Current Topics in Innate Immunity

Advances in Experimental Medicine and Biology

Recent Volumes in this Series

Volume 590
CROSSROADS BETWEEN INNATE AND ADAPTIVE IMMUNITY
Edited by Peter D. Katsikis, Bali Pulendran and Stephen P. Schoenberger

Volume 591
SOMATIC CELL NUCLEAR TRANSFER
Edited by Peter Sutovsky

Volume 592
REGULATORY MECHANISMS OF STRIATED MUSCLE CONTRACTION
Edited by Setsuro Ebashi and Iwao Ohtsuki

Volume 593
MICROARRAY TECHNOLOGY AND CANCER GENE PROFILING
Edited by Simone Mocellin

Volume 594
MOLECULAR ASPECTS OF THE STRESS RESPONSE
Edited by Peter Csermely and Laszlo Vigh

Volume 595
THE MOLECULAR TARGETS AND THERAPEUTIC USES OF CURCUMIN IN HEALTH AND DISEASE
Edited by Bharat B. Aggarwal, Yung-Joon Surh and Shishir Shishodia

Volume 596
MECHANISMS OF LYMPHOCYTE ACTIVATION AND IMMUNE REGULATION XI
Edited by Sudhir Gupta, Frederick Alt, Max Cooper, Fritz Melchers and Klaus Rajewsky

Volume 597
TNF RECEPTOR ASSOCIATED FACTORS (TRAFs)
Edited by Hao Wu

Volume 598
CURRENT TOPICS IN INNATE IMMUNITY
Edited by John D. Lambris

Current Topics in Innate Immunity

Edited by

JOHN D. LAMBRIS
Department of Pathology & Laboratory Medicine
University of Pennsylvania
Philadelphia, Pennsylvania

Springer

Aegean Conferences

Editor:
John D. Lambris, Ph.D.
University of Pennsylvania
Philadelphia, PA 19104

Library of Congress Control Number: 2007926254

Proceedings from the 4th International Conference on Innate Immunity held in Corfu, Greece, June 4-9, 2006.

ISBN-13: 978-0-387-71765-4 e-ISBN-13: 978-0-387-71767-8

Springer Aegean Conferences

Printed on acid-free paper.

10 9 8 7 6 5 4 3 2 1

springer.com

Preface

Innate Immunity has long been regarded as the non-specific arm of immune response, acting immediately and in a generic way, to defend the host from infections. In the post genomic era, our knowledge of the innate immune system is enriched by findings on the specificity of innate immune reactions as well as to novel functions that do not strictly correlate with immunological defense and surveillance, immune modulation or inflammation. Several studies indicate that molecules involved in innate immunity exert functions that are either more complex than previously thought, or go well beyond the innate immune character of the system.

The advent of high-throughput platforms for genome and proteome-wide profiling, together with the enormous amount of raw genetic information that has accumulated in the databases, have stirred new expectations in biomedical research. They have led scientists to revisit established biological systems from a global and integrative perspective. Innate Immunity research is now faced with the challenge of trying to integrate isolated biochemical pathways into complex gene and protein regulatory circuits. In this respect, scientists from around the world convened at the 4th International Conference on Innate Immunity (June 4 - 9, 2006), in Corfu, Greece to discuss recent advances in this fast evolving field. This volume represents a collection of topics on natural killer cells, mast cells, phagocytes, toll like receptors, complement, host defense in plants and invertebrates, evasion strategies of microorganisms, pathophysiology, protein structures, design of therapeutics, and experimental approaches discussed during the conference.

I am grateful to the contributing authors for the time and effort they have devoted to writing, what I consider exceptionally informative chapters in a book that will have a significant impact on the Innate Immunity field. I am grateful to Rodanthi Lambris, for her assistance in formatting the text. I also gratefully acknowledge the generous help provided by Dimitrios Lambris in managing the organization of this meeting. Finally, I also thank Andrea Macaluso and Lisa Tenaglia of Springer Publishers for their supervision in this book's production.

John D. Lambris, Ph.D.

CONTENTS

LIST OF CONTRIBUTORS

Hafiz Ahmed
Center of Marine Biotechnology
University of Maryland
Biotechnology Institute
701 East Pratt Street
Baltimore, MD 21202

Hydar Ali
Department of Pathology
School of Dental Medicine
University of Pennsylvania
Philadelphia, PA, 19104

Ali A. Ashkar
Centre for Gene Therapeutics,
Department of Pathology and
Molecular Medicine
McMaster University Health
Sciences Centre
Hamilton, Ontario

Péter Balogh
Department of Immunology
and Biotechnology
University of Pécs, Faculty of Medicine
Szigeti út 12
7643 Pécs, Hungary

Christopher Benner
Graduate Program in Bioinformatics
and Systems Biology
University of California at San Diego
La Jolla, CA 92093-0412

Anna M. Blom
Department of Laboratory Medicine
Malmo, The Wallenberg Laboratory
Malmö University Hospital
Lund University
S-205 02 Malmö, Sweden

Thomas Boller
Basel-Zurich-Plant Science Center,
Botanical Institute
University Basel
Hebelstrasse 1
4056 Basel, Switzerland

Natalie A. Borg
Protein Crystallography Unit,
Department of Biochemistry
and Molecular Biology
Monash University
3800 Victoria, Australia

Andrew G. Brooks
Department of Microbiology
and Immunology
The University of Melbourne
3010 Victoria, Australia

Robert R. Cary
Los Alamos National Laboratory
Biosciences Division
Mail Stop M888, HRL-1, TA-43
Los Alamos, NM 87545

Hui Chen
Department of Pathology and
Laboratory Medicine
University of Pennsylvania
403 Stellar Chance Labs
Philadelphia, PA 19104-6100

Delphine Chinchilla
Basel-Zurich-Plant Science Center
Botanical Institute
University Basel
Hebelstrasse 1
4056 Basel, Switzerland

Victor R. DeFilippis
Oregon Health and Science University,
Vaccine and Gene Therapy Institute
Portland, Oregon

Mariella Della Chiesa
Dipartimento di Medicina Sperimentale
Università degli Studi di Genova
Via L.B. Alberti 2
Genova 16132, Italy

Vadim Deyev
Department of Microbiology and
Immunology, Miller School of Medicine
University of Miami
Miami, FL 33136

Christina Ekerfelt
Department of Molecular and
Clinical Medicine
Linköping University
SE-581 83, Linköping, Sweden

Anna Erdei
Department of Immunology
Eötvös Loránd University
Pázmány P.s.1/C
1117 Budapest, Hungary

Jan Ernerudh
Department of Molecular and Clinical
Medicine
Linköping University
SE-581 83, Linköping, Sweden

Terje Espevik
Institute of Cancer Research and
Molecular Medicine, NTNU
N-7489 Trondheim, Norway

Bruna Ferranti
Dipartimento di Medicina Sperimentale,
Università degli Studi di Genova
Via L.B. Alberti 2
Genova 16132, Italy

Pia Forsberg
Department of Molecular and
Clinical Medicine
Linköping University
SE-581 83 Linköping, Sweden

Teizo Fujita
Department of Immunology
Fukushima Medical University
School of Medicine
1-Hikarigaoka
Fukushima 960-1295, Japan

Berhane Ghebrehiwet
Department of Medicine
Stony Brook University
Stony Brook, NY 11794

Stefano Giacomelli
Laboratory of Cell Biology
Stazione Zoologica "Anton Dohrn",
Napoli, Italy

Christopher Glass
Department of Cellular and
Molecular Medicine
University of California at San Diego
La Jolla, CA 92093-0651.

Eric L. Greidinger
Division of Rheumatology
Miami Department of Veterans Affairs
Medical Center and University of Miami
Miller School of Medicine
Miami, FL 33136

Goutam Gupta
Los Alamos National Laboratory
Biosciences Division
Mail Stop M888, HRL-1, TA-43
Los Alamos, NM 87545

Øyvind Halaas
Institute of Cancer Research and
Molecular Medicine, NTNU
N-7489 Trondheim, Norway

Anna J. Henningsson
Ryhov County Hospital
Department of Infectious Diseases,
Jönköpig, Sweden

Andrea Isaák
Department of Immunology
Eötvös Loránd University
Pázmány P.s.1/C
1117 Budapest, Hungary

Takeshi Iwata
National Institute of Sensory Organs,
National Hospital Organization Tokyo
Medical Center, 2-5-1 Higashigaoka,
Meguro-ku, Tokyo 152-8902 Japan

Kay Johswich
Department of Medical Microbiology
Medical School Hannover
30625 Hannover, Germany

Lars Kjer-Nielsen
Department of Microbiology
and Immunology
The University of Melbourne
3010 Victoria, Australia

Andreas Klos
Department of Medical Microbiology
Medical School Hannover
30625 Hannover, Germany

Magda Kucia
James Graham Brown Cancer Center
Stem Cell Biology Program
University of Louisville
Louisville, KY 40202

John D. Lambris
Department of Pathology and
Laboratory Medicine
University of Pennsylvania
401 Stellar Chance Labs
Philadelphia, PA 19104-6100

Nicole M. Lauzon
Centre for Gene Therapeutics,
Department of Pathology and
Molecular Medicine
McMaster University Health
Sciences Centre
Hamilton, Ontario, Canada

Liwu Li
Department of Biology
Virginia Tech
Blacksburg, VA 24061-0346

Martin Malmsten
Department of Pharmacy
Uppsala University
P.O. Box 580,
SE-751 23 Uppsala, Sweden

Emanuela Marcenaro
Dipartimento di Medicina Sperimentale,
Università degli Studi di Genova
Via L.B. Alberti 2
Genova 16132, Italy

Linda Mark
Department of Laboratory Medicine
Malmo
The Wallenberg Laboratory
Malmö University Hospital
Lund University,
S-205 02 Malmö, Sweden

Mano Ram Maurya
Department of Bioengineering,
University of California at San Diego,
La Jolla, CA 92093-0412

James McCluskey
Department of Microbiology
and Immunology
The University of Melbourne
3010 Victoria, Australia

Daniela Melillo
Laboratory of Cell Biology
Stazione Zoologica "Anton Dohrn",
Napoli, Italy

Firoz Mian
Centre for Gene Therapeutics,
Department of Pathology and
Molecular Medicine
McMaster University Health
Sciences Centre
Hamilton, Ontario, Canada

Michael Mihlan
Department of Infection Biology,
Leibniz Institute for Natural
Product Research and Infection
Biology- Hans Knöll Institute for Natural
Products Research
Jena, Germany,

Eszter Molnár
Department of Immunology
Eötvös Loránd University
Pázmány P.s.1/C
1117 Budapest, Hungary

Alessandro Moretta
Dipartimento di Medicina Sperimentale,
Università degli Studi di Genova,
Via L.B. Alberti 2
Genova 16132, Italy

Shuichi Mori
Fukushima Medical University
School of Medicine
1-Hikarigaoka
Fuhushima 960-1295, Japan

Kristina Nilsson Ekdahl
Department of Oncology, Radiology and Clinical Immunology
University of Uppsala,
Uppsala, Sweden

Eric W. Odom
Center of Marine Biotechnology
University of Maryland
Biotechnology Institute
701 East Pratt Street
Baltimore, MD 21202

Barbara J. Panessa-Warren
Department Materials Science
Brookhaven National Laboratory
Upton, NY 11973

Krisztián Papp
Department of Immunology
Eötvös Loránd University
Pázmány P.s.1/C,
1117 Budapest, Hungary

Paige E. Pardington
Los Alamos National Laboratory
Biosciences Division
Mail Stop M888, HRL-1, TA-43
Los Alamos, NM 87545

Danielle L. Pickett
Department of Microbiology and Immunology
The University of Melbourne
3010 Victoria, Australia

Maria Rosaria Pinto
Laboratory of Cell Biology
Stazione Zoologica "Anton Dohrn",
Napoli, Italy

Eckhard R. Podack
Department of Microbiology and Immunology, Miller School of Medicine
University of Miami
Miami, Florida 33136

Sylvain Pradervand
Department of Bioengineering,
University of California at San Diego,
La Jolla, CA 92093-0412

József Prechl
Research Group of the Hungarian Academy of Sciences at the Department of Immunology
Eötvös Loránd University
Budapest, Hungary

Mariusz Z. Ratajczak
James Graham Brown Cancer Center
Stem Cell Biology Program
University of Louisville
Louisville, KY 40202

Janina Ratajczak
James Graham Brown Cancer Center
Stem Cell Biology Program
University of Louisville
Louisville, KY 40202

Patrick C. Reading
Department of Microbiology and Immunology
The University of Melbourne
3010 Victoria, Australia

Ryan Reca
James Graham Brown Cancer Center
Stem Cell Biology Program
University of Louisville
Louisville, KY 40202

Daniel Ricklin
Department of Pathology and Laboratory Medicine
University of Pennsylvania
402 Stellar Chance Labs,
422 Curie Blvd.
Philadelphia, PA 19104-6100

Silke Robatzek
Max-Planck-Institute for Plant Breeding Research
Carl-von-Linne-Weg 10
50829 Cologne, Germany

Jamie Rossjohn
Protein Crystallography Unit
Department of Biochemistry and Molecular Biology
Monash University
3800 Victoria, Australia

Kerstin Sandholm
Department of Chemistry and
Biomedical Sciences
University of Kalmar
Kalmar, Sweden

J. Vidya Sarma
Department of Pathology
The University of Michigan
Medical School
1150 West Medical Center Drive
Ann Arbor, MI 48109-0602

Artur Schmidtchen
Section of Dermatology and
Venereology, Department of
Clinical Sciences, Biomedical Center,
Lund University
Tornavägen 10
SE-221 84 Lund, Sweden

Michael C. Schuster
Department of Pathology
and Laboratory Medicine
University of Pennsylvania
402 Stellar Chance Labs
Philadelphia, PA 19104-6100

Georgia Sfyroera
Department of Pathology
and Laboratory Medicine
University of Pennsylvania
402 Stellar Chance Labs
Philadelphia, PA 19104-6100

Shiro Shigeta
Department of Microbiology
Fukushima Medical University
School of Medicine
1-Hikarigaoka
Fuhushima 960-1295, Japan

Motoaki Shiratsuchi
Department of Microbiology and
Immunology, Miller School of Medicine
University of Miami
Miami, FL 33136

Christine Skerka
Department of Infection Biology,
Leibniz Institute for Natural
Product Research and Infection
Biology- Hans Knöll Institute for Natural
Products Research
Jena, Germany,

O. Brad Spiller
Department. of Child Health, 5th floor
University Hospital
Cardiff University, Heath Park
Cardiff, CF14 4XN, United Kingdom

Jianmin Su
Department of Biology
Virginia Tech
Blacksburg, VA 24061-0346,

Shankar Subramaniam
Department of Bioengineering
University of California at San Diego
La Jolla, CA 92093-0412.

Zsuzsanna Szekeres
Department of Immunology
Eötvös Loránd University
Pázmány P.s.1/C
1117 Budapest, Hungary

Minoru Takahashi
Fukushima Medical University
School of Medicine
1-Hikarigaoka
Fuhushima 960-1295, Japan

Lee Tantral
Department of Medicine
Stony Brook University
Stony Brook, NY 11794

Satoshi Tasumi
University of Maryland Biotechnology
Institute, Center of Marine
Biotechnology
701 East Pratt Street
Baltimore, MD 21202

Michelle D. Tate
Department of Microbiology
and Immunology
The University of Melbourne
3010 Victoria, Australia

Mathew A. Titmus
Department of Medicine
Stony Brook University
Stony Brook, NY 11794

George T. Tortora
Clinical Microbiology Laboratory
Stony Brook University
Stony Brook, NY 11794

Jennifer C. van Velkinburgh
Los Alamos National Laboratory
Biosciences Division
Mail Stop M888, HRL-1, TA-43
Los Alamos, NM 87545

Gerardo R. Vasta
Center of Marine Biotechnology
University of Maryland
Biotechnology Institute,
701 East Pratt Street
Baltimore, MD 21202.

Peter A. Ward
Department of Pathology
The University of Michigan
Medical School
1150 West Medical Center Drive
Ann Arbor, MI 48109-0602

John B. Warren
Instrumentation Division Bldg.535B,
Brookhaven National Laboratory,
Upton, NY 11973

Stanislaus S. Wong
Department of Chemistry
Stony Brook University
Stony Brook, NY 11794

Marcin Wysoczynski
James Graham Brown Cancer Center
Stem Cell Biology Program
University of Louisville
Louisville, KY 40202

Qifa Xie
Department of Biology
Virginia Tech
Blacksburg, VA 24061-0346

Asifa K. Zaidi
Department of Pathology
University of Pennsylvania School
of Dental Medicine
Philadelphia, PA, 19104

Ahmet Zeytun
Los Alamos National Laboratory
Biosciences Division,
Mail Stop M888, HRL-1, TA-43
Los Alamos, NM 87545

Peter F. Zipfel
Department of Infection Biology,
Leibniz Institute for Natural Product
Research and Infection Biology
Hans Knöll Institute
Jena, Germany

Toll-like Receptors, Natural Killer Cells and Innate Immunity

Nicole M. Lauzon[1], Firoz Mian[1] and Ali A. Ashkar[1]

[1] Centre for Gene Therapeutics, Department of Pathology and Molecular Medicine, McMaster University Health Sciences Centre, Hamilton, Ontario, Canada, ashkara@mcmaster.ca

1 Introduction

The innate immune response relies on the rapid recognition of microbial pathogens to provide the first phase of protection against infections. Innate immune cells such as macrophages, dendritic cells and natural killer (NK) cells respond to pathogen-associated molecular patterns (PAMPs) as well as host-derived factors. In the last decade receptors for many of the PAMPs have been discovered. Much of the focus in the field of Toll-like receptors (TLRs) has been given to recognition of pathogens via TLRs expressed on antigen presenting cells (APCs). NK cells are ancient lymphocytes that have very close crosstalk with APCs. In recent years, there has been an interest in the avtivation of NK cells, similar to APCs, via direct stimulation with microbial PAMPs. This review focuses on TLRs and activation of NK cells by TLR ligands. We will review the evidence for direct and/or indirect activation of NK cells by TLR ligands/agonists.

2 Toll-like Receptors and Innate Immunity

As the first line of defense against pathogens and infections, the innate immune system has evolved to respond rapidly and precisely to a variety of different challenges. Reacting quickly against non-self and avoiding responses to self, the innate immune system has developed a complex balance between sensitivity and specificity (Aderem and Ulevitch 2000; Akira and Takeda 2004). The innate immune system provides an important link between the initial responses to infection and the long term protection provided by the adaptive immune system (Akira and Takeda 2004). Unlike the adaptive immune system, the innate immune system is activated quickly after infection (Blach-Olszewska 2005). Specialized innate immune cells perform their functions while pathogens are continually evolving, adapting to host defense mechanisms via mutations. Innate immunity has therefore evolved the ability to react to a series of highly conserved molecular sequences amongst pathogens, known as pathogen-associated molecular patterns (PAMPs). These PAMPs are specific to microbial pathogens, are essential for their survival, and are not found within the host (Janeway and Medzhitov 1999; Medzhitov and Janeway 2000; Akira and Takeda 2004). Innate immune cells, including natural killer (NK) cells, recognize these PAMPs via pattern recognition receptors (PRRs) and are either induced directly or indirectly to produce appropriate responses. Subsequent populations of innate and adaptive cells are then activated via cytokine and chemokine secretion, contributing to the induction of the inflammatory response (Bowie and Haga 2005). Similar receptors have been found in plants, invertebrates, and vertebrates including mice and humans (Magor and Magor 2001; Blach-Olszewska 2005). NK cells are particularly important in the early response against pathogens, with a series of specified functions, known for their cytotoxic abilities and cytokine production. Activated NK cell-derived IFN-γ plays an important role in the early innate response, ultimately leading to inflammation and linking the innate and adaptive immune responses (Lieberman and Hunter 2002; Blach-Olszewska 2005). Different subsets of NK cells express varying degrees of activating receptors, have different intensities of cytokine production, with varying abilities to kill target cells (Fehniger, Cooper, Nuovo, Cella, Facchetti, Colonna and Caligiuri 2003; Yokoyama and Plougastel 2003; Blach-Olszewska 2005). $CD56^{dim}CD16^{+}$ NK cells are mostly found in peripheral blood, while the $CD56^{bright}CD16^{-}$ NK subset are mostly found within the lymph nodes (Fehniger, Cooper et al. 2003; Blach-Olszewska 2005).

Innate immune cells, including NK cells, perform their functions due to the varying types of innate receptors. NK cells have a range of innate receptors, including TLRs, mannose receptors, scavenger receptors and NK activating receptors, such as CD16, NKG2D, NKp30, NKp46, 2B4, and NTBA (Magor and Magor 2001; Blach-Olszewska 2005). Of these receptor types, TLRs are thought to be the most important. Toll receptors were originally discovered in the genome of *Drosophila melanogaster*, and were later found to be responsible for antifungal resistance in the fly (Lemaitre 2004; Blach-Olszewska 2005). These TLRs are classified as PRRs and are a family of transmembrane receptors that play a pivotal role in our first line of defense against invading pathogens. TLRs recognize pathogens on the basis of motifs or PAMPs displayed on the invading organisms ranging from bacterial and viral components, to fungal and protozoal molecules. All TLRs contain a highly variable extracellular leucine rich repeat (LRR) domain involved in ligand binding, and a cytoplasmic tail containing a highly conserved region, the Toll/IL-1 receptor (TIR) domain. The TIR domain transmits downstream signals by recruiting one or more TIR-containing adapter proteins. These adapters include MyD88 (myeloid differentiation primary-response protein 88), and TIR domain containing adapter proteins (TIRAP) inducing IFN-β (TRIF) (Burns, Martinon, Esslinger, Pahl, Schneider, Bodmer, Di Marco, French and Tschopp 1998; Aderem and Ulevitch 2000; Blach-Olszewska 2005). Signaling through TLRs can be broadly categorized into two pathways: the MyD88 dependent pathway and MyD88 independent pathway. Both pathways are subsequently linked to the activation of the NF-κB and MAPK pathways. Recent evidence reveal that several IRF family members such as IRF3 and IRF7 are also activated by the MyD88-dependent and independent pathways (Honda and Taniguchi 2006).

Currently 10 functional TLRs in human and 12 functional TLRs in mice have been identified. TLRs 1, 2, 4, 5, 6 and 10 are localized on the cell surface and the TLRs 3, 7, 8 and 9 are localized in the endosomal compartments (Miggin and O'Neill L 2006; Pandey and Agrawal 2006). Upon ligand binding, all TLRs, except TLR3, signal through the MyD88-dependent pathway, whereas TLR3 and TLR4 activate the MyD88 independent pathway (Honda and Taniguchi 2006). A series of molecular genetic studies have revealed the respective ligands for the TLRs which have been reviewed by a number of authors (Jiang, Mak, Sen and Li 2004; Gorski, Waller, Bjornton-Severson, Hanten, Riter, Kieper, Gorden, Miller, Vasilakos, Tomai and Alkan 2006; Kawai and Akira 2006; Miggin and O'Neill L 2006; Pandey and Agrawal 2006). LPS from Gram negative bacteria induces responses via TLR4 ligation, whereas TLR2 in association with TLR1 or TLR6 recognize peptidoglycan (PG), lipopeptide and lipoprotein of Gram positive bacteria and mycoplasma lipopeptide (Miggin and O'Neill L 2006). TLR1/2 and TLR2/6 differentiate triacyl lipopeptide and diacyl lipopeptide respectively (Miggin and O'Neill L 2006). TLR5 mediates the induction of immune responses by bacterial flagellin, while TLR9 has been shown to recognize bacterial and viral oligodeoxynucletides containing unmethylated CpG dinucleotide motifs (Takeshita, Gursel, Ishii, Suzuki, Gursel and Klinman 2004). TLR3 responds to dsRNA, a byproduct of viral replication, and its synthetic mimetic polyionosinic-polycytidylic acid (PolyI:C). TLR7 and 8 recognize synthetic imidazoquinolines (R-848) and ssRNA derived from viruses (Miggin and O'Neill L 2006). TLR11, which recognizes uropathogenic *Escherichia coli* in mice, is found in the human genome, yet it may not be expressed due to the presence of stop codons found within the TLR11 open reading frame (Zhang, Zhang, Hayden, Greenblatt, Bussey, Flavell and Ghosh 2004). Upon recognition of PAMPs, a cascade of intracellular signaling is activated, which culminate in the production of cytokines and antiviral type I interferons (IFN) in order to clear the pathogens with subsequent adaptive immune responses. TLRs, although important for the activation of other innate cell subsets, have not traditionally been viewed as being responsible for the activation of NK cells.

3 Natural Killer Cells and Toll-like Receptors

Until recent studies examining the role of TLRs in the activation of NK cells, the focus within the field remained on the balance between NK cell activating and inhibitory receptors in response to tumors and virally-infected cells (Cerwenka and Lanier 2001; Johansson, Berg, Hall and Hoglund 2005). Due to their many invariant, inducible and constitutively expressed receptors, NK cells play a key role in initial innate host defense responses (Lanier 2005). This explains their ability to rapidly produce cytokines, including IFN-γ and TNF-α, within hours after initial infections arise (Aderem and Ulevitch 2000; Chalifour, Jeannin, Gauchat, Blaecke, Malissard, N'Guyen, Thieblemont and Delneste 2004; Kalinski, Giermasz, Nakamura, Basse, Storkus, Kirkwood and Mailliard 2005).

As previously stated, TLR1-10 have been discovered in humans, while TLR1-9 and 11-13 have been identified in mice (Cerwenka and Lanier 2001; Johansson, Berg et al. 2005; Beutler, Jiang, Georgel, Crozat, Croker, Rutschmann, Du and Hoebe 2006). TLR11 in humans has been identified as a non-functional pseudogene (Gorski, Waller et al. 2006). There are discrepancies in the literature about NK cell TLR expression. Our observations (data not shown) validate that murine NK cells have been found to express mRNA for TLRs1-9 and 11 (Baratin, Roetynck, Lepolard, Falk, Sawadogo, Uematsu, Akira, Ryffel, Tiraby, Alexopoulou, Kirschning, Gysin, Vivier and Ugolini 2005; Lauzon, Mian, MacKenzie and Ashkar 2006) while human NK cells have been shown to express mRNA for all currently known human TLRs, which are expressed at varying levels (Hornung, Rothenfusser, Britsch, Krug, Jahrsdorfer, Giese, Endres and Hartmann 2002; Chalifour, Jeannin et al. 2004; Hart, Athie-Morales, O'Connor and Gardiner 2005; Lauzon, Mian et al. 2006). Human NK cells express the highest levels of TLR3 when compared to plasmocytoid dendritic cells (pDCs), B cells, monocytes and T cells (Hornung, Rothenfusser et al. 2002), while TLR1 has the highest expression levels and TLR10 the lowest (Hornung, Rothenfusser et al. 2002). TLR10 is not consistently expressed, yet several reports indicate the presence of TLR10 mRNA in small amounts (Hornung, Rothenfusser et al. 2002; Chalifour, Jeannin et al. 2004; Gorski, Waller et al. 2006; Lauzon, Mian et al. 2006). TLR2, 3, 5 and 6 are also more highly expressed than TLR4, 7, 8 and 9 (Hornung, Rothenfusser et al. 2002; Gorski, Waller et al. 2006). In support of these findings, our data indicate that purified $CD56^{+}$ cells express all known human TLRs, however this analysis was not performed on individual NK cell subsets, and was not quantitative (Lauzon, Mian et al. 2006).

4 Natural Killer Cells and Pathogen Sensing

It is widely accepted that NK cells are primed to respond quickly and effectively to pathogens. Until recently, it was believed that their activation was antigen presenting cell (APC)-dependent when faced with PAMPs (Colucci, Di Santo and Leibson 2002; Walzer, Dalod, Robbins, Zitvogel and Vivier 2005). It was determined that PAMPs were acting on DCs and macrophages, and their subsequent cytokine production and cell-cell contact elicited responses from NK cells (Watford, Hissong, Bream, Kanno, Muul and O'Shea 2004; Andoniou, van Dommelen, Voigt, Andrews, Brizard, Asselin-Paturel, Delale, Stacey, Trinchieri and Degli-Esposti 2005; Zanoni, Foti, Ricciardi-Castagnoli and Granucci 2005). Currently it is well characterized that an important link exists between APCs and NK cells where both populations act upon each other, inducing cytokine production, proliferation and maturation (Lieberman and Hunter 2002; Raulet 2004; Hart, Athie-Morales et al. 2005). Although many TLR agonists have been identified, their direct or indirect effects on NK cells have not been widely studied. Some confusion remains surrounding recent evidence suggesting that because NK cells mount rapid responses and express a variety of TLRs, therefore they are perfectly suited to respond to TLR agonists and can be activated directly in the presence of TLR ligands alone, even in the absence of activated accessory cells (Colucci, Di Santo et al. 2002; Chalifour, Jeannin et al. 2004; Schmidt, Leung, Kwong, Zarember, Satyal, Navas, Wang and Godowski 2004; Sivori, Falco, Della Chiesa, Carlomagno, Vitale, Moretta and Moretta 2004).

5 Natural Killer Cell Activation via TLR ligation: Direct or Indirect?

Despite the confusion, evidence in support of direct NK cell stimulation by TLR ligands is mounting. Becker *et al.* (2003) first characterized the membrane expression of TLR2 in human NK cells (Becker, Salaiza, Aguirre, Delgado, Carrillo-Carrasco, Kobeh, Ruiz, Cervantes, Torres, Cabrera, Gonzalez, Maldonado and Isibasi 2003). A cell surface-derived protein, LPG, from the obligate intracellular pathogen Leishmania, inhibits IL-12 production from the host macrophages. This parasite-derived LPG stimulates human NK cells via TLR2, inducing NF-κB nuclear translocation, TNF-α production and subsequently inducible nitric oxide synthase (iNOS) activation from macrophages, killing the pathogen. This process is also synergized by IFN-γ (Becker, Salaiza et al. 2003). NK cells respond through TLR2 in the absence of macrophage-derived IL-12 to produce IFN-γ, which is important for host innate defense against pathogens (Lauwerys, Garot, Renauld and Houssiau 2000; Lieberman and Hunter 2002; Strengell, Matikainen, Siren, Lehtonen, Foster, Julkunen and Sareneva 2003; Chalifour, Jeannin et al. 2004; Kalinski, Giermasz et al. 2005).

It is understandable that various cell types respond differently to the PAMPs, due to the variety of TLRs and their various expression levels. This may be dependent on the TLR mRNA expression levels or the number of functional receptors found amongst the different cell subtypes. For example, it has been previously reported that the TLR9 agonist CpG ODN cannot induce the production of cytokines or increase cytotoxicity towards tumors in the absence of inflammatory cytokines such as IL-12 or IL-18 (Sivori, Falco et al. 2004; Sivori, Carlomagno, Moretta and Moretta 2006). Furthermore, CpG did not upregulate NK cell activation markers CD69 and CD25 in the absence of these cytokines (Chace, Hooker, Mildenstein, Krieg and Cowdery 1997; Sivori, Falco et al. 2004; Sivori, Carlomagno et al. 2006). The Sivori group recognized that purified NK cells did not respond to TLR3 ligands, dsRNA and PolyI:C, without the synergistic effects of inflammatory cytokines (Sivori, Falco et al. 2004; Sivori, Carlomagno et al. 2006). It was noted that all NK cell subsets expressed CD69 which was up-regulated after TLR ligand and cytokine stimulation, suggesting that all NK cells are therefore capable of responding to TLR stimulation to some degree (Sivori, Falco et al. 2004; Sivori, Carlomagno et al. 2006). Their recent study also reported that a more potent response occurred after treatment with CpG ODN C than previously reported results with CpG ODN A and ODN B, and that NK cell responses were synergized in the presence of IFN-α (Sivori, Carlomagno et al. 2006). Our recent data support the observations that CpG ODN B does not induce significant increases in IFN-γ production after NK cell stimulation for 24-72hrs (Lauzon,

Mian et al. 2006). However, while evaluating the TLR3 ligand PolyI:C, we observed significantly increased IFN-γ and TNF-α production, as well as increased cytolytic effects from a purified NK cell population, in the absence of contaminating IL-12 (Lauzon, Mian et al. 2006).

The direct activation of NK cell populations upon TLR3 ligation has also been observed by other investigators. Schmidt *et al.* (2004) claimed an APC-independent mechanism was responsible for the activation of NK cells through TLR3 (Schmidt, Leung et al. 2004). Upon stimulation with PolyI:C, TLR3 mRNA was up-regulated; increased cytotoxicity and CD69 receptor expression were also observed. Moreover, nuclear translocation of nuclear factor NF-κB and the production of proinflammatory cytokines IL-6, IL-8 and IFN-γ were also observed. This stimulation was independent of type I IFNs, and neutralizing anti-INF-α antibodies did not block the up-regulation of CD69 from PolyI:C stimulated NK cells (Schmidt, Leung et al. 2004). This mechanism of activation is not consistent between the human and murine NK cell populations, since Poly I:C induced IFN-γ production in mice is dependent on type I IFNs and did not enhance cytolytic activity of murine NK cells (Schmidt, Leung et al. 2004). Another group found that PolyI:C is able to directly induce the cytotoxic functions of NK cells, while on the other hand it did not induce a significant increase in cytokine production (Hart, Athie-Morales et al. 2005). Pisegna *et al.* (2004) also observed that PolyI:C significantly up-regulated CD16-mediated cytotoxicity and the production of a chemokine, CXCL10, which synergized IFN-γ production (Pisegna, Pirozzi, Piccoli, Frati, Santoni and Palmieri 2004).

The effects of TLR7/8 agonist R-848 on NK cell IFN-γ production was mediated primarily via accessory cell stimulation of APC-produced IL-12-dependent fashion (Hart, Athie-Morales et al. 2005). Further evidence demonstrated that TLR7, 8 or 7/8 agonists did not result in direct NK cell activation, and any activation using these agonists was IL-18 or IL-12 dependent (Gorski, Waller et al. 2006). These two findings suggest that APC-induced IL-12 and /or IL-18 activation of NK cell functions are in contrary to the findings of Schmidt *et al.* (2004), Gorski *et al.* (2006) and our data that PolyI:C directly induced IFN-γ production from NK cells, and increased the expression of CD69 from $CD56^{dim}$ populations (Schmidt, Leung et al. 2004; Sivori, Falco et al. 2004; Hart, Athie-Morales et al. 2005; Gorski, Waller et al. 2006; Lauzon, Mian et al. 2006). In addition, MALP-2, a TLR2/6 agonist, was found to upregulate the same receptor in both $CD56^{bright}$ and $CD56^{dim}$ populations (Gorski, Waller et al. 2006). Another study examined a second TLR2 ligand, using an outer membrane protein from *Klebsiella pneumoniae* (KpOmpA) (Chalifour, Jeannin et al. 2004). The investigators determined that NK cell stimulation via TLR2 and TLR5 with the appropriate ligand flagellin resulted in a rapid release of *a*-defensins, which are antimicrobial peptides that aide in the killing of pathogens via membrane disruption (Chalifour, Jeannin et al. 2004). Tsujimoto *et al.* (2005) confirmed that both flagellin and TLR4 ligand LPS directly induced NK cell proliferation (Tsujimoto, Uchida, Efron, Scumpia, Verma, Matsumoto, Tschoeke, Ungaro, Ono, Seki, Clare-Salzler, Baker, Mochizuki, Ramphal and Moldawer 2005). As previously described with other ligands, LPS increased the expression of the CD69 marker and the response was greater than that seen with flagellin. They did not, however, find any significant production of IFN-γ after LPS stimulation, and only moderate levels were apparent after flagellin stimulation (Tsujimoto, Uchida et al. 2005). We also observed the direct activation of NK cells after treatment with PG, a TLR2 agonist, LPS and a TLR5 ligand recombinant flagellar protein FliC, with significantly higher levels observed in the NK cell population stimulated with FliC (Lauzon, Mian et al. 2006). Not everyone is in agreement however, for Eriksson *et al.* (2006) considered human uterine NK cells that expressed TLR2-4 and were unresponsive to the direct stimulation with their corresponding agonists (Eriksson, Meadows, Basu, Mselle, Wira and Sentman 2006). This suggests that TLR-dependent NK functions not only depend on different cell subtypes that respond to PAMPs in varying capacities, but that the site of interaction between pathogens and NK cells might also be important. There may be endogenously derived environmental signals, present before contact with the PAMPs, dependent on the origin of the NK cells responding to the PAMPs.

Clearly TLR ligands are important for the induction of cytokines, the up-regulation of activating receptors and are involved in increasing the cytolytic abilities of NK cells. Despite acting directly or indirectly, key functions in NK cell biology and innate protection are modulated due to TLR-TLR ligand interactions.

6 The Role of Natural Killer Cells in Early Innate Responses

One of the most important findings with regards to NK cells and their role in early responses to pathogens as described by Thale *et al.* (2005), is that they are the primary and likely only source of IFN-γ production mere hours after infection occurs (Thale and Kiderlen 2005). Earlier studies have described IFN-γ profiles after infection, excluding NK cells as the primary source of IFN-γ, without investigating the IFN-γ secretion during the early hours post infection. These models described the activation of macrophages and DCs and their subsequent production of IL-12 and IL-18 after infection, which in turn induced IFN-γ production from macrophages, not from NK cells (Kamath, Sheasby and Tough 2005; Chan, Crafton, Fan, Flook, Yoshimura, Skarica, Brockstedt, Dubensky, Stins, Lanier, Pardoll and Housseau 2006). A recent study raises questions pertaining to isolation techniques and the purity of traditional myeloid cell populations and the concept that macrophages are potent producers of IFN-γ after stimulation with IL-12 and IL-18, demonstrating autocrine macrophage activation (Schleicher, Hesse and Bogdan 2005). The authors described small amounts of $CD11b^+CD11c^+CD31^+DX5^+NK1.1^+$ NK cells or $CD3^+CD8^+TCRB^+$ T cells were responsible for the IFN-γ secretion. While macrophages expressed IFN-γ mRNA, the nuclear translocation of signal transducer and activator of transcription 4 (STAT4) or functional IFN-γ were not detectable (Schleicher, Hesse et al. 2005). This supports the observations of Thale *et al.* (2005) who described a new model where NK cells play the predominant role in early IFN-γ production. RAG knockout mice lacking T-cells were compared to wild-type mice after intravenous injection with *Listeria monocytogenes*, and IFN-γ producing cells from the liver and spleen were analyzed by flow cytometry at 9hrs, 19hrs and 29hrs post injection without any re-stimulation or cytokine treatment. MHC class II$^+$ macrophages and DCs, and NK1.1$^+$ cells were compared (Thale and Kiderlen 2005). The first detectable levels of IFN-γ were found within the NK1.1$^+$ CD3$^-$ population, which was the main source of innate IFN-γ after 19hrs and 29hrs. In the knockout mice, no MHC class II$^+$ leukocytes from the spleen or liver produced IL-12, and NK cells were the sole source of IFN-γ in these mice (Thale and Kiderlen 2005). We have recently shown that NK cell stimulation with Poly I:C for 18hrs and subsequent intracellular cytokine staining (ICCS) for IFN-γ revealed similar results in our negatively selected NK cells, confirming that 93% of the IFN-γ was produced from CD56$^+$CD3$^-$ cells, while less than 1% of the IFN-γ$^+$ cells were NKT or T cells (Lauzon, Mian et al. 2006). These recent observations unequivocally demonstrate that NK cells are important in the production of early IFN-γ.

7 Concluding Remarks and Future Directions

Despite the debate surrounding direct vs. indirect NK cell activation via TLR ligation, the evidence supports the fact that NK cells are highly sensitive to the presence of pathogens and their various PAMPs, and can be easily stimulated to respond efficiently and effectively to tumors and virus infected cells. Different ligands have varying abilities to influence the cytolytic functions, cytokine induction, receptor up-regulation and proliferation of NK cells, with varying capacities within NK cell subsets, as well as between species. Additional cytokines or chemokines may be required in many instances for successful NK cell functions, however, they may not induce responses on their own in a physiologically relevant manner. Many of these NK cell functions can be up-regulated or synergized when coupled with one or more relevant cytokine. To determine the effects of the TLR ligands on NK cells, it is absolutely important to examine several aspects of NK cell activity, cell phenotypes and to consider the source of the cells as well. In order to further describe the role of NK cells within the context of the innate immune system and how they link the innate and adaptive immune systems, further studies are required to evaluate the NK cell TLR-TLR ligand interactions. Perhaps more in depth investigations should be conducted to validate the immediate early responses, as revealed by Thale *et al.* (2005), in order to better understand and characterize the immediate and downstream consequences of pathogen or tumor associated interactions (Thale and Kiderlen 2005). Continuing investigations in the search for novel ligands, as well as novel clinical applications, for the targeted stimulation of NK cells and their responses against viral pathogens and tumors, may eventually open new avenues for combination therapies for cancer treatment. Roda *et al.* describe an example where the concurrent administration of CpG ODN with therapeutic monoclonal antibodies against tumors significantly modulate the effects of NK cell and their abilities to target the tumor cells (Roda, Parihar and Carson 2005). Other considerations may include the mode of delivery of potential therapeutic TLR ligands and how to induce desired responses from NK cells without over stimulating the immune system as a whole. NK cells remain key players in the first line of defense with an impressive system of PRRs and cytolytic abilities. Further characterization of their abilities to respond to PAMPs and subsequent interactions with pathogens and tumors will perhaps lead to vaccines against current and emerging pathogens and clinically effective therapies for a wide range of cancers.

References

Aderem, A. and Ulevitch, R. J. (2000) Toll-like receptors in the induction of the innate immune response. Nature. 406(6797), 782-7.

Akira, S. and Takeda, K. (2004) Toll-like receptor signalling. Nat Rev Immunol. 4(7), 499-511.

Andoniou, C. E., van Dommelen, S. L., Voigt, V., Andrews, D. M., Brizard, G., Asselin-Paturel, C., Delale, T., Stacey, K. J., Trinchieri, G. and Degli-Esposti, M. A. (2005) Interaction between conventional dendritic cells and natural killer cells is integral to the activation of effective antiviral immunity. Nat Immunol. 6(10), 1011-9.

Baratin, M., Roetynck, S., Lepolard, C., Falk, C., Sawadogo, S., Uematsu, S., Akira, S., Ryffel, B., Tiraby, J. G., Alexopoulou, L., Kirschning, C. J., Gysin, J., Vivier, E. and Ugolini, S. (2005) Natural killer cell and macrophage cooperation in MyD88-dependent innate responses to Plasmodium falciparum. Proc Natl Acad Sci U S A. 102(41), 14747-52.

Becker, I., Salaiza, N., Aguirre, M., Delgado, J., Carrillo-Carrasco, N., Kobeh, L. G., Ruiz, A., Cervantes, R., Torres, A. P., Cabrera, N., Gonzalez, A., Maldonado, C. and Isibasi, A. (2003) Leishmania lipophosphoglycan (LPG) activates NK cells through toll-like receptor-2. Mol Biochem Parasitol. 130(2), 65-74.

Beutler, B., Jiang, Z., Georgel, P., Crozat, K., Croker, B., Rutschmann, S., Du, X. and Hoebe, K. (2006) Genetic analysis of host resistance: Toll-like receptor signaling and immunity at large. Annu Rev Immunol. 24, 353-89.

Blach-Olszewska, Z. (2005) Innate immunity: cells, receptors, and signaling pathways. Arch Immunol Ther Exp. 53, 245-253.

Bowie, A. G. and Haga, I. R. (2005) The role of Toll-like receptors in the host response to viruses. Mol Immunol. 42(8), 859-67.

Burns, K., Martinon, F., Esslinger, C., Pahl, H., Schneider, P., Bodmer, J. L., Di Marco, F., French, L. and Tschopp, J. (1998) MyD88, an adapter protein involved in interleukin-1 signaling. J Biol Chem. 273(20), 12203-9.

Cerwenka, A. and Lanier, L. L. (2001) Natural killer cells, viruses and cancer. Nat Rev Immunol. 1(1), 41-9.

Chace, J. H., Hooker, N. A., Mildenstein, K. L., Krieg, A. M. and Cowdery, J. S. (1997) Bacterial DNA-induced NK cell IFN-gamma production is dependent on macrophage secretion of IL-12. Clin Immunol Immunopathol. 84(2), 185-93.

Chalifour, A., Jeannin, P., Gauchat, J. F., Blaecke, A., Malissard, M., N'Guyen, T., Thieblemont, N. and Delneste, Y. (2004) Direct bacterial protein PAMP recognition by human NK cells involves TLRs and triggers alpha-defensin production. Blood. 104(6), 1778-83.

Chan, C. W., Crafton, E., Fan, H. N., Flook, J., Yosh imura, K., Skarica, M., Brockstedt, D., Dubensky, T. W., Stins, M. F., Lanier, L. L., Pardoll, D. M. and Housseau, F. (2006) Interferon-producing killer dendritic cells provide a link between innate and adaptive immunity. Nat Med. 12(2), 207-13.

Colucci, F., Di Santo, J. P. and Leibson, P. J. (2002) Natural killer cell activation in mice and men: different triggers for similar weapons? Nat Immunol. 3(9), 807-13.

Eriksson, M., Meadows, S. K., Basu, S., Mselle, T. F., Wira, C. R. and Sentman, C. L. (2006) TLRs mediate IFN-gamma production by human uterine NK cells in endometrium. J Immunol. 176(10), 6219-24.

Fehniger, T. A., Cooper, M. A., Nuovo, G. J., Cella, M., Facchetti, F., Colonna, M. and Caligiuri, M. A. (2003) CD56bright natural killer cells are present in human lymph nodes and are activated by T cell-derived IL-2: a potential new link between adaptive and innate immunity. Blood. 101(8), 3052-7.

Gorski, K. S., Waller, E. L., Bjornton-Severson, J., Hanten, J. A., Riter, C. L., Kieper, W. C., Gorden, K. B., Miller, J. S., Vasilakos, J. P., Tomai, M. A. and Alkan, S. S. (2006) Distinct indirect pathways govern human NK-cell activation by TLR-7 and TLR-8 agonists. Int Immunol. 18(7), 1115-26.

Hart, O. M., Athie-Morales, V., O'Connor, G. M. and Gardiner, C. M. (2005) TLR7/8-mediated activation of human NK cells results in accessory cell-dependent IFN-gamma production. J Immunol. 175(3), 1636-42.

Honda, K. and Taniguchi, T. (2006) IRFs: master regulators of signalling by Toll-like receptors and cytosolic pattern-recognition receptors. Nat Rev Immunol. 6(9), 644-58.
Hornung, V., Rothenfusser, S., Britsch, S., Krug, A., Jahrsdorfer, B., Giese, T., Endres, S. and Hartmann, G. (2002) Quantitative expression of toll-like receptor 1-10 mRNA in cellular subsets of human peripheral blood mononuclear cells and sensitivity to CpG oligodeoxynucleotides. J Immunol. 168(9), 4531-7.
Janeway, C. A., Jr. and Medzhitov, R. (1999) Lipoproteins take their toll on the host. Curr Biol. 9(23), R879-82.
Jiang, Z., Mak, T. W., Sen, G. and Li, X. (2004) Toll-like receptor 3-mediated activation of NF-kappaB and IRF3 diverges at Toll-IL-1 receptor domain-containing adapter inducing IFN-beta. Proc Natl Acad Sci U S A. 101(10), 3533-8.
Johansson, S., Berg, L., Hall, H. and Hoglund, P. (2005) NK cells: elusive players in autoimmunity. Trends Immunol. 26(11), 613-8.
Kalinski, P., Giermasz, A., Nakamura, Y., Basse, P., Storkus, W. J., Kirkwood, J. M. and Mailliard, R. B. (2005) Helper role of NK cells during the induction of anticancer responses by dendritic cells. Mol Immunol. 42(4), 535-9.
Kamath, A. T., Sheasby, C. E. and Tough, D. F. (2005) Dendritic cells and NK cells stimulate bystander T cell activation in response to TLR agonists through secretion of IFN-alpha beta and IFN-gamma. J Immunol. 174(2), 767-76.
Kawai, T. and Akira, S. (2006) TLR signaling. Cell Death Differ. 13(5), 816-25.
Lanier, L. L. (2005) NK cell recognition. Annu Rev Immunol. 23, 225-74.
Lauwerys, B. R., Garot, N., Renauld, J. C. and Houssiau, F. A. (2000) Cytokine production and killer activity of NK/T-NK cells derived with IL-2, IL-15, or the combination of IL-12 and IL-18. J Immunol. 165(4), 1847-53.
Lauzon, N. M., Mian, F., MacKenzie, R. and Ashkar, A. A. (2006) The direct effects of Toll-like receptor ligands on human NK cell cytokine production and cytotoxicity. Cellular Immunology. doi:10.1016/j.cellimm.2006.08.004 (in press).
Lemaitre, B. (2004) The road to Toll. Nat Rev Immunol. 4(7), 521-7.
Lieberman, L. A. and Hunter, C. A. (2002) Regulatory pathways involved in the infection-induced production of IFN-gamma by NK cells. Microbes Infect. 4(15), 1531-8.
Magor, B. G. and Magor, K. E. (2001) Evolution of effectors and receptors of innate immunity. Dev Comp Immunol. 25(8-9), 651-82.
Medzhitov, R. and Janeway, C., Jr. (2000) Innate immune recognition: mechanisms and pathways. Immunol Rev. 173, 89-97.
Miggin, S. M. and O'Neill L, A. (2006) New insights into the regulation of TLR signaling. J Leukoc Biol. 80(2), 220-6.
Pandey, S. and Agrawal, D. K. (2006) Immunobiology of Toll-like receptors: emerging trends. Immunol Cell Biol. 84(4), 333-41.
Pisegna, S., Pirozzi, G., Piccoli, M., Frati, L., Santoni, A. and Palmieri, G. (2004) p38 MAPK activation controls the TLR3-mediated up-regulation of cytotoxicity and cytokine production in human NK cells. Blood. 104(13), 4157-64.
Raulet, D. H. (2004) Interplay of natural killer cells and their receptors with the adaptive immune response. Nat Immunol. 5(10), 996-1002.
Roda, J. M., Parihar, R. and Carson, W. E., 3rd (2005) CpG-containing oligodeoxynucleotides act through TLR9 to enhance the NK cell cytokine response to antibody-coated tumor cells. J Immunol. 175(3), 1619-27.
Schleicher, U., Hesse, A. and Bogdan, C. (2005) Minute numbers of contaminant CD8+ T cells or CD11b+CD11c+ NK cells are the source of IFN-gamma in IL-12/IL-18-stimulated mouse macrophage populations. Blood. 105(3), 1319-28.
Schmidt, K. N., Leung, B., Kwong, M., Zarember, K. A., Satyal, S., Navas, T. A., Wang, F. and Godowski, P. J. (2004) APC-independent activation of NK cells by the Toll-like receptor 3 agonist double-stranded RNA. J Immunol. 172(1), 138-43.
Sivori, S., Carlomagno, S., Moretta, L. and Moretta, A. (2006) Comparison of different CpG oligodeoxynucleotide classes for their capability to stimulate human NK cells. Eur J Immunol. 36(4), 961-7.
Sivori, S., Falco, M., Della Chiesa, M., Carlomagno, S., Vitale, M., Moretta, L. and Moretta, A. (2004) CpG and double-stranded RNA trigger human NK cells by Toll-like receptors: induction of cytokine

release and cytotoxicity against tumors and dendritic cells. Proc Natl Acad Sci U S A. 101(27), 10116-21.

Strengell, M., Matikainen, S., Siren, J., Lehtonen, A., Foster, D., Julkunen, I. and Sareneva, T. (2003) IL-21 in synergy with IL-15 or IL-18 enhances IFN-gamma production in human NK and T cells. J Immunol. 170(11), 5464-9.

Takeshita, F., Gursel, I., Ishii, K. J., Suzuki, K., Gursel, M. and Klinman, D. M. (2004) Signal transduction pathways mediated by the interaction of CpG DNA with Toll-like receptor 9. Semin Immunol. 16(1), 17-22.

Thale, C. and Kiderlen, A. F. (2005) Sources of interferon-gamma (IFN-gamma) in early immune response to Listeria monocytogenes. Immunobiology. 210(9), 673-83.

Tsujimoto, H., Uchida, T., Efron, P. A., Scumpia, P. O., Verma, A., Matsumoto, T., Tschoeke, S. K., Ungaro, R. F., Ono, S., Seki, S., Clare-Salzler, M. J., Baker, H. V., Mochizuki, H., Ramphal, R. and Moldawer, L. L. (2005) Flagellin enhances NK cell proliferation and activation directly and through dendritic cell-NK cell interactions. J Leukoc Biol. 78(4), 888-97.

Walzer, T., Dalod, M., Robbins, S. H., Zitvogel, L. and Vivier, E. (2005) Natural-killer cells and dendritic cells: "l'union fait la force". Blood. 106(7), 2252-8.

Watford, W. T., Hissong, B. D., Bream, J. H., Kanno, Y., Muul, L. and O'Shea, J. J. (2004) Signaling by IL-12 and IL-23 and the immunoregulatory roles of STAT4. Immunol Rev. 202, 139-56.

Yokoyama, W. M. and Plougastel, B. F. (2003) Immune functions encoded by the natural killer gene complex. Nat Rev Immunol. 3(4), 304-16.

Zanoni, I., Foti, M., Ricciardi-Castagnoli, P. and Granucci, F. (2005) TLR-dependent activation stimuli associated with Th1 responses confer NK cell stimulatory capacity to mouse dendritic cells. J Immunol. 175(1), 286-92.

Zhang, D., Zhang, G., Hayden, M. S., Greenblatt, M. B., Bussey, C., Flavell, R. A. and Ghosh, S. (2004) A toll-like receptor that prevents infection by uropathogenic bacteria. Science. 303(5663), 1522-6.

2

In the Thick of the Fray: NK Cells in Inflamed Tissues

Emanuela Marcenaro[1], Mariella Della Chiesa[1], Bruna Ferranti[1] and Alessandro Moretta[1,2]

[1] Dipartimento di Medicina Sperimentale, Università degli Studi di Genova, Via L.B. Alberti 2, 16132 Italy
[2] Centro di Eccellenza per le Ricerche Biomediche, Università degli Studi di Genova, V.le Benedetto XV, 16132 Genova, Italy

1 Introduction

Although NK cells are best known for their cytolytic function, they have been shown also to display regulatory capabilities, which are mediated by various cytokines that are released upon engagement of triggering NK receptors or signalling by other cytokines. This function appears of particular relevance during the early phases of inflammatory responses. Following the pioneering study by Fernandez et al. (Fernandez, Lozier, Flament, Ricciardi-Castagnoli, Bellet, Suter, Perricaudet, Tursz, Maraskovsky and Zitvogel 1999), who first showed that dendritic cells (DCs) can induce NK activity against tumors, a number of data have highlighted recently the role of interactions occurring between NK cells and other cells of the innate immune system during the early phases of acute inflammation secondary to infection (Moretta, Marcenaro, Sivori, Della Chiesa, Vitale and Moretta 2005). Various studies have focused on the interactions between NK cells and myeloid DCs and, more recently, on the crosstalk between NK cells, plasmocytoid DC (pDC) and other innate immunity cells. These various cellular interactions can occur after the recruitment of these cells in response to various pro-inflammatory chemokines and cytokines into inflamed tissues where they can interact each other as well as with cells that are resident within peripheral tissues (Moretta et al. 2005). These different interactions can greatly impact on both effector and regulatory mechanisms occurring not only in innate immunity, but also in down-stream adaptive immune responses.

A classical example is represented by the observation that during the earliest stages of an inflammatory response, NK cells, recruited into inflamed tissues, may interact with monocyte-derived DC. Different studies have described NK cells in close contact with DC in lesions of atopic eczema/dermatitis syndrome, as well as in

Gleevec-induced lichenoid dermatitis in gastrointestinal stromal tumors (GIST)-bearing patients (Buentke, Heffler, Wilson, Wallin, Lofman, Chambers, Ljunggren and Scheynius 2002; Borg, Terme, Taieb, Menard, Flament, Robert, Maruyama, Wakasugi, Angevin, Thielemans, Le Cesne, Chung-Scott, Lazar, Tchou, Crepineau, Lemoine, Bernard, Fletcher, Turhan, Blay, Spatz, Emile, Heinrich, Mecheri, Tursz and Zitvogel 2004).

Early studies on the interactions occurring between human NK cells and monocyte-derived DCs (MDDCs) have shown that NK cell activation is required in order to observe an effective cross-talk between the two cell types (Carbone, Terrazzano, Ruggiero, Zanzi, Ottaiano, Manzo, Karre and Zappacosta 1999; Wilson, Heffler, Charo, Scheynius, Bejarano and Ljunggren 1999; Spaggiari, Carosio, Pende, Marcenaro, Rivera, Zocchi, Moretta and Poggi 2001; Ferlazzo, Tsang, Moretta, Melioli, Steinman and Munz 2002; Gerosa, Baldani-Guerra, Nisii, Marchesini, Carra and Trinchieri 2002; Piccioli, Sbrana, Melandri and Valiante 2002). On the other hand, NK cells recruited from blood into inflamed peripheral tissues may not necessarily be activated and would thus require appropriate triggering signals in order to interact with both immature and mature MDDCs (iDCs and mDCs respectively). A possible source of such signals could be represented by tumors or virus-infected cells susceptible to NK-mediated lysis (Moretta, Bottino, Vitale, Pende, Biassoni, Mingari and Moretta 1996; Lopez-Botet, Perez-Villar, Carretero, Rodriguez, Melero, Bellon, Llano and Navarro 1997; Long 1999; Vilches and Parham 2002). In this context, it has been shown that NK cells kill certain virus-infected or tumor cells expressing low levels of MHC class I molecules and subsequently prime DCs to promote highly protective CD8+ T cell memory responses (Mocikat, Braumuller, Gumy, Egeter, Ziegler, Reusch, Bubeck, Louis, Mailhammer, Riethmuller, Koszinowski and Rocken 2003). These studies suggested that NK-DC interaction might be relevant also for the generation of protective CTL responses against non-cytopathic viruses. However, in most instances, tumors are resistant to circulating, non-activated NK cells. Their killing requires exposure of NK cells to cytokines such as IL2, IL12, IL15 or IFN-alpha capable of up-regulating their cytolytic activity (Tomasello, Blery, Vely and Vivier 2000; Moretta, Bottino, Vitale, Pende, Cantoni, Mingari, Biassoni and Moretta 2001; Bottino, Castriconi, Moretta and Moretta 2005). Along this line, the NK-mediated killing of tumors, virus-infected cells or bystander iDC would provide necrotic material and heat-shock proteins to MDDC present at the same tissue site. Indeed, MDDC have been shown to process necrotic material derived from the above sources for subsequent presentation to T cells (Moretta 2002; Zitvogel 2002). In turn MDDC, upon Ag uptake, undergo maturation and release a series of cytokines that are likely to deeply influence the functional behavior of recruited NK cells. For example, IL12 is crucial for the induction of IFN-gamma release by NK cells (Byrne, Mcguirk, Todryk and Mills 2004; Moretta 2005). The formation of “stimulatory synapses” between NK and DC would promote a polarized secretion of IL12 (present in DC preassembled stores) toward NK cells (Borg, Jalil, Laderach, Maruyama, Wakasugi, Charrier, Ryffel, Cambi, Figdor, Vainchenker, Galy, Caignard and Zitvogel 2004). Another mechanism by which NK cells could become activated within pathogen-invaded tissues involves pattern recognition receptors (Della Chiesa, Sivori, Castriconi,

Marcenaro and Moretta 2005). Indeed circulating as well as in vitro-activated human NK cells express Toll-like receptors (TLRs) (Pisegna, Pirozzi, Piccoli, Frati, Santoni and Palmieri 2004; Schmidt, Leung, Kwong, Zarember, Satyal, Navas, Wang and Godowski 2004; Sivori, Falco, Della Chiesa, Carlomagno, Vitale, Moretta and Moretta 2004) that provide an alternative mode of NK-cell activation, independent on the recognition of NK-susceptible target cells. Human NK cells express functional TLR3 and TLR9 (while they do not seem to express other TLRs), thus enabling them to respond both to viral and bacterial products (dsRNA and unmetylated CpG). In particular, the simultaneous engagement of TLR3 on both NK cells and MDDC would be sufficient to initiate a series of events characteristic of the early phases of innate immune responses (Sivori et al. 2004). Mailliard and colleagues have recently shown that, in addition to IL12, other cytokines, such as IL18, released by MDDCs or by macrophages in response to TLR stimulation, can influence the "helper" activity of NK cells (Mailliard, Alber, Shen, Watkins, Kirkwood, Herberman and Kalinski 2005). IL18 does not enhance the cytolytic activity of NK cells but induces a distinct "helper" pathway of their differentiation characterized by the acquisition of the CD56+/CD83+/CCR7+/CD25+ phenotype. These IL18-induced NK cells display an high migratory responsiveness to LN-produced chemokines, a distinctive ability to support IL12 production by MDDCs and to promote Th1 polarization of CD4+ T cells responses. The important effect of IL18 would be to render also this NK subset capable of migrating to secondary lymphoid compartments (SLCs) where they can directly influence Th1 polarization.

2 NK-Cell Mediated Editing of Myeloid Dendritic Cells

The NK-mediated killing of autologous immature MDDC (but not of the mature ones) is one of the most remarkable events occurring during the NK-DC cross-talk (Moretta 2002; Zitvogel 2002). This process is based on the capability of NK cells to discriminate between myeloid iDCs (that typically under-express HLA-class I molecules) and DCs that, after Ag uptake, up-regulate MHC expression while they undergo maturation (Ferlazzo, Morandi, D'agostino, Meazza, Melioli, Moretta and Moretta 2003). During maturation, DCs up-regulate not only HLA molecules but also chemokine receptors such as CCR7, and co-stimulatory molecules belonging to the B7 molecular family. These events are crucial for the subsequent NK cell migration to lymph nodes and priming of T lymphocytes (Moretta 2002). But how can we explain this unexpected NK cell-mediated killing of normal autologous cells? A possible interpretation could be that this may serve to keep in check the quality of DCs undergoing maturation (editing process) (Moretta 2002; Moretta 2005). Thus DCs, characterized by insufficient up-regulation of MHC molecules, would be removed. Thanks to this mechanism NK cells would avoid the survival of faulty DCs. These, after the expression of CCR7 and their migration to lymph nodes, could generate un-appropriate low affinity T cell priming eventually resulting either in Th2 responses or in a state of tolerization (Della Chiesa et al. 2005; Moretta 2005). In this context, in the absence of NK cells, the in vivo default development pathway of

CD4+ T cells has been shown to be strongly biased toward the acquisition of Th2 phenotype (Scharton and Scott 1993; Tseng and Rank 1998; Coudert, Coureau and Guery 2002; Byrne et al. 2004). Alternatively, by this mechanism, NK cells could limit the number of iDCs available for generating suitable numbers of mDCs. Thus, the lack of NK-cell-mediated of killing iDC could result in altered numbers of DCs reaching complete maturation (Moretta 2002; Zitvogel 2002).

3 IL4 and IL12 Promote Opposite Effects on the NK-MDDC Crosstalk

In the early stages of an inflammatory response in peripheral tissues, the engagement of TLRs may not be confined to NK and DCs but it could rather involved other cell types including resident mast cells (Supajatura, Ushio, Nakao, Akira, Okumura, Ra and Ogawa 2002; Kulka, Alexopoulou, Flavell and Metcalfe 2004), eosinophils (Bjerke, Gaustadnes, Nielsen, Nielsen, Schiotz, Rudiger, Reimert, Dahl, Christensen and Poulsen 1996; Nagase, Okugawa, Ota, Yamaguchi, Tomizawa, Matsushima, Ohta, Yamamoto and Hirai 2003) or plasmocytoid DCs (pDCs) (Gerosa, Gobbi, Zorzi, Burg, Briere, Carra and Trinchieri 2005). These cells, through the release of cytokines other than IL12 (e.g. IL4 or IFN-alpha), may differentially modulate the function of bystander NK and MDDCs. In this context, as recently proposed, the early exposure of NK cells to IL4 could deviate the subsequent adaptive response towards a state of tolerization or generation of either Th2 or unpolarized T cells (Marcenaro, Della Chiesa, Bellora, Parolini, Millo, Moretta and Moretta 2005). In particular, while short-term NK-cell exposure to IL12 promoted the release of high levels of both IFN-gamma and TNF-alpha and the acquisition of cytolytic activity, the exposure to IL4 resulted in poor cytokine production and cytolytic activity. Importantly, only NK cells exposed to IL12 could promote efficient DC maturation (Marcenaro et al. 2005). These data further support the notion that IL12-primed NK cells contribute, via their editing program, to the selection of mature MDDCs that are capable of promoting optimal priming of Th1 responses (Scharton-Kersten, Afonso, Wysocka, Trinchieri and Scott 1995; Mailliard, Son, Redlinger, Coates, Giermasz, Morel, Storkus and Kalinski 2003; Marcenaro, Chiesa, Dondero, Ferranti and Moretta 2007). In addition they also indicate that NK-cell priming in the presence of IL4 results in abnormal MDDC maturation (failure of the editing program) characterized by both qualitative and quantitative alterations. In conclusion, depending on the type of cytokines released during the early stages of an inflammatory response, by either resident or recruited cells (Moretta et al. 2005), NK cells differentially contribute to the quality and magnitude of innate immune responses (i.e. killing of tumor cells, DC editing and cytokine release). This, eventually, can deeply impact on the type of down-stream adaptive T cell responses (Mailliard et al. 2003; Mailliard et al. 2005; Marcenaro et al. 2005; Marcenaro et al. 2007).

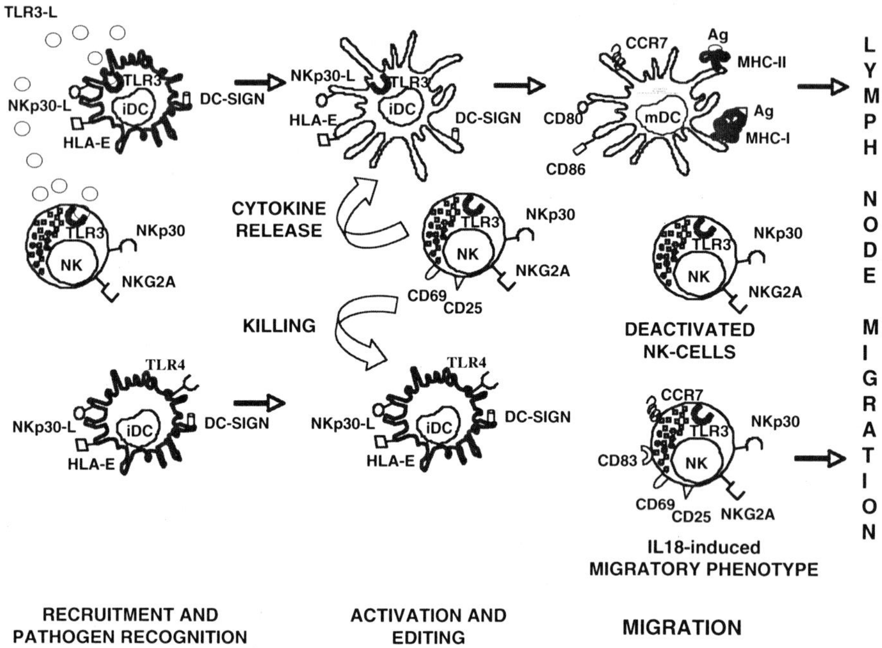

Fig. 1. The NK-mediated editing process. NK and DC are first recruited to inflammatory sites in response to cytokine gradients. Then both cell types are activated upon encounter with pathogen-derved products that are recognized via TLRs. It is likely that heterogeneity may exist among NK and iDC in the expression of given TLR. Thus only certain iDC (and NK) would undergo the process of pathogen-induced maturation that is induced by given TLR ligand (in this case TLR3 ligand). Once activated by the simultaneous exposure to both TLR ligands and IL12 (released by DCs) NK cells acquire the ability to kill DCs that have not been suitably activated by the invading pathogen.These DCs, different from those undergoing appropriate maturation, do not up-regulate MHC molecules and, similar to iDCs, would remain susceptible to NK-mediated killing. In the meanwhile NK cells start releasing TNF-alpha and IFN-gamma that have been demonstrated to promote DC maturation. These cytokines would then facilitate the progression of pathogen-responsive DCs towards a full maturation. The latter is characterized by a strong up-regulation of MHC expression by the de-novo expression of CCR7 and by the up-regulation of co-stimulatory molecules such as CD80 and CD86. At this stage mature DCs are ready for migration to secondary lymphoid compartments where they promote Th1 polarization. On the other hand, NK cells would remain within the inflamed tissue and, in the absence of further stimulation, may return to their original resting condition. Alternatively, upon exposure to massive doses of IL18, they may acquire surface CCR7 and CD83 and migrate to secondary lymphoid organs together with mature DCs.

4 Acknowledgements

This work was supported by grants awarded by Associazione Italiana per la Ricerca sul Cancro (A.I.R.C.), Istituto Superiore di Sanità (I.S.S.), Ministero della Salute - RF 2002/149, Ministero dell'Istruzione dell'Università e della Ricerca (M.I.U.R.), FIRB-MIUR progetto – cod.RBNE017B4C; Ministero dell'Università e della

Ricerca Scientifica e Tecnologica (M.U.R.S.T.), European Union FP6, LSHB-CT-2004-503319-Allostem and Compagnia di San Paolo.

References

Bjerke, T., Gaustadnes, M., Nielsen, S., Nielsen, L. P., Schiotz, P. O., Rudiger, N., Reimert, C. M., Dahl, R., Christensen, I. and Poulsen, L. K. (1996) Human blood eosinophils produce and secrete interleukin 4. Respir Med. 90, 271-277.

Borg, C., Jalil, A., Laderach, D., Maruyama, K., Wakasugi, H., Charrier, S., Ryffel, B., Cambi, A., Figdor, C., Vainchenker, W., Galy, A., Caignard, A. and Zitvogel, L. (2004) NK cell activation by dendritic cells (DCs) requires the formation of a synapse leading to IL-12 polarization in DCs. Blood. 104, 3267-3275.

Borg, C., Terme, M., Taieb, J., Menard, C., Flament, C., Robert, C., Maruyama, K., Wakasugi, H., Angevin, E., Thielemans, K., Le Cesne, A., Chung-Scott, V., Lazar, V., Tchou, I., Crepineau, F., Lemoine, F., Bernard, J., Fletcher, J. A., Turhan, A., Blay, J. Y., Spatz, A., Emile, J. F., Heinrich, M. C., Mecheri, S., Tursz, T. and Zitvogel, L. (2004) Novel mode of action of c-kit tyrosine kinase inhibitors leading to NK cell-dependent antitumor effects. J Clin Invest. 114, 379-388.

Bottino, C., Castriconi, R., Moretta, L. and Moretta, A. (2005) Cellular ligands of activating NK receptors. Trends Immunol. 26, 221-226.

Buentke, E., Heffler, L. C., Wilson, J. L., Wallin, R. P., Lofman, C., Chambers, B. J., Ljunggren, H. G. and Scheynius, A. (2002) Natural killer and dendritic cell contact in lesional atopic dermatitis skin--Malassezia-influenced cell interaction. J Invest Dermatol. 119, 850-857.

Byrne, P., Mcguirk, P., Todryk, S. and Mills, K. H. (2004) Depletion of NK cells results in disseminating lethal infection with Bordetella pertussis associated with a reduction of antigen-specific Th1 and enhancement of Th2, but not Tr1 cells. Eur J Immunol. 34, 2579-2588.

Carbone, E., Terrazzano, G., Ruggiero, G., Zanzi, D., Ottaiano, A., Manzo, C., Karre, K. and Zappacosta, S. (1999) Recognition of autologous dendritic cells by human NK cells. Eur J Immunol. 29, 4022-4029.

Coudert, J. D., Coureau, C. and Guery, J. C. (2002) Preventing NK cell activation by donor dendritic cells enhances allospecific CD4 T cell priming and promotes Th type 2 responses to transplantation antigens. J Immunol. 169, 2979-2987.

Della Chiesa, M., Sivori, S., Castriconi, R., Marcenaro, E. and Moretta, A. (2005) Pathogen-induced private conversations between natural killer and dendritic cells. Trends Microbiol. 13, 128-136.

Ferlazzo, G., Morandi, B., D'agostino, A., Meazza, R., Melioli, G., Moretta, A. and Moretta, L. (2003) The interaction between NK cells and dendritic cells in bacterial infections results in rapid induction of NK cell activation and in the lysis of uninfected dendritic cells. Eur J Immunol. 33, 306-313.

Ferlazzo, G., Tsang, M. L., Moretta, L., Melioli, G., Steinman, R. M. and Munz, C. (2002) Human dendritic cells activate resting natural killer (NK) cells and are recognized via the NKp30 receptor by activated NK cells. J Exp Med. 195, 343-351.

Fernandez, N. C., Lozier, A., Flament, C., Ricciardi-Castagnoli, P., Bellet, D., Suter, M., Perricaudet, M., Tursz, T., Maraskovsky, E. and Zitvogel, L. (1999) Dendritic cells directly trigger NK cell functions: cross-talk relevant in innate anti-tumor immune responses in vivo. Nat Med. 5, 405-411.

Gerosa, F., Baldani-Guerra, B., Nisii, C., Marchesini, V., Carra, G. and Trinchieri, G. (2002) Reciprocal activating interaction between natural killer cells and dendritic cells. J Exp Med. 195, 327-333.

Gerosa, F., Gobbi, A., Zorzi, P., Burg, S., Briere, F., Carra, G. and Trinchieri, G. (2005) The reciprocal interaction of NK cells with plasmacytoid or myeloid dendritic cells profoundly affects innate resistance functions. J Immunol. 174, 727-734.

Kulka, M., Alexopoulou, L., Flavell, R. A. and Metcalfe, D. D. (2004) Activation of mast cells by double-stranded RNA: evidence for activation through Toll-like receptor 3. J Allergy Clin Immunol. 114, 174-182.

Long, E. O. (1999) Regulation of immune responses through inhibitory receptors. Annu Rev Immunol. 17, 875-904.

Lopez-Botet, M., Perez-Villar, J. J., Carretero, M., Rodriguez, A., Melero, I., Bellon, T., Llano, M. and Navarro, F. (1997) Structure and function of the CD94 C-type lectin receptor complex involved in recognition of HLA class I molecules. Immunol Rev. 155, 165-174.

Mailliard, R. B., Alber, S. M., Shen, H., Watkins, S. C., Kirkwood, J. M., Herberman, R. B. and Kalinski, P. (2005) IL-18-induced CD83+CCR7+ NK helper cells. J Exp Med. 202, 941-953.

Mailliard, R. B., Son, Y. I., Redlinger, R., Coates, P. T., Giermasz, A., Morel, P. A., Storkus, W. J. and Kalinski, P. (2003) Dendritic cells mediate NK cell help for Th1 and CTL responses: two-signal requirement for the induction of NK cell helper function. J Immunol. 171, 2366-2373.

Marcenaro, E., Chiesa, M. D., Dondero, A., Ferranti, B. and Moretta, A. (2007) It's only innate immunity but I like it. Adv Exp Med Biol. 590, 89-101.

Marcenaro, E., Della Chiesa, M., Bellora, F., Parolini, S., Millo, R., Moretta, L. and Moretta, A. (2005) IL-12 or IL-4 prime human NK cells to mediate functionally divergent interactions with dendritic cells or tumors. J Immunol. 174, 3992-3998.

Mocikat, R., Braumuller, H., Gumy, A., Egeter, O., Ziegler, H., Reusch, U., Bubeck, A., Louis, J., Mailhammer, R., Riethmuller, G., Koszinowski, U. and Rocken, M. (2003) Natural killer cells activated by MHC class I(low) targets prime dendritic cells to induce protective CD8 T cell responses. Immunity. 19, 561-569.

Moretta, A. (2002) Natural killer cells and dendritic cells: rendezvous in abused tissues. Nat Rev Immunol. 2, 957-964.

Moretta, A. (2005) The dialogue between human natural killer cells and dendritic cells. Curr Opin Immunol. 17, 306-311.

Moretta, A., Bottino, C., Vitale, M., Pende, D., Biassoni, R., Mingari, M. C. and Moretta, L. (1996) Receptors for HLA class-I molecules in human natural killer cells. Annu Rev Immunol. 14, 619-648.

Moretta, A., Bottino, C., Vitale, M., Pende, D., Cantoni, C., Mingari, M. C., Biassoni, R. and Moretta, L. (2001) Activating receptors and coreceptors involved in human natural killer cell-mediated cytolysis. Annu Rev Immunol. 19, 197-223.

Moretta, A., Marcenaro, E., Sivori, S., Della Chiesa, M., Vitale, M. and Moretta, L. (2005) Early liaisons between cells of the innate immune system in inflamed peripheral tissues. Trends Immunol. 26, 668-675.

Nagase, H., Okugawa, S., Ota, Y., Yamaguchi, M., Tomizawa, H., Matsushima, K., Ohta, K., Yamamoto, K. and Hirai, K. (2003) Expression and function of Toll-like receptors in eosinophils: activation by Toll-like receptor 7 ligand. J Immunol. 171, 3977-3982.

Piccioli, D., Sbrana, S., Melandri, E. and Valiante, N. M. (2002) Contact-dependent stimulation and inhibition of dendritic cells by natural killer cells. J Exp Med. 195, 335-341.

Pisegna, S., Pirozzi, G., Piccoli, M., Frati, L., Santoni, A. and Palmieri, G. (2004) p38 MAPK activation controls the TLR3-mediated up-regulation of cytotoxicity and cytokine production in human NK cells. Blood. 104, 4157-4164.

Scharton, T. M. and Scott, P. (1993) Natural killer cells are a source of interferon gamma that drives differentiation of CD4+ T cell subsets and induces early resistance to Leishmania major in mice. J Exp Med. 178, 567-577.

Scharton-Kersten, T., Afonso, L. C., Wysocka, M., Trinchieri, G. and Scott, P. (1995) IL-12 is required for natural killer cell activation and subsequent T helper 1 cell development in experimental leishmaniasis. J Immunol. 154, 5320-5330.

Schmidt, K. N., Leung, B., Kwong, M., Zarember, K. A., Satyal, S., Navas, T. A., Wang, F. and Godowski, P. J. (2004) APC-independent activation of NK cells by the Toll-like receptor 3 agonist double-stranded RNA. J Immunol. 172, 138-143.

Sivori, S., Falco, M., Della Chiesa, M., Carlomagno, S., Vitale, M., Moretta, L. and Moretta, A. (2004) CpG and double-stranded RNA trigger human NK cells by Toll-like receptors: induction of cytokine release and cytotoxicity against tumors and dendritic cells. Proc Natl Acad Sci U S A. 101, 10116-10121.

Spaggiari, G. M., Carosio, R., Pende, D., Marcenaro, S., Rivera, P., Zocchi, M. R., Moretta, L. and Poggi, A. (2001) NK cell-mediated lysis of autologous antigen-presenting cells is triggered by the engagement of the phosphatidylinositol 3-kinase upon ligation of the natural cytotoxicity receptors NKp30 and NKp46. Eur J Immunol. 31, 1656-1665.

Supajatura, V., Ushio, H., Nakao, A., Akira, S., Okumura, K., Ra, C. and Ogawa, H. (2002) Differential responses of mast cell Toll-like receptors 2 and 4 in allergy and innate immunity. J Clin Invest. 109, 1351-1359.

Tomasello, E., Blery, M., Vely, F. and Vivier, E. (2000) Signaling pathways engaged by NK cell receptors: double concerto for activating receptors, inhibitory receptors and NK cells. Semin Immunol. 12, 139-147.

Tseng, C. T. and Rank, R. G. (1998) Role of NK cells in early host response to chlamydial genital infection. Infect Immun. 66, 5867-5875.

Vilches, C. and Parham, P. (2002) KIR: diverse, rapidly evolving receptors of innate and adaptive immunity. Annu Rev Immunol. 20, 217-251.

Wilson, J. L., Heffler, L. C., Charo, J., Scheynius, A., Bejarano, M. T. and Ljunggren, H. G. (1999) Targeting of human dendritic cells by autologous NK cells. J Immunol. 163, 6365-6370.

Zitvogel, L. (2002) Dendritic and natural killer cells cooperate in the control/switch of innate immunity. J Exp Med. 195, F9-14.

3

Structural Insight into Natural Killer T Cell Receptor Recognition of CD1d

Natalie A. Borg[1], Lars Kjer-Nielsen[2], James McCluskey[2] and Jamie Rossjohn[1]

[1] Natalie A. Borg and Jamie Rossjohn, Protein Crystallography Unit, Department of Biochemistry and Molecular Biology, Monash University, Victoria, Australia, 3800.
[2] Lars Kjer-Nielsen and James McCluskey, Department of Microbiology and Immunology, The University of Melbourne, Victoria, Australia, 3010.

1 Introduction

CD1d is a non-polymorphic antigen-presenting molecule that is structurally related to the MHC class I molecule. CD1d molecules are heterodimers that are composed of a heavy chain that is non-covalently associated with β2-microglobulin (β_2m). The heavy chain is composed of three extracellular domains designated α1, α2 and α3 as well as a transmembrane and an intracytoplasmic domain. The antigen-binding cavity is formed from the α1 and α2 domains of the heavy chain. These domains comprise two α-helices which flank the opening of the cleft and a β-sheet floor (see Fig. 1a and b). The cleft is hydrophobic in nature and composed of an A′ and an F′ pocket (sometimes called C′ pocket). Despite the structural similarities between MHC class I and CD1d molecules, the unique properties of their antigen-binding cavities enable them to present distinct antigens to the immune system for immune surveillance. MHC class I molecules bind and present peptides, whereas CD1d molecules bind and present a diverse range of hydrophobic lipids from self and foreign sources (Fischer et al., 2004; Giabbai et al., 2005; Kinjo et al., 2005; Mattner et al., 2005; Sriram et al., 2005; Wu et al., 2005; Wu et al., 2006; Wu et al., 2003; Zhou et al., 2004). The localisation of CD1d in late endosomes or lysosomes is ideal for the uptake of foreign lipid antigens, which accumulate in these endocytic compartments during a microbial infection (Kang & Cresswell, 2004). CD1d is a key player in our ability to overcome microbial infections because it is able to present foreign lipid antigens to the subset of T cells termed Natural Killer T (NKT) cells.

NKT cells express an αβ T cell receptor (TcR) and markers usually associated with NK cells such as NK1.1, members of the Ly-49 and DX5 receptor family, CD122 and CD16.

Although previously viewed as a single population of cells, NKT cells are heterogeneous and can be divided into functionally distinct subsets (Crowe et al.,

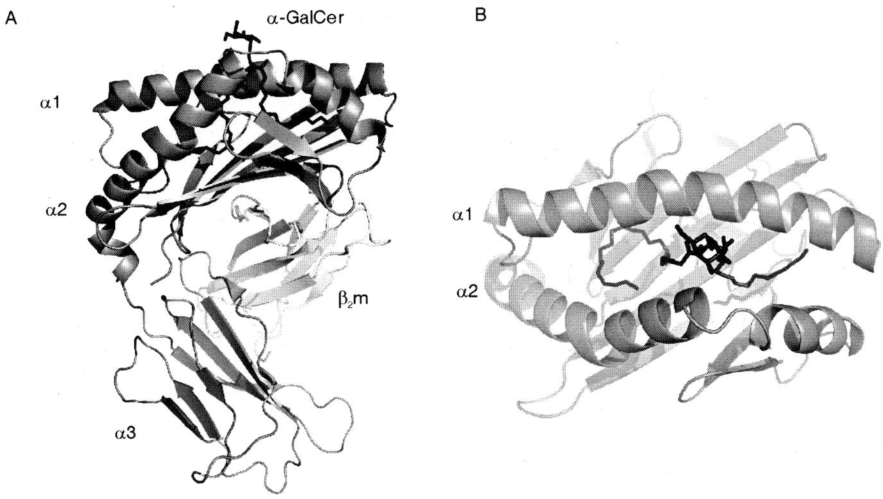

Fig. 1. Overview of the three-dimensional structure of hCD1d. Cartoon representation of hCD1d from a) a side-view b) an aerial view of the antigen-binding domain.

2005; Hammond et al., 1999). This review will focus on two NKT cell receptor subsets that are stimulated by the potent agonist α-galactosylceramide (α-GalCer) in the context of CD1d (Benlagha et al., 2000; Kawano et al., 1997; Lantz & Bendelac, 1994). These include the invariant NKT (iNKT) cell receptor and Vα24 independent TcRs. The iNKT cell receptor is the most studied subset of NKT cell receptors to date and is semi-invariant in nature. The Vα24/Jα18 (TRAV10/TRAJ18) and the Vβ11-gene segment (TRBV25-1) is preferentially selected in human iNKT cell receptors, whereas in mice the invariant α-chain Vα14/Jα18-gene segment is associated with one of three β-chain combinations (Vβ2, Vβ7, Vβ8.2). The β-chain of iNKT cell receptors typically exhibit a wide diversity of CDR3 junctions such that the CDR3β loop differs in its composition and length and generally exhibits a high degree of natural variability (Matsuda et al., 2001). CD1d/α-GalCer specific T cell clones that do not use the Vα24 gene segment (Brigl et al., 2006; Gadola et al., 2002; Gadola et al., 2006) will be referred to as Vα24 independent TcRs. Interestingly, the Vα segment of some of these TcRs has been determined to recombine with the Jα18 segment that is selected in iNKT cell receptors (Brigl et al., 2006; Gadola et al., 2006) (see Table 1). Furthermore, although the Vα24 independent TcRs use a selection of Vβ segments, they have a bias for the Vβ11 chain that is most commonly selected in human iNKT cell receptors (Brigl et al., 2006; Gadola et al., 2002; Gadola et al., 2006) (see Table 2). iNKT cell receptors and Vα24 independent TcRs have been shown to 'respond similarly' to α-GalCer (Brigl et al., 2006) and have a similar affinity for CD1d/α-GalCer as do other αβ TcRs for peptide/MHC class I (Clements et al., 2006; Ely et al., 2005). Affinities between 4-9 μM (Gadola et al., 2006; Kjer-Nielsen et al., 2006) have recently been reported for human NKT cell receptors,

Table 1. Amino acid residues composing the CDR3α loop in select iNKT and Vα24 independent (Vα24-) T cell receptors.

TcR				
iNKT	CDR3α			
NKT15	**Vα24/TRAV10**	ICVVS	**Jα18/TRAJ18**	DRGSTL
NKT12	**Vα24/TRAV10**	ICVVS	**Jα18/TRAJ18**	DRGSTL
iNKT	**Vα24/TRAV10**	ICVVS	**Jα18/TRAJ18**	DRGSTL
Vα24-				
5B	**Vα3.1/TRAV17**	FCAPF	**Jα18/TRAJ18**	DRGSTL
5E	**Vα10.1/TRAV27**	LCAGA	**Jα18/TRAJ18**	DRGSTL

Table 2. Amino acid residues composing the CDR3β loop in select iNKT and Vα24 independent (Vα24-) T cell receptors.

TcR					
iNKT	CDR3β		**TRBD + N**		
NKT15	**Vβ11/TRBV25-1**	CASS	GLRDRGL	**TRBJ2-7*01**	YEQYFGPG
NKT12	**Vβ11/TRBV25-1**	CAS	TSRRG	**TRBJ2-7*01**	SYEQYFGP
iNKT	**Vβ11/TRBV25-1**	CASS	ENIGT	**TRBJ2-7*01**	AYEQYF
Vα24-					
5B	**Vβ11/TRBV25-1**	CASS	ESRTGI	**TRBJ1-2*01**	NYGYTFGSGTRL
5E	**Vβ11/TRBV25-1**	CASS	EFRDG	**TRBJ1-4*01**	NEKLFF

while affinities of 0.1-0.3 μM have been reported for mouse NKT cell receptors (Cantu et al., 2003; Sidobre et al., 2002).

Upon activation thymus-derived NKT cells secrete large quantities of T helper type 1 and T helper type 2 cytokines like IL-2, IFN-γ, tumour necrosis factor A and IL-4. The rapidly secreted cytokines are then able to influence 'downstream' immune responses hence acting as intermediates of the innate and adaptive immune response (Nieda et al., 2004). The ability of NKT cells to release both proinflammatory and immunoregulatory cytokines, in addition to their prominent role in tumour immunity, immune surveillance, transplantation and autoimmune diseases has drawn attention to the therapeutic potential of NKT cells (Mercer et al., 2005). Manipulation of the NKT cell response may provide an opportunity to modulate cytokine patterns to achieve a more desired outcome. Paramount to this objective is the information obtained from the combined approach of structural and functional analysis that offers useful insights into the presentation of ligands by CD1d and the subsequent influence on NKT cell recognition.

2 CD1d Ligand Presentation Influences NKT Cell Recognition

CD1d is able to bind and present a diverse array of foreign and self-lipids, some of which can stimulate an NKT cell response. For example, iNKT cell receptors can be stimulated by CD1d in complex with mycobacterial phosphatidylinositol mannosides (Fischer et al., 2004); microbial cell wall glycuronosylceramide antigens such as α-galacturonosylceramide and analogues thereof (Kinjo et al., 2005; Mattner et al., 2005; Sriram et al., 2005; Wu et al., 2005; Wu et al., 2006); glycuronosylceramide homologues such as α-galactosylceramide and analogues thereof; the tumor derived disialoganglioside GD3 (Wu et al., 2003); a diacylglycerol glycolipid from *Borella burgdorferi* (Kinjo et al., 2006); the self-lipid isoglobotrihexosylceramide (iGb3) (Zhou et al., 2004) and the self glycerophospholipid phosphatidylcholine (PC) (Giabbai et al., 2005). Although iNKT cell receptors are able to recognise diverse ligands presented by CD1d they are able to selectively discriminate between analogues that have slight structural modifications (Goff et al., 2004; Wu et al., 2006; Yu et al., 2005; Zajonc et al., 2005a). This raises the questions: How can iNKT cell receptors with a semi-invariant TcR recognise diverse lipid antigens with structurally distinct head groups? How do iNKT cell receptors differentiate between structurally similar ligands?.

The three-dimensional structure of mouse or human CD1d has been determined in complex with various self and foreign lipids. The self-glycolipid/CD1d complexes include, the glycerophospholipid phosphatidylcholine (PC) (Giabbai et al., 2005) and the glycosphingolipid 3′ sulfogalactosylceramide (sulfatide) (Zajonc et al., 2005b). The foreign glycolipids include 1) α-galactosylceramide (α-GalCer), a glycosphingolipid from the marine sponge *Agelas mauritanius* (see Fig. 2a) (Koch et al., 2005); 2) PBS-25, a short-chain variant of α-GalCer (see Fig. 2b) (Zajonc et al., 2005a); 3) α-glucuronosylceramide (GalA-GSL), a microbial glycosphingolipid from *Sphingomonas* (see Fig. 2c) (Wu et al., 2006); 4) the synthetic glycerophospholipid dipalmitoyl-PIM_2 (see Fig. 2d) (Zajonc et al., 2006). The growing number of structures provide invaluable insight into 1) the nature of the antigen-binding cavity 2) the common features of glycolipids that can bind to CD1d 3) generalities in the mode of binding of foreign glycolipids in the antigen-binding cavity and 4) CD1d residues likely to interact with the NKT cell receptor. Some general trends have been observed:

1) The ligands are composed of two long hydrophobic acyl chains, one bound in the A′ and one bound in the F′ pocket of the antigen-binding cavity. The larger A′ pocket can accommodate up to 26 carbons (Koch et al., 2005), however shorter length acyl chains can still be accommodated and are often stabilised by the binding of spacer-lipids in the A′ pocket (Wu et al., 2006; Zajonc et al., 2005a).

2) The acyl chain in the A′ pocket has been reported as being either blocked (Zajonc et al., 2006) or accessible (Zajonc et al., 2005b) to TcR access depending on the glycolipid that is presented. The F′ pocket of the antigen-binding cavity is optimal for an 18-carbon length chain (Koch et al., 2005) and accommodates the sphingoid base of a glycosphingolipid. This pocket is capped in one structure such that the sphingosine tail is denied access to the CD1d surface (Zajonc et al., 2005a).

3) As yet there is no consensus as to whether the *sn*1- and *sn*2-fatty acids (eg. see Fig. 2d) of the glycerophospholipids will bind in the A′ or the F′ pocket (Giabbai et al., 2005; Zajonc et al., 2006). However, the glycerolipid tails of the glycerophospholipids do appear to be more flexible than the acyl tails of the glycosphingolipids (Zajonc et al., 2006).

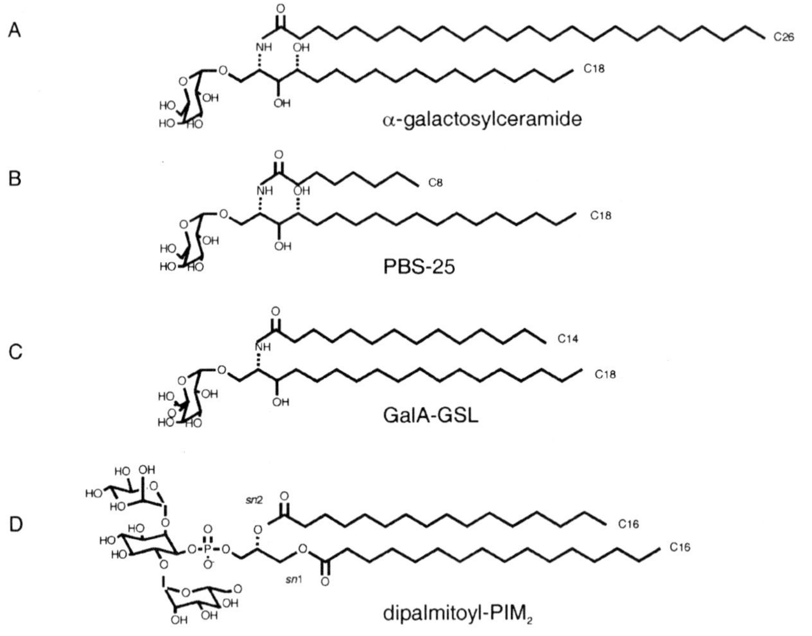

Fig. 2. Schematic representation of foreign glycolipids that have been determined in complex with hCD1d.

4) The hydrophobic tail of the glycolipid is linked via an α- or β-anomeric link to a hydrophilic head group. The composition of the head group is a single or multiple-linked sugar moiety with or without modifications. The head group of the glycolipid protrudes out of the antigen-binding cleft where it is available for TcR recognition (Koch et al., 2005; Wu et al., 2006; Zajonc et al., 2006; Zajonc et al., 2005a; Zajonc et al., 2005b).

5) Stabilising interactions are formed between residue Asp80 on the α1 helix of CD1d and the 3′-hydroxyl and in some cases the 4′-hydroxyl groups of the sphingoid base (Koch et al., 2005; Wu et al., 2006; Zajonc et al., 2005a) of a glycosphingolipid. Glycosphingolipids with a sphinganine backbone, such as GalA-GSL, lack the 4′-hydroxyl and only make contact with CD1d through the 3′-hydroxyl of the sphinganine backbone. The fewer contacts lead to a less rigid head group conformation (Wu et al., 2006) which may also alter the conformation of TcR exposed residues like Arg79 on the α1 helix of CD1d. The extent of these interactions also subsequently determines the depth that the glycolipid sits in the groove as well as the lateral disposition, orientation and the rigidity of the sugar moiety (Wu et al., 2006).

6) An aspartate on the α2 helix of CD1d (D151 in human CD1d; D153 in mouse CD1d) forms hydrogen bonds with the 2′-hydroxyl and/or the 3′-hydroxyl of the sugar moiety (Koch et al., 2005; Wu et al., 2006; Zajonc et al., 2005a). The configuration of the 2′-hydroxyl is correlated with the lack of stimulatory activity of α-mannosylceramide (α-ManCer) compared to α-GalCer, therefore such subtle differences are crucial to TcR recognition (Kawano et al., 1997).

7) The substitution of CD1d residues (R79, D80 and D153) involved in hydrogen bonds with the short chain α-GalCer ligand resulted in abolished NKT cell stimulation, most likely because the conformation of the sugar moiety was severely disrupted (Kamada et al., 2001).

8) The importance of the conformation of the sugar moiety on NKT cell recognition is reinforced by the inability (Kawano et al., 1997) or minimal ability (Parekh et al., 2004) of β-galactosyl-ceramide (β-GalCer) to induce NKT cell proliferation and cytokine release (Ortaldo et al., 2004; Parekh et al., 2004) compared to the highly potent α-GalCer. β-GalCer is predicted to make fewer contacts with the α2 helix of CD1d such that the sugar moiety adopts an alternate perpendicular orientation to the CD1d binding groove (Zajonc et al., 2005a), once again influencing TcR recognition and cytokine release.

9) It was previously considered that the length of the lipid tail in the binding groove had an impact on the antigenic potency. However, α-GalCer (C26) and the identical, but otherwise shorter tailed PBS-25 (C8) have identical antigenic potency (Brossay et al., 1998; Zajonc et al., 2005a). The cytokine response has been shown however to be altered for other α-GalCer analogues with both a shortened lipid tail and the introduction of unsaturations (Yu et al., 2005) and also with other truncated, saturated lipids (Goff et al., 2004; Miyamoto et al., 2001).

3 NKT Cell Receptors are Structurally Biased

In addition to the structural information available for CD1d with various lipids, the three-dimensional structure of CD1d/α-GalCer restricted NKT cell receptors is now available and offers a structural insight into 1) the selection of the semi-invariant iNKT cell receptor in the context of CD1d; 2) the highly selected nature of the Jα18 and the Vβ11 gene segments in iNKT cell receptors and some Vα24 independent TcRs; 3) cross-species reactivity; 4) the ability of iNKT cell receptors to recognise a diverse array of lipids in the context of CD1d.

The five NKT cell receptors for which there is structural information available include 1) three different semi-invariant iNKT cell receptors, NKT12, NKT15 and iNKT (PDB accession codes 2EYR, 2EYS and 2CDE respectively) which are $CD4^-$ $CD8^-$ (DN) (Gadola et al., 2006; Kjer-Nielsen et al., 2006); 2) two Vα24 independent TcRs that are also CD1d/α-GalCer restricted but either $CD4^+$ (5B TcR, PDB accession code 2CDF) or $CD8\alpha\beta^+$ (5E TcR, PDB accession code 2CDG) (Gadola et al., 2006). Interestingly, although the 5B and 5E TcRs are Vα24 independent they recombine with the same Jα18 segment to create their CDR3α loop as do iNKT cell

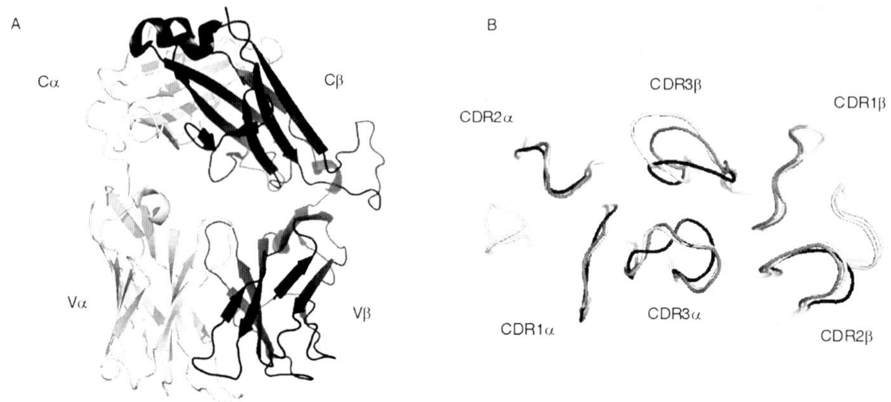

Fig. 3. Overview of the structure of the NKT cell receptor. a) overall view of NKT12. b) superposition of the three iNKT cell receptors for which structural information is available, NKT12 (grey), NKT15 (white) and iNKT (black). View of the conformation of the CDR loops at the antigen-binding interface.

receptors and in addition are also Vβ11[+] (see Table 1). Therefore the residues comprising the CDR3α loops of iNKT cell receptors and these Vα24 independent TcRs are for the most part the same.

All five NKT cell receptors are structurally homologous to MHC class I restricted αβ TcRs, being composed of an α and β-chain, each with a constant (C) and variable (V) domain. The three-complementarity determining regions (CDR) reside on the variable domain of each chain and include CDR1α, CDR2α, CDR3α, CDR1β, CDR2β and CDR3β (see Fig. 3a). The six CDRs comprise the antigen-binding site of the TcR. A superposition of the CDR loops of the three iNKT cell receptors, iNKT, NKT12, and NKT15 shows that the Cα trace of the CDR1α, CDR2α, CDR1β, and CDR2β loops is almost the same (see Fig. 3b). The differences however lie in the conformation of the CDR3α and CDR3β loops. While each of these TcRs use the invariant α-chain, and thus have an identical composition of their CDR3α loops, the composition and length of their CDR3β loops differ by up to four residues due to diversity in the junctional regions. This indicates that the conformation of the CDR3α loop is dependent to some extent on the conformation of the CDR3β loop. The conformation of three CDR loops of the Vα24 independent TcRs differ remarkably from that of the iNKT cell receptors (figure not shown). These loops include CDR2α and more significantly, CDR3α and CDR3β.

One of the three iNKT cell receptors solved (NKT12) has a markedly electropositive cavity on the surface of the antigen-binding domain (see Fig. 4a) and this is lined by residues from the CDR1α, CDR3α, CDR1β, and CDR3β loops. This cavity appears to be significant as one of the residues lining the base of the cavity (R33α) is unique to the Vα24 gene segment that is highly selected in human iNKT cell receptors, whilst three residues that line the cavity (G96α, S97α, and L104α) are unique to the Jα18 gene segment. The importance of the residues from the Jα18 gene

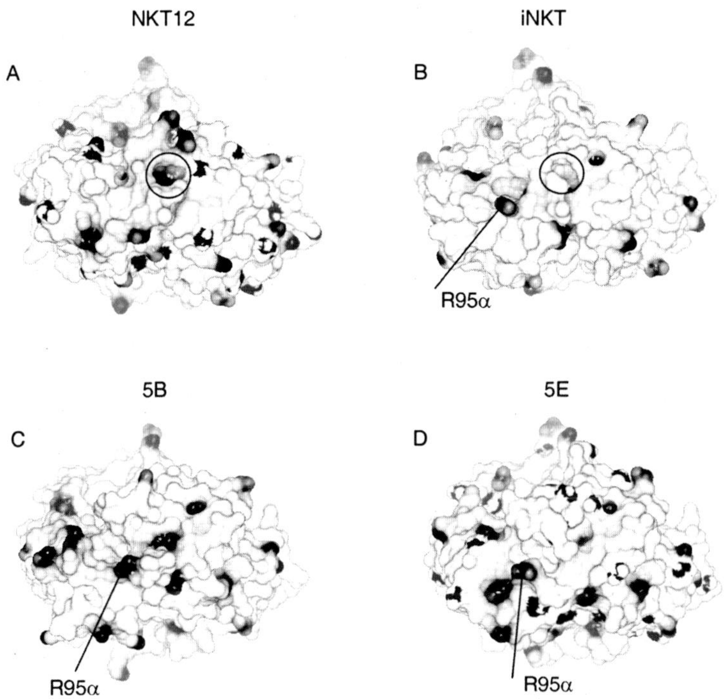

Fig. 4. Surface representations of the antigen-binding domain of select iNKT and Vα24 independent NKT cell receptors. Electropositive residues are highlighted in black. iNKT cell receptors a) NKT12, with the electropositive cavity circled b) iNKT and the Vα24 independent NKT cell receptors c) 5B and d) 5E.

segment to ligand recognition is also highlighted by the selection of this segment in the Vα24 independent TcRs 5B and 5E.

The electropositive cavity of NKT12 is approximately 120 $Å^3$ in volume and large enough to accommodate a carbohydrate moiety, such as the polar galactose moiety of α-GalCer which protrudes out of the CD1d antigen-binding cleft (Koch et al., 2005). However, NKT cell receptors are also activated by ligands with diverse carbohydrate groups such as the terminal Gal α1,3 Gal β1-4 Glc of iGb3 (Zhou et al., 2004) and the four mannose moieties of PIM_4 (Fischer et al., 2004). Due to 1) The flexible nature of the CDR3β loop in our crystal structures; 2) our analysis of hybrid TcRs with switched or mutated CDR3β regions; 3) the structural differences in the CDR3β loop of iNKT compared to NKT12 and NKT15 (see Fig. 3b); we propose that the function of the CDR3β loop is to fine tune antigen recognition. Movement of the CDR3β loop may alter the volume of the proposed antigen-binding cavity and enable more diverse head group/s of CD1d ligands to be accommodated.

Although the residues lining the electropositive cavity are conserved between NKT12, NKT15, and iNKT, the cavity is only observed in NKT12. The cavity of NKT15 is obstructed by the bulky Y103β on the CDR3β loop. However, this residue

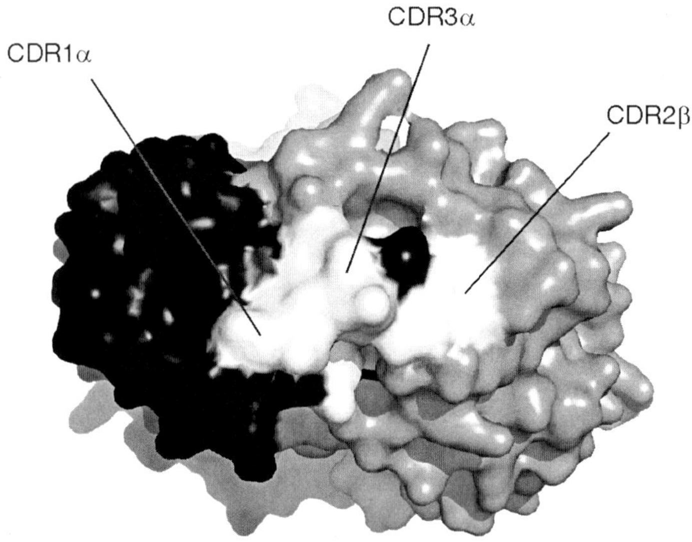

Fig. 5. Surface representation of the antigen-binding domain of the iNKT cell receptor, NKT12. Conserved residues between human and mouse iNKT cell receptors are coloured white. The α-chain is coloured black and the β-chain is coloured grey.

is highly mobile in the 2.2 Å crystal structure and could potentially be displaced. In the remaining iNKT cell receptor structure (iNKT) a distinct electropositive cavity is not observed on the surface of the antigen-binding domain, but instead, this region is flanked by the electropositive R95α from the CDR3α loop (see Fig. 4b). This prominent electropositive arginine is conserved and also protrudes from the antigen-binding domain of the Vα24 independent 5B and 5E TcRs which also lack an electropositive cavity (Gadola et al., 2006) (see Fig. 4c and d). Therefore, it may act as an electrostatic guide for docking onto CD1d/α-GalCer.

4 Cross-Species Reactivity

The mouse homologue of the human iNKT cell receptor (Vα24/Jα18 and Vβ11) is composed of the Vα14/Jα18 and Vβ8.2 gene segments. Human and mouse iNKT cell receptors share greater than 50% sequence identity in their α and β-chains and exhibit remarkable cross-species reactivity (Benlagha et al., 2000; Brossay et al., 1998; Schumann et al., 2003). Thus, a human iNKT cell receptor is able to recognise α-GalCer in the context of both human and mouse CD1d and vice versa for the mouse homologue of the iNKT cell receptor. This cross-species reactivity is indicative of evolutionary conservation and is unlike the highly restricted syngeneic TcR recognition of MHC class I molecules (Brossay et al., 1998). The three-dimensional

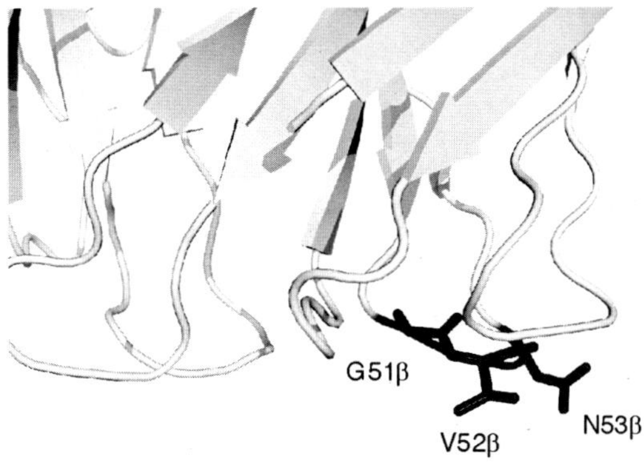

Fig. 6. Residues 51-53 on the CDR2β loop of mouse iNKT cell receptors have an impact on the recognition of CD1d/α-GalCer. In the human iNKT cell receptor structure shown above, residues 51-53 are a glycine, valine and asparagine respectively (^{51}GVN^{53}), whereby G51β is buried and both V52β and N53β are surface exposed.

structure of the iNKT cell receptor provided an opportunity to view the location of surface exposed amino acids that are conserved in the mouse and human iNKT cell receptor. These residues are located on the CDR1α, CDR3α, and CDR2β loops and form a contiguous conserved surface on the TcR antigen-binding domain (see Fig. 5) (Kjer-Nielsen et al., 2006). Therefore these regions are likely to contribute to cross-species reactivity between human and mouse CD1d presenting α-GalCer. Interestingly, three of the species-conserved residues from the CDR3α loop also line the afore-mentioned electropositive cavity of NKT12 (G96α, S97α, and L104α).

The CDR2β loop of the mouse iNKT cell receptor has previously been implicated in cross-species reactivity with rat iNKT cell receptors and the binding to mouse CD1d/α-GalCer (Pyz et al., 2006). This was investigated using chimeric and mutated mouse iNKT cell receptors composed of mouse or rat β-chains and these were subsequently tested for their ability to bind mouse CD1d/α-GalCer complexes and their ability to be stained using mouse CD1d/α-GalCer tetramers. Interestingly, the substitution of three amino acids (^{51}GAG^{53} to ^{51}DVN^{53}) in the CDR2β chain of the TcR abolished binding to mouse CD1d/α-GalCer, whereas a single amino acid substitution at position 53 (G53N) was tolerated even though the three-dimensional structure of the human iNKT cell receptor shows residue 53 is solvent exposed (see Fig. 6). This indicates that residues at position 51-52 contribute to the binding of α-GalCer even though only residue 52 is surface exposed (see Fig. 6). This might indicate that the substitution of residues larger than glycine or alanine cause a steric impact on nearby CDR loops and hence ligand recognition.

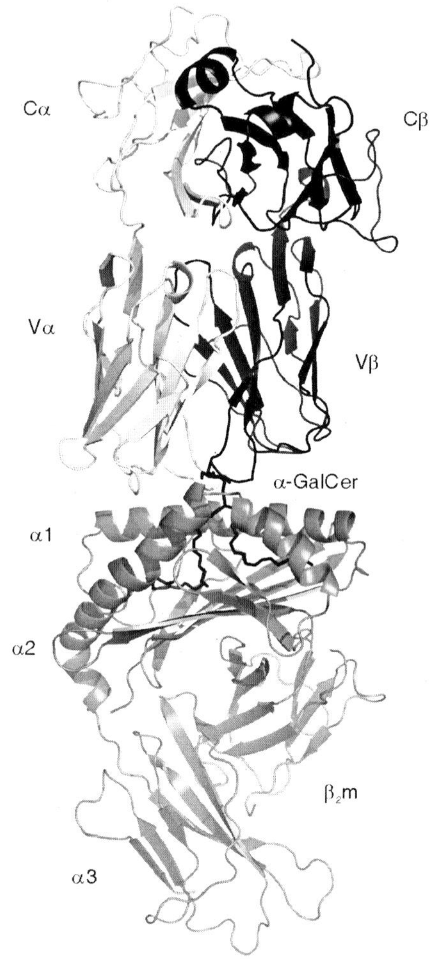

Fig. 7. Proposed docking of the iNKT cell receptor onto hCD1d presenting α-GalCer according to Kjer-Nielsen et al., 2006. The model was based on the HLA-B8/FLR/LC13 peptide MHC/TcR complex structure (PDB accession code 1MI5) as 1) the prominent feature of the FLR epitope was in a similar position compared to the head-group of α-GalCer 2) the LC13 TcR provided the best model for docking of the central TcR cavity over the α-GalCer head-group.

5 TcR Specificity and Binding Model

The ternary structure of CD1d/α-GalCer/NKT has not as yet been determined by X-ray crystallography. However, information obtained from 1) CD1d mutagenesis studies and the corresponding effect on NKT cell activation (Kamada et al., 2001); 2) structural information from the unliganded NKT cell receptor structures; 3) structural information from the various CD1d/lipid complexes available; offer yet another tool for modelling NKT cell receptor docking (see Fig. 7). Various docking models have been proposed in the literature (Gadola et al., 2006; Giabbai et al., 2005; Kamada et al., 2001; Kjer-Nielsen et al., 2006) and there are some common predictions between them. These include 1) the NKT cell receptor assumes the diagonal docking mode onto CD1d that is typical of an αβ TcR onto a MHC class I (Gadola et al., 2006; Kjer-Nielsen et al., 2006); 2) Predict that the CDR3α and/or CDR3β loops flank or are in close proximity to the polar galactose head group of α-GalCer (Gadola et al., 2006; Kjer-Nielsen et al., 2006) and recognise it directly (Kamada et al., 2001) or other head groups directly (Giabbai et al., 2005). We further predict that the CDR1α and CDR2α loops are likely to interact with the α2 helix of CD1d, whereas the CDR1β and CDR2β loops are likely to interact with the α1 helix of CD1d (Kjer-Nielsen et al., 2006). Likewise, the CDR1β and/or CDR2β loops have been attributed to the avidity of binding of mouse iNKT cell receptors to CD1d-IgG1 dimers (Schumann et al., 2003). Furthermore, we consider it likely that the conserved residues within the CDR3α loop mediate interactions with conserved residues on CD1d. The prominent role of CDR3α in interacting with CD1d may explain NKT cell cross reactivity between mouse and human CD1d, analogous to the way that the CDR3α loop of the LC13 TcR is suggested to dictate its known alloreactivity to HLA-B*4402 (Borg et al., 2005; Macdonald et al., 2003). Functional analysis on mutant mouse CD1d transfectants (Burdin et al., 2000; Kamada et al., 2001) or human CD1d show residues equivalent to R79, D80, and E83 (human equivalent numbering) on the α1 helix play an important role in NKT cell activation as do residues on the α2 helix including D151 (Brossay et al., 1998; Kamada et al., 2001) and T154 (Burdin et al., 2000) (human equivalent numbering).

6 Conclusions

We propose a structural basis for the selection of the invariant α-chain in iNKT cell recognition of CD1d-glycolipid complexes through the identification of a putative pre-formed electropositive antigen-binding cavity that is lined by unique and conserved residues. Amongst others, the cavity is lined by some residues from the Jα18 gene segment and the marked bias for this gene segment in Vα24 independent TcRs fortifies its importance in ligand recognition. Although not all NKT cell receptors have an electropositive cavity on the surface of their antigen-binding domain, the CDR3β loop has been shown to be flexible and may partially cause the occlusion of the cavity. Interestingly, those NKT cell receptors without an electropositive cavity have a prominent electropositive arginine residue on their antigen-binding domain that may serve for electrostatic steering onto CD1d.

While the individual structures of the iNKT cell receptor and hCD1d in complex with antigens that activate iNKT cell receptors have been determined, they only enable one to predict the interaction between an NKT cell receptor and CD1d. Knowledge of the precise interaction requires the three-dimensional structure.

A complex crystal structure will highlight the docking mode of the NKT cell receptor and enable the precise interactions of the invariant α-chain to be determined. This will shed light on what makes the invariant α-chain so highly selected. Furthermore, it will be interesting to note if conformational changes in the CDR loops and/or α-GalCer occur upon TcR ligation, potentially as a means of accommodating various ligands. Overcoming this challenge will enable us to determine precisely how NKT cell receptors bind to CD1d presenting diverse glycolipids and in turn how this specific interaction enables signalling to be altered upon activation. It will also be crucial in exploiting the immunotherapeutic potential of α-GalCer with regards to the activation of NKT cells.

7 Acknowledgements

J. Rossjohn is supported by an Australian Research Council Federation Fellowship and N.A. Borg is supported by a National Health and Medical Research Council (NHMRC) Peter Doherty Fellowship. This work was also supported in part by the NHMRC Australia, the Australian Research Council, the Cancer Council Victoria, and the Roche Organ Transplantation Research Foundation.

References

Benlagha, K., Weiss, A., Beavis, A., Teyton, L. & Bendelac, A., 2000, In vivo identification of glycolipid antigen-specific T cells using fluorescent CD1d tetramers, *J Exp Med.* **191**:1895-903.

Borg, N.A., Ely, L.K., Beddoe, T., Macdonald, W.A., Reid, H.H., Clements, C.S., Purcell, A.W., Kjer-Nielsen, L., Miles, J.J., Burrows, S.R., McCluskey, J. & Rossjohn, J., 2005, The CDR3 regions of an immunodominant T cell receptor dictate the 'energetic landscape' of peptide-MHC recognition, *Nat Immunol.* **6**:171-80.

Brigl, M., van den Elzen, P., Chen, X., Meyers, J.H., Wu, D., Wong, C.H., Reddington, F., Illarianov, P.A., Besra, G.S., Brenner, M.B. & Gumperz, J.E., 2006, Conserved and heterogeneous lipid antigen specificities of CD1d-restricted NKT cell receptors, *J Immunol.* **176**:3625-34.

Brossay, L., Chioda, M., Burdin, N., Koezuka, Y., Casorati, G., Dellabona, P. & Kronenberg, M., 1998, CD1d-mediated recognition of an alpha-galactosylceramide by natural killer T cells is highly conserved through mammalian evolution, *J Exp Med.* **188**:1521-8.

Burdin, N., Brossay, L., Degano, M., Iijima, H., Gui, M., Wilson, I.A. & Kronenberg, M., 2000, Structural requirements for antigen presentation by mouse CD1, *Proc Natl Acad Sci U S A.* **97**:10156-61.

Cantu, C., 3rd, Benlagha, K., Savage, P.B., Bendelac, A. & Teyton, L., 2003, The paradox of immune molecular recognition of alpha-galactosylceramide: low affinity, low specificity for CD1d, high affinity for alpha beta TCRs, *J Immunol.* **170**:4673-82.

Clements, C.S., Dunstone, M.A., Macdonald, W.A., McCluskey, J. & Rossjohn, J., 2006, Specificity on a knife-edge: the alphabeta T cell receptor, *Curr Opin Struct Biol.*

Crowe, N.Y., Coquet, J.M., Berzins, S.P., Kyparissoudis, K., Keating, R., Pellicci, D.G., Hayakawa, Y., Godfrey, D.I. & Smyth, M.J., 2005, Differential antitumor immunity mediated by NKT cell subsets in vivo, *J Exp Med.* **202**:1279-88.

Ely, L.K., Kjer-Nielsen, L., McCluskey, J. & Rossjohn, J., 2005, Structural studies on the alphabeta T-cell receptor, *IUBMB Life.* **57**:575-82.

Fischer, K., Scotet, E., Niemeyer, M., Koebernick, H., Zerrahn, J., Maillet, S., Hurwitz, R., Kursar, M., Bonneville, M., Kaufmann, S.H. & Schaible, U.E., 2004, Mycobacterial phosphatidylinositol mannoside is a natural antigen for CD1d-restricted T cells, *Proc Natl Acad Sci U S A.* **101**:10685-90.

Gadola, S.D., Dulphy, N., Salio, M. & Cerundolo, V., 2002, Valpha24-JalphaQ-independent, CD1d-restricted recognition of alpha-galactosylceramide by human CD4(+) and CD8alphabeta(+) T lymphocytes, *J Immunol.* **168**:5514-20.

Gadola, S.D., Koch, M., Marles-Wright, J., Lissin, N.M., Shepherd, D., Matulis, G., Harlos, K., Villiger, P.M., Stuart, D.I., Jakobsen, B.K., Cerundolo, V. & Jones, E.Y., 2006, Structure and binding kinetics of three different human CD1d-alpha-galactosylceramide-specific T cell receptors, *J Exp Med.* **203**:699-710.

Giabbai, B., Sidobre, S., Crispin, M.D., Sanchez-Ruiz, Y., Bachi, A., Kronenberg, M., Wilson, I.A. & Degano, M., 2005, Crystal structure of mouse CD1d bound to the self ligand phosphatidylcholine: a molecular basis for NKT cell activation, *J Immunol.* **175**:977-84.

Goff, R.D., Gao, Y., Mattner, J., Zhou, D., Yin, N., Cantu, C., 3rd, Teyton, L., Bendelac, A. & Savage, P.B., 2004, Effects of lipid chain lengths in alpha-galactosylceramides on cytokine release by natural killer T cells, *J Am Chem Soc.* **126**:13602-3.

Hammond, K.J., Pelikan, S.B., Crowe, N.Y., Randle-Barrett, E., Nakayama, T., Taniguchi, M., Smyth, M.J., van Driel, I.R., Scollay, R., Baxter, A.G. & Godfrey, D.I., 1999, NKT cells are phenotypically and functionally diverse, *Eur J Immunol.* **29**:3768-81.

Kamada, N., Iijima, H., Kimura, K., Harada, M., Shimizu, E., Motohashi, S., Kawano, T., Shinkai, H., Nakayama, T., Sakai, T., Brossay, L., Kronenberg, M. & Taniguchi, M., 2001, Crucial amino acid residues of mouse CD1d for glycolipid ligand presentation to V(alpha)14 NKT cells, *Int Immunol.* **13**:853-61.

Kang, S.J. & Cresswell, P., 2004, Saposins facilitate CD1d-restricted presentation of an exogenous lipid antigen to T cells, *Nat Immunol.* **5**:175-81.

Kawano, T., Cui, J., Koezuka, Y., Toura, I., Kaneko, Y., Motoki, K., Ueno, H., Nakagawa, R., Sato, H., Kondo, E., Koseki, H. & Taniguchi, M., 1997, CD1d-restricted and TCR-mediated activation of valpha14 NKT cells by glycosylceramides, *Science.* **278**:1626-9.

Kinjo, Y., Tupin, E., Wu, D., Fujio, M., Garcia-Navarro, R., Benhnia, M.R., Zajonc, D.M., Ben-Menachem, G., Ainge, G.D., Painter, G.F., Khurana, A., Hoebe, K., Behar, S.M., Beutler, B., Wilson, I.A., Tsuji, M., Sellati, T.J., Wong, C.H. & Kronenberg, M., 2006, Natural killer T cells recognize diacylglycerol antigens from pathogenic bacteria, *Nat Immunol.* **7**:978-86.

Kinjo, Y., Wu, D., Kim, G., Xing, G.W., Poles, M.A., Ho, D.D., Tsuji, M., Kawahara, K., Wong, C.H. & Kronenberg, M., 2005, Recognition of bacterial glycosphingolipids by natural killer T cells, *Nature.* **434**:520-5.

Kjer-Nielsen, L., Borg, N.A., Pellicci, D.G., Beddoe, T., Kostenko, L., Clements, C.S., Williamson, N.A., Smyth, M.J., Besra, G.S., Reid, H.H., Bharadwaj, M., Godfrey, D.I., Rossjohn, J. & McCluskey, J., 2006, A structural basis for selection and cross-species reactivity of the semi-invariant NKT cell receptor in CD1d/glycolipid recognition, *J Exp Med.* **203**:661-73.

Koch, M., Stronge, V.S., Shepherd, D., Gadola, S.D., Mathew, B., Ritter, G., Fersht, A.R., Besra, G.S., Schmidt, R.R., Jones, E.Y. & Cerundolo, V., 2005, The crystal structure of human CD1d with and without alpha-galactosylceramide, *Nat Immunol.* **6**:819-26.

Lantz, O. & Bendelac, A., 1994, An invariant T cell receptor alpha chain is used by a unique subset of major histocompatibility complex class I-specific CD4+ and CD4-8- T cells in mice and humans, *J Exp Med* **180**:1097-106.

Macdonald, W.A., Purcell, A.W., Mifsud, N.A., Ely, L.K., Williams, D.S., Chang, L., Gorman, J.J., Clements, C.S., Kjer-Nielsen, L., Koelle, D.M., Burrows, S.R., Tait, B.D., Holdsworth, R., Brooks, A.G., Lovrecz, G.O., Lu, L., Rossjohn, J. & McCluskey, J., 2003, A naturally selected dimorphism within the HLA-B44 supertype alters class I structure, peptide repertoire, and T cell recognition, *J Exp Med* **198**:679-91.

Matsuda, J.L., Gapin, L., Fazilleau, N., Warren, K., Naidenko, O.V. & Kronenberg, M., 2001, Natural killer T cells reactive to a single glycolipid exhibit a highly diverse T cell receptor beta repertoire and small clone size, *Proc Natl Acad Sci U S A.* **98**:12636-41.

Mattner, J., Debord, K.L., Ismail, N., Goff, R.D., Cantu, C., 3rd, Zhou, D., Saint-Mezard, P., Wang, V., Gao, Y., Yin, N., Hoebe, K., Schneewind, O., Walker, D., Beutler, B., Teyton, L., Savage, P.B. & Bendelac, A., 2005, Exogenous and endogenous glycolipid antigens activate NKT cells during microbial infections, *Nature.* **434**:525-9.

Mercer, J.C., Ragin, M.J. & August, A., 2005, Natural killer T cells: rapid responders controlling immunity and disease, *Int J Biochem Cell Biol.* **37**:1337-43.

Miyamoto, K., Miyake, S. & Yamamura, T., 2001, A synthetic glycolipid prevents autoimmune encephalomyelitis by inducing TH2 bias of natural killer T cells, *Nature.* **413**:531-4.

Nieda, M., Okai, M., Tazbirkova, A., Lin, H., Yamaura, A., Ide, K., Abraham, R., Juji, T., Macfarlane, D.J. & Nicol, A.J., 2004, Therapeutic activation of Valpha24+Vbeta11+ NKT cells in human subjects results in highly coordinated secondary activation of acquired and innate immunity, *Blood.* **103**:383-9.

Ortaldo, J.R., Young, H.A., Winkler-Pickett, R.T., Bere, E.W., Jr., Murphy, W.J. & Wiltrout, R.H., 2004, Dissociation of NKT stimulation, cytokine induction, and NK activation in vivo by the use of distinct TCR-binding ceramides, *J Immunol.* **172**:943-53.

Parekh, V.V., Singh, A.K., Wilson, M.T., Olivares-Villagomez, D., Bezbradica, J.S., Inazawa, H., Ehara, H., Sakai, T., Serizawa, I., Wu, L., Wang, C.R., Joyce, S. & Van Kaer, L., 2004, Quantitative and qualitative differences in the in vivo response of NKT cells to distinct alpha- and beta-anomeric glycolipids, *J Immunol.* **173**:3693-706.

Schumann, J., Voyle, R.B., Wei, B.Y. & MacDonald, H.R., 2003, Cutting edge: influence of the TCR V beta domain on the avidity of CD1d:alpha-galactosylceramide binding by invariant V alpha 14 NKT cells, *J Immunol.* **170**:5815-9.

Sidobre, S., Naidenko, O.V., Sim, B.C., Gascoigne, N.R., Garcia, K.C. & Kronenberg, M., 2002, The V alpha 14 NKT cell TCR exhibits high-affinity binding to a glycolipid/CD1d complex, *J Immunol.* **169**:1340-8.

Sriram, V., Du, W., Gervay-Hague, J. & Brutkiewicz, R.R., 2005, Cell wall glycosphingolipids of Sphingomonas paucimobilis are CD1d-specific ligands for NKT cells, *Eur J Immunol.* **35**:1692-701.

Wu, D., Xing, G.W., Poles, M.A., Horowitz, A., Kinjo, Y., Sullivan, B., Bodmer-Narkevitch, V., Plettenburg, O., Kronenberg, M., Tsuji, M., Ho, D.D. & Wong, C.H., 2005, Bacterial glycolipids and analogs as antigens for CD1d-restricted NKT cells, *Proc Natl Acad Sci U S A.* **102**:1351-6.

Wu, D., Zajonc, D.M., Fujio, M., Sullivan, B.A., Kinjo, Y., Kronenberg, M., Wilson, I.A. & Wong, C.H., 2006, Design of natural killer T cell activators: structure and function of a microbial glycosphingolipid bound to mouse CD1d, *Proc Natl Acad Sci U S A.* **103**:3972-7.

Wu, D.Y., Segal, N.H., Sidobre, S., Kronenberg, M. & Chapman, P.B., 2003, Cross-presentation of disialoganglioside GD3 to natural killer T cells, *J Exp Med.* **198**:173-81.

Yu, K.O., Im, J.S., Molano, A., Dutronc, Y., Illarionov, P.A., Forestier, C., Fujiwara, N., Arias, I., Miyake, S., Yamamura, T., Chang, Y.T., Besra, G.S. & Porcelli, S.A., 2005, Modulation of CD1d-restricted NKT cell responses by using N-acyl variants of alpha-galactosylceramides, *Proc Natl Acad Sci U S A.* **102**:3383-8.

Zajonc, D.M., Ainge, G.D., Painter, G.F., Severn, W.B. & Wilson, I.A., 2006, Structural characterization of mycobacterial phosphatidylinositol mannoside binding to mouse CD1d, *J Immunol.* **177**:4577-83.

Zajonc, D.M., Cantu, C., 3rd, Mattner, J., Zhou, D., Savage, P.B., Bendelac, A., Wilson, I.A. & Teyton, L., 2005a, Structure and function of a potent agonist for the semi-invariant natural killer T cell receptor, *Nat Immunol.* **6**:810-8.

Zajonc, D.M., Maricic, I., Wu, D., Halder, R., Roy, K., Wong, C.H., Kumar, V. & Wilson, I.A., 2005b, Structural basis for CD1d presentation of a sulfatide derived from myelin and its implications for autoimmunity, *J Exp Med.* **202**:1517-26.

Zhou, D., Mattner, J., Cantu, C., 3rd, Schrantz, N., Yin, N., Gao, Y., Sagiv, Y., Hudspeth, K., Wu, Y.P., Yamashita, T., Teneberg, S., Wang, D., Proia, R.L., Levery, S.B., Savage, P.B., Teyton, L. & Bendelac, A., 2004, Lysosomal glycosphingolipid recognition by NKT cells, *Science.* **306**:1786-9.

4

The Journey of Toll-like Receptors in the Cell

Øyvind Halaas[1,2], Harald Husebye[1] and Terje Espevik[1]

[1] Institute of Cancer Research and Molecular Medicine, NTNU, N-7489 Trondheim, Norway. Terje.Espevik@ntnu.no
[2] St. Olavs Hospital, N-7006 Trondheim, Norway

1 Introduction

Multicellular organisms are constantly challenged by microbes that are threatening to invade the host and causing genome or tissue destruction and pathology. In order to fight back attacks it is essential for the host to detect pathogens early before any tissue damage has occurred. For this purpose the Toll-like receptors (TLR) emerged as conserved microbial recognition proteins in species as different as worms (*Caenorhabditis elegans*) and humans. To date, 12 different TLRs have been found in mammals. TLR1, 2, 4, and 6 are found on the plasma membrane of immune cells and recognize lipoproteins and lipoglycans found on the surface of microbes. TLR5 is also on the plasma membrane and recognizes the motility apparatus protein flagellin.

TLR3, 7, 8 and 9 are found intracellularly in immune cells and recognize a variety of nucleotides and nucleoside analogues found more frequently in microbes than in vertebrates. Lack of TLR signaling may results in severe loss of anti-microbial defense, and erroneous TLR signaling may result in allergies, auto-immunity or cancer (Bohnhorst, Rasmussen, Moen, Flottum, Knudsen, Borset, Espevik and Sundan 2006).

This chapter will focus on localization and trafficking of TLRs and the reasons there may be for this compartmentalization.

2 The Toll-like receptors in immune reactions

TLRs recognize molecules more frequently found in microbes than in the host. TLRs are type I transmembrane proteins with an intracellular signaling domain and a leucine-rich repeat (LRR) ectodomain responsible for ligand binding. The crystal structure for the TLR3 ectodomain has been determined and shows extensive glycosylation, except at one face that is thought to mediate ligand binding and dimerization (Choe, Kelker and Wilson 2005). TLR1/2 and 2/6 heterodimers recognize lipoproteins, peptidoglycans and glycans from bacteria, fungi and some parasites. TLR4 in complex with soluble MD2 recognizes the prevalent lipopolysaccharides from gram-negative bacterial cell walls. TLR5 recognizes the flagellin monomer of the flagellum propulsion apparatus. TLR11 recognizes components from uropathogenic bacteria and a profilin-like protein from protozoa (Yarovinsky, Zhang, Andersen, Bannenberg, Serhan, Hayden, Hieny, Sutterwala, Flavell, Ghosh and Sher 2005). The closely interrelated TLR3, 7, 8, and 9, are located in intracellular compartments and recognize nucleic acids. TLR3 recognizes double stranded RNA, a result of viral replication. TLR7 and 8, nucleoside analogues and single stranded RNA from virus. TLR9 recognizes unmethylated CpG rich dsDNA, which is far less common in vertebrate than in microbial genomes. The TLRs contain a cytoplasmic domain homologues to the IL-1/IL-18 receptor and has been termed the Toll/Interleukin-1 receptor intracellular domain, TIR. The TIR domain is also found on the cytosolic TIR adaptors (MyD88, MAL, TRAM, and TRIF), that are recruited to the TLR TIR domains and mediate signaling (for review see (Uematsu and Akira 2006)).

All TLRs, except TLR3, signal through the MyD88 pathway, leading to activation of the NFκB gene-transcription program and production of pro-inflammatory cytokines. The TRIF adaptor is more restricted and signals from TLR4 and TLR3 leading to activation of interferon-regulatory factors (IRF), and production of anti-viral type I interferons. In addition, non-transcriptional activation programs are initiated from both, such as phosphoinositide kinase activity, which affects endocytosis, phagocytosis, trafficking, motility and antigen presentation.

TLR signaling is initiated when TLRs are oligomerized. TLR4 recruits TRAM and MAL, TLR1/2 and TLR2/6 recruit MAL. TLR5, 7, 8, 9 directly recruit MyD88, TLR1/2 and 2/6 heterodimers and TLR4 homodimers recruit MyD88 through MAL. TLR3 directly recruits TRIF and TLR4 recruits TRIF through TRAM. Thus, all TLRs, except TLR3, recruit MyD88 whereas only TLR4 and TLR3 recruit TRIF.

The MyD88-dependent pathway proceeds through phosphorylation of various interleukin-1 receptor-associated kinases (IRAK1, 2 and 4), subsequent activation of TRAF6, phosphorylation of IKK, and release and nuclear translocation of NFκB. In parallel, the mitogen-activated protein kinases (MAPK) mediates activation of activator protein 1 (AP-1). NFκB then promotes production of proinflammatory cytokines and microbicidal proteins, modulation of the phagocytic, endocytic and exocytic machinery, motility and homing preferences.

The MyD88-independent pathway from TLR3 and 4 proceeds through activation of TRIF, c-Src, TBK1/IKKε and the transcription factors IRF-3 and STAT-1 (Johnsen, Nguyen, Ringdal, Tryggestad, Bakke, Lien, Espevik and Anthonsen 2006; Uematsu et al. 2006) to yield IFNβ.

This activation is negatively regulated by a number of proteins.

2.1 Negative Regulatory Mechanisms of TLR Signaling

TLR signaling promotes activities that, if uncontrolled, can be detrimental to the host. So it comes as no surprise that TLR signaling is under extensive negative control (for review see; (Liew, Xu, Brint and O'Neill 2005)).

IRAK-M, IRAK2c/d splice variants, SOCS1, Tollip and SIGIRR all interfere with IRAK activation in different ways. Phosphorylation of IRAK is essential for downstream events and modulation of this phosphorylation is used for controlling TLR signaling. Further downstream, A20 de-ubiquitinates TRAF6 preventing NFκB-activation. The surface proteins STL2 and SIGIRR sequester MyD88 or MAL, preventing recruitment to the TLRs. The short truncated MyD88 splice variant, MyD88s, binds to the same domain as MyD88, but does not recruit downstream effectors. Class I Ptdlns 3 kinase also inhibits TLR signaling, although the precise mechanism remains obscure. The activated receptors themselves must be degraded to shut off the signal. TLRs are ubiquitinated and targeted for degradation by the E3 ubiquitin ligase Triad3A (Chuang and Ulevitch 2004). The events following the ubiquitination will be described in detail.

2.2 Trafficking and Membrane Dynamics

Transmembrane proteins are translated in endoplasmic reticulum (ER) and are modified in ER and the *Golgi* complex before they reach their assigned destination. At the plasma membrane there is constant re-circulation of membrane and non-activated membrane proteins, whereas activated receptors are targeted for destruction.

The lysosomal targeting of transmembrane proteins is determined by tyrosine phosphorylation, di-leucin motifs and ubiquitination (Bonifacino and Traub 2003). Tyrosine based YXXØ motifs can be found in TLR4 (587YDAF, 622YRDF, and 707YLEW) as can a di-leucin motif [DE]XXXL[LI] (69ELYRLL). These motifs are recognized by clathrin and endocytosis adaptor proteins, and mediate the internalization by promoting import of the stretch of membrane containing the cargo. In the surrounding membranes there are phospholipids that attract effector proteins with organelle specific activity.

The integral membrane phopspholipid phosphatidylinositol (Ptdlns) has been implicated in almost all aspects of cell physiology by providing the basis for signal transduction, cytoskeletal organization and membrane trafficking (for review see (Di Paolo and De Camilli 2006; Gruenberg and Stenmark 2004). The Ptdlns are modified phospholipids that carry an attached inositol carbohydrate ring. Reversible phosphorylation of the inositol ring at positions 3, 4 and 5 results in the generation of seven species of Ptdlns. The precursor of Ptdlns is produced in the ER and then distributed to the different membrane compartments throughout the cells by vesicle transport (or specific carrier proteins). The different cellular membrane compartments vary in their specific Ptdlns content. The *Golgi* apparatus is highly enriched in Ptdlns(4)P, early/sorting endosomes in Ptdlns(3)P, the late endsomes in Ptdlns(3,5)P_2 and the plasma membrane Ptdlns(4,5)P_2. The Ptdlns are important for the recruitment of different effector proteins having organelle specific tasks such as endocytic adaptors during endocytosis and endosomal sorting factors controlling recycling and degradation.

2.3 Distribution of TLRs in resting cells

As mentioned, the TLRs are compartmentalized so that TLR1, 2, 4, 5 and 6 are found at the plasma membrane, whereas TLR3, 7, 8, and 9 are expressed intracellularly.

In unstimulated cells, TLR4 is located as a large pool in the *Golgi* area, on endsomal structures and at the plasma membrane both in transfected cells and in primary immune cells (Husebye, Halaas, Stenmark, Tunheim, Sandanger, Bogen, Brech, Latz and Espevik 2006; Latz, Visintin, Lien, Fitzgerald, Monks, Kurt-Jones, Golenbock and Espevik 2002). LPS and CD14 have also been found in *Golgi* (Thieblemont and Wright 1999). The LPS receptor complex rapidly cycles between the plasma membrane and the *Golgi*, but the *Golgi* localization is not required for TLR4 signaling(Latz et al. 2002). Non-immune cells may have only intracellular TLR4, but are still activated by LPS (Guillot, Medjane, Le-Barillec, Balloy, Danel, Chignard and Si-Tahar 2004; Hornef, Normark, Vandewalle and Normark 2003). TLR2 is exposed at the plasma membrane in resting cells, and is enriched in phagocytic cups containing zymosan particles (Nilsen, Nonstad, Khan, Knetter, Akira, Sundan, Espevik and Lien 2004).

TLR9 and 3 are found in the ER in unstimulated cells but not on the plasma membrane. Addition of ligands somehow transfer TLR9 and TLR3 from ER to early and late endosomes, resulting in co-localization with the ligand (Johnsen et al. 2006; Latz, Schoenemeyer, Visintin, Fitzgerald, Monks, Knetter, Lien, Nilsen, Espevik and Golenbock 2004). The sorting signals for newly synthesized TLR9 has been studied in detail. Either the ectodomain or the cytoplasmic domain is sufficient to retain TLR9 at an intracellular location (Leifer, Brooks, Hoelzer, Lopez, Kennedy, Mazzoni and Segal 2006). The cytoplasmic domain of TLR3 and the transmembrane domain of TLR7 contains the necessary information to keep the receptors cytoplasmic (Nishiya, Kajita, Miwa and Defranco 2005). Further analysis showed that a cytoplasmic linker between the membrane and the TIR domain of TLR3 is responsible for the intracellular localization (Funami, Matsumoto, Oshiumi, Akazawa, Yamamoto and Seya 2004). The export signals for the extracellular TLRs have not been described, but could simply be a result of lack of retention signals.

The TIR adaptors are recruited to the TLRs at the location of ligand binding. TLR4 signals from the plasma membrane. The TLR4 adaptors, MAL and TRAM, are located at the inner side of the plasma membrane even in resting cells. MAL binds to plasma membrane PtdIns(4,5)P_2 (Kagan and Medzhitov 2006), whereas TRAM is myristoylated and thereby embedded in the plasma membrane (Rowe, McGettrick, Latz, Monks, Gay, Yamamoto, Akira, O'Neill, Fitzgerald and Golenbock 2006). Localization of MAL or TRAM to the plasma membrane is required for MyD88 or TRIF signaling, respectively.

An obvious deduction for the differentiated localization is that the surface TLRs (TLR1, 2, 4, 5, and 6) recognize molecules that are on the surface of pathogens (such as cell walls). The cytosolic TLRs (TLR3, 7, 8, and 9), on the other hand, recognize molecules that are on the inside of pathogens (such as genetic material), requiring some degradation to obtain access. The TLRs are primarily expressed in antigen-presenting cells, possibly pointing to a specific function for these cells beyond the detection of microbes. Non-immune cells indeed do have mechanisms for detecting and dealing with cytoplasmic pathogens (Uematsu et al. 2006), but these cells can not present extracellular antigens or prime T cells. One important consequence of TLR stimulation in antigen-presenting cells is the post-activation migration to the lymph nodes, the primary site for activation of T cells. This activity should indeed be limited to a subset of cells to avoid cell crowding in the lymph nodes. Thus, one can view the TLR-expressing phagocytic cells as sentinels detecting pathogens, engulfing them, degrading them, analyzing their contents, migrating to the lymph node with partially digested pathogenic material and presenting these fragments to passing T cells. The physiological process of antigen presentation is thus intimately coupled to the TLRs.

2.4 Activity of TLRs at the Plasma Membrane

The LPS receptor complex encounters LPS at the plasma membrane. There, TLR4 undergoes activation, oligomerization (Saitoh, Akashi, Yamada, Tanimura, Matsumoto, Fukase, Kusumoto, Kosugi and Miyake 2004) and tyrosine phosphorylation (Chen, Zuraw, Zhao, Liu, Huang and Pan 2003). The activated receptors recruit effector proteins carrying Src-homology 2 (SH2) domains binding to phosphorylated tyrosine residues contained in the intracellular domain. Among proteins recruited are the ubiquitin ligases. Ubiquitin (Ub) is a small 76aa protein that can be conjugated to the ε-aminogroup of lysine residues within a substrate. Ubiquitination of lysines in the intracellular part of transmembrane receptors has been implicated both in ligand-mediated endocytosis and endosomal sorting. E3 ubiquitin ligases attach one or more ubiquitins to lysines of the target protein. Ubiquitinated transmembrane proteins are sorted into multivesicular endosomes, which subsequently fuse with lysosomes resulting in degradation.

It has been shown that a single ubiquitin is sufficient for directing EGFR for degradation (Haglund, Sigismund, Polo, Szymkiewicz, Di Fiore and Dikic 2003). However, recent published work shows that Lys63-linked polyubiquitin chains also serve a role as sorting signals for lysosomal degradation (Huang, Kirkpatrick, Jiang, Gygi and Sorkin 2006), contrasting the traditional view that poly-ubiquitination only serves a role for directing proteins for proteasomal degradation.

TLRs, with the exception of TLR2, has been shown to be ubiquitinated by E3 ligase Triad3A (Chuang et al. 2004). This study did, however, suggest proteasomal degradation of the ubiquitinated receptor, which has previously been shown to be important mostly for misfolded newly synthesized proteins in the ER (Meusser, Hirsch, Jarosch and Sommer 2005). The most prominent role of ubiquitin for transmembrane protein outside the ER is the segregation of obsolete signaling receptors at the early/sorting endosome, which will be discussed below.

As mentioned, the plasma membrane contains mainly PtdIns(4,5)P_2. Class I phosphatidylinositol 3-kinase (Class I PI3-kinase) uses PtdIns(4,5)P_2 as their main substrate yielding the product PtdIns(3,4,5)P_3. PI3 kinase activity is controlled by stimulatory signals, resulting in dramatic fluctuations in the level of PtdIns(3,4,5)P_3

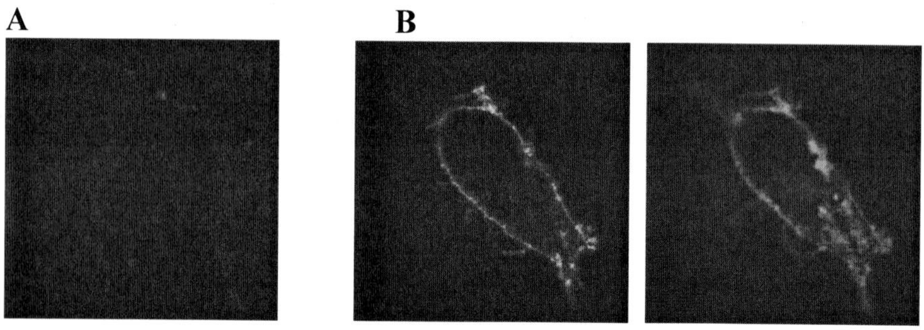

Fig. 1. LPS induces the formation of PtdIns(3,4,5)P_3 at the plasma membrane of TLR4 expressing cells. HEK293 cells stably expressing human TLR4 were transfected 24 hours with MD2, CD14 and the ARNO derived plextrin homology domain fused to YFP (PH_{ARNO}-YFP) and serum starved for 24 hours. The cells were either left untreated (**A**) or stimulated with 250 ng/ml CY5-LPS for 80 minutes (**B**). LPS stimulation results in a massive recruitment of PH_{ARNO}-YFP (Fig. 1B, left panel) to LPS containing regions of the plasma membrane (Fig. 1B, right panel).

at the plasma membrane. The plextrin homology (PH) domain from ADP-ribosylation factor nucleotide-binding site opener (ARNO) has been shown to work as a probe for cellular PtdIns(3,4,5)P_3 when fused to GFP (Venkateswarlu, Oatey, Tavare and Cullen 1998), We fused the ARNO-PH-domain with CFP to detect formation of PtdIns(3,4,5)P_3 in TLR4/MD2/CD14 expressing cells. Indeed, PtdIns(3,4,5)P_3 was generated at the plasma membrane following LPS stimulation (Fig. 1). Stimuli-dependent generation of PtdIns(3,4,5)P_3 has been coupled to proliferation, migration, phagocytosis, macropinocytosis, differentiation, survival and metabolic changes (Di Paolo et al. 2006; Gruenberg et al. 2004).

Invagination of membranes is caused by recruitment of clathrin to microdomains through the clathrin adaptors, pulling the membrane into the cell. If this process is stopped, the receptor will stay longer at the plasma membrane and continue signaling. Indeed, when the clathrin heavy chain is knocked down by a specific siRNA, the LPS uptake is reduced (Husebye et al. 2006) but the NFκB signal increases (Halaas, Husebye and Espevik, unpublished results), suggesting that internalization of the receptor attenuates signaling. Also the TLR9 ligand CpG and the TLR3 ligand dsRNA enters cells through the clathrin-mediated endocytic pathway (Johnsen et al. 2006; Latz et al. 2004), but internalization of these ligands are a pre-requisite for signaling through the intracellular TLRs.

Cells expressing a dominant negative version of dynamin I, should also be expected to have reduced internalization of plasma membrane receptors. Like the siRNA directed towards clathrin heavy chain, over-expression of the dominant negative version of dynamin I increases LPS-induced NFκB activation. This also results in a dramatic reduction of LPS internalization suggesting the endocytosis of the activated LPS receptor complex is dependent both on clathrin and dynamin. After the clathrin coated transport vesicles are pinched in the endocytotic process are transported to the early (or sorting) endosomes where they fuse with the endosomal limiting membrane (Husebye et al. 2006).

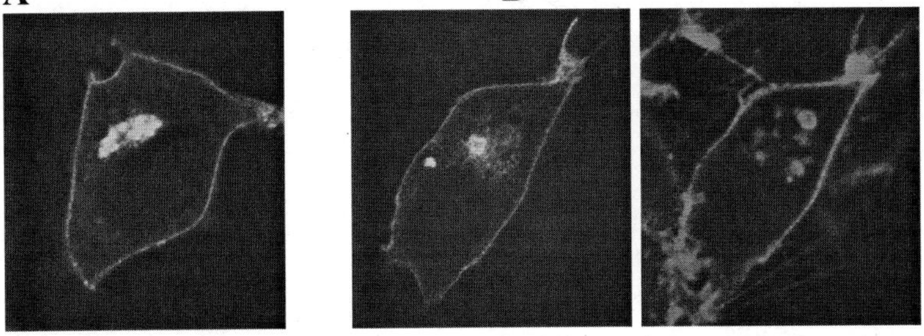

Fig. 2. The TIR signaling adapter TRAM appears on endosomal structures containing LPS following stimulation. HEK293 cells expressing TLR4YFP/MD2 were transfected with CD14 and TRAM fused to CFP (TRAM-CFP) for 48 hours. The cells were either left untreated (A) or stimulated with 250 ng/ml CY5-LPS for 50 minutes (B). In unstimulated cells, TRAM-CFP localizes to the plasma membrane and the perinuclear region (Fig. 2A). LPS stimulation results in the recruitment of TRAM-CFP to the limiting membrane of endosomes (Fig. 2B, left panel) overlapping with LPS (Fig. 2B, right panel)

2.5 Activity of TLRs at the Endosomes

Early/sorting endosomes are highly enriched in PtdIns(3)P, which is essential for the function of sorting endosomes in eukaryotic cells. PtdIns(3)P is generated by the action of the class III PI3-kinase, hVPS34. Endosomal proteins, such as hepatocyte growth factor-regulated tyrosine kinase substrate (Hrs) and the early endosomal marker 1 (EEA1), carry the PtdIns(3)P affinity domain FYVE (for conserved in Fab1, YOTB, Vac1 and EEA1) (Stenmark, Aasland and Driscoll 2002). A tandem FYVE domain (2×FYVE) derived from Hrs has been extensively used as PtdIns(3)P probe (Gillooly, Morrow, Lindsay, Gould, Bryant, Gaullier, Parton and Stenmark 2000). This probe is excellent for the detection of PtdIns(3)P in living cells. At the early/sorting endosome a decision is made whether an internalized receptors is a transport protein or unactivated receptor that should be recycled to the surface, or an activated signaling receptor which should be directed towards lysosomes for degradation.

Upon stimulation, several TLRs are found together with their cognate ligands in PtdIns(3)P positive early endosomes (Husebye et al. 2006; Johnsen et al. 2006; Latz et al. 2004; Nilsen et al. 2004). In this compartment, TLR4 co-localizes with Hrs in defined microdomains in the limiting membrane. Hrs has in addition to the FYVE domain, an ubiquitin-interaction motif (UIM) and a clathrin-binding domain. The flat clathrin-coat at the endosomal membrane acts as a detention scaffold for ubiquitinated cargo. Hrs complexes with Eps15, STAM1, and STAM2, which also contain UIMs. Through cooperative binding this complex delivers the ubiquitinated receptor cargo to the endosomal complexes required for transport (ESCRT complex). In turn, the ESCRT complex internalizes the receptor embedded in microvesicles into the lumen of the endosome, turning it into a multivesicular endsome (MVE) (Gruenberg et al. 2004).

Both plasma membrane and ER-residing TLRs are translocated to endosomes upon stimulation with their cognate ligands. Thus, TLR4/MD2/CD14 co-localize with LPS, TLR9 with CpG, and TLR3 with dsRNA in early endosomes. In fact, both TLR3 and TLR9 first encounter their ligands there. The endocytosis of both CpG and dsRNA seem to be independent of TLR signaling. Contrary to the situation with TLR4 and LPS, when dsRNA endocytosis is inhibited by PtdIns(3)P kinase inhibitors (Sarkar, Peters, Elco, Sakamoto, Pal and Sen 2004) or dominant negative Eps15 (Johnsen et al. 2006), signaling is decreased. This is most likely because ligands are inefficiently internalized and thus unavailable for TLR3 and 9. A mutation in the ER-resident protein Unc93b1 results in defective TLR3, 7, and 9 signaling as well as defective exogenous antigen-presentation (Tabeta, Hoebe, Janssen, Du, Georgel, Crozat, Mudd, Mann, Sovath, Goode, Shamel, Herskovits, Portnoy, Cooke, Tarantino, Wiltshire, Steinberg, Grinstein and Beutler 2006). The precise function of this protein is still not known, but is assumed to affect incidences in the communication between the ER and the endo-lysosomal system.

TLR4 is probably ubiquinated at the surface, and LPS increases the ubiquitination. Ubiquitin does not seem to interfere with LPS internalization but decrease NFκB signaling. The reason for this is that ubiquitin acts as a garbage tag at the early endosome, directing the ubiquitinated TLR4 down the lysosomal degradation pathway. Depletion of Hrs or ESCRT by the use of siRNA technology or over-expression of dominant negative ubiquitin not capable of lysine ligation resulted in significantly elevated LPS-induced NFκB activation.

The increased signaling observed by the inhibition of ubiquitination of TLR4 or by insufficient targeting into the lumen of multivesicular endosomes was taken as evidence that TLR4 is still able to signal from the endosomal surface. Indeed, the TRAM signaling adaptor is recruited to LPS positive endosomes following LPS stimulation (Fig. 2). In contrast, cells with moderate over-expression of Hrs show a significantly faster reduction in the TLR4 levels upon LPS stimulation (Husebye et al. 2006), probably due to more efficient delivery of TLR4 to the lumen of the endosome.

The PtdIns(3) kinase/Akt pathway has been implicated as negative regulator of TLR4 activation (Guha and Mackman 2002), possibly because inhibition of early endosome formation allows TLR4 prolonged signaling. When PtdIns-3-Kinase is inhibited by the addition of PtdIns-3-kinase inhibitor LY294002, the TLR4 levels increase both in the presence and absence of LPS. This suggests that the PtdIns-3-kinase is involved both in constitutive and induced degradation of TLR4.

Concomitant with acidification of the early endosome, interaction with lysosomes promotes continued maturation to late endosomes.

2.6 Activity of TLRs in the Late Endosomes

The lysosomes are preformed organelles with pH ~4.5 containing catabolic enzymes. Through fusion events or in a "kiss-and-run/linger"-fashion it delivers its contents to the maturing endosome (Bright, Gratian and Luzio 2005). The lysosomes and late endosomes are positive for LAMP-1 and will co-localize with fluid-phase markers such as dextran or BSA. The late multivesicular endosomes are the organelles responsible for presentation of exogenous peptides on MHC class II (HLA-DR/Q/P) and are often referred to as class II loading compartments. Acidity in these organelles promotes dissociation of the CLIP-peptide of the invariant chain Ii from the peptide-binding cleft on MHC class II, opening it for loading with endocytosed peptides. It has been suggested that TLRs are directly involved in sorting of antigens and generation of T-cell receptor ligands based on discrimination of TLR-ligand containing phagosomes from phagosomes containing only senescent host cells (Blander and Medzhitov 2006), although others have reported TLR-independent antigen presentation (Tabeta et al. 2006).

TLR4 and LPS can be found in LAMP-1-positive compartments (Husebye et al. 2006), as can TLR2 (Nilsen et al. 2004), TLR9 (Ahmad-Nejad, Hacker, Rutz, Bauer, Vabulas and Wagner 2002) and TLR3 (Johnsen et al. 2006). The late endosome is the final destination for TLR4, and activated TLR4 is degraded here by proteolysis. Inhibition of endo/lysosomal acidification with chloroquine prevented degradation of TLR4 (Husebye et al. 2006). The TLR ligands LPS (Husebye, H. and Espevik, T., unpublished), CpG (Latz et al. 2004) and dsRNA (Johnsen et al. 2006) can also be found together with dextran in dynamic tubular lysosomes within hours of ligand stimulation. These dynamic tubulae have been reported to act as transport expressways for antigen-loaded MHC II from late endosomes to the surface, and may significantly increase the rate and number of presented antigens.

In the late endosomes, peptide-loading of HLA-DR (MHC class II) takes place on the inner vesicles in the MVEs, a process facilitated by HLA-DM in the limiting membrane. This suggests that at some point MHC II and TLR4 are on intraluminal vesicles in the same MVE, TLR4 ending the mission of detecting invading microbes, MHC II starting the mission of instructing T helper cells what protein epitopes should be considered dangerous. We obtained proof that TLR4 and MHC II somehow cross paths by showing that antigens (i.e. antibodies) bound to TLR4, MD2, or CD14, were degraded and loaded onto HLA-DR and presented to T helper cells (Husebye et al. 2006). So either material must be transported from TLR4 containing intraluminal vesicles to MHC II containing vesicles, or more easily these vesicles may fuse to deliver TLR4 associated molecules to MHC II.

Erroneous antigen-presentation may result in auto-immunity. It has been shown that endogenous ligands for TLR7 and TLR9 induce and exacerbate autoimmunity (reviewed in (Marshak-Rothstein 2006).

It has also been shown that in animal strains susceptible to autoimmune diseases, administration of TLR-ligands can trigger the onset of disease. In experimental systems, TLR-ligands have induced arthritis, multiple sclerosis, myocarditis, diabetes and atherosclerosis (Marshak-Rothstein 2006). One explanation given is that TLR signaling overcomes the immuno-suppressive effect of regulatory T cells (Pasare and Medzhitov 2004). Another explanation is that the target organ itself is converted to an inflammatory state by the TLR-ligands, promoting self-destructing activities (Lang, Recher, Junt, Navarini, Harris, Freigang, Odermatt, Conrad, Ittner, Bauer, Luther, Uematsu, Akira, Hengartner and Zinkernagel 2005). In the case of systemic lupus erythematosis (SLE), immune comlexes of endogenous CpG DNA, autoantigens, anti-nuclear antibodies and TLR9 are directly involved in progression of the disease (Means, Latz, Hayashi, Murali, Golenbock and Luster 2005).

2.7 Significance of TLRs in Detection of Intracellular Bacteria

Intracellular bacteria can be found at diverse locations in the cell. Mycoplasmae adhere to the plasma membrane, mycobacteria and ehrlichiae dwell in early endosomes, *Salmonella* and *Histoplasma* in late endosomes, coxiellea in lysosomes, and leishmania in lysosomes and autophagosomes. Listeriae and rickettsiae escape phagosomes and live in the cytoplasm. Legionellae and brucellae associate with ER and chlamydiae and toxoplasmae with mitochondria (Gruenberg and van der Goot 2006; Schaible and Kaufmann 2004). Pathogens can either enter phagocytes and antigen presenting cells as hosts, or as passengers in senescent infected cells. Either way, the surface can be recognized by TLR1, 2, 4, 5, or 6 on the *outside* of the cell or in phagosomes (which topographically also can be considered to be the outside the cell). To gain access to the intracellular TLR ligands, either the intact genetic material of pathogens or nucleotides that has entered the cytoplasm, the phagocyte much first degrade the surrounding material. Whether the TLRs are directly responsible for the phagocytosis or antigen presentation or merely alter the general phenotype of the cells is still debated. No matter what one chose to believe, the TLRs must still travel through the endocytic compartments to search for intruders.

The TLR pathway is crucial for the innate defense against several intracellular bacteria, but MyD88 deficient mice are still able to mount adaptive responses. No single TLR seems to be sufficient for innate defense against intracellular bacteria as single TLR knockout mice are relatively resistant to infections whereas MyD88 knock out mice are highly susceptible (Edelson and Unanue 2002; Feng, Scanga, Collazo-Custodio, Cheever, Hieny, Caspar and Sher 2003; Uematsu et al. 2006). Even though TLRs patrol intracellular compartments in phagocytes in search for microbes, it is tempting to speculate that stealth inhabitance in cellular compartments is used by the bacteria for evading the immune cells.

3 Conclusions

TLR family members are essential in many aspects of immunology. The primary role is to detect microbes and to promote anti-microbial activities, such as inflammation and antigen presentation. A detailed description of cellular TLR localization and trafficking is helpful in understanding immunobiology, and is summarized in Fig 3. TLR1, 2, 4, 5, and 6 are located at the plasma membrane in unstimulated cells, whereas TLR3, 7, 8, and 9 are located in the endoplasmic reticulum. TLR4 is activated by LPS at the surface, where the receptor recruits signaling adaptors MAL/MyD88 and TRAM/TRIF and becomes phosphorylated and ubiquitinated. Clathrin and associated proteins form invaginations at the plasma membrane and the resulting transport vesicle is pinched off by dynamin I. The vesicle merges with early/sorting endosomes positive for PtdIns(3)P. At the limiting membrane of the endosome, ubiquitinated TLR4 associates with Hrs. Hrs and associated effector proteins sort ubiquitinated receptors embedded in microvesicles into the lumen of the endosomes, now called a multivesicular endosome. Lysosomal delivery of catabolic

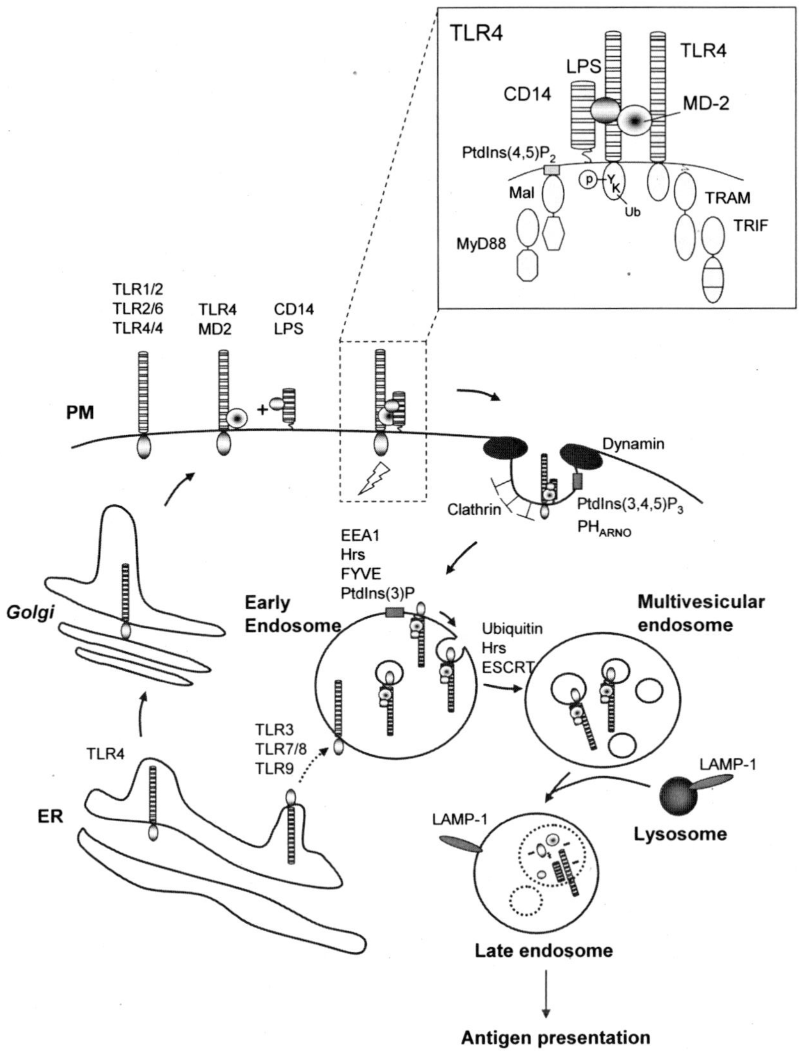

Fig 3. Overview over TLR trafficking. Plasma membrane TLR4/MD2 receives LPS from CD14 and initiates signaling. Clathrin and dynamin transport TLR4 to early endosomes positive for PtdIns(3)P. Hrs and ESCRT deliver TLR4 to the lumen of multivesicular endosomes. Through fusion with lysosomes, TLR4 is degraded, and co-delivered antigens are processed for class II presentation. TLR3, 7, 8, and 9 are recruited directly from ER to ligand containing early endosomes. See text for details.

enzymes to the multivesicular endosomes results in formation of a late endosome, which is responsible for degradation of TLR4 and termination of signaling. Antigens co-delivered to the late endosomes are subsequently presented by MHC class II (HLA-DR/Q/P) to T helper cells. The intracellular TLR3, 7, 8, and 9, on the other hand, are recruited from ER to endosomes containing the cognate ligands where they recruit the signaling adaptors. TLR3 and 9 and the ligands can later be found in tubular lysosomes. The activation of innate immune cells through TLRs and the subsequent presentation of antigens is important for fighting infections, but imperfections in the discrimination of self from non-self may also result in autoimmunity, which has been extensively documented in relation to TLR9.

4 Acknowledgements

This work was supported by the Norwegian Research Council, The Norwegian Cancer Society, and the Commission of the European Communities, LSMH-CT-2004-512093, AMIS.

References

Ahmad-Nejad, P., Hacker, H., Rutz, M., Bauer, S., Vabulas, R.M. and Wagner, H. (2002) Bacterial CpG-DNA and lipopolysaccharides activate Toll-like receptors at distinct cellular compartments. Eur. J. Immunol. 32, 1958-1968.

Blander, J.M. and Medzhitov, R. (2006) Toll-dependent selection of microbial antigens for presentation by dendritic cells. Nature 440, 808-812.

Bohnhorst, J., Rasmussen, T., Moen, S.H., Flottum, M., Knudsen, L., Borset, M., Espevik, T. and Sundan, A. (2006) Toll-like receptors mediate proliferation and survival of multiple myeloma cells. Leukemia 20, 1138-1144.

Bonifacino, J.S. and Traub, L.M. (2003) Signals for sorting of transmembrane proteins to endosomes and lysosomes. Annu. Rev Biochem. 72, 395-447.

Bright, N.A., Gratian, M.J. and Luzio, J.P. (2005) Endocytic delivery to lysosomes mediated by concurrent fusion and kissing events in living cells. Curr. Biol. 15, 360-365.

Chen, L.Y., Zuraw, B.L., Zhao, M., Liu, F.T., Huang, S. and Pan, Z.K. (2003) Involvement of protein tyrosine kinase in Toll-like receptor 4-mediated NF-kappa B activation in human peripheral blood monocytes. Am J Physiol Lung Cell Mol Physiol 284, L607-L613.

Choe, J., Kelker, M.S. and Wilson, I.A. (2005) Crystal structure of human toll-like receptor 3 (TLR3) ectodomain. Science 309, 581-585.

Chuang, T.H. and Ulevitch, R.J. (2004) Triad3A, an E3 ubiquitin-protein ligase regulating Toll-like receptors. Nat Immunol 5, 495-502.

Di Paolo, G. and De Camilli, P. (2006) Phosphoinositides in cell regulation and membrane dynamics. Nature 443, 651-657.

Edelson, B.T. and Unanue, E.R. (2002) MyD88-dependent but Toll-like receptor 2-independent innate immunity to Listeria: no role for either in macrophage listericidal activity. J. Immunol. 169, 3869-3875.

Feng, C.G., Scanga, C.A., Collazo-Custodio, C.M., Cheever, A.W., Hieny, S., Caspar, P. and Sher, A. (2003) Mice lacking myeloid differentiation factor 88 display profound defects in host resistance and immune responses to Mycobacterium avium infection not exhibited by Toll-like receptor 2 (TLR2)- and TLR4-deficient animals. J. Immunol. 171, 4758-4764.

Funami, K., Matsumoto, M., Oshiumi, H., Akazawa, T., Yamamoto, A. and Seya, T. (2004) The cytoplasmic 'linker region' in Toll-like receptor 3 controls receptor localization and signaling. Int. Immunol. 16, 1143-1154.

Gillooly, D.J., Morrow, I.C., Lindsay, M., Gould, R., Bryant, N.J., Gaullier, J.M., Parton, R.G. and Stenmark, H. (2000) Localization of phosphatidylinositol 3-phosphate in yeast and mammalian cells. EMBO J 19, 4577-4588.

Gruenberg, J. and Stenmark, H. (2004) The biogenesis of multivesicular endosomes. Nat Rev Mol Cell Biol 5, 317-323.

Gruenberg, J. and van der Goot, F.G. (2006) Mechanisms of pathogen entry through the endosomal compartments. Nat. Rev. Mol. Cell Biol. 7, 495-504.

Guha, M. and Mackman, N. (2002) The phosphatidylinositol 3-kinase-Akt pathway limits lipopolysaccharide activation of signaling pathways and expression of inflammatory mediators in human monocytic cells. J. Biol. Chem. 277, 32124-32132.

Guillot, L., Medjane, S., Le-Barillec, K., Balloy, V., Danel, C., Chignard, M. and Si-Tahar, M. (2004) Response of human pulmonary epithelial cells to lipopolysaccharide involves Toll-like receptor 4 (TLR4)-dependent signaling pathways: evidence for an intracellular compartmentalization of TLR4. J. Biol. Chem. 279, 2712-2718.

Haglund, K., Sigismund, S., Polo, S., Szymkiewicz, I., Di Fiore, P.P. and Dikic, I. (2003) Multiple monoubiquitination of RTKs is sufficient for their endocytosis and degradation. Nat. Cell Biol. 5, 461-466.

Hornef, M.W., Normark, B.H., Vandewalle, A. and Normark, S. (2003) Intracellular recognition of lipopolysaccharide by toll-like receptor 4 in intestinal epithelial cells. J. Exp. Med. 198, 1225-1235.

Huang, F., Kirkpatrick, D., Jiang, X., Gygi, S. and Sorkin, A. (2006) Differential regulation of EGF receptor internalization and degradation by multiubiquitination within the kinase domain. Mol. Cell 21, 737-748.

Husebye, H., Halaas, O., Stenmark, H., Tunheim, G., Sandanger, O., Bogen, B., Brech, A., Latz, E. and Espevik, T. (2006) Endocytic pathways regulate Toll-like receptor 4 signaling and link innate and adaptive immunity. EMBO J. 25, 683-692.

Johnsen, I.B., Nguyen, T.T., Ringdal, M., Tryggestad, A.M., Bakke, O., Lien, E., Espevik, T. and Anthonsen, M.W. (2006) Toll-like receptor 3 associates with c-Src tyrosine kinase on endosomes to initiate antiviral signaling. EMBO J. 25, 3335-3346.

Kagan, J.C. and Medzhitov, R. (2006) Phosphoinositide-mediated adaptor recruitment controls Toll-like receptor signaling. Cell 125, 943-955.

Lang, K.S., Recher, M., Junt, T., Navarini, A.A., Harris, N.L., Freigang, S., Odermatt, B., Conrad, C., Ittner, L.M., Bauer, S., Luther, S.A., Uematsu, S., Akira, S., Hengartner, H. and Zinkernagel, R.M. (2005) Toll-like receptor engagement converts T-cell autoreactivity into overt autoimmune disease. Nat. Med. 11, 138-145.

Latz, E., Schoenemeyer, A., Visintin, A., Fitzgerald, K.A., Monks, B.G., Knetter, C.F., Lien, E., Nilsen, N.J., Espevik, T. and Golenbock, D.T. (2004) TLR9 signals after translocating from the ER to CpG DNA in the lysosome. Nat. Immunol. 5, 190-198.

Latz, E., Visintin, A., Lien, E., Fitzgerald, K.A., Monks, B.G., Kurt-Jones, E.A., Golenbock, D.T. and Espevik, T. (2002) Lipopolysaccharide rapidly traffics to and from the golgi apparatus with the toll-like receptor 4-MD-2-CD14 complex in a process that is distinct from the initiation of signal transduction. J. Biol. Chem. 277, 47834-47843.

Leifer, C.A., Brooks, J.C., Hoelzer, K., Lopez, J.L., Kennedy, M.N., Mazzoni, A. and Segal, D.M. (2006) Cytoplasmic targeting motifs control localization of toll-like receptor 9. J. Biol. Chem.

Liew, F.Y., Xu, D., Brint, E.K. and O'Neill, L.A. (2005) Negative regulation of toll-like receptor-mediated immune responses. Nat. Rev. Immunol. 5, 446-458.

Marshak-Rothstein, A. (2006) Toll-like receptors in systemic autoimmune disease. Nat. Rev. Immunol. 6, 823-835.

Means, T.K., Latz, E., Hayashi, F., Murali, M.R., Golenbock, D.T. and Luster, A.D. (2005) Human lupus autoantibody-DNA complexes activate DCs through cooperation of CD32 and TLR9. J. Clin. Invest 115, 407-417.

Meusser, B., Hirsch, C., Jarosch, E. and Sommer, T. (2005) ERAD: the long road to destruction. Nat. Cell Biol. 7, 766-772.

Nilsen, N., Nonstad, U., Khan, N., Knetter, C.F., Akira, S., Sundan, A., Espevik, T. and Lien, E. (2004) Lipopolysaccharide and double-stranded RNA up-regulate toll-like receptor 2 independently of myeloid differentiation factor 88. J. Biol. Chem. 279, 39727-39735.

Nishiya, T., Kajita, E., Miwa, S. and Defranco, A.L. (2005) TLR3 and TLR7 are targeted to the same intracellular compartments by distinct regulatory elements. J. Biol. Chem. 280, 37107-37117.

Pasare, C. and Medzhitov, R. (2004) Toll-dependent control mechanisms of CD4 T cell activation. Immunity. 21, 733-741.

Rowe, D.C., McGettrick, A.F., Latz, E., Monks, B.G., Gay, N.J., Yamamoto, M., Akira, S., O'Neill, L.A., Fitzgerald, K.A. and Golenbock, D.T. (2006) The myristoylation of TRIF-related adaptor molecule is essential for Toll-like receptor 4 signal transduction. Proc. Natl. Acad. Sci. U. S. A 103, 6299-6304.

Saitoh, S., Akashi, S., Yamada, T., Tanimura, N., Matsumoto, F., Fukase, K., Kusumoto, S., Kosugi, A. and Miyake, K. (2004) Ligand-dependent Toll-like receptor 4 (TLR4)-oligomerization is directly linked with TLR4-signaling. J. Endotoxin. Res. 10, 257-260.

Sarkar, S.N., Peters, K.L., Elco, C.P., Sakamoto, S., Pal, S. and Sen, G.C. (2004) Novel roles of TLR3 tyrosine phosphorylation and PI3 kinase in double-stranded RNA signaling. Nat. Struct. Mol. Biol. 11, 1060-1067.

Schaible, U.E. and Kaufmann, S.H. (2004) Iron and microbial infection. Nat. Rev. Microbiol. 2, 946-953.

Stenmark, H., Aasland, R. and Driscoll, P.C. (2002) The phosphatidylinositol 3-phosphate-binding FYVE finger. Febs Letters 513, 77-84.

Tabeta, K., Hoebe, K., Janssen, E.M., Du, X., Georgel, P., Crozat, K., Mudd, S., Mann, N., Sovath, S., Goode, J., Shamel, L., Herskovits, A.A., Portnoy, D.A., Cooke, M., Tarantino, L.M., Wiltshire, T., Steinberg, B.E., Grinstein, S. and Beutler, B. (2006) The Unc93b1 mutation 3d disrupts exogenous antigen presentation and signaling via Toll-like receptors 3, 7 and 9. Nat. Immunol. 7, 156-164.

Thieblemont, N. and Wright, S.D. (1999) Transport of bacterial lipopolysaccharide to the golgi apparatus. J. Exp. Med. 190, 523-534.

Uematsu, S. and Akira, S. (2006) Toll-like receptors and innate immunity. J. Mol. Med.

Venkateswarlu, K., Oatey, P.B., Tavare, J.M. and Cullen, P.J. (1998) Insulin-dependent translocation of ARNO to the plasma membrane of adipocytes requires phosphatidylinositol 3-kinase. Curr. Biol. 8, 463-466.

Yarovinsky, F., Zhang, D., Andersen, J.F., Bannenberg, G.L., Serhan, C.N., Hayden, M.S., Hieny, S., Sutterwala, F.S., Flavell, R.A., Ghosh, S. and Sher, A. (2005) TLR11 activation of dendritic cells by a protozoan profilin-like protein. Science 308, 1626-1629.

5

Differential Regulation of Key Signaling Molecules in Innate Immunity and Human Diseases

Liwu Li[1], Jianmin Su and Qifa Xie

[1] Department of Biology, Virginia Tech, Blacksburg, VA 24061-0346, lwli@vt.edu

1 Introduction

Following the discovery of Toll-Like-Receptors (TLRs) and related downstream signaling molecules, the field of innate immunity and inflammation has drawn immense interest. Human hosts can specifically respond to distinct non-self molecules via TLRs and illicit complex yet specific responses through the expression of diverse cellular mediators such as cytokines, chemokines, complement factors, and co-stimulatory molecules (Li 2004) . Interleukin-1 receptor associated kinases (IRAK-1, 2, M, and 4) are intracellular kinases that can be recruited to the TLR complex and mediate diverse downstream signaling. Molecular and cellular analyses reveal that distinct IRAK proteins are differentially regulated and play unique roles in mediating downstream signaling processes (Yamin and Miller 1997; Kollewe, Mackensen, Neumann, Knop, Cao, Li, Wesche and Martin 2004; Su, Richter, Zhang, Gu and Li 2007). In particular, IRAK-4 appears to be the critical kinase necessary for the classical NFκB activation pathway (Li, Strelow, Fontana and Wesche 2002; Suzuki, Suzuki, Suzuki, Duncan, Millar, Wada, Mirtsos, Takada, Wakeham, Itie, Li, Penninger, Wesche, Ohashi, Mak and Yeh 2002a; Suzuki, Suzuki and Yeh 2002b). In contrast, IRAK-1 is selectively involved in enhancing transcriptional activities of p65/RelA as well as IRF3/5/7 by facilitating their phosphorylation (Uematsu, Sato, Yamamoto, Hirotani, Kato, Takeshita, Matsuda, Coban, Ishii, Kawai, Takeuchi and Akira 2005; Song, Jen, Soni, Kieff and Cahir-McFarland 2006; Oganesyan, Saha, Guo, He, Shahangian, Zarnegar, Perry and Cheng 2006; Hacker, Redecke, Blagoev, Kratchmarova, Hsu, Wang, Kamps, Raz, Wagner, Hacker, Mann and Karin 2006). Sumoylated IRAK-1 is not related to NFκB regulation, and is primarily responsible for Stat3 activation in cell nucleus (Huang, Li, Sane and Li 2004; Su et al. 2007). On the other hand, IRAK-M deactivates TLR-mediated NFκB activation and may help to prevent excessive expression of cellular inflammatory mediators (Kobayashi, Hernandez, Galan, J aneway Jr., Medzhitov and Flavell 2002). Recent advances further reveal that IRAK family molecules are not limited to TLR-mediated innate immunity signaling processes. Intriguingly, selected IRAK molecules can also associate with protein partners involved in T cell and B cell receptor-mediated signaling pathways, indicating that IRAK proteins are critical for both innate and adaptive immunity signaling (Ohnuma, Yamochi, Uchiyama, Nishibashi, Iwata, Hosono, Kawasaki, Tanaka, Dang and Morimoto 2005; Suzuki, Suzuki, Millar, Unno, Hara, Calzascia, Yamasaki, Yokosuka, Chen, Elford, Suzuki, Takeuchi, Mirtsos, Bouchard, Ohashi, Yeh and Saito 2006). Studies employing transgenic mice as well as human population-based studies have revealed that genetic variations in various IRAK genes are linked with diverse diseases such as infection, atherosclerosis, sepsis, auto-immune diseases, and cancer (Huang et al. 2004, Cardenes, von Bernuth, Garcia-Saavedra, Santiago, Puel, Ku, Emile, Picard, Casanova, Colino, Bordes, Garfia and Rodriguez-Gallego 2006; Arcaroli, Silva, Maloney, He, Svetkauskaite, Murphy and Abraham 2006; Deng, Radu, Diab, Tsen, Hussain, Cowdery, Racke and Thomas 2003; Sun, Wiklund, Hsu, Balter, Zheng, Johansson, Chang, Liu, Li, Turner, Li, Li, Adami, Isaacs, Xu and Gronberg 2006). We intend to summarize updated understanding of the differential regulation and function of IRAK family proteins and their involvements in the pathogenesis of various human diseases.

2 Regulation and Function of IRAK-1

IRAK-1 was first identified by Cao *et al.* through biochemical purification of the IL-1 dependent kinase activity that co-immunoprecipitates with the IL-1 type 1 receptor (Cao, Henzel and Gao 1996). Micropeptide sequencing and subsequent cDNA library screening yielded a full length cDNA clone encoding a protein with 712 amino acids and a predicted molecular size of ~76KD. IRAK-1 message is expressed ubiquitously in diverse human tissues. By radiation hybrid analysis, Thomas *et al.* mapped the murine IRAK-1 gene to Xq29.52-q29.7 and human IRAK-1 gene to Xq28 (Thomas, Allen, Tsen, Dubnicoff, Danao, Liao, Cao and Wasserman 1999). IRAK-1 protein contains an N-terminal death domain, a central serine/threonine kinase domain, and a C-terminal serine/threonine rich region. There are a putative nuclear localization sequence (NLS at aa 503-508) and a nuclear exit sequence (NES at aa 518-526) (Su et al. 2007). Using human THP-1 cells, primary blood mononuclear cells, as well as mice splenocytes, we have demonstrated that there are two signature forms of IRAK-1; the unmodified 80KD form, and the modified 100KD form (Li, Cousart, Hu, and McCall 2000). IRAK-1 modification consists of phosphorylation, ubiquitination , and sumoylation (Yamin et al. 1997; Kollewe et al. 2004; Su et al. 2007). Depending upon the nature of its modification, IRAK-1 may perform distinct functions including activation of NFκB (Knop, H. Wesche, D. Lang, and M.U. Martin 1998; Li, Commane, Burns, Vithalani, Cao and Stark 1999; Song et al. 2006), IRF3/5/7 (Uematsu et al. 2005; Oganesyan et al. 2006; Schoenemeyer, Barnes, Mancl, Latz, Goutagny, Pitha, Fitzgerald and Golenbock 2005), and Stat1/3 (Nguyen, Chatterjee-Kishore, Jiang, Qing, Ramana, Bayes, Commane, Li and Stark 2003; Huang et al. 2004).

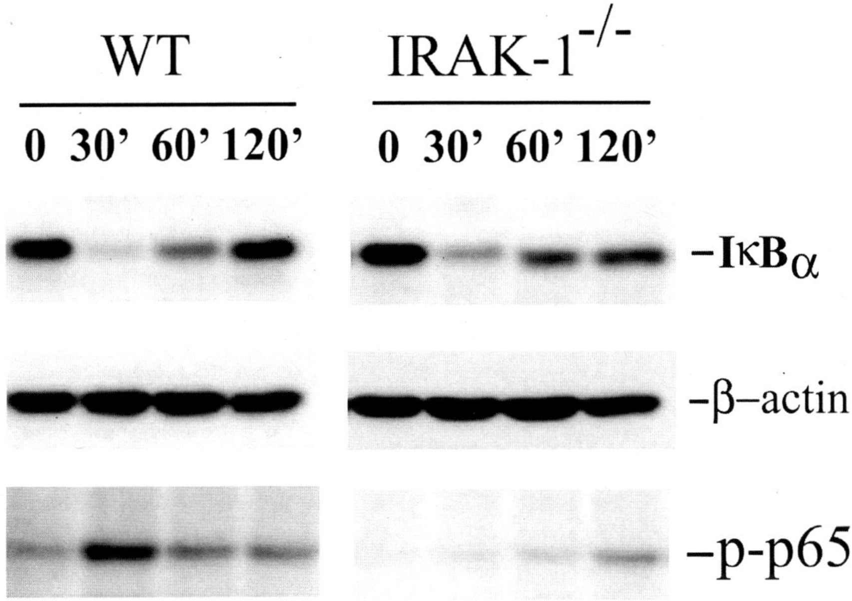

Fig. 1. IRAK-1 is selectively involved in Pam_3CSK_4-mediated p65 phosphorylation, instead of the classical pathway leading to IκBα degradation. Bone marrow derived macrophages (BMDM) from wild type and IRAK-$1^{-/-}$ mice were stimulated with 500ng/ml Pam_3CSK_4 for various time periods. IκBα and p536-p65 levels were detected through Western blot.

There are multiple pathways and steps leading to fully activated NFκB (Hacker and Karin 2006). The first step involves the classical pathway causing the activation of IKKα/β complex, which contributes to IκBα phosphorylation and degradation, and subsequent nuclear translocation of p65/RelA. The second step involves IKKε/TBK-1 dependent p65/RelA phosphorylation, which is independent of the classical pathway and IκBα degradation. Recent evidence indicates that IRAK-1 contributes to NFκB activation by facilitating p65/RelA phosphorylation, instead of the classical pathway leading to IκBα degradation and p65 nuclear translocation (Song et al. 2006). In agreement with this argument, we have also demonstrated that IRAK-1 deficient cells exhibit compromised p65/RelA phosphorylation, yet normal IκBα degradation following TLR2-ligand Pam_3CSK_4 challenge (Figure 1). It is likely that IRAK-1 also contributes to IRF5/7 activation through IKKε/TBK-1 (Oganesyan et al. 2006; Hacker and Karin 2006).

It appears that the unmodified IRAK-1 form is primarily involved in the phosphorylation of NFκB p65/RelA and IRF5/7. Several lines of evidence support such claim. First, the kinase-dead IRAK-1 mutant (K239A), which can not undergo

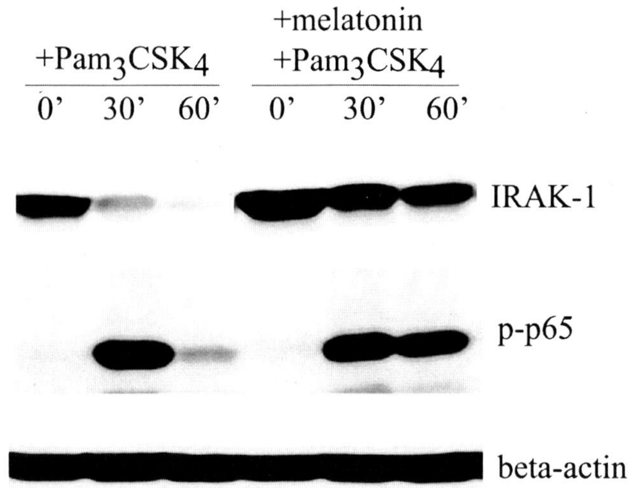

Fig. 2. Melatonin-pretreatment preserves the un-modified IRAK-1 form, and prolongs Pam_3CSK_4-mediated p65 phosphorylation. Human monocytic THP-1 cells were pretreated with or without 2 μM melatonin for 20 minute, followed by challenge with 500 ng/ml Pam_3CSK_4 for various time periods. IRAK-1 and p536-p65 levels were detected through Western blot.

any modification upon challenge, can still potently activate NFκB activity (Li et al. 1999; Knop and Martin 1999). Second, we have observed that melatonin pretreatment, which can alleviate Pam_3CSK_4-triggered IRAK-1 modification/degradation and preserve IRAK-1 in its unmodified form, leads to prolong p65 phosphorylation (figure 2).

Intriguingly, we observed that IRAK-1 gets sumoylated following either LPS or Pam_3CSK_4 challenge (Huang et al. 2004; Su et al. 2007). Sumoylated IRAK-1 is not involved in NFκB activation. Rather, it enters the nucleus and contributes to Stat3 activation and selected gene expression (Huang et al. 2004). TLR ligation also induces IRAK-1 ubiquitination, which subsequently leads to proteosome-mediated degradation (Yamin et al. 1997). IRAK-1 degradation correlates with reduced host response to endotoxin, and has been correlated with endotoxin tolerance observed in septic leukocytes (Li et al. 2000; Ferlito, Romanenko, Ashton, Squadrito, Halushka and Cook 2001; Moors, Li and Mizel 2001; Adib-Conquy and Cavaillon 2002; Sato, Takeuchi, Fujita, Tomizawa, Takeda and Akira 2002; Jacinto, Hartung, McCall and Li 2002). The dynamic balance of IRAK-1 sumoylation and ubiquitination may therefore regulate cellular IRAK-1 protein levels and contributes to its diverse yet distinct functions (Figure 3).

Besides mediating TLR signaling, IRAK-1 also participates in the regulation of adaptive immune response. For example, T cell co-stimulatory molecule CD26 can trigger the association of IRAK-1 with caveolin on antigen presenting monocytes,

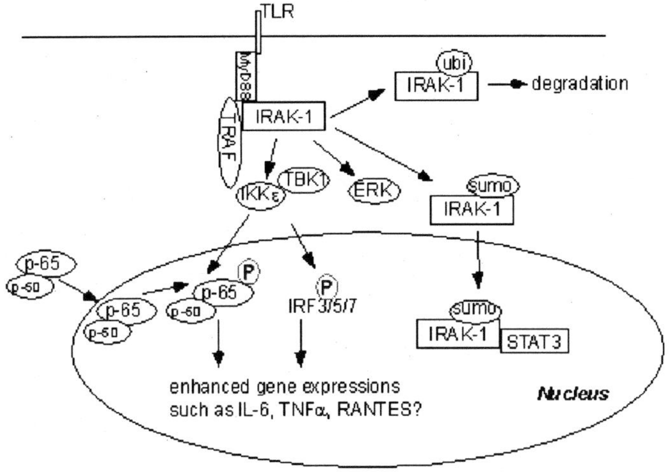

Fig. 3. Schematic illustration of IRAK-1 regulation and function.

which is responsible for the subsequent expression of co-stimulatory molecule CD86 (Ohnuma et al. 2005).

Differential splicing events in human cells give rise to two additional IRAK-1 variants, IRAK-1b and IRAK-1c. IRAK-1b derives from alternative splicing and deletion of 90bp within exon 12, which yields an in-frame deletion of 30 amino acids (residues 514-543) (Jensen and Whitehead 2001). IRAK-1c is due to alternative splicing and deletion of exon 11 and part of exon 12 (Rao, Nguyen, Ngo and Fung-Leung 2005). IRAK-1b exists in minute amount (less then 1% of IRAK-1) in most human cells and tissues with unknown function. On the other hand, the full length IRAK-1 and IRAK-1c are abundantly expressed in human leukocytes and most tissues (Rao et al. 2005; Su et al. 2007). In contrast to IRAK-1, both IRAK-1b and IRAK-1c are stable and do not undergo covalent modification following various stimulations (Li et al. 2000; Jensen et al. 2001; Su et al. 2007). Overexpression of IRAK-1c blocks IL-1β induced MAP kinase activation, suggesting that IRAK-1c may serve as a negative regulator of inflammation. Intriguingly, IRAK-1 is absent and IRAK-1c is the predominant form in young human brain tissues (Rao et al. 2005; Su et al. 2007). The absence of full length IRAK-1 may help keeping human brain in an immune-privileged state. In contrast to young humans, we recently found that both IRAK-1 and IRAK-1c are equally present in brain tissues obtained from aged humans (Su et al. 2007). This may bear significant implication in terms of aging. Increased chronic inflammation is a hallmark of aging process as evidenced by local infiltration of inflammatory cells such as macrophages, and higher circulatory levels of pro-inflammatory cytokines, complement components and adhesion molecules. Consequently, aging is often accompanied by increasing incidences of chronic inflammatory diseases such as Alzheimer's or Parkinson's disease. The molecular mechanisms contributing to the chronic inflammatory state during cellular senescence and aging process is not clearly understood. Our finding that the full-length IRAK-1 and IRAK-1c are equally present in aged human brains may provide at least a partial explanation for the aging process. Future studies determining the mechanism of IRAK-1 mRNA differential splicing and the functions of different IRAK-1 splice forms are warranted.

Given the significant and diverse roles IRAK-1 play in mediating both the innate and adaptive immune responses, it is not surprising that variations in IRAK-1 gene will lead to diverse inflammatory diseases. Indeed, deletion of IRAK-1 gene in mice decreases the risk of Experimental Autoimmune Encephalomyelitis (EAE) (Deng et al. 2003). We have found that IRAK-1 protein in leukocytes from human atherosclerosis patients is constitutively activated/sumoylated and localizes in cell nucleus (Huang et al. 2004). Furthermore, our human population-based study indicates that genetic variation in IRAK-1 gene correlates with the severity of atherosclerosis and serum C reactive protein levels (Larkoski, Li, Langfield, Liu, Howard, Xu, Bowden and Herrington 2006). There are two IRAK-1 haplotypes and the rare variant haplotype (~10% of human population) contains three exon single nucleotide polymorphisms (SNPs). Humans harboring the variant IRAK-1 gene tend to have higher serum CRP levels and higher risk for diabetes and hypertension (Larkoski et al. 2006). IRAK-1 gene variation is also linked to the risk of sepsis. Arcaroli et al. recently demonstrated that sepsis patients with the rare variant IRAK-1 haplotype have increased incidence of shock, prolonged requirement for mechanical ventilatory support, and greater 60-day mortality (Arcaroli 2006).

3 Regulation and Function of IRAK-2

IRAK-2 was initially identified by Dixit's group based on the search of the human expressed sequence tag (EST) database for sequences homologous to IRAK-1(Muzio, Ni, Feng and Dixit 1997). Subsequent screening of a human umbilical vein endothelial cell cDNA library resulted in the isolation of a full-length cDNA clone which encodes a 590-amino acid protein with a predicted size of 65 KD. The human IRAK-2 gene is mapped to chromosome 3 at position 3p25.3-3p24.1. Upon overexpression, IRAK-2 can associate with MyD88 as well as TRAF6, and activate NFκB -dependent reporter gene expression. Intriguingly, IRAK-2, instead of IRAK-1 can also interact with another distinct TLR intracellular adaptor molecule Mal/TIRAP (Fitzgerald, Palsson-McDermott, Bowie, Jefferies, Mansell, Brady, Brint, Dunne, Gray, Harte, McMurray, Smith, Sims and O'Neill 2001). Dominant negative IRAK-2 can block Mal/TIRAP-induced signaling while dominant negative IRAK-1 fails to do so. These studies suggest that IRAK-2 may selectively be recruited by Mal/TIRAP and participate in subsequent NFκB activation. Besides activating NFκB, IRAK-2 also participates in the regulation of cellular apoptosis (Ruckdeschel, O. Mannel, and P. Schrottner 2002). Dominant negative IRAK-2 can diminish LPS-induced macrophage apoptosis (Ruckdeschel et al. 2002).

O'Neil's group has identified the murine IRAK-2 gene which locates at chromosome 6 position E3 (Hardy, and O'Neill 2004). In contrast to its human counterpart which only encodes one single transcript, the murine IRAK-2 gene can generate four alternatively spliced isoforms (designated as IRAK-2a, 2b, 2c, and 2d) that have various N-terminal deletions. Upon overexpression, IRAK-2a and IRAK-2b could activate, while IRAK-2c and IRAK-2d inhibit NFκB activation. Alternative splicing of the IRAK-2 gene in mice instead of humans reflects the difference between human and murine TLR signaling processes and innate immunity regulations. The physiological function of IRAK-2 is poorly defined. Intriguingly, it was reported that several cases of human liver tumors harbor Hepatitis B Virus (HBV) DNA insertion near the IRAK-2 gene, which implies that IRAK-2 and related cellular signaling pathways may regulate human carcinogenesis (Paterlini-Brechot, Saigo, Murakami, Chami, Gozuacik, Mugnier, Lagorce and Brechot 2003). Further studies are needed to determine the biochemical regulation of IRAK-2 and its participation in various cellular signaling pathways.

4 Regulation and Function of IRAK-M

Using the similar EST search, Wesche *et al.* identified a murine EST sequence which encodes a polypeptide sharing significant homology with IRAK-1 (Wesche, Gao, Li, Kirschning, Stark and Cao 1999). Human IRAK-M gene is mapped to chromosome 12 at position 12q14.1-12q15, and its murine homolog is mapped to chromosome 10. Screening of human peripheral blood leukocyte library with this EST sequence resulted in the isolation of a full length cDNA clone that encodes a protein with 596 amino acids and a calculated molecular mass of 68 kDa. Northern blot analysis revealed that IRAK-M transcript is primarily present in the peripheral blood leukocytes and monocytic cell lines. Initial studies revealed that IRAK-M overexpression can activate NFκB activity (Wesche et al. 1999). Strikingly, later studies using IRAK-$M^{-/-}$ cells indicate otherwise. IRAK-$M^{-/-}$ macrophages exhibit enhanced NFκB activity and elevated expression of various inflammatory cytokines upon stimulation with several TLR ligands, indicating that IRAK-M may actually attenuate NFκB activation (Kobayashi et al. 2002). Phenotypically, IRAK-$M^{-/-}$ mice develop severe osteoporosis, which is associated with the accelerated differentiation of osteoclasts, an increase in the half-life of osteoclasts, and their activation (Li, Cuartas, Cui, Choi, Crawford, Ke, Kobayashi, Flavell and Vignery 2005). These studies indicate that IRAK-M may help to attenuate TLR signaling and prevent excessive inflammation.

Although IRAK-M may play a pivotal role in preventing excessive activation of NFκB and subsequent inflammatory response, such mechanism may also be exploited by tumor cells or bacteria to evade active immune surveillance. Sepsis syndrome is initiated by dissemination of bacteria or bacterial products (endotoxin) in blood circulation (Ishii, and Akira 2004). The host develops an endotoxin-tolerant state in which blood leukocytes can no longer exhibit inducible NFκB activation and expression of selected inflammatory cytokines, such as TNF-alpha and IL-6 (West, and Heagy 2002). Suppressed expression of inflammatory cytokines further puts the host in danger of secondary infection. The suppressed state is caused by deactivation of innate immunity signaling process, including persistent degradation of IRAK-1 and elevated IRAK-M protein levels in blood leukocytes (Li et al. 2000; Cuschieri,

Bulmus, Gourlay, Garcia, Hoffman, Stayton and Maier 2004; Deng, Cheng, Newstead, Zeng, Kobayashi, Flavell and Standiford 2006; Nakayama, Okugawa, Yanagimoto, Kitazawa, Tsukada, Kawada, Kimura, Hirai, Takagaki and Ota 2004). Analogously, deactivation of innate immunity may also tolerate tumor growth and progression. For example, through a yet-to-be-determined mechanism, tumor-associated macrophages fail to express pro-inflammatory cytokines, such as IL12p40 and TNFα, and tolerate/facilitate tumor growth instead of mounting an efficient tumor cell-killing defense (Mytar, Woloszyn, Szatanek, Baj-Krzyworzeka, Siedlar, Ruggiero, Wieckiewicz and Zembala 2003; Gallina, Dolcetti, Serafini, De Santo, Marigo, Colombo, Basso, Brombacher, Borrello, Zanovello, Bicciato and Bronte 2006; Dirkx, Egbrink, Wagstaff and Griffioen 2006). A recent study indicates that deactivated macrophages incubated with tumor cells exhibit increased IRAK-M protein levels (del Fresno, Otero, Gomez-Garcia, Gonzalez-Leon, Soler-Ranger, Fuentes-Prior, Escoll, Baos, Caveda, Garcia, Arnalich and Lopez-Collazo 2005). Tumor hosts also have compromised antigen presentation, decreased function of effector T cells, and exacerbated inhibitory function of negative T regulatory cells in both circulating leukocytes as well as within tumor tissues, which is reflective of suppressed adaptive immunity. Given the evidence indicating the role of IRAK-M in deactivating both innate and adaptive immunity signaling, we hypothesize that cancer cells may exploit the inhibitory function of IRAK-M to evade host immune surveillance.

5 Regulation and Function of IRAK-4

Lastly, yet another EST search yielded the human IRAK-4 cDNA sequence that encodes a distinct polypeptide sharing significant homology with the other IRAKs (Li et al. 2002). Human IRAK-4 gene is mapped to chromosome 12 at position 12p11.22, and its murine homolog is mapped to chromosome 15. Full-length IRAK-4 cDNA encodes a protein with 460 amino acids and a calculated molecular mass of 52 kDa. In contrast to IRAK-1 or IRAK-M deficient mice, IRAK-4$^{-/-}$ mice exhibit severe impairment in NFκB activation and expression of various inflammatory cytokines upon challenges with several TLR ligands (Suzuki et al. 2002a). Overexpression of kinase-dead IRAK-4 mutant strongly diminishes IL-1/LPS induced NFκB activation, pointing to the essential role of its kinase activity (Li et al. 2002). MyD88 is critically involved in recruiting IRAK-4 into the TLR4 complex (Li et al. 2002). These studies indicate that IRAK-4 is the primary kinase in the TLR signaling process essential for mediating NFκB activation. Recently, Suzuki *et al.* reported that IRAK-4 is also critically involved in T cell receptor (TCR)-induced T cell proliferation through NF-κB activation (Suzuki et al. 2006). T cell responses *in vivo* are significantly impaired in IRAK-4 deficient mice. Upon TCR stimulation, IRAK-4 is recruited to T cell lipid rafts, where it can associate with Zap70 and activate protein kinase C. This finding indicates that there is an intricate connection between innate and adaptive immune system activation, and IRAK-4 may be directly involved in the cross talk between the two systems.

Because of the central role IRAK-4 plays in mediating NFκB activation and innate immunity signaling, humans carrying IRAK-4 gene variation may be prone to microbial infections. Indeed, a study by Picard *et al.* revealed that IRAK-4 mutations are present in three children suffering from persistent pyogenic bacteria infection and poor inflammatory responses (Picard, Puel, Bonnet, Ku, Bustamante, Yang, Soudais, Dupuis, Feinberg, Fieschi, Elbim, Hitchcock, Lammas, Davies, Al-Ghonaium, Al-Rayes, Al-Jumaah, Al-Hajjar, Al-Mohsen, Frayha, Rucker, Hawn, Aderem, Tufenkeji, Haraguchi, Day, Good, Gougerot-Pocidalo, Ozinsky and Casanova 2003). These patients did not respond to IL-1β, IL-18, or any of the TLR1-6 or 9 ligands, as assessed by activation of NF-κB and p38-MAPK, and induction of IL-1β, IL-6, IL-12, TNFα, and IFN-γ. It is intriguing that the spectrum of infections was relatively narrow, with most infections caused by Gram-positive bacteria *Staphylococcus aureus* and *Streptococcus pneumoniae.* The patients showed poor sign of inflammatory response and the frequency of infection decreased with age, potentially due to the compensatory action of adaptive immunity. In a separate study, a patient was identified who suffered from recurrent bacterial infections and failed to respond to gram-negative LPS *in vivo*, and whose leukocytes were profoundly hyporesponsive to LPS and IL-1 *in vitro* (Medvedev, Thomas, Awomoyi, Kuhns, Gallin, Li, and Vogel 2005). This patient also exhibits deficient responses in a skin blister model of aseptic inflammation. Cloning and sequencing of IRAK-4 gene revealed that this patient expresses a "compound heterozygous" genotype, with a point mutation (C877T in cDNA) and a two-nucleotide, AC deletion (620-621del in cDNA) encoded by distinct alleles of the IRAK-4 gene. Both mutations encode proteins with an intact death domain, but a truncated kinase domain, therefore precluding expression of full-length IRAK-4. Recently, a case of invasive, systemic, extraintestinal Gram-negative *Shigella* infection was reported in a patient with inherited IRAK-4 deficiency (Cardenes et al. 2006). This case indicates that although the pyogenic Gram-positive bacteria *Staphylococcus* and *Pneumococcus* remain the most frequent pathogens associated with IRAK-4 deficiency, Gram-negative bacteria such as *Shigella* may threaten humans with IRAK-4 deficiency and cause severe illness and mortality.

6 Concluding remarks

In summary, collective research efforts regarding gene expression, message splicing, protein structure and function, signal transduction, cellular activation of human and murine immune cells, as well as genetic variation in humans, have led to extensive understanding of the complex roles and regulations of IRAK family proteins. It is apparent that although IRAK proteins share some common structural features, they are clearly not redundant and each has unique and distinct function. Furthermore, individual member may have multiple roles depending upon its modification, cellular and tissue distribution. Besides mediating various TLR signaling pathways, IRAKs are also involved in other signaling networks including T cell receptor and B cell

Table 1. IRAK family genes and related diseases.

	Chromosome location		Human gene variation and related diseases		Transgenic mice model
	Human	Mouse	Single nucleotide Polymorphisms (SNPs)	Related Diseases	
IRAK-1	Xq28	Xq29.5-q29.7	***Intron*** rs3027898T→G rs731642 G→A rs2239673T→C rs5945174A→G rs7061789A→G ***exon*** rs1059701C→T rs1059702C→T rs1059703T→C	Sepsis Diabetes and heart diseases	IRAK-1$^{-/-}$ mice are resistant to experimental autoimmune encephalomyelitis
IRAK-2	3p25.3-3p24.1	Chr 6	Not reported	Not reported	Not reported
IRAK-M	12q14.1 -12q15	Chr10	Not reported	Not reported	IRAK-M$^{-/-}$ mice develop Osteoporosis
IRAK-4	12p11.22	Chr 15	Point mutation (C877T in mRNA) leading to expression of truncated protein	microbial infections	IRAK-4$^{-/-}$ mice have increased mortality upon bacterial infection

receptor mediated signaling. Conceivably, these molecules pose as viable targets for designing new therapeutic strategies for various human inflammatory diseases.

7 Acknowledgements

This work is supported by NIH grants to L.L. We thank Andrew Becker in our group for proof-reading and formatting of the manuscript.

References

Adib-Conquy, M. and Cavaillon, J.M. (2002) Gamma interferon and granulocyte/monocyte colony-stimulating factor prevent endotoxin tolerance in human monocytes by promoting interleukin-1 receptor-associated kinase expression and its association to MyD88 and not by modulating TLR4 expression. J. Biol. Chem. 277, 27927-27934.

Arcaroli, J., Silva, E., Maloney, J. P., He, Q., Svetkauskaite, D., Murphy, J.R. and Abraham, E. (2006) Variant IRAK-1 haplotype is associated with increased nuclear factor-kappaB activation and worse outcomes in sepsis. Am. J. Respir. Crit. Care. Med. 173, 1335-1341.

Cao, Z., Henzel, W.J. and Gao, X. (1996) IRAK: a kinase associated with the interleukin-1 receptor. Science 271, 1128-1131.

Cardenes, M., von Bernuth, H., Garcia-Saavedra, A., Santiago, E., Puel, A., Ku, C.L., Emile, J.F., Picard, C., Casanova, J.L., Colino, E., Bordes, A., Garfia, A. and Rodriguez-Gallego, C. (2006) Autosomal recessive interleukin-1 receptor-associated kinase 4 deficiency in fourth-degree relatives. J. Pediatr. 148, 549-551.

Cuschieri, J., Bulmus, V., Gourlay, D., Garcia, I., Hoffman, A., Stayton, P. and Maier, R.V. (2004) Modulation of macrophage responsiveness to lipopolysaccharide by IRAK-1 manipulation. Shock 21, 182-188.

del Fresno, C., Otero, K., Gomez-Garcia, L., Gonzalez-Leon, M.C., Soler-Ranger, L., Fuentes-Prior, P., Escoll, P., Baos, R., Caveda, L., Garcia, F., Arnalich, F. and Lopez-Collazo, E. (2005) Tumor cells deactivate human monocytes by up-regulating IL-1 receptor associated kinase-M expression via CD44 and TLR4. J. Immunol. 174, 3032-3040.

Deng, C., Radu, C., Diab, A., Tsen, M. F., Hussain, R., Cowdery, J. S., Racke, M.K. and Thomas, J.A. (2003) IL-1 receptor-associated kinase 1 regulates susceptibility to organ-specific autoimmunity. J. Immunol. 170, 2833-2842.

Deng, J.C., Cheng, G., Newstead, M.W., Zeng, X., Kobayashi, K., Flavell, R.A. and Standiford, T.J. (2006) Sepsis-induced suppression of lung innate immunity is mediated by IRAK-M. J. Clin. Invest. 116, 2532-2542.

Dirkx, A. E., Egbrink, M.G., Wagstaff, J. and Griffioen, A.W. (2006) Monocyte/macrophage infiltration in tumors: modulators of angiogenesis. J. Leukoc. Biol. 80, 1183-96.

Ferlito, M., Romanenko, O.G., Ashton, S., Squadrito, F., Halushka, P.V. and Cook, J.A. (2001) Effect of cross-tolerance between endotoxin and TNF-alpha or IL-1beta on cellular signaling and mediator production. J. Leukoc. Biol. 70, 821-829.

Fitzgerald, K.A., Palsson-McDermott, E. M., Bowie, A.G., Jefferies, C.A., Mansell, A.S., Brady, G., Brint, E., Dunne, A., Gray, P., Harte, M.T., McMurray, D., Smith, D.E., Sims, J.E. and O'Neill, L.A. (2001) Mal (MyD88-adapter-like) is required for Toll-like receptor-4 signal transduction. Nature 413, 78-83.

Gallina, G., Dolcetti, L., Serafini, P., De Santo, C., Marigo, I., Colombo, M.P., Basso, G., Brombacher, F., Borrello, I., Zanovello, P., Bicciato, S. and Bronte, V. (2006) Tumors induce a subset of inflammatory monocytes with immunosuppressive activity on CD8+ T cells. J. Clin. Invest. 116, 2777-2790.

Hacker, H., Redecke, V., Blagoev, B., Kratchmarova, I., Hsu, L.C., Wang, G.G., Kamps, M. P., Raz, E., Wagner, H., Hacker, G., Mann, M. and Karin, M. (2006) Specificity in Toll-like receptor signalling through distinct effector functions of TRAF3 and TRAF6. Nature 439, 204-207.

Hacker, H., and Karin, M. (2006) Regulation and function of IKK and IKK-related kinases. Sci STKE, re13.

Hardy, M.P. and O'Neill, L.A. (2004) The murine IRAK2 gene encodes four alternatively spliced isoforms, two of which are inhibitory. J. Biol. Chem. 279, 27699-27708.

Huang, Y., Li, T., Sane, D.C., and Li, L. (2004) IRAK1 serves as a novel regulator essential for lipopolysaccharide-induced interleukin-10 gene expression. J. Biol. Chem. 279, 51697-51703.

Ishii, K. J. and Akira, S. (2004) Toll-like Receptors and Sepsis. Curr. Infect. Dis. Rep. 6, 361-366.

Jacinto, R., Hartung, T., McCall, C. and Li, L. (2002) Lipopolysaccharide- and lipoteichoic acid-induced tolerance and cross-tolerance: distinct alterations in IL-1 receptor-associated kinase. J. Immunol. 168, 6136-6141.

Jensen, L.E. and Whitehead, A.S. (2001) IRAK1b, a novel alternative splice variant of interleukin-1 receptor-associated kinase (IRAK), mediates interleukin-1 signaling and has prolonged stability. J. Biol. Chem. 276, 29037-29044.

Knop, J. and Martin, M.U. (1999) Effects of IL-1 receptor-associated kinase (IRAK) expression on IL-1 signaling are independent of its kinase activity. FEBS Lett. 448, 81-85.

Knop, J., Wesche, H., Lang, D. and Martin, M.U. (1998) Effects of overexpression of IL-1 receptor-associated kinase on NFkappaB activation, IL-2 production and stress-activated protein kinases in the murine T cell line EL4. Eur. J. Immunol. 28, 3100-3109.

Kobayashi, K., Hernandez, L.D., Galan, J.E., Janeway, C.A., Jr., Medzhitov, R. and Flavell, R.A. (2002) IRAK-M is a negative regulator of Toll-like receptor signaling. Cell 110, 191-202.

Kollewe, C., Mackensen, A.C., Neumann, D., Knop, J., Cao, P., Li, S., Wesche, H. and Martin, M.U. (2004) Sequential autophosphorylation steps in the interleukin-1 receptor-associated kinase-1 regulate its availability as an adapter in interleukin-1 signaling. J. Biol. Chem. 279, 5227-5236.

Larkoski, S. G., Li, L., Langfield, C.D., Liu, Y.M., Howard, T.D., Xu, J.F., Bowden, D.W. and Herrington, D. M. (2005) Variation in innate immunity gene (IRAK1) is associated with C-reactive protein in Caucasian women in the Diabetes Heart Study (DHS) Circulation 112, U851.

Li, H., Cuartas, E., Cui, W., Choi, Y., Crawford, T.D., Ke, H.Z., Kobayashi, K.S., Flavell, R.A. and Vignery, A. (2005) IL-1 receptor-associated kinase M is a central regulator of osteoclast differentiation and activation. J. Exp. Med. 201, 1169-1177.

Li, L. (2004) Regulation of innate immunity signaling and its connection with human diseases. Curr. Drug Targets Inflamm. Allergy 3, 81-86.

Li, L., Cousart, S., Hu, J. and McCall, C.E. (2000) Characterization of interleukin-1 receptor-associated kinase in normal and endotoxin-tolerant cells. J. Biol. Chem. 275, 23340-23345.

Li, S., Strelow, A., Fontana, E. J. and Wesche, H. (2002) IRAK-4: a novel member of the IRAK family with the properties of an IRAK-kinase. Proc. Natl. Acad. Sci. U. S. A. 99, 5567-5572.

Li, X., Commane, M., Burns, C., Vithalani, K., Cao, Z. and Stark, G.R. (1999) Mutant cells that do not respond to interleukin-1 (IL-1) reveal a novel role for IL-1 receptor-associated kinase. Mol. Cell. Biol. 19, 4643-4652.

Medvedev, A.E., Thomas, K., Awomoyi, A., Kuhns, D.B., Gallin, J.I., Li, X. and Vogel, S.N. (2005) Cutting edge: expression of IL-1 receptor-associated kinase-4 (IRAK-4) proteins with mutations identified in a patient with recurrent bacterial infections alters normal IRAK-4 interaction with components of the IL-1 receptor complex. J. Immunol. 174, 6587-6591.

Moors, M.A., Li, L. and Mizel, S.B. (2001) Activation of interleukin-1 receptor-associated kinase by gram-negative flagellin. Infect. Immun. 69, 4424-4429.

Muzio, M., Ni, J., Feng, P. and Dixit, V.M. (1997) IRAK (Pelle) family member IRAK-2 and MyD88 as proximal mediators of IL-1 signaling. Science 278, 1612-1615.

Mytar, B., Woloszyn, M., Szatanek, R., Baj-Krzyworzeka, M., Siedlar, M., Ruggiero, I., Wieckiewicz, J. and Zembala, M. (2003) Tumor cell-induced deactivation of human monocytes. J. Leukoc. Biol. 74, 1094-1101.

Nakayama, K., Okugawa, S., Yanagimoto, S., Kitazawa, T., Tsukada, K., Kawada, M., Kimura, S., Hirai, K., Takagaki, Y. and Ota, Y. (2004) Involvement of IRAK-M in peptidoglycan-induced tolerance in macrophages. J. Biol. Chem. 279, 6629-6634.

Nguyen, H., Chatterjee-Kishore, M., Jiang, Z., Qing, Y., Ramana, C.V., Bayes, J., Commane, M., Li, X. and Stark, G.R. (2003) IRAK-dependent phosphorylation of Stat1 on serine 727 in response to interleukin-1 and effects on gene expression. J. Interferon Cytokine Res. 23, 183-192.

Oganesyan, G., Saha, S.K., Guo, B., He, J.Q., Shahangian, A., Zarnegar, B., Perry, A. and Cheng, G. (2006) Critical role of TRAF3 in the Toll-like receptor-dependent and -independent antiviral response. Nature 439, 208-211.

Ohnuma, K., Yamochi, T., Uchiyama, M., Nishibashi, K., Iwata, S., Hosono, O., Kawasaki, H., Tanaka, H., Dang, N.H. and Morimoto, C. (2005) CD26 mediates dissociation of Tollip and IRAK-1 from caveolin-1 and induces upregulation of CD86 on antigen-presenting cells. Mol. Cell. Biol. 25, 7743-7757.

Paterlini-Brechot, P., Saigo, K., Murakami, Y., Chami, M., Gozuacik, D., Mugnier, C., Lagorce, D. and Brechot, C. (2003) Hepatitis B virus-related insertional mutagenesis occurs frequently in human liver cancers and recurrently targets human telomerase gene. Oncogene 22, 3911-3916.

Picard, C., Puel, A., Bonnet, M., Ku, C.L., Bustamante, J., Yang, K., Soudais, C., Dupuis, S., Feinberg, J., Fieschi, C., Elbim, C., Hitchcock, R., Lammas, D., Davies, G., Al-Ghonaium, A., Al-Rayes, H., Al-Jumaah, S., Al-Hajjar, S., Al-Mohsen, I.Z., Frayha, H.H., Rucker, R., Hawn, T.R., Aderem, A., Tufenkeji, H., Haraguchi, S., Day, N.K., Good, R.A., Gougerot-Pocidalo, M.A., Ozinsky, A. and Casanova, J.L. (2003) Pyogenic bacterial infections in humans with IRAK-4 deficiency. Science 299, 2076-2079.

Rao, N., Nguyen, S., Ngo, K. and Fung-Leung, W.P. (2005) A novel splice variant of interleukin-1 receptor (IL-1R)-associated kinase 1 plays a negative regulatory role in Toll/IL-1R-induced inflammatory signaling. Mol. Cell Biol. 25, 6521-6532.

Ruckdeschel, K., Mannel, O. and Schrottner, P. (2002) Divergence of apoptosis-inducing and preventing signals in bacteria-faced macrophages through myeloid differentiation factor 88 and IL-1 receptor-associated kinase members. J. Immunol. 168, 4601-4611.

Sato, S., Takeuchi, O., Fujita, T., Tomizawa, H., Takeda, K. and Akira, S. (2002) A variety of microbial components induce tolerance to lipopolysaccharide by differentially affecting MyD88-dependent and -independent pathways. Int. Immunol. 14, 783-791.

Schoenemeyer, A., Barnes, B. J., Mancl, M.E., Latz, E., Goutagny, N., Pitha, P. M., Fitzgerald, K.A. and Golenbock, D.T. (2005) The interferon regulatory factor, IRF5, is a central mediator of toll-like receptor 7 signaling. J. Biol. Chem. 280, 17005-17012.

Song, Y.J., Jen, K.Y., Soni, V., Kieff, E. and Cahir-McFarland, E. (2006) IL-1 receptor-associated kinase 1 is critical for latent membrane protein 1-induced p65/RelA serine 536 phosphorylation and NF-kappaB activation. Proc. Natl. Acad. Sci. U. S. A. 103, 2689-2694.

Su, J., Richter, K., Zhang, C., Gu, Q. and Li, L. (2007) Differential regulation of interleukin-1 receptor associated kinase 1 (IRAK1) splice variants. Mol. Immunol. 44, 900-905.

Sun, J., Wiklund, F., Hsu, F. C., Balter, K., Zheng, S.L., Johansson, J.E., Chang, B., Liu, W., Li, T., Turner, A. R., Li, L., Li, G., Adami, H.O., Isaacs, W.B., Xu, J. and Gronberg, H. (2006) Interactions of sequence variants in interleukin-1 receptor-associated kinase4 and the toll-like receptor 6-1-10 gene cluster increase prostate cancer risk. Cancer Epidemiol. Biomarkers Prev. 15, 480-485.

Suzuki, N., Suzuki, S., Duncan, G.S., Millar, D.G., Wada, T., Mirtsos, C., Takada, H., Wakeham, A., Itie, A., Li, S., Penninger, J. M., Wesche, H., Ohashi, P.S., Mak, T.W. and Yeh, W.C. (2002a) Severe impairment of interleukin-1 and Toll-like receptor signalling in mice lacking IRAK-4. Nature 416, 750-756.

Suzuki, N., Suzuki, S., Millar, D.G., Unno, M., Hara, H., Calzascia, T., Yamasaki, S., Yokosuka, T., Chen, N.J., Elford, A.R., Suzuki, J. , Takeuchi, A., Mirtsos, C., Bouchard, D., Ohashi, P.S., Yeh, W.C. and Saito, T. (2006) A critical role for the innate immune signaling molecule IRAK-4 in T cell activation. Science 311, 1927-1932.

Suzuki, N., Suzuki, S. and Yeh, W.C. (2002b) IRAK-4 as the central TIR signaling mediator in innate immunity. Trends Immunol. 23, 503-506.

Thomas, J.A., Allen, J.L., Tsen, M., Dubnicoff, T., Danao, J., Liao, X.C., Cao, Z. and Wasserman, S.A. (1999) Impaired cytokine signaling in mice lacking the IL-1 receptor-associated kinase. J. Immunol. 163, 978-984.

Uematsu, S., Sato, S., Yamamoto, M., Hirotani, T., Kato, H., Takeshita, F., Matsuda, M., Coban, C., Ishii, K.J., Kawai, T., Takeuchi, O. and Akira, S. (2005) Interleukin-1 receptor-associated kinase-1 plays an essential role for Toll-like receptor (TLR)7- and TLR9-mediated interferon-{alpha} induction. J. Exp. Med. 201, 915-923.

Wesche, H., Gao, X., Li, X., Kirschning, C. J., Stark, G.R. and Cao, Z. (1999). IRAK-M is a novel member of the Pelle/interleukin-1 receptor-associated kinase (IRAK) family. J. Biol. Chem. 274, 19403-19410.

West, M.A. and Heagy, W. (2002) Endotoxin tolerance: A review. Crit .Care. Med. 30, S64-S73.

Yamin, T.T., and Miller, D.K. (1997) The interleukin-1 receptor-associated kinase is degraded by proteasomes following its phosphorylation. J. Biol. Chem. 272, 21540-21547.

6

Systems Biology of Macrophages

Mano Ram Maurya[1], Christopher Benner[2], Sylvain Pradervand[1], Christopher Glass[2,3] and Shankar Subramaniam[1,2]

[1] Department of Bioengineering, University of California at San Diego, La Jolla, CA 92093-0412.
[2] Graduate Program in Bioinformatics and Systems Biology, University of California at San Diego, La Jolla, CA 92093-0412.
[3] Department of Cellular and Molecular Medicine, University of California at San Diego, La Jolla, CA 92093-0651.

Abstract. Cells and tissues function in context. Under a given growth or survival medium they perform tasks, replicate and die. Given a stimulus they respond by invoking myriad biomolecular networks that result in a specified cellular outcome. At any given instant it can be argued that the cell is in a "state" defined by its components – their concentrations and locations, the interactions between components – that are modulated in space and time, and the complex circuitry – that involves a large number of interacting networks and a snapshot of the dynamical processes – such as gene expression, cell cycle, transport of components, etc. At present, we can measure, using high and low throughput methods, several cellular components in a context-dependent manner and obtain a partial picture of cellular networks and dynamical processes. Are these measurements sufficient to answer important biological questions and help reconstruct a systems-level understanding of a mammalian cell? This chapter will address systems biology strategies developed to address this question and demonstrate the power of integration of diverse cellular data for answering interesting biological questions in macrophages. We will use this systems biology approach to address the following questions: (1) How good are macrophage cell lines in addressing phenotypic biology of primary macrophages? (2) How do signals associated with inflammatory molecules regulate gene transcription in macrophages? (3) How can we combine proteomic and other cellular measurements to characterize the repertoire of upstream signaling networks invoked by macrophages? (4) How do designed knockdowns of proteins influence cellular phenotypes?

1 Introduction

Macrophages are a key cell type in the study of the immune system, serving as one of the first lines of defense and functioning as key mediators of the transition from innate to adaptive immunity. Starting as haematopoietic stem cells in the bone marrow, macrophage precursors proliferate and differentiate in response to a variety of cytokines and growth factors and are released into the blood stream as monocytes

(Valledor, Borras, Cullell-Young, Celada 1998). These monocytes eventually find their way into the peripheral tissue where they undergo a final program of differentiation into macrophages. Macrophage functions include antigen presentation to lymphocytes, phagocytosis of foreign antigens and apoptotic cells, synthesis of anti-microbial products, and the expression of a diverse group of cytokines and other hormone-like molecules that influence the initiation and evolution of the adaptive immune responses. (Linton and Fazio 2003; Lucas and Greaves 2001).

Because of their importance in biology and medicine, and their diverse range of activity, macrophages have served as important model systems for studies of signal transduction, cellular proliferation, differentiation, motility, gene expression and other cellular phenomena. Serving many functions of innate immunity, macrophages allow the study of innate pattern recognition of foreign antigens and other harmful molecules (like dsRNA) and the resulting physiological response (Beutler, Hoebe, Du, and Ulevitch 2003; Gordon, 2002). Macrophages have become a focal point for lipidomics, lipid homeostasis and lipid signaling research (Lee and Evans 2002). Macrophages have also been used in several models of disease, such as atherosclerosis, where macrophages circulating in the blood stream develop into foam cells contributing to arterial stenosis (Plutzky 2003; Linton and Fazio 2003). More recently macrophages have been identified serving important roles in obesity and type II diabetes (Skurk, Herder, Kraft, Muller-Scholze, Hauner, and Kolb 2004).

Several large-scale projects have focused on macrophages with a view to deciphering the molecular components that play a role in response to stimulation by ligands. These include the Alliance for Cellular Signaling [http://www.cellularsignaling.org], the LIPID MAPS [http://www.lipidmaps.org], the Inflammation and Response to Injury Project [http://www.gluegrant.org/] and the Innate Immunity Project [http://www.systemsbiology.org/]. The vast number of measurements carried out on primary macrophages and macrophage derived cell lines under different stimuli can answer important questions about macrophage biology. The Alliance for Cellular Signaling studied the response of macrophage cells to 23 distinct extracellular ligands binding to a variety of receptors including the Toll receptors, G-protein coupled receptors, and others by themselves and in pairs, by measuring select phosphoproteins, transcribed genes, second messengers and cytokines as a function of time after applying stimulus. The LIPID MAPS project has measured changes in hundreds of distinct lipid molecules as function of time in macrophages in response to activation of TLR4 receptor by KDO2 Lipid A. The Inflammation and Innate Immunity projects have carried out extensive proteomic and transcriptomic measurements in primary macrophage cells after activation. And recent ChIP-chip measurements of transcriptional regulation in macrophages upon activation have shed considerable light on mechanisms of activation and repression in gene transcriptional regulation. These exhaustive and focused measurements on macrophages spanning metabolic, signaling and regulatory biology of macrophages provide a unique opportunity to obtain a comprehensive view of the biochemical pathways and networks invoked in macrophages and to build a systems perspective on inflammation and innate immunity.

In this Chapter, we will address key biological questions concerning macrophages using a “systems biology” approach. In this approach, we will use cellular components, their interactions and biochemical networks of these components derived from experimental data to answer significant questions concerning macrophage biology. Systems biology brings constructionist, quantitative and engineering perspectives to cell biology and helps develop testable mechanistic hypotheses for complex cellular phenotypes.

2 In Vitro Measurements in Macrophages and Cell Lines

It has become common practice in cell biology to use cell lines that are expected to display phenotypes similar to primary cells in vitro and in vivo to study cellular behavior. The RAW264.7 (RAW) cell line is a macrophage-like cell line originating from murine macrophages transformed by the Abelson Leukemia Virus (Raschke, Baird, Ralph, Nakoinz 1978) and has been used extensively to study behavior of macrophages under many different stimuli. Despite its use as a model for primary macrophages, very little work has been done comparing RAW cells with the primary macrophages cell types. Most publications that reference both RAW cells and primary macrophages use RAW cells to validate the results of the main study or perform a follow up experiment that is more easily performed in the cell line. How good are macrophage cell lines in addressing phenotypic biology of primary macrophages? System-wide gene expression changes serve as convenient indicators of "cell state" for highlighting the similarities and differences between the primary cells and the cell line.

We have compared the resting state of the macrophages and the response of the macrophage cell types to lipopolysaccharide (LPS) a compound found in the cell wall of gram-negative bacteria, and commonly used as an antigen for studies of innate immunity. Macrophages stimulated by LPS become cytokine factories that prime the immune system for a response, ultimately surrendering to apoptosis. LPS binds to Toll-like receptor 4 (Tlr4) in a complex with LBP, CD14, and MD2, on the surface of the plasma membrane of macrophages. The Tlr4 protein belongs to a family of Toll-like receptor proteins involved in pattern recognition that share a common toll domain and play a large role in the innate immune system (for review, see (Palsson-McDermott and O'Neill 2004)). We used microarray technology to compare the global expression profiles of primary macrophages, thioglycollate-elicited macrophages and bone marrow derived macrophages and a cell line RAW 264.7, in a resting state and when treated with LPS. We have also included the expression profiles of mouse embryonic stem cells as a non-macrophage cell type control in this study.

2.1 Comparison of Untreated Cell Types

The untreated expression profiles of each cell type was compared against each of the other cell types, identifying differentially expressed genes that exhibit elevated expression levels in one cell type or the other (Table 1). There are roughly 650 genes that are differentially expressed between the primary cell types. RAW cells have about 1400 genes that are differentially expressed relative to the primary cells, and the

Table 1. Comparison of functions of genes differentially expressed between different macrophage cell types (RAW 267.4 cell line, bone marrow derived macrophages (BM), thioglycollate-elicited macrophages (TM) and murine embryonic stem cells (ES)).

High In	Low In	# genes	Most Significant group from Biological Process (GO)
BM	RAW	553	Transition metal ion transport (3.5e-5)
BM	TM	362	cell cycle (6.9e-17)
BM	ES	923	inflammatory response (2.3e-5)
RAW	BM	383	negative regulation of organismal physiological process (1.4e-3)
RAW	TM	631	DNA replication and chromosome cycle (1.2e-6)
RAW	ES	1075	cell surface receptor linked signal transduction (4.6e-7)
TM	BM	311	antigen processing (2.9e-4)
TM	RAW	743	immune cell chemotaxis (2.3e-5)
TM	ES	868	immune response (2.4e-8)
ES	BM	900	cellular metabolism (9.15e-23)
ES	RAW	1066	cellular metabolism (5.01e-13)
ES	TM	1003	RNA processing (2.7e-27)

ES cells have roughly 150% that number of differentially expressed genes between each of the macrophage cell types.

Analyzing these lists of differentially expressed genes for functional enrichment recovered several phenotypic differences between the cell types. In general, the macrophage cell types preferentially express signal transduction and defense response genes compared to the ES cells. Conversely, the ES cells preferentially express genes associated with metabolism, development, and cell cycle.

The comparison between the RAW and primary cell types revealed several surprising results. Genes with higher expression in the primary cell types showed enrichment for immune cell migration (4.1e-5) and the positive regulation of phagocytosis (1e-5). In addition, the primary cells were enriched for genes encoding proteins residing in the lysosome (4.5e-11) and Golgi apparatus (4.5e-5). The RAW cells had higher relative expression of cell cycle genes than the TM cells, in particular genes associated with mitosis (1.4e-5) and DNA metabolism (6.7e-5).

Close analysis of the genes that were preferentially expressed in RAW cells relative to the primary cell types revealed several targets of the Sonic Hedgehog pathway (SHH) / Gli transcription factors. Targets of the pathway have been identified from a variety of sources and are included in Fig. 1.

Nearly every known target is up regulated in RAW cells compared to the primary cell types. In addition, all three of the Gli transcription factors (Gli, Gli2, Gli3), including Gli5, display similar patterns of up-regulation in the RAW cells.

The differences between the two primary macrophages are of particular interest since they are frequently used interchangeably in macrophage studies. Analysis of

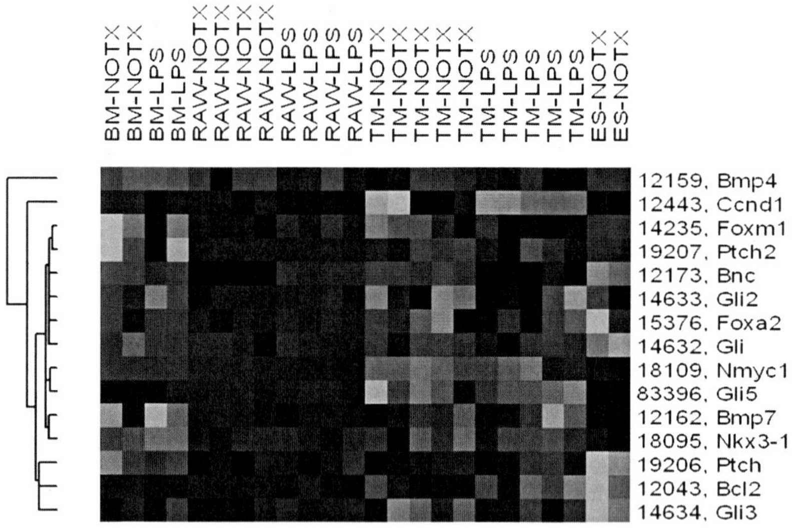

Fig. 1. Heat map of genes differentially expressed in RAW 264.7 cells relative to primary macrophage cells.

the untreated BM and TM expression profiles revealed 311 genes with higher expression in TM and 362 genes with higher expression in BM. Genes with higher expression in TM were enriched for defense response functions (9.6e-3), and in particular antigen processing and presentation (1.0e-4). This is most likely a byproduct of TM activation resulting from the thioglycolate priming. Genes with expression higher in BM were enriched for cell cycle (8.1e-16) and DNA replication (1.8e-10), consistent with the fact that BM cells are still proliferating while the TM cells are post mitotic.

2.2 Comparison of Genes Induced by LPS

Stimulation of macrophages by LPS results in a robust transcriptional response inducing roughly 500-600 genes depending on the cell type. Figure 2 shows the overlap of genes induced by LPS in all three macrophage cell types. In addition to the 243 genes that are commonly induced between the 3 cell types, 151 genes are shared between the two primary cell types, which is 3 fold more than the number of genes shared uniquely between the RAW cells and either of the primary cells (TM – 55, BM – 38).

Most commonly studied LPS-inducible genes are robustly induced in all three macrophage cell types. There appears to be no systematic difference between the cell types in the response of genes induced by Myd88 dependent and independent arms of the Tlr4 pathway. Common inflammatory and antiviral genes like Il1b, Il6, Tnf, iNOS, Cox-2, IP-10, and Il10 are all strongly up regulated in all cell types. Induction of the autoregulatory NFkB and Nfkbia genes are also induced across the cell types. A major exception to this is the induction of Rela (and to a lesser extent Relb), with a

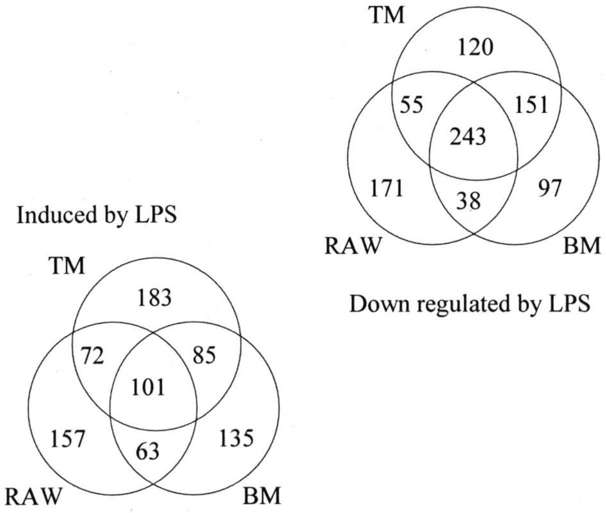

Fig. 2. Venn diagram displaying genes induced and downregulated, common and distinct to different cell types.

magnitude of induction in RAW cells well below that of the primary cells. The type II interferon gamma is only expressed in TM cells.

Several commonly studied genes are not induced in RAW cells. Il12a and Il12b, Gro1 (Cxcl1), Ccl8 (MCP-2), Gem, GR, and M-CSF (Csf1) fail to induce in response to LPS. Conversely, RAW cells show the preferential induction of several genes with important immune functions that are not induced in the primary cell types. C/EBP delta (NF-IL6 beta), Oct-2 (Pou2f2) and Ets1 are all strongly up regulated in RAW cells.

Functional analyses of the genes induced by LPS are similar for each of the cell types. Defense and inflammatory response have enrichment p-values of roughly 1e-15. Other categories closely related such as cytokine biosynthesis are also highly significant. The regulation of apoptosis (~1e-7 in RAW, 1e-4 in BM/TM) is also enriched. Genes belonging to the auto-regulatory NFkB pathway are induced, as are down stream targets of type I interferon signaling. Analysis with transcription factor target genes from Transfac showed targets of NFkB p65 (Rela), NFkB p50/p65 heterodimers, C/EBP beta and delta, Stat1/3/5, and IRF1 to be enriched in the LPS induced genes, all of which are known to be activated by LPS either within the primary or secondary responses.

Although functional categories can be readily assigned to genes commonly induced by LPS among the macrophage cell types, the functional assignment of genes that are specific to only one or two of the cell types proves a more difficult problem. Genes induced by both TM and BM cells show enrichment for the same major categories found in genes common to all three. They also show specific enrichment in genes belonging to Positive regulation of phagocytosis (Ptx3, C3, Ptpns1, Mfge8). For genes that are only induced in RAW cells, several genes involved in growth arrest (genes such as Gas1, Ddit3, Gadd45a) and anti-apoptotic genes (Eef1a2, Pim2, Nr4a1, Pik3ca) are enriched in the group.

Although the two primary cells and the RAW cell line are generally referred to as roughly equivalent in most research, there are clear differences between the cell types. For example, preparations of TM are generally post-mitotic while BM continue to proliferate when cultured in serum rich conditions. TM are also slightly activated as a result from the thioglycolate priming, secreting various cytokines at elevated levels. RAW cells are a transformed cell line locked in a state of continual proliferation, much like a cancer. Most of these large physiological differences are easily recovered by examination of the differences in gene expression and analyzing for functional enrichment from the Gene Ontology. It is still unclear to what extent this transformation has affected the macrophage-like function of the RAW cells.

Only a handful of publications actually report differences in the behavior of the cell types. Pparg, a nuclear receptor with anti-inflammatory functions, is not expressed in RAW cells. IL-12 p70 and Interferon γ, two important cytokines that function at the barrier of innate and adaptive immunity, are not secreted by RAW cells (Mastsuura, Saito, Hirai, Okamura 2003). Dectin-1 (Clecsf12), a plasma membrane protein involved in phagocytosis, was shown to have low expression in RAW cells relative to primary macrophages (Taylor, Brown, Herre, Williams, Willment, Gordon 2004). Cd36 expression also appears to be absent in RAW cells but present in the primary cell types (Nicholson, 2004).

Gro1 and Gem are strongly induced in the primary cell types by LPS but fail to change in RAW cells. A recent study comparing RAW and peritoneal macrophages in ACT signaling found these two genes failed to express in RAW cells (Galindo, Fadl, Sha, Chopra 2004), suggesting a common faulty mechanism with LPS signaling. The lack of induction of Il12a and Il12b has been observed before (Mastsuura, Saito, Hirai, Okamura 2003), however several studies have demonstrated activation of the Il12a and Il12b promoters by LPS in various expression constructs. This could suggest the epigenetic remodeling of the Il12a and Il12b genomic regions during the Abelson virus transformation process.

Caveolin-1 is upregulated in peritoneal macrophages in response to LPS and down regulated in RAW cells, which has been demonstrated independently (Lei and Morrison 2000; Frank and Lisanti 2004). Caveolin-1 is a major component of the caveolae membrane and is negatively regulated by the Ras-p42/44 Map Kinase pathway, which is involved in proliferation and activated by v-abl during Abelson virus transformation (Engelman, Zhang, Razani, Pestell, and Lisanti 1999). For a review of abelson virus pathways, see (Shore, Tantravahi, and Reddy 2002). Caveolin-1 has been show to play an important role in Atherosclerosis. Werb and Chin (1983) (Meir and Leitersdorf 2004) have shown that another protein important in Atherosclerosis, apoprotein E, is secreted in TM and BM but not in macrophage-like cell lines, including J774.2 and RAW264.7. With so many important genes involved in Atherosclerosis exhibiting stark differences between RAW and primary cell types, it would be advised to carefully consider any experiment using RAW cells as a model for Atherosclerosis.

The general structure of the Tlr4/LPS signaling pathway appears to be conserved between the RAW and primary macrophages. Traditional downstream targets of Tlr4 signaling are nearly all up regulated across the cell types with few exceptions. However, on the global scale there are a disproportional number of genes only induced in one or two of the macrophage cell types. Part of this may be due to the fact that RAW cells have played an important part in macrophage research over the past 20 years and have been used to verify studies performed on primary macrophages. For this reason our contemporary view of Tlr4 signaling in macrophages may be biased as a common subset of the pathway between well studied cell types. The large overlap of induced genes common only to TM and BM may be the signature of a macrophage specific arm of the LPS pathway that is yet to be discovered.

3 Reconstruction of Signaling Modules in Stimulated Macrophages

Signaling network reconstruction involves the integration of several sources of data to describe the biochemical transformations that occur in a given network. Contextual specificity is a crucial consideration in answering five questions for signaling network reconstruction: What proteins and other network components participate? What are the ligand–receptor interactions? What are the receptor–intracellular-component interactions? What are the intracellular-component–intracellular component interactions? What are the intracellular-component–DNA interactions? Genome annotation, biochemical experimentation, cell physiology characterizations, expression arrays, and other such data sources each provide different types of datum that answer these questions and contribute to the reconstruction of a given cellular signaling network.

A main component of the inflammatory response is the production and release of immuno-regulatory cytokines and chemokines by macrophages. Among them, pro-inflammatory cytokines, such as tumor necrosis factor (TNF)α, interleukins IL-1, IL-6, IL-12, granulocyte macrophage colony stimulating factor GM-CSF and interferon (IFN)γ, induce both acute and chronic inflammatory responses; chemokines, such as MIP-1a and RANTES, are involved in the chemotaxis of leucocytes; and anti-inflammatory cytokines, such as IL-4, IL-10 and transforming growth factor (TGF)β, limit the magnitude and the extent of inflammation (Oppenheim, Feldman, Durum, Hirano, and Vilcek 2000; Oppenheim and Feldmann 2000). Upon macrophage activation, cytokine synthesis and release are triggered (Ma, Chen, Mandelin, Ceponis, and Miller 2003). This process is mainly regulated transcriptionally although posttranscriptional and translational mechanisms may also take place (Shaw and Kamen 1986; Pasparakis, Alexopoulou, Douni, Kolias 1996; Kontoyiannis, Pasparakis, Pizarro, Cominelli, Kollias 1999). Several pathways transmit the signals that trigger cytokine production. Nuclear factor kappa B (NF-κB), mitogen-activated protein kinases (MAPK), signal transducer and activator of transcription (STAT), cAMP - protein kinase A (PKA), interferon regulatory factor (IRF) or CAAT/enhancer-binding proteins (C/EBP) pathways have been described to be involved in cytokine production in macrophages (Oppenheim et al. 2000; Ozato, Tsujimura, and Tamura 2002). These pathways are not distinct entities, but are part of a general network whose different signals are produced by multiple stimuli that generate different cytokine responses. We do not have sufficient knowledge to provide a complete and detailed picture of these signaling pathways. However, using specific markers of signaling pathways, one can develop a coarse-grained description/model that will allow (1) elucidate what are the common and different signaling modules required for the release of different cytokines and, (2) predict quantitative estimate of the cytokines release to given signaling pathway activations.

General strategies have been developed for the reconstruction of signaling networks (Pradervand, Maurya, and Subramaniam, 2006; Papin, Hunter, Palsson, and Subramaniam 2005). Using a systematic profiling of signaling responses and cytokine release in RAW 264.7 macrophages made available by the Alliance for Cellular Signaling, we developed a strategy that integrates principal component regression (PCR) and exhaustive-search-based model-reduction to identify required signaling factors necessary and sufficient to predict the release of seven cytokines (G-CSF, IL-1α, IL-6, IL-10, MIP-1α, RANTES, TNFα) in response to selected ligands (Pradervand et al. 2006). This study provided a model-based quantitative estimate of cytokine release and identified ten signaling components involved in cytokine production. The models captured many of the known signaling pathways involved in cytokine release and predict potentially important novel signaling components, like p38 MAPK for G-CSF release, IFNγ and IL-4 specific pathways for IL-1a release, and a M-CSF specific pathway for TNFα release (Fig. 3).

Evaluation of our models using literature data shows a good agreement with current knowledge of stimulation of RAW264.7 macrophages. Our minimal model covers all known mechanisms of activation of G-CSF and highlights a potential role for p38 in its posttranscriptional regulation. For IL-1α release, besides all known activators, IFNγ and IL-4 were identified as potential novel independent activators. For IL-6 release, four predictors were confirmed by literature data. For RANTES release, all known mechanisms of activation were found. Finally, all known signaling pathways with the exception of ERK were found for TNFα release. Cytokines mediate pathogenesis of many diseases (chronic inflammatory diseases, autoimmune diseases, cancer, etc.). With increasing quantitative knowledge about the important pathways in the production of cytokines, model building using a systems biology approach, as discussed here, will help identify novel targets in order to maximize the efficacy of a drug for affecting one or few cytokines while minimizing the effect on the homeostasis of other cytokines.

4 Kinetic Modeling of Calcium Signaling Networks in Macrophages

A major goal of systems biology is to be able to quantitatively model cellular signaling networks. Cytosolic calcium is a second messenger and plays an important role in intracellular signaling (Carafoli 2002). Dynamic changes in intracellular

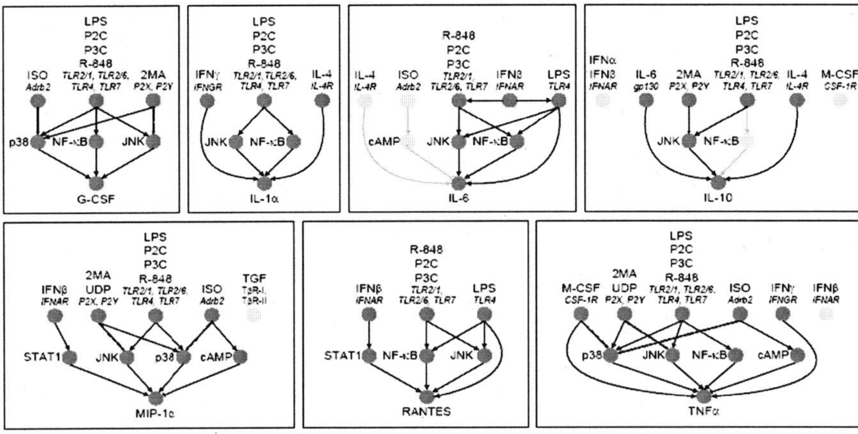

Fig. 3. Topologies of signaling networks leading to cytokine releases derived from principal component regression (PCR) minimal models and ANOVA analysis (Pradervand et al. 2006). In each panel, nodes in the upper row represent ligands that significantly regulate respective cytokines (ANOVA). Nodes in the middle row represent significant pathways identified by PCR minimal models. Edges between top and middle rows represent significant signaling pathway regulation by the given ligands (ANOVA). Edges between top and bottom rows, or middle and bottom rows, represent significant participation identified by PCR minimal models. Weak activations of signaling pathways are indicated by dashed edges. Light gray: pathways demonstrated in the literature to not play any role (false positives).

calcium serve both as an important indicator of cellular events and a quantitative measure of cellular response to stimuli. A simplified model for calcium signaling in RAW 264.7 cells that incorporates most of the relevant mechanisms included in the published models has been developed in our laboratory (shown in Fig. 4). In the calcium model for RAW 264.7 cells, the stimulus is the complement protein C5a (activates G-protein Gi). Most of the mechanisms are generic and they can be easily tailored for another stimulus in a non-excitable cell. We have included some mechanisms explicitly so that knockdown of important proteins, such as GPCR kinase (GRK), Arrestin, Gβγ and Gα,i, can be modeled quantitatively. Below we concisely present a schematic and mathematical model and the main results. A detailed description of the model and the related results are presented elsewhere (Maurya and Subramaniam 2007a,b).

4.1 Schematic Model

Figure 4 shows an overall schematic of ligand-induced release of calcium from endoplasmic reticulum (ER) into cytosol, binding of calcium (Ca_i) to proteins (Pr) in the cytosol (shown) and in the ER (not shown) and other calcium exchange fluxes to/from ER, the extra-cellular space and mitochondria. Figure 4B shows the reactions that have been modeled explicitly (either as shown or in a lumped manner).

Receptor module: Reaction 1 is activation through ligand binding. Reactions 3-5 lead to desensitization through phosphorylation of the ligand-bound active receptor

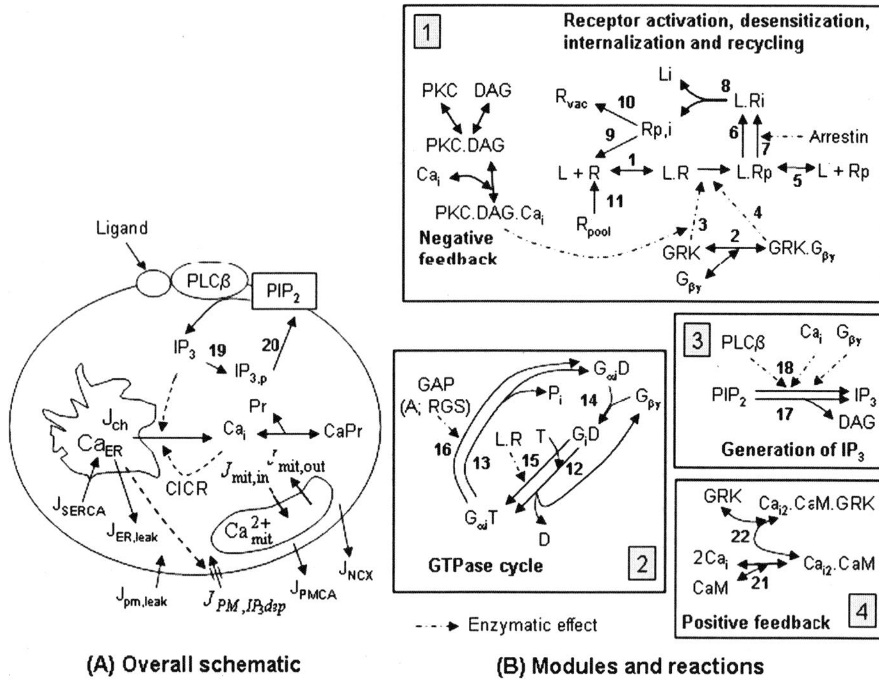

Fig. 4. A simplified model for calcium signaling including calcium influx, ER and mitochondrial exchange and storage, used in the conceptual model based computation.

(reaction 3 is catalyzed by GRK and reaction 4 is catalyzed by GRK.$G_{\beta\gamma}$) (Riccobene, Omann and Linderman 1999; Woolf and Linderman 2003; Lemon, Gibson and Bennett 2003). Reactions 6 and 7 are internalization (reaction 7 is catalyzed by Arrestin). Reactions 8-10 are for recovery of the receptor (L.Ri: internalized ligand-bound phosphorylated receptor (Sitaramayya and Bunnett 1999)). Hydrolysis of phosphate of $R_{p,i}$ (internalized phosphorylated receptor) is lumped in both reactions 9 and 10 (Hoffman, Linderman and Omann 1996, Lemon et al. 2003). Reaction 11 is fresh receptor generation (Yi, Kitano and Simon 2003). In most existing models, GRK and Arrestin have not been explicitly included. Reaction 2 represents binding of GRK to $G_{\beta\gamma}$ and is very fast. Rate constant for reaction 4 is much higher (about 100-fold) than for reaction 3 so that most phosphorylation occurs due to GRK.$G_{\beta\gamma}$ when ligand is present and due to GRK during basal state. Similarly, internalization (reactions 6 and 7) is enhanced substantially by Arrestin (buffered). Reactions 3, 4 and 7 are lumped-enzymatic reactions.

GTPase cycle module: Reactions 12-16 depict GTPase cycle. Reactions 12 and 13 are in the absence of active receptor and GAP, respectively. Reaction 15 is similar to reaction 12 but is catalyzed by L.R (ligand-bound active receptor) and reaction 16 is similar to 13 but is catalyzed by GAP (A, RGS). Reactions 15 and 16 model the G-protein activation (binding of the alpha subunit to GTP) and GTP hydrolysis catalyzed by GTPase activating protein, respectively (T is GTP, D is GDP, A is GAP (RGS)).

IP_3 module: During basal state, most of the IP_3 is generated due to slow hydrolysis of PIP2 (reaction 17) since free $G_{\beta\gamma}$ is present in very small amounts. Upon G protein activation, dissociated $G_{\beta\gamma}$ binds to phospholipase-C (PLC) β. Each of PLCβ.$G_{\beta\gamma}$, PLCβ.Ca_i and PLCβ.$G_{\beta\gamma}$.Ca_i catalyze hydrolysis of PIP2 but PLCβ.$G_{\beta\gamma}$.Ca_i is the most potent. In our model, for simplification, a lumped-enzymatic reaction is used to model the enhancement due to PLCβ.$G_{\beta\gamma}$.Ca_i (reaction 18). Reactions 19 and 20 are simplified and highly lumped representations of IP_3 metabolism (degradation/conversion to/from other inositol-phosphates and back to PIP2) with only one intermediate pseudo-species ($IP_{3,p}$, IP_3 product (Lemon et al. 2003)). This representation is sufficient both for excitable and non-excitable cells since the main mechanism responsible for calcium oscillation in muscle cells is the interaction between fluxes through IP_3 receptor (IP_3R), RyR channels, sarco(endo)plasmic reticulum (SR) calcium ATPase (SERCA) pump and the voltage-operated channels (VOC) and store-operated channels (SOC). Marhl et al. (Marhl, Haberichter, Brumen, and Heinrich 2000) have shown that interactions with the mitochondria (included in our model) can also result in oscillations specially at cytosolic [Ca^{2+}] above 0.5 uM.

Feedback effects from calmodulin and PKC: As shown in Fig. 4B, calmodulin (CaM) binds with intracellular Ca^{2+} (Ca_i) (reaction 21) and the resulting complex binds with GRK (reaction 22) reducing the effective amount of free GRK that can bind $G_{\beta\gamma}$. The result is reduced phosphorylation of the active-receptor and thus this constitutes a functional positive feedback. On the contrary, PKC-DAG-Ca_i complex enhances the activity of GRK and thus, promotes phosphorylation resulting in a negative feedback. This effect is modeled as an enzymatic-activation by calcium to avoid explicit modeling of the complexes.

4.2 Mathematical Representation of the Model for Macrophages

In the model, the state variables are described by a set of ordinary differential equations ODEs (Schuster, Marhl and Hofer 2002) involving the Ca^{2+} fluxes between different cellular compartments and other fluxes due to reactions. Apart from the state variables used to model the details of ligand-induced generation of IP_3, the four main state variables are $[Ca^{2+}]_i$, $[Ca^{2+}]_{ER}$, $[Ca^{2+}]_{mit}$, and [IP_3], which represent the Ca^{2+} concentrations in cytosolic, endoplasmic reticulum (ER), mitochondrial compartment, and IP_3 concentration in cytosol, respectively. The differential equations for the state variables related to the reactions (including [IP_3]) are automatically derived using a reaction parser developed in our laboratory. The expressions for other state variables and related fluxes, specified manually, are given below:

$$\frac{d[Ca^{2+}]_i}{dt} = \beta_i (J_{ch} + J_{ER,leak} + J_{SOC} - J_{SERCA} - (J_{PMCA} + J_{NCX} - J_{PM,leak}) + (J_{mit,out} - J_{mit,in}) - 2v_{21}) \tag{1}$$

$$\frac{d[Ca^{2+}]_{ER}}{dt} = \frac{\beta_{ER}}{\rho_{ER}}(J_{SERCA} - J_{ch} - J_{ER,leak}) \tag{2}$$

$$\frac{d[Ca^{2+}]_{mit}}{dt} = \frac{\beta_m}{\rho_m}(J_{mit,in} - J_{mit,out}) \tag{3}$$

$$\frac{dh}{dt} = k_{on}(Q - ([Ca^{2+}]_i + Q)h) \tag{4}$$

where, β_i, β_{ER}, and β_m are the ratio of free calcium to free and bound calcium in the cytosol, ER and mitochondria, respectively, assuming fast-buffering (equilibrium) with calcium binding proteins. The expressions for other terms are as follows:

$$J_{ch} = v_{\max,ch}\left(\left(\frac{[IP_3]}{[IP_3]+K_{IP3}}\right)\left(\frac{[Ca^{2+}]_i}{[Ca^{2+}]_i + K_{act}}\right)h\right)^3 \left([Ca^{2+}]_{ER} - [Ca^{2+}]_i\right) \tag{5}$$

$$J_{SERCA} = V_{\max}\left([Ca^{2+}]_i^2 / \left([Ca^{2+}]_i^2 + K_P^2\right)\right) \tag{6}$$

$$J_{ER,leak} = k_{ER,leak}([Ca^{2+}]_{ER} - [Ca^{2+}]_i) \tag{7}$$

$$J_{PM,IP_3dep} = V_{\max,PM,IP_3dep}[IP_3]^2 / \left(K_{m,PM,IP_3dep}{}^2 + [IP_3]^2\right) \tag{8}$$

$$J_{PM,leak} = v_{pm,leak} \tag{9}$$

$$J_{PMCA} = \frac{V_{\max,PMCA,l}[Ca^{2+}]_i^2}{[Ca^{2+}]_i^2 + K_{M,PMCA,l}^2} + \frac{V_{\max,PMCA,h}[Ca^{2+}]_i^5}{[Ca^{2+}]_i^5 + K_{M,PMCA,h}^5}$$
$$J_{NCX} = \frac{V_{\max,NCX}[Ca^{2+}]_i}{[Ca^{2+}]_i + K_{M,NCX}} \tag{10}$$

$$J_{mit,out} = (k_{out}\left([Ca^{2+}]_i^2 / \left(K_1^2 + [Ca^{2+}]_i^2\right)\right) + k_m)[Ca^{2+}]_{mit} \tag{11}$$

$$J_{mit,in} = k_{in}[Ca^{2+}]_i^4 / \left(K_2^4 + [Ca^{2+}]_i^4\right) \tag{12}$$

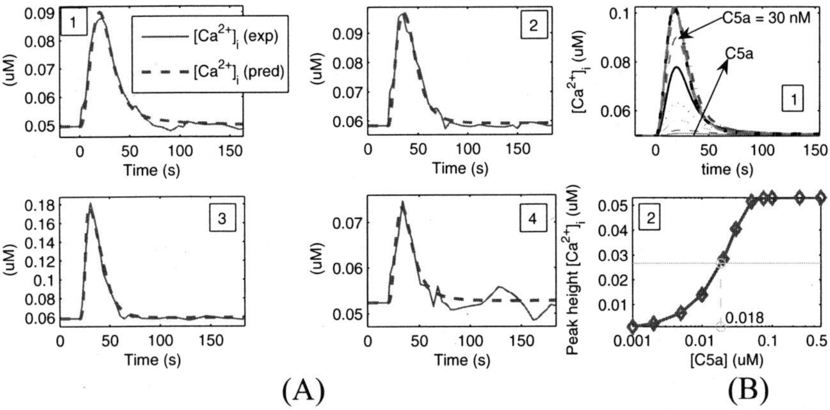

Fig. 5. Results on calcium modeling: (A) Fit to four data sets: panel 1: control for knockdown of RGS with 30 nM C5a; panel 2: control for knockdown of $G_{\alpha i}$ with 100 nM C5a; panel 3: 85% knockdown of $G_{\alpha i}$ with 100 nM C5a; panel 4: 83% knockdown of PLCβ with 100 nM C5a, (B) Predicted dose response curve for C5a. EC_{50} is ~18 nM.

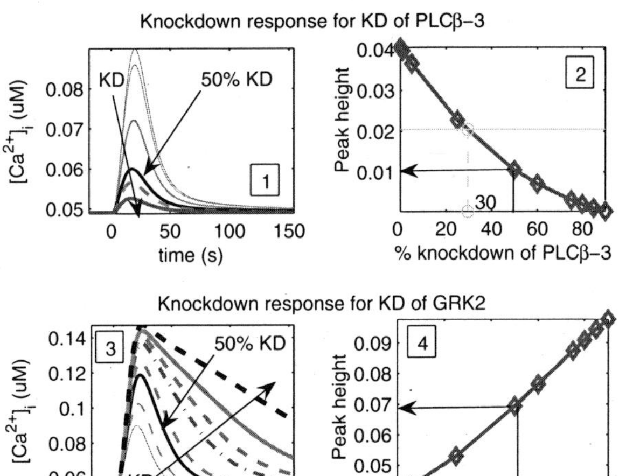

Fig. 6. Predicted knockdown response for varying knockdowns of PLCβ-3 (panel 1 and 2) and GRK2 (panel 3 and 4). KD; KnockDown.

$$Q = K_{inh}([IP_3]+d_1)/([IP_3]+d_3) \tag{13}$$

and v_{21} is the rate of reaction 21 (Fig. 4B).

We note that, (1) J_{ch}, Ca^{2+} flux rate from ER to cytosol through the Ca^{2+} releasing channel, is based on the IP_3R model of De Young and Keizer (De Young and Keizer 1992) and Li and Rinzel (Li and Rinzel, 1994) (see also, (Fink, Slepchenko, Moraru, Watras, Schaff and Loew 2000)), by implementing the calcium induced calcium release (CICR) mechanism of the Ca^{2+} releasing channel (*h* is the fraction of IP_3R in which Ca^{2+} is not bound to the inhibitory site); (2) J_{SERCA}, Ca^{2+} pumping rate by

SERCA pump, is expressed as Hill-type equation with Hill constant of 2 as observed in many cell types (Gill and Chueh 1985, Lytton, Westlin, burk, Shull and MacLennan 1992); (3) $J_{mit,in}$ and $J_{mit,out}$, Ca^{2+} uptake and release by mitochondria is from (Haberichter, Marhl and Heinrich 2001) except that for $J_{mit,in}$, a Hill-coefficient of 4 is used instead of 8; (4) J_{PMCA} and J_{NCX}, extrusions of Ca^{2+} to the extra-cellular space by plasma-membrane calcium ATPase (PMCA) pump and Na^+/Ca^{2+} exchanger (NCX) on plasma membrane are modeled as in (Wiesner, Bernk and Nerem 1996) and $J_{PM,leak}$, leakage flux from extra-cellular space to cytosol, is treated as in (Hofer, Venance and Giaume 2002); (5) The expression for J_{PM,IP_3dep} which includes SOC flux is from (Hofer et al. 2002).

4.3 Results for Stimulation of RAW 264.7 Cells with C5a

We have used 4 datasets (a control dataset, $G_{\alpha,i}2,3$ knockdown data, the corresponding control data and PLCβ-3 knockdown data) to constrain the parameters. The independent control/basic dataset is later used for prediction of dose-response of C5a and knockdown-response for several proteins. The results are shown in Fig. 5. The optimizer is able to find parameter-values that satisfy the data from all the four datasets (Fig. 5A, panels 1-4) and hence, we have confidence in the structure/mechanisms of the model. The sigmoidal shape of the dose-response curve (Fig. 5B, panel 2) is as expected. The predicted knockdown response for various proteins such as the PLCβ-3 (Fig. 6, panels 1 and 2), GRK2 (Fig. 6, panels 3 and 4), receptor, Arrestin, $G_{\beta\gamma}$ and RGS10 (not shown) is accurate. For example, upon knockdown of PLCβ-3 (panels 1 and 2 in Fig. 6), rate of IP_3 generation through hydrolysis of PIP_2 is reduced which results in lower channel flux (J_{ch}) and hence a lower peak is observed. Along similar lines, knockdown of GRK2 leads to higher amount of free $G_{\beta\gamma}$ for longer time resulting in larger IP_3 and hence larger peak and slower return to the basal state (panels 3 and 4 in Fig. 6). More details are presented by Maurya and Subramaniam (2007a,b).

5 Conclusions

The main objective of this review is to emphasize the role systems biology plays in the analysis of complex mammalian cell phenotypes. Diverse phenotypic questions such as how similar are cell lines in portraying in vivo phenotypes, how can one reconstruct inflammatory pathways in macrophages, and how can one quantitatively assess a cellular phenotype are addressed from a systems biology perspective. With the availability of more high and medium throughput measurements, it will be possible to carry out systemic analysis of related components including the complement and other cells associated with innate immunity. This will permit us to move beyond causal descriptions of cellular phenotypes to mechanistic and quantitative analysis of networks and systems.

6 Acknowledgements

The authors would like to acknowledge grants from that National Institutes of Health and from the National Science Foundation.

References

Beutler B, Hoebe K, Du X and Ulevitch RJ. (2003) How we detect microbes and respond to them: the Toll-like receptors and their transducers. J Leukoc Biol., **74**(4):479-85.

Carafoli, E. (2002) Calcium signaling: a tale for all seasons. Proc Natl Acad Sci U S A. **99**(3): 1115-22.

De Young, G.W. and Keizer, J. (1992) A single-pool inositol 1,4,5-trisphosphate-receptor-based model for agonist-stimulated oscillations in Ca2+ concentration', *Proc Natl Acad Sci U S A*. **89**(20), pp. 9895-9.

Engelman JA, Zhang XL, Razani B, Pestell RG and Lisanti MP. (1999) p42/44 MAP kinase-dependent and -independent signaling pathways regulate caveolin-1 gene expression. Activation of Ras-MAP kinase and protein kinase a signaling cascades transcriptionally down-regulates caveolin-1 promoter activity. J Biol Chem. 1999 Nov 5;274(45):32333-41.

Fink, C.C., Slepchenko, B., Moraru, II, Watras, J., Schaff, J.C. and Loew, L.M. (2000) An image-based model of calcium waves in differentiated neuroblastoma cells, *Biophys J*. **79**(1), pp. 163-83.

Frank PG and Lisanti MP. (2004) Caveolin-1 and caveolae in atherosclerosis: differential roles in fatty streak formation and neointimal hyperplasia. Curr Opin Lipidol. 15(5):523-9.

Galindo CL, Fadl AA, Sha J, Chopra AK. (2004) Microarray analysis of Aeromonas hydrophila cytotoxic enterotoxin-treated murine primary macrophages. Infect Immun. 72(9):5439-45.

Gill, D.L. and Chueh, S.H. (1985) An intracellular (ATP + Mg2+)-dependent calcium pump within the N1E-115 neuronal cell line' *J Biol Chem* **260**(16), pp. 9289-97.

Gordon S. (2002) Pattern recognition receptors: doubling up for the innate immune response. Cell. 111(7):927-30.

Haberichter, T., Marhl, M. and Heinrich, R. (2001) Birhythmicity, trirhythmicity and chaos in bursting calcium oscillations, *Biophys Chem*. **90**(1), pp. 17-30.

Hofer, T., Venance, L. and Giaume, C. (2002) Control and plasticity of intercellular calcium waves in astrocytes: a modeling approach, *J Neurosci*. **22**(12), pp. 4850-9.

Hoffman, J.F., Linderman, J.J. and Omann, G.M. (1996) Receptor up-regulation, internalization, and interconverting receptor states. Critical components of a quantitative description of N-formyl peptide-receptor dynamics in the neutrophil, *J Biol Chem*. **271**(31), pp. 18394-404.

Kontoyiannis D, Pasparakis M, Pizarro TT, Cominelli F, Kollias G (1999) Impaired on/off regulation of TNF biosynthesis in mice lacking TNF AU-rich elements: implications for joint and gut-associated immunopathologies. Immunity. 10: 387-398.

Lee CH, Evans RM. (2002) Peroxisome proliferator-activated receptor-gamma in macrophage lipid homeostasis. Trends Endocrinol Metab. 13(8):331-5.

Lei MG, Morrison DC. (2000) Differential expression of caveolin-1 in lipopolysaccharide-activated murine macrophages. Infect Immun. 68(9):5084-9.

Lemon, G., Gibson, W.G. and Bennett, M.R. (2003) Metabotropic receptor activation, desensitization and sequestration-I: modelling calcium and inositol 1,4,5-trisphosphate dynamics following receptor activation, *Journal of Theoretical Biology*. **223**(1), pp. 93-111.

Li, Y.X. and Rinzel, J. (1994) Equations for InsP3 receptor-mediated [Ca2+]i oscillations derived from a detailed kinetic model: a Hodgkin-Huxley like formalism, *J Theor Biol*. **166**(4), pp. 461-73.

Linton MF, Fazio S. (2003) Macrophages, inflammation, and atherosclerosis. Int J Obes Relat Metab Disord. Suppl 3:S35-40.

Lucas AD, Greaves DR. (2001) Atherosclerosis: role of chemokines and macrophages. Expert Rev Mol Med. pp. 1-18.

Lytton, J., Westlin, M., Burk, S.E., Shull, G.E. and MacLennan, D.H. (1992) Functional comparisons between isoforms of the sarcoplasmic or endoplasmic reticulum family of calcium pumps, *J Biol Chem*. **267**(20), pp. 14483-9.

Ma J, Chen T, Mandelin J, Ceponis A, Miller NE, Hukkanen M, Ma GF, Konttinen YT (2003) Regulation of macrophage activation. *Cell Mol Life Sci,* **60**, pp. 2334-46.

Marhl, M., Haberichter, T., Brumen, M. and Heinrich, R. (2000) Complex calcium oscillations and the role of mitochondria and cytosolic proteins', *Biosystems.* **57**(2), pp. 75-86.

Matsuura M, Saito S, Hirai Y, Okamura H. (2003) A pathway through interferon-gamma is the main pathway for induction of nitric oxide upon stimulation with bacterial lipopolysaccharide in mouse peritoneal cells. Eur J Biochem. 270(19):4016-25.

Maurya, M. R. and S. Subramaniam (2007a). A Kinetic Model for Calcium Dynamics in RAW 264.7 Cells: 1. Mechanisms, Parameters and Sub-populational Variability. Biophysical Journal, In print, Vol. 93, August.

Maurya, M. R. and S. Subramaniam (2007b). A Kinetic Model for Calcium Dynamics in RAW 264.7 Cells: 2. Knockdown Response and Long-Term Response. Biophysical Journal, In print, Vol. 93, August.

Meir KS, Leitersdorf E. (2004) Atherosclerosis in the apolipoprotein-E-deficient mouse: a decade of progress. Arterioscler Thromb Vasc Biol. 24(6):1006-14.

Nicholson AC. (2004) Expression of CD36 in macrophages and atherosclerosis: the role of lipid regulation of PPARgamma signaling. Trends Cardiovasc Med. 14(1):8-12.

Oppenheim JJ, Feldman M, Durum SK, Hirano T, Vilcek J, et al., editors (2000) Cytokine reference: A compendium of cytokines and other mediators of host defense. 1st ed. San Diego: Academic Press.

Oppenheim JJ, Feldmann M (2000) Introduction to the Role of Cytokines in Innate Host Defense and Adaptive Immunity. In: Oppenheim JJ, Feldman M, Durum SK, Hirano T, Vilcek J et al., editors. Cytokine reference: A compendium of cytokines and other mediators of host defense. 1st ed. San Diego: Academic Press. pp. 3-20.

Ozato K, Tsujimura H, Tamura T (2002) Toll-like receptor signaling and regulation of cytokine gene expression in the immune system. Biotechniques Suppl: 66-68, 70, 72 passim.

Palsson-McDermott, E. M. and L. A. O'Neill (2004). Signal transduction by the lipopolysaccharide receptor, Toll-like receptor-4. Immunology **113**(2): 153-62.

Papin, J.A., Hunter, T., Palsson, B.O. and Subramaniam, S. (2005) Reconstruction of cellular signalling networks and analysis of their properties, *Nat Rev Mol Cell Biol.* **6**(2), pp. 99-111.

Pasparakis M, Alexopoulou L, Douni E, Kollias G (1996) Tumour necrosis factors in immune regulation: everything that's interesting is.new! Cytokine Growth Factor Rev 7: 223-229.

Plutzky J. (2003) The vascular biology of atherosclerosis. Am J Med. 115 Suppl 8A:55S-61S.

Pradervand, S., Maurya, M.R. and Subramaniam, S. (2006) Identification of signaling components required for the prediction of cytokine release in RAW 264.7 macrophages, *Genome Biol.* **7**(2), pp. R11.1-R11.14.

Raschke WC, Baird S, Ralph P, Nakoinz I. (1978) Functional macrophage cell lines transformed by Abelson leukemia virus. Cell. 15(1):261-7.

Riccobene, T.A., Omann, G.M. and Linderman, J.J. (1999) Modeling activation and desensitization of G-protein coupled receptors - Provides insight into ligand efficacy, *Journal of Theoretical Biology.* **200**(2), pp. 207-222.

Schuster, S., Marhl, M. and Hofer, T. (2002) 'Modeling of simple and complex calcium oscillations, from single-cell responses to intercellular signaling', *Eur. J. Biochem.* **269**, pp. 1333-1355.

Shaw G, Kamen R (1986) A conserved AU sequence from the 3' untranslated region of GM-CSF mRNA mediates selective mRNA degradation. Cell 46: 659-667.

Shore SK, Tantravahi RV, Reddy EP. (2002) Transforming pathways activated by the v-Abl tyrosine kinase. Oncogene. 21(56):8568-76.

Sitaramayya, A. and Bunnett, N.W. (1999) Cell surface receptors: Mechanisms of signaling and activation, in *Introduction to Cellular Signal Transduction,* (Birkhauser, Boston, MA, USA), pp. 7-28.

Skurk T, Herder C, Kraft I, Muller-Scholze S, Hauner H, Kolb H. (2004) Production and Release of Macrophage Migration Inhibitory Factor from Human Adipocytes. Endocrinology. 146(3):1006-11.

Taylor PR, Brown GD, Herre J, Williams DL, Willment JA, Gordon S. (2004) The role of SIGNR1 and the beta-glucan receptor (dectin-1) in the nonopsonic recognition of yeast by specific macrophages. J Immunol. 172(2):1157-62.

Valledor AF, Borras FE, Cullell-Young M, Celada A. (1998) Transcription factors that regulate monocyte/macrophage differentiation. J Leukoc Biol. 63(4):405-17.

Werb Z, Chin JR. (1983) Onset of apoprotein E secretion during differentiation of mouse bone marrow-derived mononuclear phagocytes. J Cell Biol. 97(4):1113-8.

Wiesner, T.F., Berk, B.C. and Nerem, R.M. (1996) A mathematical model of cytosolic calcium dynamics in human umbilical vein endothelial cells, *American Journal of Physiology-Cell Physiology*. **39**(5), pp. C1556-C1569

Woolf, P.J. and Linderman, J.J. (2003) Untangling ligand induced activation and desensitization of G-protein-coupled receptors', *Biophys J.* **84**(1), pp. 3-13.

Yi, T.M., Kitano, H. and Simon, M.I. (2003) A quantitative characterization of the yeast heterotrimeric G protein cycle', *Proc Natl Acad Sci USA*, **100**(19), pp. 10764-10769.

7

The Alternative Pathway of Complement: a Pattern Recognition System

Peter F. Zipfel [1,2], Michael Mihlan [1] and Christine Skerka [1]

[1] Department of Infection Biology, Leibniz Institute for Natural Product Research and Infection Biology- Hans Knöll Institute for Natural Products Research, Jena, Germany, Peter.Zipfel@hki-jena.de
[2] Friedrich Schiller University Jena

1 Pattern recognition a central aspect of immune defence

Discrimination between self and non self is central for the integrity of the host (Janeway and Medzhitov 2002). In order to maintain integrity, the host has evolved a complex immune system, which is able to discriminate between self and non self and which mounts a toxic response towards a foreign, non self invader, i.e. microbes. The immune system acts on multiple levels and based on the time required upon first contact with a foreign invader the response is divided into the innate response, which is immediately and directly activated and an adaptive response which requires longer time for initiation. Complement particularly when initiated by the alternative pathway represents an immediately acting and highly selective part of innate immunity (Walport 2001). Here we summarize how alternative pathway activation relates to innate immunity and describe a common strategy used by pathogens for immune evasion.

Higher vertebrates have established effective and highly sophisticated immune response in order to combat foreign microbes and to maintain tissue integrity. The immune system aims in the elimination of a foreign invader: once activated the reaction cascades generate highly toxic products that damage any kind of cell. Based on the toxic potential and activity the discrimination between host cells (self) and microbes (foreign) is important. Thus the immune system serves four central functions (i) recognition of self or of foreign cells, (ii) discrimination between self and foreign, (iii) protection of self cells, and (iv) killing or elimination of foreign cells and microbes.

Complement is a very ancient defence system that has the capacity to generate toxic activation products that kill and eliminate microbes and foreign particles (Walport 2001; Nordahl, Rydengard, Nyberg, Nitsche, et al. 2004). The newly generated, highly toxic activation products must be delivered in time to the site of the target.

Based on the toxic activity of the activated complement system the individual reactions of the cascade are tightly balanced and activation must be focused to the surface of the foreign invader. At the same time adjacent, neighbouring self cells, i.e. bystander cells must be protected. Thus for the host activation and effector function is desired on the surface of a microbe but simultaneously the same reactions must be inhibited on the surface of any host cells, particularly those cells that are located in direct vicinity of the microbe. Consequently host cells must actively downregulate the immune response and the inflammatory reactions. The limitation in terms of time and space is a central aspect of the immune response and of inflammatory reactions, and inappropriate control results in autoimmunity. The recent research has highlighted the central role of the innate immunity (Medzhitov and Janeway 1997). The innate immune system acts independently, but also in combination with the adaptive response mediated by T and B cells. The innate immune system is the evolutionary older part (Kimbrell and Beutler 2001) and acts immediately upon entry of a microbe into an immunocompotent host.

The innate and adaptive immune responses employ unique but different mechanisms for sensing foreign structures and further recognition of self cells. The innate immune response forms the fist defence line against microbes and it is of interest to understand, how (i) this system acts, (ii) it recognizes foreign microbes, (iii) it differentiates between foreign and self cells, and (iv) control ensures to limit destructive action on the surface of foreign microbes. In addition to Toll like receptors that form one major part of the innate immune response (Uematsu and Akira 2006), the complement system is a further central part of innate immunity. The alternative pathway of complement, that serves as a continuous immune surveillance system and forms one of the first barriers for invading microbes. The alternative pathway is an evolutionary old defence system and has been presented in invertebrates at the level of *C. elegans* (Schulenburg, Kurz and Ewbank 2004) and early vertebrates as the bony fish barred sand bass. A defective control of this defence system results in diverse human diseases (Zipfel, Heinen, Jozsi and Skerka 2006; ASGHAR 1995).

2 The Alternative Pathway

2.1 Introduction complement : Focus on the Alternative Pathway

Complement is an important immune effector system of host defence (Walport 2001; Volanakis 2002). It consists of over 40 proteins that circulate as plasma proteins and are associated with cellular membranes (Lambris, Reid and Volanakis 1999). Similar to the blood clotting system the individual components represent either inactive proforms of cascade proteins or regulators that control activation and the formation of the cascade. In the normal setting the cascade components are inactive and require activation to gain biological activity. Activation converts an inactive proform (zymogen) into an active protein, which either gains enzymatic activity or forms protein-protein complexes with additional components. The combined activity of complement

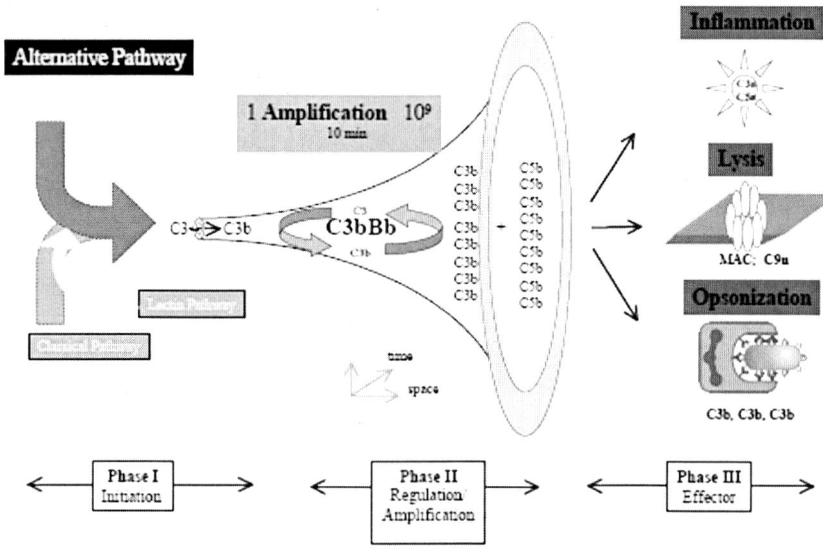

Fig. 1. Alternative complement activation and control
Activation of the alternative complement pathway proceeds in two steps. **Phase I : The Initiation Phase** is initiated by the spontaneous activation of C3. The native C3 molecule undergoes a spontaneous conformational change, thereby exposing its reactive thioester. The thioester reacts with any molecule in its direct neighbourhood. -this initial step is indiscrimatory and occurs on host cells as well as invading microbes. The next step, **second** phase is however discriminatory and actually this step decides on the nature of the loop. In its default setting the reaction proceeds and results in amplification. As microbes lack any regulators consequently the activation proceeds unrestricted and eventuates in a powerful **amplification loop**. Thus activation is by default and is favoured on targets such as microbes and foreign particles. The default setting of the cascade is aimed for activation and particularly surface bound C3b is relevant. Activation includes the formation of a convertase C3bBb. The surface bound C3b binds Factor B, that in consequence is cleaved by Factor D to form the active C3bBb complex. This complex has a half life of milliseconds and can be stabilized by properdin. All these reactions are continuous, constant and indiscriminatory. As a consequence the active C3 convertase causes the deposition of a multiple C3b molecules at the surface. This opsonization favours phagocytosis. In contrast to foreign microbes, self cells utilize regulators to control and regulate the amplification loop. To this end the cells are equipped with regulators and attach regulators from the fluid phase to their surface. As a consequence the cascade is inhibited at a very early step and the amplification reaction is blocked. This step represents the regulatory phase and blocks the disastrous effect of an activated complement. Inhibition is performed by both membrane bound regulators, such as DAF, MCP, CR1 and soluble surface attached regulators Factor H and FHL-1. The control steps are highly discriminatory between self cells and foreign cells such as microbes. If activation proceeds the third phase, the effector response is initiated which results in (i) the release of anaphylactic peptides C3a and C5a, (ii) pore formation, i.e. membrane attack complexes (MAC) insertion into the cell membrane and (iii) opsonization of the foreign surface with C3b which enhances phagocytosis.

is necessary to eliminate microorganisms and potentatiate the production of specific antibodies (Dempsey, Allison, Akkaraju, Goodnow and Fearon 1996).

Based on its activity complement activation, via the alternative pathway can be divided into distinct phases. The first, the initiation phase is indiscriminatory and occur continuously and on any type of surface, on host cell and on foreign particles, such as microbes. Thus in its default setting low level activation occurs in the fluid phase and on any kind of surface. This initial indiscriminatory phase is slow and allows ample time for control and regulation and is followed by a discriminatory phase (or decision phase). Regulators which are present in the fluid phase and on the surface of host cells inhibit activation of complement cascade. Microbes lack such endogenous regulators or binding moieties on their surface. Thus complement is activated on the microbial surface, the amplification phase is initiated and the activated complement system performs its toxic activity that results in the elimination of the microbe.

2.2 Activation and control

2.2.1 Initiation-The First Phase

The alternative pathway of complement provides the first or a very early shield for invading microbes. This part of the innate immune defence system is immediately activated, within less than a second once a microbe is in contact with the host.

In its normal setting the alternative pathway of complement is continuously and constantly activated (Thurman and Holers 2006). Activation is initiated by the spontaneous conformational change of the central component C3. This tick over occurs constantly and continuously, however at low levels. The spontaneous conformational change of C3 exposes the highly reactive thioester bond that can attach to any molecule in its vicinity. The newly generated $C3_{H2O}$ molecule can bind Factor B and with the help of Factor D form an alternative pathway convertase termed $C3b_{H2O}Bb$. This convertase converts C3 and generates C3b, and the anaphylatoxin C3a which is released (Fig. 1) (Pangburn 1998). This activation occurs in the fluid phase and the active enzyme generates additional and multiple C3b molecules which can deposit on any surface and form a cluster. This reaction is central for the alternative pathway and actually the ‘milieu’ where the reaction occurs, i.e. the type of surface determines whether the reaction proceeds or whether it is inhibited. This particular step is crucial for alternative complement activation. It represents the major decision of these reactions: progressions vs inactivation. Progression is favoured for foreign microbe, which needs to be eliminated and inactivation is favoured on self cells, that need to be protected.

2.2.2 C3b formation

In native C3 the thioester bond appears protected within a hydrophobic pocket and is exposed in the C3b fragment upon cleavage of C3 by the C3 convertase. The transiently exposed thioester bond has a half life of about 100 μs and can then transacylate reactions with nucleophilic groups present on cell surfaces or with complex carbohydrates or immune complexes (Sahu and Lambris 2001). Proteolytic cleavage of C3 between Arg726 and Ser727 by the alternative pathway convertase results in the generation of C3a, a small (9 kDa) anaphylactic peptide and a large C3b fragment (185 kDa) that remains covalently attached to the surfaces. The attached and cleaved C3b exposes binding sites for Factor B and other complement proteins. Binding of Factor B in the presence of Factor D leads to the generation of a C3bBb complex that displays enzymatic activity and can generate more C3b molecules. Consequently the step amplified activation of the alternative pathway initiates the amplification loop. This step has the capacity to generate up to 10^{10} C3b molecules in a period of ca. 10 – 15 min (Pangburn 1998).

The recognition mechanism of the alternative pathway of complement is intriguing both by its simplicity and its efficiency. From a conceptual point of view this defence mechanism is distinct in terms of activation and control from the adaptive immune response. The initiation phase, (phase I) is by default, is indiscriminatory and occurs everywhere, at any place, at any time and everywhere in the body. Although aimed at being activated specifically at the surface of foreign microbes, in this phase the activation products are formed on virtually every cell and tissue, i.e. this phase does not discriminate self and self cells.

However further activation is controlled by regulators such as the fluid phase regulators Factor H and FHL-1, and also by membrane bound regulators, such as membrane cofactor protein 1 (MCP/CD46), decay accelerating factor (CD55) complement receptor 1 (CR1/CD35) and protectin (CD59), that all inhibit progression progression of activation and or shut off the cascade (Morgan and Harris, 1999).

2.2.3 Amplification / Regulation Loop-The second phase decides on amplification vs inhibition. This phase is discriminatory and thus can differentiate between dangerous foreign particles/cells and non dangerous, self cells. A spontaneously formed C3 convertase is stabilized by properdin, and complexed with properdin the half life of the convertase is increased about 10fold. The amplification cycle generates large amounts of C3b that deposit on the surface in direct vicinity of the initial enzyme. This newly formed cluster of C3b leads to a propagation of the complement activation pathway and results in formation of the C3 convertases C3bBb that generates anaphylactic C3a and C5a peptides, which initiate an inflamematory response, forms multiple C3b, and consequently further C3bBb convervases. This increases enzymatic activity and results in the opsonization of a particle, and/or the initiation of the lytic activity of complement. Once activated on the surface of a microbe this cascade results in elimination of a microbe (Figure 1).

From the side of the host these outlined reactions are favourable and highly advantageous when they occur on the surface of a microbe, modified self cells and/or immune complexes. However because of their toxic nature these events are highly unfavourable on the surface of tissue cells, particularly bystander cells that are in direct vicinity of microbes and sites of inflammation. Consequently multiple internal control regulators exist that inhibit their activity on host cells and leave the reactions unrestricted to the site of the foreign particles.

2.2.4 Inactivation by regulators

Regulator proteins which are either distributed in the fluid phase or are expressed on the membrane of host cells are responsible for the inactivation of the initial short lived intermediate and consequently for shutting down the reaction. This shut off is essential and several proteins with largely redundant activities exists, that control alternative pathway activation. These regulators have overlapping, redundant activity but are differently distributed. They are present as soluble plasma proteins, i.e. Factor H and the Factor H like protein 1 (FHL-1) or are integrated within the cellular membrane i.e. MCP (CD46) CR1 (CD35) and DAF (CD55) (Kim, and Song 2006). These five regulators have redundant and highly overlapping activities and are structurally related, (i) they show the same domain structure, (ii) are composed exclusively or almost exclusively of complement control modules (also termed short consensus repeats) and (iii) their genes are located on the regulators of complement activation gene cluster on human chromosome 1q32 (Zipfel and Skerka 1994; de Cordoba, Esparza-Gordillo, de Jorge, Lopez-Trascasa, and Sanchez-Corral 2004). Each of these proteins acts upon C3b inactivates the protein irreversibly or they target and inactivate the C3 convertases themselves. The fact that five distinct proteins control the same phase of the reaction highlights the central importance of these reactions. In addition protectin (CD59), which controls the terminal complement components membrane attack complex is also expressed on the cell surface.

The redundant activities of these regulators include

- **(i)** **cofactor activity,** as they allow the inactivation of C3b by the serine protease Factor I
- **(ii)** **inhibition of convertase formation**, and
- **(iii)** **decay accelerating activity, i.e.** dissociation of the preformed C3bBb (or C3bBbProperidn) convertases.

Despite their overlapping activities, the individual regulators also show difference in their activity. The soluble regulators are specific for the alternative complement pathway and have no (or a very minor) activity in the classical pathway. In contrast all three membrane bound regulators act in both the alternative and the classical pathway of complement. The two soluble regulators Factor H and FHL-1, and also the membrane bound protein CR1 perform all three activities, i.e. cofactor activity, i.e. inactivation of C3b, (ii) inhibition of convertase formation, and once a convertase is formed, (iii) the decay accelerating of the convertase. The activity of MCP and DAF is restricted to either cofactor or decay accelerating activity.

2.2.5 Expression, distribution and expression profiles

Factor H and FHL-1 are present at high levels (ca. 500 and ca. 30 ug/ml) in plasma and in several body fluids including synovial fluid (Schlaf, Demberg, Beisel, Schieferdecker and Gotze 2001). Similar to other soluble complement proteins Factor H and FHL-1 are synthesized in the liver but these soluble regulators are also synthesized locally by activated monocytes, fibroblast, synovial fibroblasts etc. Factor H and particularly FHL-1 expression is induced during inflammation and by inflammatory mediators (Friese, Hellwage, Jokiranta, Meri, et al. 1999).

MCP and DAF are widely distributed in human cells (Seya, Hirano, Matsumoto and Ueda 1999; Liszewski and Atkinson, 1992; Lublin and Atkinson 1989). MCP is expressed in all peripheral blood mononuclear cells, epithelial and endothelial cells. On leucocytes the number of copies is 5 000 to 15 000 molecules per cell. DAF shows a broad tissue distribution. The protein is expressed on all peripheral blood mononuclear cells, on epithelial and endothelial cells and is also present on erythrocytes (3500 copies per cell) and neutrophils (85 000). CR1 has a more restricted distribution, it is present on erythrocytes (200 – 1000 copies per cell), neutrophils (5000 – 20 000), eosinophils, B lymphocytes and some T lymphocytes, tissue macrophages, glomerular podocytes and follicular dendritic cells, (Krych-Goldberg and Atkinson 2001).

Although most of the membrane bound alternative pathway regulators are expressed in several cell types it is currently unclear whether all cells of a body express any or a specific combination of the membrane regulators. It is apparent that expression of specific surface proteins stop the cascade and protect the host cell. Recent experimental evidence shows that in addition to the membrane bound regulators also plasma regulators, like Factor H and FHL-1 attach to host/self cells and host surfaces and mediate protection (Jokiranta, Cheng, Seeberger, Jozsi, Heinen, et al 2005). In addition extracellular structures and surfaces like extracellular matrixes and the glomerular basement membrane of the kidney, which do not express endogenous regulators and require the attachment of soluble regulators like Factor H and FHL-1 for immune protection (Zipfel 2001).

Thus the soluble regulators act in close vicinity to the cell surface (Jozsi, Heinen and Zipfel 2006). In addition to tissue cells certain structures of the body like the basal membrane of the kidney glomerulus or extracellular matrices serve as activation sites for complement and need to be controlled at these particular sites. However these host surfaces do not express endogenous membrane attached regulators and dependent on attached fluid phase regulators for protection.

3 Bystander cells

Particularly any host cells that is located or positioned in the vicinity of a microbe is in contact with the activated system and these cells form an active border to limit the action of the activated complement system. Therefore control of the system at these sites is central for the integrity of host cells.

4 Microbial Evasion Strategies Pattern Recognition

Pathogenic microbes have found means to circumvent attack mediated by this safeguard system of their host. Pathogens and viruses either express surface proteins that bind host regulators or they express endogenous regulators, e.g. in form of proteases that

Table 1: Complement and Immune Evasion of Pathogens by Binding soluble host Alternative Pathway Inhibitors.

Pathogen	Bound Host Alternative pathway Regulator	Microbial Protein	Reference
Streptococcus pneumonia	Factor H, FHL-1	PspC	MacDaniel
Streptococcus pyogenes	Factor H, FHL-1 FHR-3	M Protein, Fba	
Streptococcus agalacticae	Factor H	ND	
Borrelia Species[1)]	Factor H, FHL-1, FHR-1	BbCRASP-1 - BbCRASP-5	
Leptrospira interogans	Factor H, FHL-1	ND	
Neisseria gonorroheae	Factor H, FHL-1	LOS, Por1A	
Neisseria meningitidis	Factor H, FHL-1	GNA1870	
Pseudomonas aeruginosa	Factor H, FHL-1	PaCRASP-1	Kunert et al. submitted
Candida albicans	Factor H, FHL-1	CaCRASP-1	Meri et al
Aspergillus fumigatus	Factor H	ND	
Onchocerca volvulus	Factor H	ND	
Virally encoded regulators West Nile virus	Factor H	NS1	Chung, PNAS
HIV	Factor H		

[1)] Borrelia species include *B. afzelii B. burgdorferii, B. hermissi, B. spielmanni, B.recurrentis* and *B. duttonii*

inactivate complement components or endogenous complement regulators. (Fig. 2) The detailed understanding of both the activation reactions of the alternative complement pathway and the microbial complement evasion strategies is a fundamental aspect of innate immunity and of microbial immune evasion. A wide range of microbes Gram-negative, Gram-positive bacteria, pathogenic yeast and parasites express surface molecules that bind host complement regulators (Kraiczy and Wurzner 2006; Lindahl, Sjobring, and Johnsson 2000). (Table 1)

Microbes are recognized by the immune system and as outlined here the major role of the alternative complement pathway is the identification of the foreign invaders, to target microbes and eliminate the invader. The alternative pathway of complement is initiated immediately upon entry of a microbe. In order to survive microbes have found means to inactivate or circumvent complement attack and these features make a microbe a pathogen. Inhibition of complement activation, e.g. by acquisition of host regulators is a major step for immune evasion of pathogenic microbes. (Fig. 3) An increasing number of microbes is identified that bind specifically host alternative pathway regulators Factor H, FHL-1 and FHR-1 (Thurman et. al. 2006) and also classical pathway regulator C4BP to their surface

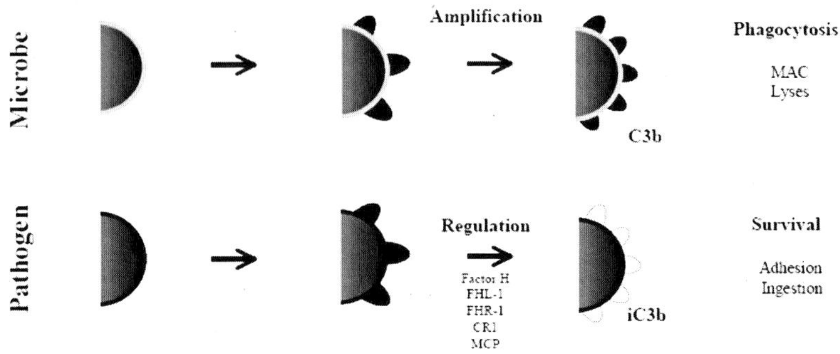

Fig. 2. The surface of a foreign particle mediates complement control. Microbes do not restrict complement activation, thus the default setting of the alternative pathway results in amplification, C3b deposition, which results in opsonization and MAC formation and as a consequence elimination of the microbe. In contrast pathogens express specific surface proteins, generally termed CRASP (complement regulator acquiring surface proteins) which attach host fluid phase alternative pathway regulators that inhibit complement activation. This limitation of complement control results in survival and damaging of the host.

(Jenkins, Mark, Ball, Persson, Lindahl, et al 2006; Blom, Rytkonen, Vasquez, Lindahl, Dahlback et al 2001). This is reported for gram positive, gram-negative bacteria, for pathogenic yeast and multicellular parasites.

Alternative Pathway is a true pattern recognition mechanisms as the combined systems senses foreign and self cells and ensures that activation proceeds specifically on the surface of a target, i.e. a microbe. The activated system can proceed through opsonization and or lysis, i.e. Mac formation. Thus pattern recognition to identify microbes is actually the constant attack and targeting of any cell, structure and surface. The system is active by default and every structure and surface serves as a target. In order to survive, self cells permanently inactivate the attack and downregulate activation. Based on the multiple regulators and the high concentration of the fluid phase regulators and as normal downregulation occurs at the very initial phase. Downregulation of the initial reaction is a continuous, steady process for self cells and tissues. In these cases the cofactor activity is required to inactivate any newly generated C3 convertases. In the normal way it is the cofactor activity that inactivates the enzyme and inhibits assembly of the convertase. it is getting more difficult when activation has proceeded and in the case of bystander cells, cells that are present within an inflammatory process and are facing a inflammatory reactions. Even in these cases the cells and surfaces are equipped with control proteins that interfere also at a later stage, when an active convertase has already been formed and needs to be dissociated.

The system has to ensure that active convertases formed on the surface of self cells are controlled and inactivated, but the same convertases formed on the target can proceed unrestricted. Particularly the soluble regulators in form of Factor H and FHL-1

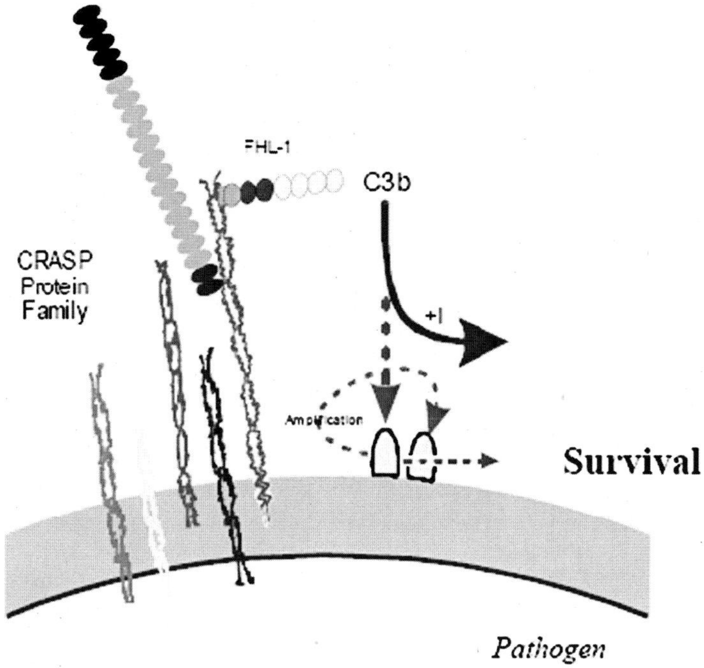

Fig. 3: Complement evasion of pathogens.
Pathogens express specific surface proteins termed CRASP (complement regulator acquiring surface proteins) which attach host fluid phase alternative pathway regulators (Factor H, FHL-1 and FHR-1) that inhibit complement activation and formation of the alternative pathway amplification convertase C3bBb. As a consequence C3b is inactivated to iC3b, does not attach to the surface and the pathogen survives complement attack.

have two functional regions, in a complement regulatory region that binds C3b and a recognition region that binds to cell surfaces (Jokiranta et al 2005).

5 Microbes utilize complement inhibitors for immune evasion

In the pattern recognition system microbes lack regulatory activities, that are required to inhibit and stop the activated system. Thus as microbes lack such inhibitors they are immediately targeted by the surveillance activation and are marked as an invader.

The relevance of these mechanisms is confirmed by the immune evasion strategies used by several microbes (Finlay and McFadden 2006).

Several viruses are known that expresses endogenous complement regulators (Favoreel, Van de Walle, Nauwynck, and Pensaert 2003; Lee, Jung and Means 2003). The vaccinia virus complement control protein is the first example. This protein is structurally related to members of the human RCA family. VCP (Mc Kenzie, Kotwal, Moss, Hammer and Frank 1992; Mullick, Bernet, Panse, Hallihosur, Singh, et al. 2005) is a 27 kDa protein that is exclusively composed of four CCP domains. VCP inactivates human complement system and inhibits the alternative (and also the classical) pathway of human complement similar to Factor H (and C4BP). Similar proteins have been found in pox virus including small pox virus, herpes virus, saimirii Karposi sarcoma associated Herpes virus and and murine herpes virus (Mullick, Kadam and Sahu 2003; Mark, Spiller, Villoutreix and Blom 2007).

An apparent similar however distinct approach is used by pathogens and is employed by Gram-negative (Kraiczy, Hellwage, Skerka, Becker, Kirschfink et al. 2004) Gram- positive Bacteria (Foster 2005) as well as pathogenic yeast (Meri, Hartmann, Lenk, Eck 2002) and multi cellular organism (Sacks and Sher 2002). The pathogenic forms express surface proteins that bind soluble host immune regulators to their surface. This kind of disguise "the wolve in sheep clothes principle" appears a common principle for immune escape. Specific binding of the soluble host complement regulators Factor H, FHL-1 and also C4BP has been shown for a number of pathogens. Although they bind the same host proteins at related sites the various pathogenic binding proteins are distinct in sequence and size. No apparent sequence motif is identified between the various Factor H, and FHL-1 binding proteins that include the five borrelial proteins (BbCRASP1, BbCRASP-2, BbCRASP-3, BbCRASP-4 and BbCRASP-5), the streptococcal M- and Fba proteins and the pneumococcal Hic, BAc, SpsA proteins.

6 Acknowledgement

The work of the author is funded by the Deutsche Forschungsgemeinschaft, the Foundation for Children with atypical HUS, the KidNeeds program (Iowa City, Iowa).

References

Asghar, S.S. (1995) Biology of Disease – Membrane Regulators of Complement Activation and their aberrant Expression in Disease. *Laboratory Investigation* 72 (3), 254-271.

Blom, A.M., Rytkonen, A., Vasquez, P., Lindahl, G., Dahlback, B. and Jonsson, A.B. (2001) A novel interaction between type IV pili of *Neisseria gonorrhoeae* and the human complement regulator C4b-binding protein. *J. Immunol.* 166 (11), 6764-6770.

de Cordoba, R.S., Esparza-Gordillo, J., de Jorge, G.E., Lopez-Trascasa, M. and Sanchez-Corral, P. (2004) The human complement factor H: functional roles, genetic variations and disease associations. *Mol. Immunol.* 41 (4), 355-67.

Dempsey, P.W., Allison, M.E., Akkaraju, S., Goodnow, C.C. and Fearon, D.T. (1996) C3d of complement as a molecular adjuvant: bridging innate and acquired immunity. *Science* 271 (5247), 348-50.

Favoreel, H.W., Van de Walle, G.R., Nauwynck, H.J. and Pensaert, M.B. (2003) Virus complement evasion strategies. *JOURNAL OF GENERAL VIROLOGY* 84, 1-15.

Finlay, B.B. and Mc Fadden, G. (2006) Anti-immunology: Evasion of the host immune system by bacterial and viral pathogens. *Cell*24 (4), 767-782.

Foster, T.J. (2005) Immune evasion by Staphylococci. *Nature Reviews Microbiology* 3 (12), 948-958.

Friese, M.A., Hellwage, J., Jokiranta, T.S., Meri, S., Peter, H.H., Eibel, H. and Zipfel, P.F. (1999) Different transcriptional regulation and expression of members of the factor H family by inflammatory mediators and in rheumatoid arthritis. *Mol. Immunol.* 36 (4-5), 281-281.

Janeway, C.A. and Medzhitov, R. (2002) Innate immune recognition. *Annual Review of Immunology* 20, 197-216.

Jenkins, H.T., Mark, L., Ball, G., Persson, J., Lindahl, G., Uhrin, D., Blom, A.M. and Barlow, P.N. (2006) Human C4b-binding protein, structural basis for interaction with streptococcal M protein, a major bacterial virulence factor. *J. Biol. Chem.* 281 (6), 3690-3697.

Jokiranta, T.S., Cheng, Z.Z., Seeberger, H., Jozsi, M., Heinen, S., Noris ,M., Remuzzi, G., Ormsby, R., Gordon, D.L., Meri, S., Hellwage, J. and Zipfel, P.F. (2005) Binding of complement factor H to endothelial cells is mediated by the carboxy-terminal glycosaminoglycan binding site. *Am. J. Pathol.* 167 (4), 1173-1181.

Jozsi, M., Heinen, S. and Zipfel, P.F. (2006) Protective role of surface-attached complement factor H for endothelial cells. *Mol. Immunol.* 43 (1-2), 143-143.

Kim, D.D. and Song, W.C. (2006) Membrane complement regulatory proteins. *Clinical. Immunol.* 118 (2-3), 127-136.

Kimbrell, D.A. and Beutler, B. (2001) The evolution and genetics of innate immunity. *Nature Reviews Genetics* 2 (4), 256-267.

Kraiczy, P. and Wurzner, R. (2006) Complement escape of human pathogenic bacteria by acquisition of complement regulators. *Mol. Immunol.* 43 (1-2), 31-44.

Kraiczy, P., Hellwage, J., Skerka, C., Becker, H., Kirschfink, M., Simon, M.M., Brade, V., Zipfel, P.F. and Wallich, R. (2004) Complement resistance of *Borrelia burgdorferi* correlates with the expression of BbCRASP-1, a novel linear plasmid-encoded surface protein that interacts with human factor H and FHL-1 and is unrelated to Erp proteins. *J. Biol. Chem* .279 (4), 2421-2429.

Krych-Goldberg, M. and Atkinson, J.P. (2001) Structure-function relationships of complement receptor type 1. *Immunol. Reviews* 180, 112-122.

Lambris, J.D., Reid, K.B.M. and Volanakis, J.E. (1999) The evolution, structure, biology and pathophysiology of complement. *Immunology Today* 20 (5), 207-211.

Lee, S.H., Jung, J.U. and Means, R.E. (2003) 'Complementing' viral infection: mechanisms for evading innate immunity. *Trends in Microbiol.* 11 (10), 449-452.

Lindahl, G., Sjobring, U. and Johnsson, E. (2000) Human complement regulators: a major target for pathogenic microorganisms. *Current Opinion in Immunol.* 12 (1), 44-51.

Liszewski, M.K. and Atkinson, J.P. (1992) Membrane cofactor protein. *Current Topics in Microbiology and Immunology* 178, 45-60.

Lublin, D.M. and Atkinson, J.P. (1989) Decay-accelerating factor: biochemistry, molecular biology, and function. *Annual Review of Immunology* 7, 35-58.

Mark, L., Spiller, O.B., Villoutreix, B.O. and Blom, A.M.(2007) Kaposi's sarcoma-associated herpes virus complement control protein: KCP - complement inhibition and more. *Molecular Immunology* 44 (1-3), 11-22.

Mc Kenzie, R., Kotwal, G.J., Moss, B., Hammer, C.H. and Frank, M.M. (1992) Regulation of Complement Activity by Vaccinia Virus Complement-Control Protein. *J. Infectious Diseases* 166 (6), 1245-1250.

Medzhitov, R. and Janeway, C.A. (1997) Innate immunity: Impact on the adaptive immune response. *Current Opinion in Immunol.* 9 (1), 4-9.

Meri, T., Hartmann, A., Lenk, D., Eck, R., Wurzner, R., Hellwage, J., Meri, S. and Zipfel, P.F. (2002) The yeast *Candida albicans* binds complement regulators factor H and FHL-1. *Infection and Immunity* 70 (9), 5185-5192.

Morgan, B.P. and Harris, C.L. (1999) Regulation in the complement system. In: Morgan, B.P., Harris, C.L., Complement Regulatory Proteins. Academic Press 32-40.

Mullick, J., Bernet, J., Panse, Y., Hallihosur, S., Singh, A.K. and Sahu, A. (2005) Identification of complement regulatory domains in vaccinia virus complement control protein. *J. Virol.* 79 (19), 12382-12393.

Mullick, J., Kadam, A. and Sahu, A. (2003) Herpes and pox viral complement control proteins: 'the mask of self''. *Trends in Immunology* 24 (9), 500-507.

Nordahl, E.A., Rydengard, V., Nyberg, P., Nitsche, DP., Morgelin, M., Malmsten, M., Bjorck, L. and Schmidtchen A. (2004) Activation of the complement system generates antibacterial peptides. *Proceedings of the Nat. Acad. Of Sciences of the USA 101 (48), 16879-16884.*

Pangburn, M.K. (1998) Alternative Pathway: Activation and Regulation. In: Rother, K., Till, G.O., Hänsch, G.M (Eds.), The complement system. Springer, Berlin 93-115.

Sacks, D. and Sher, A. (2002) Evasion of innate immunity by parasitic protozoa. *Nature Immunol. 3 (11), 1041-1047.*

Sahu, A. and Lambris, J.D. (2001) Structure and biology of complement protein C3, a connecting link between innate and acquired immunity. *Immunological Reviews 180, 35-48.*

Schlaf, G., Demberg, T., Beisel, N., Schieferdecker, H.L. and Gotze, O. (2001) Expression and regulation of complement factors H and I in rat and human cells: some critical notes. *Mol. Immunol. 38 (2-3), 231-239.*

Seya, T., Hirano, A., Matsumoto, M., Nomura, M. and Ueda, S. (1999) Human membrane cofactor protein (MCP, CD46): multiple isoforms and functions. *The International Journal of Biochemistry & Cell Biology 31(11), 1255-60.*

Schulenburg, H., Kurz, C.L. and Ewbank, J.J. (2004) Evolution of the innate immune system: the worm perspective. *Immunological Reviews 198 (1), 36-58.*

Thurman, J.M., Holers, V.M. (2006) The central role of the alternative complement pathway in human disease. *J. Immunol.* 176 (3),1305-1310.

Uematsu, S. and Akira, S. (2006) Toll-like receptors and innate immunity. *J. Mol. Med.* 84 (9), 712-725.

Volanakis, J.E. (2002) The role of complement in innate and adaptive immunity. *Current Topics in Microbiol. And Immunol.* 266, 41-56.

Walport, M.J. (2001) Advances in immunology: Complement (First of two parts). *New England J. Med.* 344 (14), 1058-1066.

Zipfel, P.F., Heinen, S., Jozsi, M. and Skerka, C. (2006) Complement and diseases: Defective alternative pathway control results in kidney and eye diseases. *Mol. Immunol.* 43, 97-106.

Zipfel, P.F. (2001) Complement factor H: Physiology and pathophysiology. *Seminars in Thrombosis and Hemostasis* 27 (3), 191-199.

Zipfel P.F. and Skerka, C. (1994) Complement Factor-H and Related Proteins – An Expanding Family of Complement-Regulatory Proteins. *Immunology Today* 15 (3), 121-126.

8

Role of MBL-associated Serine Protease (MASP) On Activation of the Lectin Complement Pathway

Minoru Takahashi[1*], Shuichi Mori[2], Shiro Shigeta[2] and Teizo Fujita[1]

[1] Department of Immunology, Fukushima Medical University School of Medicine, minolta@fmu.ac.jp
[2] Department of Microbiology, Fukushima Medical University School of Medicine

Abstract. Mannose-binding lectin (MBL) and ficolin are pattern recognition molecules in the complex with the MBL-associated serine proteases (MASPs). Three kinds of MASPs, termed as MASP-1, MASP-2 and MASP-3 have been identified. When MBL or ficolins binds to carbohydrates on the surface of microbes, conformational modifications of these molecules trigger to activate zymogens of MASPs, followed by consequential complement activation. MASP-2 cleaves C4 and C2 to make a C3 convertase, C4b2a. MASP-1 has an ability to cleave C3 directly, although this activity has not been detected in physiological conditions. Natural target molecules for MASP-3 are still discussible. To elucidate the physiological meanings of MASPs, we generated MASPs-deficient mice. Not only MASP-2-deficient mouse but also MASP-1-/MASP-3-deficient mouse reduced activities for C3 deposition on the surface of mannan and zymosan, suggesting MASP-1/3 also contribute the activation of complement by the lectin pathway. Also, MASP-1/3-deficient mice showed the susceptible to an influenza virus.

1 Introduction

The lectin complement pathway (LCP) has been identified as the third pathway of the initial activation for complement. Mannose-binding lectin (MBL) is a C-type serum lectin and acts as a pattern recognition molecule to binds to several kinds of carbohydrates on microbes, followed by complement activation. Recently, other serum lectins, ficolins (L-ficolin, H-ficolin and M-ficolin in human and ficolin-A in mouse) have been found to activate complement (Endo et al., 2005; Liu et al., 2005; Matsushita et al., 2000a; Matsushita et al., 2002). These pattern recognition molecules are associated with serine protease, termed MBL-associated serine protease (MASP). MASP was initially identified to have the ability to cleave C4 and C2 (Matsushita and Fujita, 1992). Matsushita et al. also found that MASP can cleave C3 directly

(Matsushita and Fujita, 1995). However, it is still controversial as to whether direct C3 cleavage by MASP actually acts *in vivo* or not. Sequence analysis of MASP cDNA isolated from human liver predicted that MASP has the same domain structure to those of C1r and C1s of the classical complement pathway (Sato et al., 1994).

In a further study, Thiel et al. has isolated the second MASP (termed MASP-2), and characterized as the protease that has the activity to cleave C4 (Thiel et al., 1997). The first identified MASP has been reffered as MASP-1 which is unable to cleave C4 (Matsushita et al., 2000b; Thiel et al., 1997). Subsequently, the third protease, MASP-3 has been identified in the MBL complex (Dahl et al., 2001). MASP-1 and MASP-3 have an identical heavy chain that contains five domains at N-terminal, but their serine protease domains are different. MASP-3 may act a negative regulator for the lectin pathway by competitive manner to make complexes with MBL and ficolins (Dahl et al., 2001). It was suggested a candidate of the substrates for MASP-3 might be insulin-like growth factor-binding protein 5, although the functional meanings are unknown (Cortesio and Jiang, 2006).

In addition to three MASPs, a truncated protein of MASP-2 termed as small MBL-associated protein (sMAP) or 19kDa MBL-associated protein (MAp19) is present in MBL-MASPs complex, that has no catalytic domain (Stover et al., 1999b; Takahashi et al., 1999). Domain structure of sMAP/MAp19 is identical with the two N-terminal domains, which are the first C1r/C1s/sea urchin Uegf/bone morphogenic protein (CUB) and epidermal growth factor (EGF)-like domain, except for four amino acids at its C-terminal. A recent our study reported that sMAP/MAp19 was in competition with MASP-2 to form the complex with MBL and attenuated the C4 cleavage activity of MASP-2 in normal murine serum (Iwaki et al., 2006). The LCP has been shown to participate the several tissue damage such as intestine, lung, kidney and heart after ischemia-reperfusion injury (de Vries et al., 2004; Hart et al., 2005; Moller-Kristensen et al., 2005; Walsh et al., 2005). Therefore, our recent finding propose the possibility that sMAP/MAp19 is one of the candidates for inhibitor to prevent the tissue damage after ischemia-reperfusion.

To elucidate the physiological meanings of MASPs and sMAP/MAp19, we generated MASPs and sMAP/MAp19 knockout mice.

2 Domain structures and proteolytic activation of MASP-1, MASP-2 and MASP-3

As shown in Fig. 1, MASPs, C1r and C1s share a common domain structure which consists of the first CUB, EGF-like, the second CUB, two complement control protein (CCP), also known as sushi domain or short consensus repeats (SCR) and serine protease domain. MASPs were present in the circulation as a single-chain, which are their zymogen forms. When MBL-MASPs complexes bind to the surface of microbes, some conformational changes of MBL trigger activation of MASPs, in which MASPs are cleaved at a conserved Arg-Ile linkage located between the second CCP and protease domains. Therefore, activated MASPs consist of two chains (heavy- and

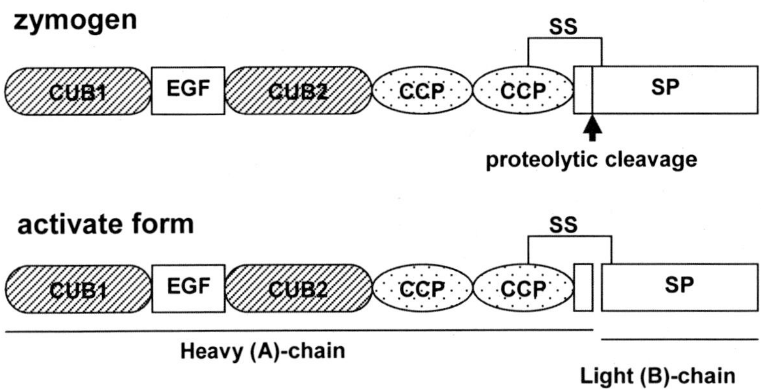

Fig. 1. A schematic representation of domains organization of the MASPs/C1r/C1s family.

light-chains, or A- and B-chains); these polypeptide chains are linked by a disulfide bridge (Fig. 1).

Recombinant MASP-1 and MASP-2 are activated by an auto-catalytic fashion during expression and purification steps (Rossi et al., 2001; Vorup-Jensen et al., 2000; Zundel et al., 2004). On the other hand, recombinant MASP-3 does not undergo spontaneous self-activation (Zundel et al., 2004).

Several truncated MASP proteins has been expressed to elucidate the functions of each domain in MASPs. Wallis et al. reported that the CUB-EGF domains contribute to form the MASP homodimer using the recombinant N-terminal two or three domains of rat MASP-1 and MASP-2 (Wallis and Dodd, 2000). They concluded that it was formed in a Ca^{2+}-independent manner, since it was not disrupted by EDTA. Human sMAP/MAp19, N-terminal CUB-EGF segments of MASP-1/3 and MASP-2 are also involved in homodimer formation in the presence of calcium ion (Thielens et al., 2001).

The N-terminal three domains, CUB-EGF-CUB of MASPs are reported to be sufficient to form complexes with MBL or L-ficolin in a Ca^{2+}-dependent manner. The CUB-EGF segment, but not CUB-EGF-CUB segment of rat MASP-1 could not bind to MBL (Wallis and Dodd, 2000). In contract, surface plasmon resonance spectrometry using human truncated MASPs segments showed that the CUB-EGF moieties of MASP-1 and MASP-2 are sufficient for binding to MBL and L-ficolin (Cseh et al., 2002; Thielens et al., 2001).

Rossi et al. expressed the truncated catalytic region of human MASP-1 and MASP-2 using a baculovirus/insect cell system (Rossi et al., 2001). In their study, the CCP1/2-SP fragments of MASP-1 and MASP-2 were recovered in partially activated forms, whereas CCP2-SP of MASP-2 was recovered in a fully proenzyme form. This data suggests the first CCP module is essential for autoactivation of MASP-2. In addition, they showed that the CCP1/2-SP fragment of MASP-2 could cleave C2 and C4 at high efficiency. On the other hand, CCP1/2-SP of MASP-1 showed marginal activity toward C3 and C2, and no activity on C4. These results were confirmed by another study using the recombinants prepared in a bacteria expression system

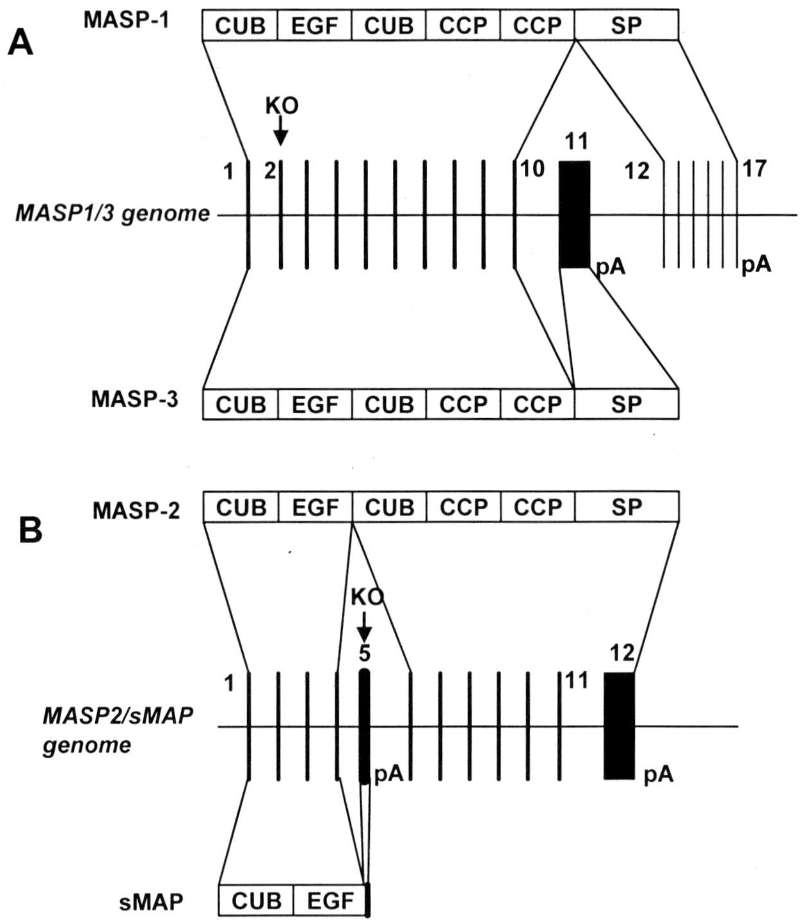

Fig. 2. Genomic organization and its transcribed products of the *MASP1/3* and *MASP2/sMAP* genes

(A) The exon-intron structure of *MASP1/3* gene. Vertical lines indicate exons. After 10 exons that encode MASP-1 and -3 heavy chain, a single exon for MASP-3 light chain and 6 split exons for MASP-1 light chain are tandemly located. Two polyadenylation signals exist in exon 11 and 17 (indicated as "pA"). In MASP1/3 knockout mice, the second exon was replaced with neomycin resistant gene (indicated as "KO" and an arrow).

(B) The exon-intron structure of *MASP2/sMAP* gene. Vertical lines indicate exons. MASP2 product is encoded from exon 1 to 12, except for exon 5, which is a specific exon for sMAP. Two polyadenylation sites exist in exon 5 and 12. In sMAP/MASP2 knockout mice, the sMAP specific exon (exon 5) was replaced with neomycin resistant gene.

(Ambrus et al., 2003). However, other studies suggested that C2 might be a natural substrate for MASP-1 (Chen and Wallis, 2004; Matsushita et al., 2000b)

3 Evolutionary aspects of the MASP/C1r/C1s family

A review well documents about the evolutionary aspects in MASPs (Fujita, 2002). *MASPs* cDNA have been identified in almost vertebrates such as lamprey, shark, carp, frog, birds, mouse, rat and human (Endo et al., 1998; Lynch et al., 2005). In addition to vertebrates, *MASPs* cDNA has been found in two invertebrates, ascidians (Ji et al., 1997) and amphioxus (Endo et al., 2003). On the basis of a phylogenetic tree of catalytic domain, the MASP/C1r/C1s family are classified into two groups, termed as AGY-type and TCN-type (Endo et al., 1998). MASP-1 belongs to TCN-type, which is characterized by TCN codon at the active site serine and split exons for protease domain. MASP-2, MASP-3, C1r and C1s belong to the AGY-type, which is characterized by AGY codon at the active site serine and a single exon for protease domain. It is consented that TCN-type is a prototype of the MASP/C1r/C1s family. Actually only TCN-type (MASPa and MASPb) was found in ascidian (Ji et al., 1997). As described below, the organization of the *MASP1/3* gene is very unique, in which a single exon that encodes the entire of MASP-3 catalytic domain locates at upstream of six split exons encoding MASP-1 catalytic domain. This gene structure of *MASP1/3* was found in an invertebrate, amphioxus (Endo et al., 2003), suggesting MASP-1 and MASP-3 might have evolved before emergence of the amphioxus lineage. On the other hand, the *MASP2* gene was only found from the amphibian lineage. It was reported that the split exon type protease domain is exceptionally absent in the MASP1/3 gene in birds (Lynch et al., 2005).

4 *MASP1/3* gene structure and its knockout mice

MASP-1 and MASP-3 are alternative spliced products from the same gene (Dahl et al., 2001). The *MASP1/3* gene is mapped to chromosome 3 (3q27-q28) in human and to chromosome 16 (16B2-B3) in mouse (Sato et al., 1994; Takada et al., 1995). The *MASP1/3* gene extends in much longer region more than 67 kbp (Takayama et al., 1999). The heavy chain of MASP-1/3 is encoded by 10 exons, followed by 6 exons that encode MASP-1 light chain (Endo et al., 1996; Takayama et al., 1999)(Fig. 2A). An additional single exon encoding the light chain of MASP-3 is present at the upstream of split exons encoding the MASP-1 light chain (Dahl et al., 2001). Northern blot analysis indicated the hepatic tissues primarily produced MASP-1 and MASP-3 (Schwaeble et al., 2002).

To generate mice that are deficient in both MASP-1 and -3 (MASP1/3KO), the common second exon was disrupted by a gene targeting strategy (Fig. 2A). No MASP-1 protein is absent in serum of MASP1/3KO, and no transcription of *MASP3* mRNA in MASP1/3KO liver. A light growth retardation was observed in the knockout mice.

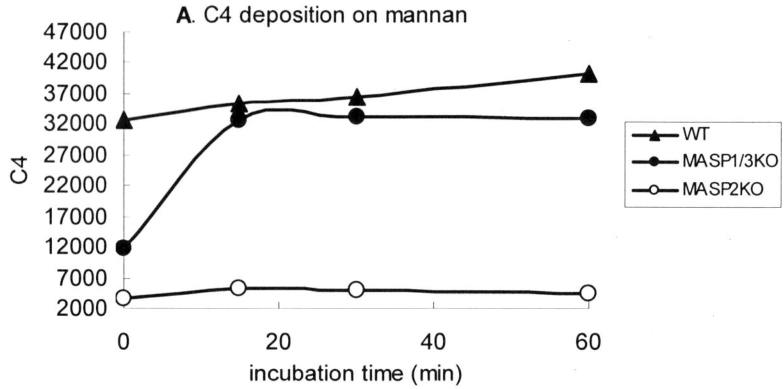

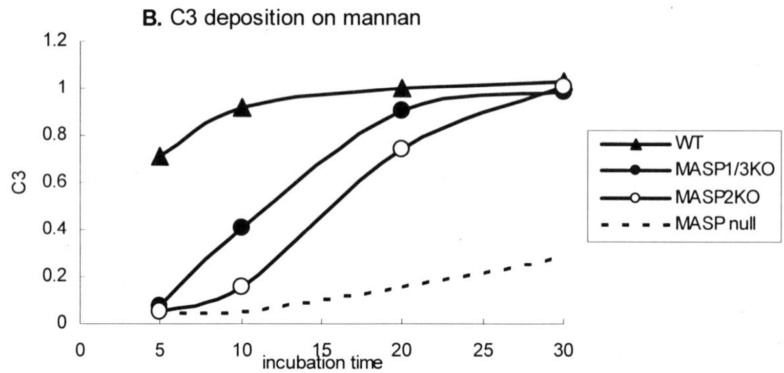

Fig. 3. C4 and C3 deposition activities on mannan-coated wells using sera from knockout mice. (A)C4 deposition activity on mannan-surface. 2 % diluted murine sera (0.1 ml) were incubated on mannan (Sigma-Aldrich: M7504)-coated wells at 37 °C, followed by incubation with human C4 at 37 °C for 30 min. Amount of deposited human C4 was detected by FITC-conjugated anti-human C4. Levels of Y axis indicate fluorescent intensities. Each point shows average for three mice. (B)C3 deposition activity on mannan-surface. 2 % diluted murine sera (0.1 ml) were incubated on mannan (Sigma-Aldrich: M7504)-coated wells at 37 °C. Levels of deposited C3 were detected by anti-murine C3 antibody and HRP-conjugated secondary antibody. Colors were developed by TMB substrates.

5 *MASP2/sMAP(MAp19)* gene structure and its knockout mice

The human *MASP2* gene was assigned to the chromosome 1 (1p36.3-p36.2) by in situ hybridization and somatic cell hybrid analysis (Stover et al., 1999a). The *MASP2* gene encodes two serum proteins, MASP-2 and sMAP/MAp19 by alternative polyadenylation/splicing (Stover et al., 1999b; Takahashi et al., 1999). As shown in Fig. 2B, *MASP2* gene encompasses 12 exons within 20 kbp in human (Stover et al., 2001). Since the *MASP2* gene has two distinct polyadenylation signals in exons 5 and 12, two different mRNAs are transcribed from a single gene. Northern blot analysis reveals 0.9 kbp transcript encoding sMAP/MAp19, that is dominantly transcribed than MASP2 transcript (2.4-4.8 kbp) in human liver (Stover et al., 1999b; Takahashi et al., 1999).

In mouse, the *MASP2* gene locates on the chromosome 4 (4E1) and its gene structure is similar to the human *MASP2* gene. To generate sMAP/MAp19 deficient mice, the exon 5 that is specific for sMAP/MAp19 was replaced with neomycin-resistant gene (Fig. 2B). These knockout mice showed undetectable level of MASP-2 in sera in addition to the absence of sMAP/MAp19, therefore named as MASP2/sMAP KO (Iwaki et al., 2006).

6 Lectin complement pathway in MASPs-deficient mice

Consistent with the previous studies (Matsushita et al., 2000b; Thiel et al., 1997), analysis of the phenotypes of MASPs-deficient mice demonstrated that only MASP-2 has an ability to cleave C4. Actually, C4 deposition activity of sMAP/MASP2 KO sera on mannan was undetectable. On the other hand, MASP1/3 KO sera showed the comparable activities of C4 deposition after 30 min incubation with wild-type sera, although it was decreased at the early phase of activation. These data supported that only MASP-2 activates C4 and that MASP-1 plays some roles in this activation (Fig. 3).

We also investigated the C3 deposition activities by the knockout mice sera on mannan. In sera of wild-type, C3 was highly activated on mannan within 5 min incubation and reached the peak at 10 min. In contrast, no C3 deposition activities were observed with 5 min incubation in sMAP/MASP2KO and MASP1/3KO. However, C3 activation was detected at the extended period in sera of both knockout mice, although the onset of activation delayed. The MASPs-null mice, which were generated by cross of both knockout mice, showed the decreased ability of C3 activation until 30 min of the incubation. These findings indicated that not only MASP-2 but also MASP-1/3 has an essential role for the initiation of the lectin complement pathway. Recently Selandar et al. demonstrated C3 and the alternative pathway were activated through an MBL-dependent C2 bypass mechanism (Selander et al., 2006). They also described MASP-1, -2 and -3 are unnecessary for the C2-bypass mechanism, but this issue may be still controversial. Altogether, it is possible that at least two routes (one is C2/C4-dependent and the other is C2/C4-independent) would be present in the lectin complement pathway (Fig. 5). Recent our findings demonstrated the possibility that MASP-1 might participate the activation of MASP-2 (manuscript in preparation).

7 MASP-2 deficiency in human

A lot of studies have shown that low serum levels of MBL in patients with immunodeficiencies, such as frequent unexplained infections and systemic lupus erythematosus. Since MBL affects the complement activation *via* activation of its

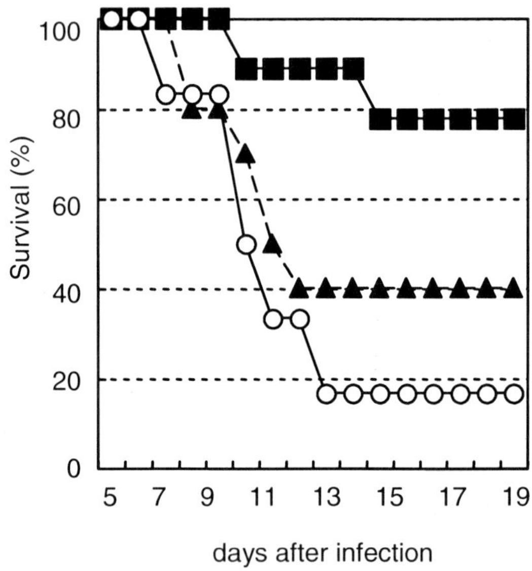

Fig. 4. Susceptibility of MASP1/3(-/-) mice against influenza A virus infection.
The infections of mice with A/PR8/34 of influenza A virus were performed. The mice were observed for at least 3 weeks. Closed squares, closed triangles and opened circles represent MASP1/3(+/+) (n=9), (+/-) (n=10) and (-/-) (n=6) mice, respectively.

associated MASPs, it is suspected that the patients who are deficient in MASPs also would suffer from similar diseases.

Recently a patient who has an inherited deficiency of MASP-2 was found (Stengaard-Pedersen et al., 2003). The patient had a history of infections and chronic inflammatory disease. In the patient, no C4b deposition activity on solid-mannan was observed despite a normal level of MBL. A point mutation that results in the substitution of an aspartic acid with a glycine at position 105 in the mature MASP-2 (D105G) was found in exon 3 of the MASP2 gene. The site of the mutation is close to the C-terminal of CUB1 domain in MASP-2 and essential for binding of calcium ion (Sorensen et al., 2005). The patient was homozygous for the mutation. It was also reported the mutant MASP-2 failed to bind to MBL (Stengaard-Pedersen et al., 2003). No patient with MASP-1-deficiency was found.

8 MASP1/3KO mice were affected to infection of influenza A (H1N1) virus

The previous reports have shown that MBL is involved in the neutralization of the influenza virus (Anders et al., 1990; Reading et al., 1997). To know whether MASP-1 or MASP-3 is involved in the virulence of the influenza virus, we performed the infection experiment with one LD50 of PR8/A/34 (H1N1) against BALB/c, although

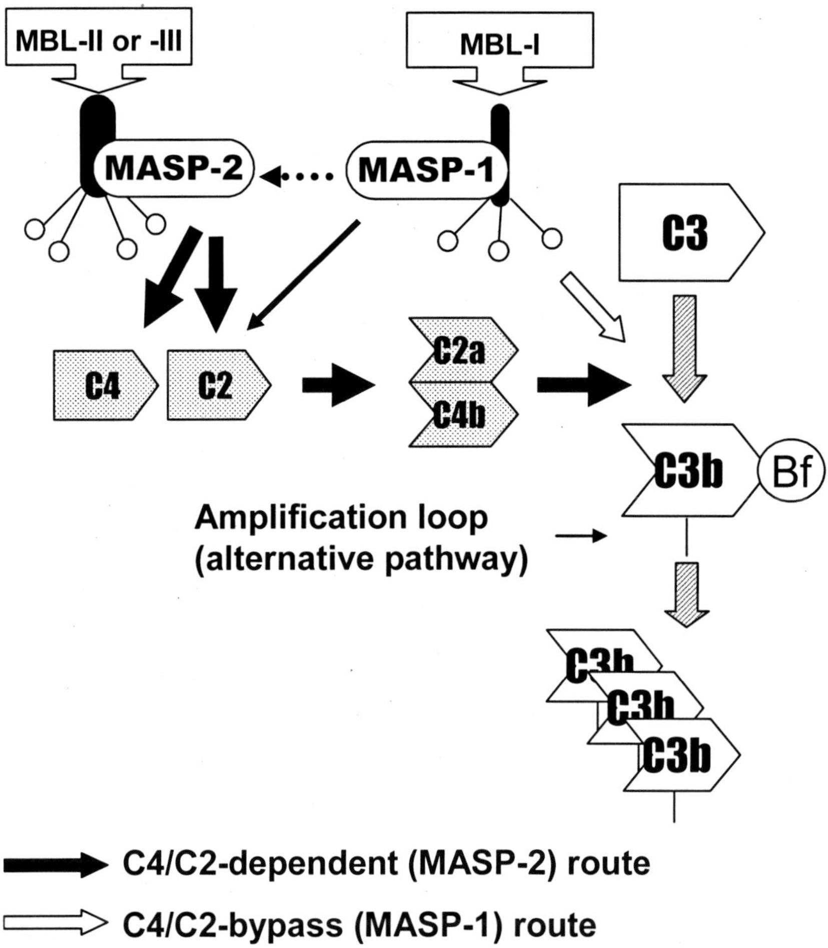

Fig. 5. Two routes are present in lectin pathway.
Different oligomerizations of MBL exist in serum. Mainly trimer of MBL, termed as MBL-I associate with MASP-1 and sMAP, whereas tetramer of MBL, termed as MBL-II associate with MASP-2 and MASP-3. MASP-1 can trigger the complement activation in absent of C4/C2 by direct activation of C3 (C4/C2-bypass, MASP-1 route). In another route, it is essential to activate C4 and C2 by MASP-2 (C4/C2-dependent, MASP-2 route). MASP-1 might be able to enhance the MASP-2 route through the activtion of C2. Also, MASP-1 plays some roles to activate MASP-2.

MBL does not bind directly to H1N1 among the influenza virus (Reading et al., 1997). In the case of MASP1/3(+/+) (n=9), 80 % of the infected mice were survival for 3 weeks (Fig. 4). However, in MASP1/3(-/-) mice, only one of six mice was survival for 2 weeks, showing a significant difference in the survival period. The heterozygoutes (MASP1/3(+/-)) also showed shorter survival periods than wild-type mouse (40 % survival for 3 weeks) although it was statistically not significant. There was no significant relationship between the survival and the body weight, eliminating the possibility that MASP1/3(-/-) mice might be easily died because of their growth retardation. This suggests that MASP-1 and/or MASP-3 are involved in host defense against influenza A virus in mouse.

9 Conclusion

It is likely that at least two distinct routes coordinate for the activation of the lectin pathway. One of them is C4/C2 dependent MASP-2 route and the other is C4/C2 independent route. In the C4/C2 independent route, it is possible that MASP-1 might cleave C3 directly (Fig. 5). In human sera, lower oligomerized forms of MBL, termed as MBL-I associate with MASP-1 and sMAP, whereas higher oligomerized forms of MBL, termed as MBL-II associate mainly with MASP-2 and MASP-3 (Dahl et al., 2001). This finding might support the above hypothesis.

MASP-2-deficiency in human is known to suffer from some immunological diseases such as recurrent infection and SLE. Furthermore, MASP1/3KO mice were susceptible to infection of influenza A virus.

10 Acknowledgement

We thank Yuichi, Endo for helpful comments and critical reading.

References

Ambrus, G., Gal, P., Kojima, M., Szilagyi, K., Balczer, J., Antal, J., Graf, L., Laich, A., Moffatt, B. E., Schwaeble, W., *et al.* (2003). Natural substrates and inhibitors of mannan-binding lectin-associated serine protease-1 and -2: a study on recombinant catalytic fragments. J Immunol 170, 1374-1382.

Anders, E. M., Hartley, C. A., and Jackson, D. C. (1990). Bovine and mouse serum beta inhibitors of influenza A viruses are mannose-binding lectins. Proc Natl Acad Sci U S A 87, 4485-4489.

Chen, C. B., and Wallis, R. (2004). Two mechanisms for mannose-binding protein modulation of the activity of its associated serine proteases. J Biol Chem 279, 26058-26065.

Cortesio, C. L., and Jiang, W. (2006). Mannan-binding lectin-associated serine protease 3 cleaves synthetic peptides and insulin-like growth factor-binding protein 5. Arch Biochem Biophys.

Cseh, S., Vera, L., Matsushita, M., Fujita, T., Arlaud, G. J., and Thielens, N. M. (2002). Characterization of the interaction between L-ficolin/p35 and mannan-binding lectin-associated serine proteases-1 and -2. J Immunol 169, 5735-5743.

Dahl, M. R., Thiel, S., Matsushita, M., Fujita, T., Willis, A. C., Christensen, T., Vorup-Jensen, T., and Jensenius, J. C. (2001). MASP-3 and its association with distinct complexes of the mannan-binding lectin complement activation pathway. Immunity 15, 127-135.

de Vries, B., Walter, S. J., Peutz-Kootstra, C. J., Wolfs, T. G., van Heurn, L. W., and Buurman, W. A. (2004). The mannose-binding lectin-pathway is involved in complement activation in the course of renal ischemia-reperfusion injury. Am J Pathol 165, 1677-1688.

Endo, Y., Nakazawa, N., Liu, Y., Iwaki, D., Takahashi, M., Fujita, T., Nakata, M., and Matsushita, M. (2005). Carbohydrate-binding specificities of mouse ficolin A, a splicing variant of ficolin A and ficolin B and their complex formation with MASP-2 and sMAP. Immunogenetics 57, 837-844.

Endo, Y., Nonaka, M., Saiga, H., Kakinuma, Y., Matsushita, A., Takahashi, M., Matsushita, M., and Fujita, T. (2003). Origin of mannose-binding lectin-associated serine protease (MASP)-1 and MASP-3 involved in the lectin complement pathway traced back to the invertebrate, amphioxus. J Immunol 170, 4701-4707.

Endo, Y., Sato, T., Matsushita, M., and Fujita, T. (1996). Exon structure of the gene encoding the human mannose-binding protein-associated serine protease light chain: comparison with complement C1r and C1s genes. Int Immunol 8, 1355-1358.

Endo, Y., Takahashi, M., Nakao, M., Saiga, H., Sekine, H., Matsushita, M., Nonaka, M., and Fujita, T. (1998). Two lineages of mannose-binding lectin-associated serine protease (MASP) in vertebrates. J Immunol 161, 4924-4930.

Fujita, T. (2002). Evolution of the lectin-complement pathway and its role in innate immunity. Nat Rev Immunol 2, 346-353.

Hart, M. L., Ceonzo, K. A., Shaffer, L. A., Takahashi, K., Rother, R. P., Reenstra, W. R., Buras, J. A., and Stahl, G. L. (2005). Gastrointestinal ischemia-reperfusion injury is lectin complement pathway dependent without involving C1q. J Immunol 174, 6373-6380.

Iwaki, D., Kanno, K., Takahashi, M., Endo, Y., Lynch, N. J., Schwaeble, W. J., Matsushita, M., Okabe, M., and Fujita, T. (2006). Small mannose-binding lectin-associated protein plays a regulatory role in the lectin complement pathway. J Immunol 177, 8626-8632.

Ji, X., Azumi, K., Sasaki, M., and Nonaka, M. (1997). Ancient origin of the complement lectin pathway revealed by molecular cloning of mannan binding protein-associated serine protease from a urochordate, the Japanese ascidian, Halocynthia roretzi. Proc Natl Acad Sci U S A 94, 6340-6345.

Liu, Y., Endo, Y., Iwaki, D., Nakata, M., Matsushita, M., Wada, I., Inoue, K., Munakata, M., and Fujita, T. (2005). Human M-ficolin is a secretory protein that activates the lectin complement pathway. J Immunol 175, 3150-3156.

Lynch, N. J., Khan, S. U., Stover, C. M., Sandrini, S. M., Marston, D., Presanis, J. S., and Schwaeble, W. J. (2005). Composition of the lectin pathway of complement in Gallus gallus: absence of mannan-binding lectin-associated serine protease-1 in birds. J Immunol 174, 4998-5006.

Matsushita, M., Endo, Y., and Fujita, T. (2000a). Cutting edge: complement-activating complex of ficolin and mannose-binding lectin-associated serine protease. J Immunol 164, 2281-2284.

Matsushita, M., and Fujita, T. (1992). Activation of the classical complement pathway by mannose-binding protein in association with a novel C1s-like serine protease. J Exp Med 176, 1497-1502.

Matsushita, M., and Fujita, T. (1995). Cleavage of the third component of complement (C3) by mannose-binding protein-associated serine protease (MASP) with subsequent complement activation. Immunobiology 194, 443-448.

Matsushita, M., Kuraya, M., Hamasaki, N., Tsujimura, M., Shiraki, H., and Fujita, T. (2002). Activation of the lectin complement pathway by H-ficolin (Hakata antigen). J Immunol 168, 3502-3506.

Matsushita, M., Thiel, S., Jensenius, J. C., Terai, I., and Fujita, T. (2000b). Proteolytic activities of two types of mannose-binding lectin-associated serine protease. J Immunol 165, 2637-2642.

Moller-Kristensen, M., Wang, W., Ruseva, M., Thiel, S., Nielsen, S., Takahashi, K., Shi, L., Ezekowitz, A., Jensenius, J. C., and Gadjeva, M. (2005). Mannan-binding lectin recognizes structures on ischaemic reperfused mouse kidneys and is implicated in tissue injury. Scand J Immunol 61, 426-434.

Reading, P. C., Morey, L. S., Crouch, E. C., and Anders, E. M. (1997). Collectin-mediated antiviral host defense of the lung: evidence from influenza virus infection of mice. J Virol 71, 8204-8212.

Rossi, V., Cseh, S., Bally, I., Thielens, N. M., Jensenius, J. C., and Arlaud, G. J. (2001). Substrate specificities of recombinant mannan-binding lectin-associated serine proteases-1 and -2. J Biol Chem 276, 40880-40887.

Sato, T., Endo, Y., Matsushita, M., and Fujita, T. (1994). Molecular characterization of a novel serine protease involved in activation of the complement system by mannose-binding protein. Int Immunol 6, 665-669.

Schwaeble, W., Dahl, M. R., Thiel, S., Stover, C., and Jensenius, J. C. (2002). The mannan-binding lectin-associated serine proteases (MASPs) and MAp19: four components of the lectin pathway activation complex encoded by two genes. Immunobiology 205, 455-466.

Selander, B., Martensson, U., Weintraub, A., Holmstrom, E., Matsushita, M., Thiel, S., Jensenius, J. C., Truedsson, L., and Sjoholm, A. G. (2006). Mannan-binding lectin activates C3 and the alternative complement pathway without involvement of C2. J Clin Invest 116, 1425-1434.

Sorensen, R., Thiel, S., and Jensenius, J. C. (2005). Mannan-binding-lectin-associated serine proteases, characteristics and disease associations. Springer Semin Immunopathol 27, 299-319.

Stengaard-Pedersen, K., Thiel, S., Gadjeva, M., Moller-Kristensen, M., Sorensen, R., Jensen, L. T., Sjoholm, A. G., Fugger, L., and Jensenius, J. C. (2003). Inherited deficiency of mannan-binding lectin-associated serine protease 2. N Engl J Med 349, 554-560.
Stover, C., Endo, Y., Takahashi, M., Lynch, N. J., Constantinescu, C., Vorup-Jensen, T., Thiel, S., Friedl, H., Hankeln, T., Hall, R., *et al.* (2001). The human gene for mannan-binding lectin-associated serine protease-2 (MASP-2), the effector component of the lectin route of complement activation, is part of a tightly linked gene cluster on chromosome 1p36.2-3. Genes Immun 2, 119-127.
Stover, C. M., Schwaeble, W. J., Lynch, N. J., Thiel, S., and Speicher, M. R. (1999a). Assignment of the gene encoding mannan-binding lectin-associated serine protease 2 (MASP2) to human chromosome 1p36.3-->p36.2 by in situ hybridization and somatic cell hybrid analysis. Cytogenet Cell Genet 84, 148-149.
Stover, C. M., Thiel, S., Thelen, M., Lynch, N. J., Vorup-Jensen, T., Jensenius, J. C., and Schwaeble, W. J. (1999b). Two constituents of the initiation complex of the mannan-binding lectin activation pathway of complement are encoded by a single structural gene. J Immunol 162, 3481-3490.
Takada, F., Seki, N., Matsuda, Y., Takayama, Y., and Kawakami, M. (1995). Localization of the genes for the 100-kDa complement-activating components of Ra-reactive factor (CRARF and Crarf) to human 3q27-q28 and mouse 16B2-B3. Genomics 25, 757-759.
Takahashi, M., Endo, Y., Fujita, T., and Matsushita, M. (1999). A truncated form of mannose-binding lectin-associated serine protease (MASP)-2 expressed by alternative polyadenylation is a component of the lectin complement pathway. Int Immunol 11, 859-863.
Takayama, Y., Takada, F., Nowatari, M., Kawakami, M., and Matsu-ura, N. (1999). Gene structure of the P100 serine-protease component of the human Ra-reactive factor. Mol Immunol 36, 505-514.
Thiel, S., Vorup-Jensen, T., Stover, C. M., Schwaeble, W., Laursen, S. B., Poulsen, K., Willis, A. C., Eggleton, P., Hansen, S., Holmskov, U., *et al.* (1997). A second serine protease associated with mannan-binding lectin that activates complement. Nature 386, 506-510.
Thielens, N. M., Cseh, S., Thiel, S., Vorup-Jensen, T., Rossi, V., Jensenius, J. C., and Arlaud, G. J. (2001). Interaction properties of human mannan-binding lectin (MBL)-associated serine proteases-1 and -2, MBL-associated protein 19, and MBL. J Immunol 166, 5068-5077.
Vorup-Jensen, T., Petersen, S. V., Hansen, A. G., Poulsen, K., Schwaeble, W., Sim, R. B., Reid, K. B., Davis, S. J., Thiel, S., and Jensenius, J. C. (2000). Distinct pathways of mannan-binding lectin (MBL)- and C1-complex autoactivation revealed by reconstitution of MBL with recombinant MBL-associated serine protease-2. J Immunol 165, 2093-2100.
Wallis, R., and Dodd, R. B. (2000). Interaction of mannose-binding protein with associated serine proteases: effects of naturally occurring mutations. J Biol Chem 275, 30962-30969.
Walsh, M. C., Bourcier, T., Takahashi, K., Shi, L., Busche, M. N., Rother, R. P., Solomon, S. D., Ezekowitz, R. A., and Stahl, G. L. (2005). Mannose-binding lectin is a regulator of inflammation that accompanies myocardial ischemia and reperfusion injury. J Immunol 175, 541-546.
Zundel, S., Cseh, S., Lacroix, M., Dahl, M. R., Matsushita, M., Andrieu, J. P., Schwaeble, W. J., Jensenius, J. C., Fujita, T., Arlaud, G. J., and Thielens, N. M. (2004). Characterization of recombinant mannan-binding lectin-associated serine protease (MASP)-3 suggests an activation mechanism different from that of MASP-1 and MASP-2. J Immunol 172, 4342-4350.

9

Viral Heparin-Binding Complement Inhibitors – A Recurring Theme

Anna M. Blom[1], Linda Mark[1] and O. Brad Spiller[2]

[1] Lund University, Department. of Laboratory Medicine Malmö, The Wallenberg Laboratory, 4th floor, Malmö University Hospital, entrance 46, S-205 02 Malmö, Sweden
[2] Cardiff University, Department. of Child Health, 5th floor University Hospital, Heath Park, Cardiff, CF14 4XN, United Kingdom

1 Introduction

The complement system is an important part of the innate immune system providing immediate protection against pathogens without requirement for previous exposure. Complement also serves an important role in priming the adaptive immune response through opsonization, leukocyte recruitment and enhancing humoral immune responses. The importance of complement is exemplified through recurrent fulminant bacterial (especially *Neiserria spp.*) infections in individuals with complement deficiencies, predisposition to autoimmune pathologies in individuals lacking the C1q component, and through the many complement evasion strategies (including acquisition of endogenous host complement inhibitors and expression of microbial homologues) identified for a wide range of viruses and bacteria. The identification of heparin binding sites within complement regulators, of both host and viral origin, is well established. Heparin (or more appropriately heparan sulfate) binding mediates appropriate localization of soluble proteins, such as factor H or vaccinia virus complement control protein (VCP), to cell surfaces in the correct orientation for maximal efficacy. The specificity of glycosaminoglycan (GAG) required to recruit factor H enables differential protection of host, but not pathogen, in most cases; while secreted soluble VCP utilizes GAG binding to concentrate the viral complement regulator on surfaces at the point of infection, where it will be most beneficial to the virus. However, it is difficult to see an equivalent benefit for viral glycoproteins that are expressed at the cell- or virion surface: Despite a complete lack of apparent structural homology for alphaherpesvirus family glycoprotein C and gammaherpesvirus open-reading frame 4 (e.g. KSHV Complement Control Protein;

KCP), heparin binding and complement component C3b binding sites overlap and compete with each other for respective function. Thus raising the question: is the presence of heparin binding sites in complement regulators non-specific and simply due to concentration of positively charged residues in conserved structural motifs required for C3b or C4b binding?

2 Complement Activation by Pathogens

Complement comprises greater than 30 soluble proteins in the serum, some present at millimolar concentrations, which are activated in a cascade with amplification at each step culminating in the final formation of a multi-protein lytic pore (called the membrane attack complex; MAC). Three linked activation pathways (reviewed in Walport 2001), ensure that the surfaces of pathogens including bacteria, viruses, fungi, and parasites are identified as hostile and trigger enzymatic cleavage of sequential complement components.

The lectin pathway (with MBL or ficolin, MASP, C4, C2, C3) acts to recognise carbohydrate residues normally not present in the host and result in complement activation by many prokaryotes, fungi, and parasites. The binding of the globular heads of MBL (or ficolin) to "foreign" carbohydrate surfaces results in MASP activation, which in turn cleaves C4, then C2 to their active forms in a coordinated manner leading to a covalently tethered C3 convertase that cleaves and activates the central complement component C3 (reviewed in Arnold, Dwek, Rudd and Sim 2006; Endo, Takahashi and Fujita 2006). Similarly the alternative pathway (factors B & D, properdin, and C3) is sensitive to prokaryotic and fungal pathogens; however, not through specific recognition of pathogens by an initiating complement component such as MBL. Alternative pathway activation is more reliant on a lack of complement regulator expression or lack of ability to recruit soluble serum complement inhibitors (eg. GAG recruitment of factor H). The alternative pathway exists due to low level spontaneous hydrolysis of the strained internal thioester bond within C3 (called "tick-over"). The resulting C3-H_20 can bind factor B, which is then cleaved by factor D. This soluble alternative C3 convertase (C3-H_2OBb) cleaves C3 molecules into highly reactive intermediates capable of non-specific covalent attachment to surface proteins and carbohydrates. The conformation of this bound C3 provides a nexus for an amplification loop (also used by the lectin and classical pathways) designed to maximise a covalent foothold on activating surfaces. The C3 and factor B complex that make up the alternative pathway C3 convertase are stabilized by properdin and cleaved to active forms by factor D. As with the lectin pathway, the end product is an activated covalently bound form of C3 on the activating surface. The presence of complement regulators expressed on, or recruited by, host cell surfaces inactivates the initial bound C3 almost immediately, limiting alternative pathway activation to non-host surfaces.

It is easy to see from these two mechanisms how prokaryotes and fungi activate complement, but since viruses use the host's own synthetic machinery to synthesize the glycoproteins used to produce progeny virions, in many cases acquiring portions of the host cell's membrane as an envelope during egress, these mechanisms probably do not readily apply to many viruses. Enveloped viruses such as cytomegalovirus, vaccinia, and HIV can also passively acquire the host's complement inhibitors as components of the envelope (Spear, Lurain, Parker, Ghassemi, Payne and Saifuddin 1995; Spiller, Hanna, Devine and Tufaro 1997; Vanderplasschen, Mathew, Hollinshead, Sim and Smith 1998). However, any virus that is transmitted by insect vectors, or acquired from an animal reservoir will have altered glycosylation patterns relative to human hosts and may activate these pathways. Further, many viruses shut down host protein synthesis as part of their infectious cycle that may also result in emergence of incompletely processed carbohydrates on glycoproteins and glycolipids as infection progresses.

Finally, the classical pathway (C1, C4, C2, C3) is mostly associated with complement activation initiated through specific immunoglobulin binding to invading pathogens, although immune complexes have also been suggested to activate the alternative and lectin pathways as well (Saevarsdottir, Vikingsdottir and Valdimarsson 2004). A single bound IgM molecule or a threshold concentration of bound IgG provides the presentation of sufficient Fc immunoglobulin regions to bind multiple arms of C1q and trigger catalytic conversion of C1r and C1s chains of the C1 complex molecule. Although this suggests complement activation is subsequent to previous humoral immune response, some virally-infected cells (e.g. cytomegalovirus (Spiller and Morgan 1998)) can directly bind C1q leading to complement activation. Protein glycosylation inconsistent with host patterns can also activate the classical pathway. Natural or spontaneous antibodies to toxins, bacteria and erythrocytes are present in the sera of normal, non-immunized humans and mice (Avrameas 1991; Herzenberg and Kantor 1993). These antibodies have been shown to neutralize herpes simplex virus (HSV) that lack their virally encoded complement inhibitor (Hook, Lubinski, Jiang, Pangburn and Friedman 2006). Furthermore, higher primates lack the required enzyme to generate terminal alpha 1-3 linked galactose molecules on their carbohydrates, in contrast to other mammals and many prokaryotes, which means that up to 1% of all circulating human IgG my be directed against this carbohydrate epitope (Galili, Mandrell, Hamadeh, Shohet and Griffiss 1988; Galili, Shohet, Kobrin, Stults and Macher 1988). While these antibodies are useful in complement activation by bacteria, yeast, and fungi, it could, for example, aid in neutralizing influenza virus that originated from pigs or fowl due to the presence of terminal alpha 1-3 galactose residues on their glycoproteins.

All of these activation pathways converge at activation of C3 and precede the final enzymatic cleavage of all pathways (C5 into C5a and C5b), the first step in the terminal lytic pathway that results in the sequential non-covalent association of C5b, C6, C7, C8, and multiple units of C9, leading to the of penetrating lipid membranes (MAC). When sufficient MAC are deposited into the membrane, osmotic balance is disrupted and cell death (or virolysis) ensues. Furthermore, C5a and C3a released during activation of any of the three pathways are strong chemoattractants and activators of phagocytes.

3 Consequence of Complement Activation for Viruses

Viruses are not alive; therefore, mediation of osmotic imbalance probably does not play a dominant role in removing them, although loss of the viral envelope and associated glycoproteins will result in neutralization. Furthermore, viruses encased in proteinaceous capsids (e.g. poliovirus) are not likely susceptible to MAC. However, lectin and classical pathway activation result in covalent C4b coating of virions, and all activation pathways lead to abundant covalent C3b attachment. Since the first requirement for virus infection is interaction with receptors expressed on host cells, disabling receptor engagement through steric hindrance is an effective method of neutralization. Incubation of herpes simplex virus (HSV) with purified C1 and C4 under activating conditions was found sufficient to significantly reduce viral infectivity (Daniels, Borsos, Rapp, Snyderman and Notkins 1970). Neutralizing anti-virus antibodies work through the same principle, although the additive effect of complement activation can enhance neutralizing antibodies by 2-16 fold (Eizuru, Ueno and Minamishima 1988; Lewis, Matzke, Albrecht and Pollard 1986; Rundell and Betts 1982). Antibodies that are not capable of neutralizing virus on their own, can also activate complement leading to indirect neutralization by C4b/C3b deposition with or without virolysis. Receptors that recognize bound C3b and the largest inactivation fragment (iC3b) are found on granulocytes, monocytes, macrophages, and dendritic cells, and covalent attachment of C3b to the surface (opsonisation) greatly enhances phagocytosis and digestion of antigens by these cells. While this phenomenon represents further versatility of complement to enhance pathogen clearance by other factions of the innate immune system, a few viruses have adapted to utilise this mechanism as a secondary mechanism of infection. HIV's primary receptor CD4 is found on a subset of T-lymphocytes, however, C3b has been found to enable infection of CD4-negative dendritic cells and macrophages (Bajtay, Speth, Erdei and Dierich 2004; Banki, Stoiber and Dierich 2005; Thieblemont, Haeffner-Cavaillon, Ledur, L'Age-Stehr, Ziegler-Heitbrock and Kazatchkine 1993) by triggering endocytosis, but escaping the endosome prior to destruction by low pH and proteolysis.

Enzymatic processing of C3b covalently bound to viral glycoproteins to its smallest form (C3d) by serum factor I (fI) and phagocyte expressed complement receptor 1 (CR1; CD35), results in a more potent antigen for the stimulation of specific B-lymphocyte response (reviewed in Morgan, Marchbank, Longhi, Harris and Gallimore 2005). The C3d receptor, complement receptor 2 (CR2; CD21), acts to lower the signaling threshold on B-lymphocytes when co-engaged with specific antigen receptor expressed on the lymphocyte surface. Expression of foreign antigens fused to multiple copies of C3d has been shown to significantly enhance the antibody response to test antigens, including influenza and HIV proteins (Green, Montefiori and Ross 2003; Watanabe, Ross, Tamura, Ichinohe, Ito, Takahashi, Sawa, Chiba, Kurata, Sata and Hasegawa 2003). Thus complement not only enhances the effectiveness of existing antibodies, it acts to enhance their production and refine the specificity of antibodies to prevalent viral proteins coated with C3b.

4 Regulation of Complement Activation

Complement regulation occurs at key points in the cascade. C1 is inhibited in an irreversible one-to-one fashion by soluble C1 inhibitor present in the serum, while CD59 is expressed on the surface of most human cells and blocks the formation of MAC through physically blocking membrane integration by multiple molecules of C9. With the exception of these regulators, complement inhibition is mediated through targeting of the two covalently bound complement components, C3b and C4b, and the C3 or C5 convertases they form. Some of these inhibitors have been briefly mentioned already above, and a detailed description of host regulators is outside the scope this review. To date, the identified host inhibitors of these components belong to a family of proteins located on chromosome 1 referred to as the regulators of complement activation (RCA). The members of the RCA family are composed of 4-35 structural complement control protein (CCP) modules, which are approximately 60 amino acids long, have 4 conserved cysteines (linked through 1-3 and 2-4 disulphide bridges), and many conserved residues that enable the formation of several anti-parallel beta-sheets. Regulation of complement at the level of C3b and C4b can be mediated in two ways: 1) accelerate the dissociation of the two components forming C3 convertase complexes (C4b2a or C3bBb) so that it cannot cleave C3b into C3a and C3b or 2) act as a cofactor to recruit serum serine protease fI into cleaving C3b or C4b into inactive fragments.

Viruses are well known to subvert host complement inhibitors for their own protection. Enveloped viruses such as cytomegalovirus, vaccinia, HIV and T-cell lymphotrophic retroviruses are known to acquire cell-bound inhibitors; (Spear et al. 1995; Spiller et al 1997; Vanderplasschen et al. 1998) cytomegalovirus further upregulates the expression of decay-accelerating factor (DAF; CD55) and membrane cofactor protein (MCP; CD46) to maximize the protection of progeny virions and infected cells (Spiller, Morgan, Tufaro and Devine 1996). Some viruses also actively recruit soluble complement inhibitors (in addition to those that bind acquired host GAGs) through binding by surface expressed viral glycoproteins (eg. HIV gp120 binding of factor H; Pinter, Beltrami, Stoiber, Negri, Titti and Clivio 2000) although this strategy is more associated with bacterial strains of Staphylococcus, Streptococcus and Borrelia (Rooijakkers and van Strijp 2007). However, two viral families (*herpesviridae* and *poxviridae*) encode their own complement regulators; most are viral RCA homologues composed of CCP domains and regulate complement at the level of C3b and/or C4b. Notable exceptions are the CD59 homologue (non-CCP terminal pathway regulator) expressed by herpesvirus saimiri and the glycoprotein C-related proteins (non-CCP C3b regulators) expressed by alphaherpesviruses.

5 Viral Complement Inhibitors Homologous to RCA Proteins

5.1 Orthopoxviruses

The first viral RCA homologue to be discovered was the CCP-containing soluble complement inhibitor encoded by vaccinia virus, VCP (Kotwal, Isaacs, McKenzie,

Frank and Moss 1990; Kotwal and Moss 1988). Initially identified as one of the most abundant proteins produced by vaccinia infected cells, VCP remains the only viral complement inhibitor whose structure has been determined by X-ray crystallography (Ganesh, Smith, Kotwal and Murthy 2004; Murthy, Smith, Ganesh, Judge, Mullin, Barlow, Ogata and Kotwal 2001). VCP regulates complement by accelerating classical C3 convertase decay and by acting as a cofactor for fI cleavage of both C3b and C4b (McKenzie, Kotwal, Moss, Hammer and Frank 1992; Sahu, Isaacs, Soulika, and Lambris 1998). Decay of the alternative convertase was noted only under some conditions and even then it was not very efficient (Liszewski, Leung, Hauhart, Buller, Bertram, Wang, Rosengard, Kotwal and Atkinson 2006; McKenzie et al. 1992; Sahu et al. 1998). The pathological contribution of VCP to vaccinia infection was reported as increased lesion size following intradermal injection into rabbit ears, when compared to a recombinant VCP-knock-out strain of vaccinia (Isaacs, Kotwal and Moss 1992). However, vaccinia was chosen for its lack of pathological consequences to eradicate the etiological agent responsible for small pox: variola virus. Variola also encodes an RCA homologue, small pox inhibitor of complement enzymes (SPICE), which only differs in amino acid sequence by 11 residues from VCP (Rosengard, Liu, Nie and Jimenez 2002). Despite sharing greater than 95% identical amino acid composition between VCP and SPICE, the latter was found to be 100-fold more potent at inactivating C3b, (Rosengard et al. 2002) and 5 to 10-fold more effective at classical C3 convertase decay-accelerating activity (Liszewski, et al. 2006). While SPICE has been found to be so much more active than VCP, it would be reckless to suggest that this is a major contributor to the difference in pathogenesis between the parent viruses, but it certainly lends itself to speculation. The RCA homologue of another member of the orthopoxvirus family, monkeypox, has also recently been described (Liszewski et al. 2006). MOPICE (monkeypox inhibitor of complement enzymes) only has 3 full CCP domains, which is the fewest found in any naturally occurring complement inhibitor (host or viral) to date. However, the relative truncation of MOPICE also correlates to a decreased functionality: fI cofactor activity was comparable to VCP, but there was a complete absence of decay-accelerating activity for both classical and alternative C3 convertases. Liszewski and colleagues (Liszewski et al. 2006) have recently compared all three of these viral complement inhibitors. Using ELISA techniques and immobilized C3b, concentrations for 50% maximum binding showed VCP (24 nM) bound C3b less efficiently than MOPICE (4 nM) and SPICE (0.27 nM); however, all of these were much lower than the host membrane cofactor protein (MCP; CD46: 0.0009 nM). Furthermore, the authors confirmed that none of these 4 proteins were able to accelerate the decay of alternative C3 convertases, limiting the mechanism of inhibition to fI cofactor activity.

5.2 Gammaherpesviruses

Complement inhibitors have also been identified in many members of the herpesvirus family. Kaposi's sarcoma-associated herpesvirus (KSHV) or human herpesvirus 8, is the most recently sequenced member of the family and classified in the rhadinovirus sub-family (Russo, Bohenzky, Chien, Chen, Yan, Maddalena, Parry, Peruzzi, Edelman, Chang, and Moore 1996). KSHV is the etiologic agent responsible for endothelial neoplasma referred to as Kaposi's sarcoma, as well as the B-lymphocyte proliferative disorder primary effusion lymphoma (Cesarman, Chang, Moore, Said and Knowles 1995) and the plasma cell variant of multicentric Castleman's disease (Soulier, Grollet, Oksenhendler, Cacoub, Cazals-Hatem, Babinet, d'Agay, Clauvel, Raphael and Degos 1995). Like VCP and SPICE, the RCA homologue encoded by KSHV contains 4 CCP domains, but in contrast is expressed as a cell-surface and virion-expressed glycoprotein (Spiller, Mark, Blue, Proctor, Aitken, Blom, and Blackbourn 2006; Spiller, Robinson, O'Donnell, Milligan, Morgan, Davison, and Blackbourn 2003). Investigation of the KSHV complement control protein (KCP) expression identified 3 alternatively spliced isoforms; all of which maintain an in-frame transmembrane region and the 4 CCP domains located at the N-terminus (Spiller, Robinson, et al. 2003). Variations in size of the glycoprotein result from removal of a large serine/threonine-rich region, likely to be highly O-glycosylated (common to many host RCA proteins), with or without an intervening short stretch of residues containing two additional cysteines that do not participate in CCP domain structure. No functional difference was found between these isoforms and no predilection for a particular isoform to be expressed on the virion has been identified as yet. Recombinant KCP regulates C3b deposition regardless of whether it is cell-surface expressed or added exogenously (Spiller, Blackbourn, Mark, Proctor, and Blom 2003; Mullick, Bernet, Singh, Lambris and Sahu 2003). KCP accelerated the decay of classical/lectin C3 convertases, but was relatively inefficient at accelerating the decay of alternative C3 convertases. KCP also had a 10-fold higher affinity for C4b compared to C3b, even though it was capable of acting as a cofactor for fI cleavage of both components to inactive forms (Spiller, Blackbourn, et al. 2003; Mullick, Singh, Panse, Yadav, Bernet and Sahu 2005).

Other members of *gammaherpesvirinae* have also been previously described to express functional RCA homologues. The first of these was identified in 1992 by Albrecht and colleagues (Albrecht, Nicholas, Biller, Cameron, Biesinger, Newman, Wittmann, Craxton, Coleman and Fleckenstein 1992), and this protein, known as the complement control protein homologue (CCPH) expressed by herpesvirus saimiri, has recently been more extensively examined (Fodor, Rollins, Bianco-Caron, Rother, Guilmette, Burton, Albrecht, Fleckenstein and Squinto 1995; Singh, Mullick, Bernet and Sahu 2006). Similar to the findings for KCP, CCPH had a 24-fold higher affinity for C4b than C3b and accelerated the decay of classical/lectin C3 convertases, as well as acting as a cofactor for fI cleavage of both C3b and C4b (Singh et al. 2006). CCPH was 10-fold better than VCP at accelerating the decay of the alternative C3 convertases, but this activity was still negligible compared to host RCA proteins. CCPH has also been found to be alternatively spliced, although this resulted in production of transmembrane-containing and secreted forms of the protein. The murine gammaherpesvirus 68 also encodes an RCA homologue (γHV68-RCA), which also contains 4 CCP domains bound to the surface by a transmembrane spanning region (Kapadia, Molina, van Berkel, Speck and Virgin 1999). A soluble form of this protein was also identified, although whether this form arose through alternative splicing or protease-assisted shedding remains to be determined. γHV68-RCA inhibited both the classical and alternative mouse complement system, but the specific mechanisms underlying this inhibition have not yet been elucidated. The role of this protein in pathogenesis has also been determined by infecting mice with recombinant viruses lacking or expressing γHV68-RCA. These studies found a complement-dependent increase in virulence and mortality for RCA homologue-containing virus (Kapadia, Levine, Speck and Virgin 2002).

5.3 Alphaherpesviruses

Alphaherpesviruses also encode complement inhibitors, and although they do not have any homology to RCA proteins, they do bind C3b. In 1984, herpes simplex virus type 1 (HSV-1) was found to contain a virally-encoded C3b binding protein (Friedman, Cohen, Eisenberg, Seidel and Cines 1984). Glycoprotein C (gC) was subsequently identified as the C3b-binding viral glycoprotein that was expressed both on cell surfaces and the virion envelope (Fries, Friedman, Cohen, Eisenberg, Hammer and Frank 1986; Chen, Wang, Cheng and Hung 2003; Friedman et al. 1984). This viral glycoprotein is approximately the same size as KCP and has a transmembrane region, but has no homology to CCP-containing proteins and exists as a single isoform. HSV-1 gC has been found to bind native C3, C3b, iC3b and the released C3c fragment, (Kostavasili, Sahu, Friedman, Eisenberg, Cohen and Lambris 1997; Tal-Singer, Seidel-Dugan, Fries, Huemer, Eisenberg, Cohen, and Friedman 1991) to accelerate the decay of the alternative C3 convertase by binding to properdin, (Fries et al. 1986; Hung, Peng, Kostavasili, Friedman, Lambris, Eisenberg, and Cohen 1994) and also to block the initiation of the terminal lytic pathway by preventing C5 binding to C3b (Kostavasili et al. 1997). Even though gC does not appear to interact with C4b and no cofactor activity has been described, it does efficiently inhibit complement-mediated virus neutralization (Friedman, Wang, Fishman, Lambris, Eisenberg, Cohen and Lubinski 1996; McNearney, Odell, Holers, Spear and Atkinson 1987) and complement-dependent lysis of infected cells (Harris, Frank, Yee, Cohen, Eisenberg and Friedman 1990). Further, deletion of gC from the HSV-1 genome was found to significantly reduce the virulence of this virus 50-100 fold in various animal models, in a complement-dependent fashion (Lubinski, Wang, Mastellos, Sahu, Lambris and Friedman 1999; Lubinski, Wang, Soulika, Burger, Wetsel, Colten, Cohen, Eisenberg, Lambris and Friedman 1998) Other herpesviruses found to express gC homologues that bind to C3b include HSV type 2, (Gerber, Belval and Herold 1995) simian herpes B virus, (Huemer, Wechselberger, Bennett, Falke and Harrington 2003) equine herpesviruses 1 and 4, (Huemer, Nowotny, Crabb, Meyer and Hubert 1995) and pseudorabies virus (Huemer, Larcher and Coe 1992; Maeda, Hayashi, Tanioka, Matsumoto and Otsuka 2002). While HSV-1 and HSV-2 have highly homologous genomes, there are several differences in the N-terminal portion of the gC between these two strains. As a result, cells transfected with gC from HSV-1 and HSV-2 bind to C3b, (Seidel-Dugan, Ponce de Leon, Friedman, Eisenberg and Cohen 1990; Seidel-Dugan, Ponce de Leon, Friedman, Fries, Frank, Cohen and Eisenberg 1988) but HSV-2 gC does not inhibit properdin binding to the alternative C3 convertase (Huemer, Wang, Garred, Koistinen and Oppermann 1993). However, HSV-2 gC does function to protect the virions from complement mediated neutralization, through an undefined mechanism (Gerber et al. 1995). Recombinant gC homologues from both simian herpes B virus and pseudorabies virus were found to regulate complement activation, (Huemer et al. 2003; Maeda et al. 2002) but indirect evidence suggested that the host complement regulators acquired by pseudorabies virus mediated far more protection than gC (Maeda et al. 2002).

It is interesting to note that despite the differences in origin, method of cell association, and structure that many, if not all, of these proteins have at least one heparin binding site including soluble host RCA proteins factor H and C4b-binding protein (C4BP), and viral proteins VCP, SPICE, MOPICE, KCP, and all gC homologues.

6 Introduction to Heparin

Sulfated polysaccharides are found throughout the animal kingdom and often have heterogeneous structures that are difficult to structurally define. It is inevitable that any basic protein will be attracted to their large, acidic sulfate substituents. For the first 30 years of clinical use of heparin as an anti-coagulant, not much was known about its structure except that it was composed of glucosamine and uronic acid, with heavy sulfate substitution (Petitou, Casu and Lindahl 2003). The main repeating structure of heparin was established as a disaccharide of alternating *N*- and 6-*O*-sulfated α-D-glucosamine and 2-*O*-sulphated α-L-D-glucuronic acid, with a minor proportion of N-acetyl glucosamine and β-D-glucuronic acid. Heparan sulfate consists of the same structural elements as heparin, although less sulfated, and is almost as ubiquitous as the GAG component of proteoglycans (such as syndecans and glypicans) on the surfaces of mammalian cells and perlecan in the extracellular matrix. In heparan sulfate, the same basic disaccharide unit listed above is interrupted by N-sulfated, iduronate-containing sequences known as S-regions.

Heparin binding is generally used as a model for binding to cell surfaces through interaction with abundant heparan sulphate and other abundant related GAGs. With the exception of the secreted viral and host complement inhibitors that supposedly utilize GAG binding to orient themselves correctly on activating surfaces, it is presumed that GAG binding by virion-expressed glycoproteins is required to localize virions non-specifically in the vicinity of true receptors and the colocalization of complement regulation to these proteins is unrelated. The length of chains that make up the linear polysaccharide heparin can range in molecular weight from 5000 to 40,000 and can attain an average negative charge of 100 per chain. Naturally occurring endogenous heparin is primarily an intracellular polysaccharide, localized to the granules of mast cells, while heparan sulfate is an extracellular polysaccharide found both in the plasma and on cell surfaces with many roles in maintenance of normal physiological activities (Linhardt 2003).

The increased relative sulfation of heparin usually mediates stronger ligand interactions than heparan sulfate, due to the increased negative charge. The number of modifications to heparan sulphate also varies greatly between different organs and many are specific for certain tissues, cell types and can be influenced by factors such as aging or disease. The pattern, composition and spacing of basic amino acids in heparin binding peptides and proteins have been investigated in detail in many cases. From these studies it is known that heparin (characterized by a high charge density) interacts tightly with peptides displaying a series of at least three to five positively charged residues and that the binding is stronger when the cluster contains arginine residues rather than lysines. Amino acid sequence patterns such as XBBXBX and XBBBXXBX (B denotes basic and X nonbasic residues; Hileman, Fromm, Weiler and Linhardt 1998) have been defined as potential heparin recognition sites but often the heparin binding site is not easily identifiable in linear sequence as the basic residues may be located far apart in the linear sequence, but be topological neighbours in the three-dimensional (3D) structure.

6.1 Heparin Binding to Host Complement Inhibitors

Heparin interacts with several complement proteins, but the affinity of heparin binding to C4BP is the highest among the complement proteins, most likely due to the multimeric nature of C4BP. C4BP binding is also relatively selective: it interacts with heparin but not with chondroitin sulfate or hyaluronan, which are also negatively charged due to abundant sulfoxyl and/or carboxyl groups. Heparin binding has been localized to CCP2 of the C4BP α-chain, with some contribution from CCP1 and CCP3 (Blom, Kask and Dahlback 2001). A patch of positively charged residues, on the interface between CCP1 and CCP2, were found to be crucial to heparin binding. Furthermore, C4BP binding to C4b interaction can be inhibited by heparin, suggesting that the C4b and heparin binding sites overlap. In comparison, three separate heparin-binding sites have been identified in factor H in CCP7, CCP13 and CCP20 (Blackmore, Hellwage, Sadlon, Higgs, Zipfel, Ward and Gordon 1998; Sharma and Pangburn 1996) Studies using a panel of monoclonal antibodies to block factor H and heparin binding led Oppermann et al. (Oppermann, Manuelian, Jozsi, Brandt, Jokiranta, Heinen, Meri, Skerka, Gotze and Zipfel 2006) to postulate that factor H circulates in a compact structure and that initial binding occurs preferentially through the CCP20 heparin-binding site. Upon first contact, the protein then unfolds and exposes the additional binding sites; thus heparin enhances binding of factor H to surface-bound C3b and subsequent inactivation of C3b by fI. This model also helps to explain the unusual association of naturally occurring mutations in CCP20 that are associated with factor H dysfunction and complement activation in patients with hemolytic uremic syndrome (Caprioli, Bettinaglio, Zipfel, Amadei, Daina, Gamba, Skerka, Marziliano, Remuzzi and Noris 2001; Perez-Caballero, Gonzalez-Rubio, Gallardo, Vera, Lopez-Trascasa, Rodriguez de Cordoba and Sanchez-Corral 2001; Richards, Buddles, Donne, Kaplan, Kirk, Venning, Tielemans, Goodship and Goodship 2001; Zipfel 2001). This contrasts with heparin's competition with C4BP complement inhibition, although it could be argued that the heptameric nature of C4BP enables 3 α-chains to mediate attachment while the other 4 α-chains would be left in close proximity to the cell surface to interact with C4b and fI. Thus, interaction of C4BP with heparin-like molecules, present at the surface of some cells, could protect them from the destructive and inflammatory consequences of complement activation. The localization of the heparin binding site at the tip of the C4BP tentacles is consistent with this hypothesis since if such a site was located closer to the central core region of C4BP, it would not be as easily accessible for binding.

6.2 Heparin binding to Viral Complement Inhibitors with CCP Domains

The amino acid sequence of VCP predicts 4 putative heparin binding sites based on conserved surface exposed K/R-X-K/R motifs and a net overall positive charge (Smith, Mullin, Parkinson, Shchelkunov, Totmenin, Loparev, Srisatjaluk, Reynolds, Keeling, Justus, Barlow and Kotwal 2000). However, only two of these binding sites (one at the C-terminal end and the other at K64, R65, R66 in CCP1 combined with the CCP1-2 linker region) had realistic heparin-binding potential based on the solved X-ray crystal structure and modeling based on fibroblast growth factor in complex with heparin (Murthy et al. 2001). Eventually the X-ray crystallographic structure for VCP bound to heparin was solved (Ganesh et al. 2004) and only found the binding site in CCP4 occupied. Binding to heparin did not alter the conformation of CCP1-3, but induced a 29.5 degree rotation of the CCP3-4 hinge region. Whether or not the heparin-induced change in CCP4 position, relative to the rest of VCP, is important to complement regulation remains to be elucidated. Given the difference in efficacy between SPICE and VCP, some point mutation studies to improve VCP's functional activity has been undertaken. Of particular interest, Ghebremariam et al. (Ghebremariam, Odunuga, Janse and Kotwal 2005) found that mutating H98 to Y, E102 to K, and E120 to K increased the potency of VCP for inhibiting the classical and alternative activation pathways 100-fold and speculated that these changes potentially created an additional putative heparin-binding site in CCP2.

The heparin-binding site has also been investigated in KCP (Mark, Lee, Spiller, Villoutreix, and Blom 2006) through site-directed mutagenesis and molecular docking models. Removal of CCP4 from KCP did not reduce the heparin binding affinity. Heparin binding was localized to the positive patch of residues stretching from CCP1 into the CCP1-2 linker region, almost identical in position to the proposed second heparin-binding site in VCP. Mutation of residue clusters R20/R33/R35 or K64/K65/K88 to uncharged residues of similar space-filling proportions resulted in significant to total loss of heparin binding (Mark et al. 2006). The contribution of these residues to heparin binding also varied: K65 played a much greater role in heparin binding than K64 and K88 by KCP. Addition of 0.3 mg/ml heparin to KCP inhibited binding to immobilized C4b, as measured by surface plasmon resonance, and inhibited the cofactor activity for C4b cleavage. However, it is important to note that heparin binding is a model of heparan sulphate binding (which has a lower affinity for most viral proteins) and native C4 and fI have been reported to directly bind heparin with affinities of 60.8 and 36.2 nM, respectively (Yu, Munoz, Edens and Linhardt 2005).

6.3 Heparin binding by HSV Glycoprotein C

Of the 10 virus envelope glycoproteins, only glycoproteins B, D, H and L are essential for in vitro infection of cultured cells, while the other six contribute to virus infectivity and spread in the host. While loss of gC was not essential for HSV-1 infection, the resultant virus was impaired in its binding to cells and slightly impaired in its penetration (Herold, WuDunn, Soltys and Spear 1991; Homa, Purifoy, Glorioso and Levine 1986; Tal-Singer, Peng, Ponce De Leon, Abrams, Banfield, Tufaro, Cohen and Eisenberg 1995). Like KSHV, there is a redundancy in heparin binding of envelope glycoproteins, as gB has also been found to bind heparin (Akula, Pramod, Wang and Chandran 2001; Herold, Visalli, Susmarski, Brandt and Spear 1994; Wang, Akula, Pramod, Zeng and Chandran 2001). However, only gC is capable of binding C3b, inhibiting complement, and removal of gC leads to an attenuation of the virus *in vivo* (Lubinski et al. 1998; Lubinski et al. 1999). HSV-1 gC binds heparin with a K_D of 13 nM, heparan sulfate with an affinity of 100 nM, and C3b with an affinity of 28 nM, and pre-incubation of gC with 100-fold molar excess of heparan sulfate only partially inhibited gC binding to immobilized C3b (Mardberg, Trybala, Glorioso and Bergstrom 2001). Complete inhibition of C3b binding by pre-incubation with 1000-fold excess of heparan sulfate suggests that heparin- and C3b-binding sites overlap in this molecule. Although gC contains 8 conserved cysteine residues, the resulting disulphide bridges are inconsistent with any remote similarity to potential CCP domain formation, (Mardberg et al. 2001) and the structure of gC has yet to be solved. However, there is substantial investigation into the regions gC that make up the primary antigenic epitopes (site I and II), heparin binding sites, and C3b, properdin and C5 binding sites (summarized in Figure 1). The first 26 amino acids constitute the signal sequence, while the most N-terminal portion of the expressed protein (residues 27-133) contains several O-glycosylation sites and the properdin and C5 binding sites (Hung et al. 1994). Site-directed mutagenesis of pairs of arginine residues found that R143, R145, R151, and R155 play a dominant role in heparan sulfate binding, while R135, R139, R147, and R160 also significantly contributed to binding (Mardberg et al. 2001). All of these residues reside within, or are in close proximity to a 16 amino acid loop formed by the first two cysteines (C127 & C144) in gC (Figure 1), and notably non-polar I142 (but not V140) was also found important suggesting that nonionic forces may also contribute to the free energy of heparin binding. Monoclonal antibody B1C1 that blocks both C3b and heparin binding to gC was also mapped to T150 in this region, as well as R143, R145, R147, R151, and R155 (Huemer, Broker, Larcher, Lambris and Dierich 1989; Mardberg et al. 2001; Olofsson, Bolmstedt, Biller, Mardberg, Leckner, Malmstrom, Trybala and Bergstrom 1999; Trybala, Bergstrom, Svennerholm, Jeansson, Glorioso and Olofsson 1994). The C3b binding site was initially investigated by using linker insertion mutation to disrupt sites through incorporation of 4-5 residues at defined sites (Hung, Srinivasan, Friedman, Eisenberg and Cohen 1992). Insertion mutations in the first 123 residues did not inhibit C3b binding (including insertions at residues 116, 122, and 123); however, insertions at residues 130 and 134 (but not 138) within the 16 amino acid loop resulted in the loss of C3b binding. Four other insertion mutations were found to disrupt C3b binding: at residue 237 and in close proximity to the next disulfide bridge (C286 & C347) at residues 288, 353 and 359. Of interest, an insertion at G247 did not disrupt C3b binding, while G247 had been identified as important to heparin binding

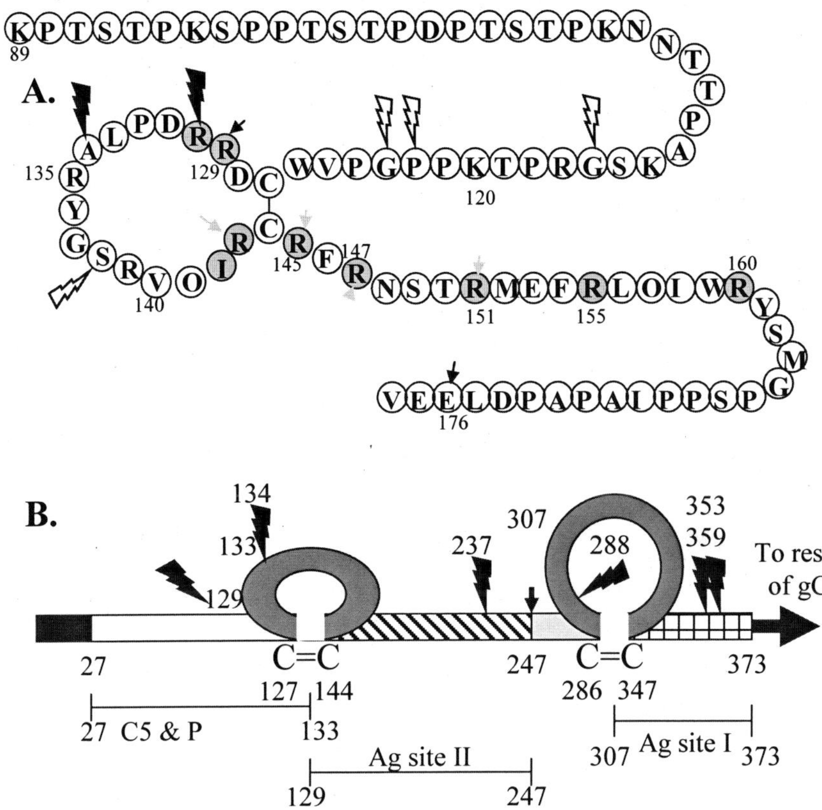

Fig. 1. **A.** Amino acid sequence of HSV gC from residues 89-179 based in part on Mardberg *et al.* (2001). **B.** A larger scale diagram depicting the N-terminal 373 residues (out of 511), showing the first two disulphide bonds and antigenic (Ag) sites I and II. Residues found to be essential to heparin binding are shown as shaded circles in (A). Linker insertion points that resulted in loss of C3b binding are indicated with a black lightning bolt (A and B), while insertions that did not abrogate C3b binding are shown as open lighting bolts (A only). Mutations in antibody resistant strains (mar) that resulted in the loss of C3b binding are shown as black arrows (including G247 in B), while sequence mutations in mar strains that retained C3b binding are shown as grey arrows. The region responsible for binding to C5 and properdin (P) are also shown in (B).

(Hung et al. 1992; Trybala, Bergstrom, Svennerholm, Jeansson, Glorioso and Olofsson 1994). Analysis of monoclonal antibody resistant (mar) HSV-1 strains (both cell-binding and C3b binding), and subsequent identification of their point mutations by DNA sequencing, has also provided information about the location of these sites (Friedman, Glorioso, Cohen, Hastings, Harris and Eisenberg 1986; Trybala et al. 1994; Wu, Levine, Homa, Highlander and Glorioso 1990). Mutations localized to antigenic site I (which partially includes the dicysteine bonds between 286 and 347) were relatively ineffective at blocking C3b binding, with the exception of P373Q (mar C2.1) that may have had significant consequences to the overall protein conformation. In comparison, mutations to antigenic site II (including the 127 to 144 loop) including E176K (mar 3.1), R129C (mar 16.1), and G247E (mar C10.1) inhibited C3b binding, while R143Q (mar C9.6), R145Q (mar13.2), R147W (mar 17.2/7.1) and R151H (mar 17.3) did not play an apparent role in C3b binding, despite their importance in heparin binding. It is also interesting to note the discrepancy in C3b binding to G247 determined by linker insertion mutagenesis and mar strain sequencing (Friedman et al. 1986; Hung et al. Cohen 1992; Wu et al. 1990). These data indicate that the two N-terminal loops are in close proximity to the intervening G247 residue and that while the heparin- and C3b-binding site reside within this region, they do not completely overlap.

6.4 Uncharacterized Heparin Binding by Other RCA Host Proteins

As part of our previous investigations, we found that KCP bound to heparin-sepharose columns and required 181 mM NaCl (greater than physiological concentration; 150 mM) to elute it from the column (Mark et al. 2006). While the soluble host complement regulators (factor H, C4BP, C1inhibitor) have been investigated for heparin binding capacity, the membrane-bound complement regulators have not been examined in detail. Figure 2 shows heparin-sepharose elution profiles (using our previously published conditions; Mark et al. 2006) for recombinant proteins, expressed and purified with immunoglobulin Fc tags (except for TP-10). CCP-containing complement inhibitors tested include human and mouse decay-accelerating factor (DAF; Harris, Spiller and Morgan 2000), soluble human CR1 (TP-10; Rioux 2001), human MCP (courtesy of Claire Harris, Cardiff University, U.K.), and rat Crry (McGrath, Wilkinson, Spiller, and Morgan 1999). Mouse CD59 (courtesy of B. Paul Morgan, Cardiff University, U.K.) was included as a non-CCP membrane-bound complement inhibitor and gave almost identical results to human

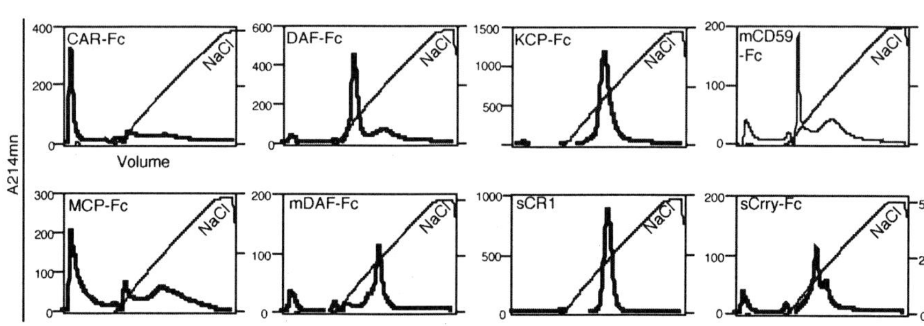

Fig. 2. Elution of complement inhibitors and control protein (expressed as Fc fusion proteins) from heparin-sepharose. NaCl gradient from 0 to 500 mM is shown and eluted protein was detected by absorbance at 214 nm, full details of protocol can be seen for experiments in reference #78 for investigations of KCP binding (shown for comparison). Proteins tested (left to right, top to bottom): Coxsackie-Adenovirus receptor, human DAF, KCP, mouse CD59, human MCP, mouse DAF, human soluble CR1 (TP-10), and rat Crry protein.

CD59 (not shown). Coxsackie-adenovirus receptor (CAR; Goodfellow, Evans, Blom, Kerrigan, Miners, Morgan and Spiller 2005) was included as a non-complement membrane-bound protein. CAR did not bind to the heparin-sepharose column at all, and mouse CD59 bound but was eluted with NaCl concentrations well below physiological levels. Similarly human DAF and rat Crry bound to heparin and were eluted just below the 150 mM NaCl threshold, while mouse DAF eluted just above this threshold. Human MCP bound with a complex profile, similar to some that we have previously seen for KCP containing mutations located within the binding region (K64, K65, K88; Mark et al. 2006), which can probably be interpreted as poor heparin binding.

Therefore, while membrane-bound CCP-containing host complement inhibitors seemed to bind heparin better than CD59 and CAR, they do not bind heparin (and therefore likely cell surface GAG) under physiological conditions, with the exception of sCR1. It is interesting to speculate that heparin binding by sCR1 may enhance its potential when utilized as a therapeutic, and further experiments localizing the heparin-binding site should be pursued.

7 Conclusion

Both soluble and membrane-bound viral complement inhibitors bind to heparin and C3b/C4b. VCP, SPICE, and MOPICE bind heparin with a greater affinity for heparin than factor H, (Liszewski et al. 2006), but SPICE and MOPICE lack the dominant C-terminal heparin-binding site identified for VCP (Mark et al. 2006). This suggests that SPICE and MOPICE, although soluble, may use the binding site located in CCP1 through the CCP1-2 linker region, homologous to that identified for KCP. The ability of CCP-containing viral complement inhibitors, including VCP, SPICE, MOPICE, CCPH, and KCP, to regulate the alternative C3 convertase is very weak to negligible, while their ability to regulate classical C3 convertases (with the exception of MOPICE) rivals many host complement inhibitors. All of these viral proteins also contain varying affinities for C3b and C4b and act as cofactors for the fI-mediated inactivation of these proteins. In sharp contrast, gC of HSV-1 contains no CCP domains, is an excellent inhibitor of the alternative C3 convertase and C5 convertase, and does not appear to have any cofactor activity. While the gC homologues identified in HSV-2 and other alphaherpesviruses retain the strong affinity for C3b and their removal appears to decrease viral ability to evade complement, they do not appear to bind C5 or properdin and, therefore, likely do not have the same alternative C3 convertase decay-accelerating activity. All of these viral proteins have conserved the colocalized binding of heparin and complement inhibition under evolutionary pressure, despite the presence of redundant heparin binding by other viral proteins. There is no question that basic residues central to C3b/C4b binding participate in the electrostatic interactions with heparin in most cases, although mutational analyses also show distinct differences between detailed requirements for these respective binding sites. Host membrane-bound RCA proteins, do not share the heparin-binding affinity, with the exception of CR1, which may suggest that heparin binding is only important to complement inhibitors when the active site is a sub-optimal distance from the cell membrane or their primary role is to inhibit on neighboring cells or as a soluble form (either virion associated or secreted/ protease released). C4BP may represent the best model of this, as virions may be viewed as larger scale versions of multi-chain repeats, and VCP, MOPICE, and SPICE were recently reported to be secreted as disulphide linked dimers (which have increased functional activity relative to monomers) when produced by mammalian cells (Liszewski et al. 2006).

8 Acknowledgements

Anna Blom is supported by Cancerfonden, the Swedish Foundation for Strategic Research (INGVAR) and the Swedish Research Council. Brad Spiller would like to thank the Wellcome Trust and Cancer Research UK for financial support.

References

Akula, S.M., Pramod, N.P., Wang, F.Z., and Chandran, B. (2001) Human herpesvirus 8 envelope-associated glycoprotein B interacts with heparan sulfate-like moieties. Virology *284*, 235-49.

Albrecht, J.C., Nicholas, J., Biller, D., Cameron, K.R., Biesinger, B., Newman, C., Wittmann, S., Craxton, M.A., Coleman, H., and Fleckenstein, B. (1992) Primary structure of the herpesvirus saimiri genome. J Virol *66*, 5047-58.

Arnold, J.N., Dwek, R.A., Rudd, P.M., and Sim, R.B. (2006) Mannan binding lectin and its interaction with immunoglobulins in health and in disease. Immunol Lett *106*, 103-10.

Avrameas, S. (1991) Natural autoantibodies: from 'horror autotoxicus' to 'gnothi seauton'. Immunol Today *12*, 154-9.

Bajtay, Z., Speth, C., Erdei, A., and Dierich, M.P. (2004) Cutting edge: productive HIV-1 infection of dendritic cells via complement receptor type 3 (CR3, CD11b/CD18) J Immunol *173*, 4775-8.

Banki, Z., Stoiber, H., and Dierich, M.P. (2005) HIV and human complement: inefficient virolysis and effective adherence. Immunol Lett *97*, 209-14.

Blackmore, T.K., Hellwage, J., Sadlon, T.A., Higgs, N., Zipfel, P.F., Ward, H.M., and Gordon, D.L. (1998) Identification of the second heparin-binding domain in human complement factor H. J Immunol *160*, 3342-8.

Blom, A.M., Kask, L., and Dahlback, B. (2001) Structural requirements for the complement regulatory activities of C4BP. J Biol Chem *276*, 27136-44.

Caprioli, J., Bettinaglio, P., Zipfel, P.F., Amadei, B., Daina, E., Gamba, S., Skerka, C., Marziliano, N., Remuzzi, G., and Noris, M. (2001) The molecular basis of familial hemolytic uremic syndrome: mutation analysis of factor H gene reveals a hot spot in short consensus repeat 20. J Am Soc Nephrol *12*, 297-307. Notes: CORPORATE NAME: Itaslian Registry of Familial and Recurrent HUS/TTP

Cesarman, E., Chang, Y., Moore, P.S., Said, J.W., and Knowles, D.M. (1995) Kaposi's sarcoma-associated herpesvirus-like DNA sequences in AIDS-related body-cavity-based lymphomas. N Engl J Med *332*, 1186-91.

Chen, Y.T., Wang, Y.H., Cheng, Y.Y., and Hung, S.L. (2003) Direct binding of herpes simplex virus type 1 virions to complement C3. Viral Immunol *16*, 347-55.

Daniels, C.A., Borsos, T., Rapp, H.J., Snyderman, R., and Notkins, A.L. (1970) Neutralization of sensitized virus by purified components of complement. Proc Natl Acad Sci U S A *65*, 528-35.

Eizuru, Y., Ueno, I., and Minamishima, Y. (1988) Evaluation of immunoglobulin G preparations for anti-cytomegalovirus antibodies with reference to neutralizing antibody in the presence of complement. J Clin Microbiol *26*, 1881-3.

Endo, Y., Takahashi, M., and Fujita, T. (2006) Lectin complement system and pattern recognition. Immunobiology *211*, 283-93.

Fodor, W.L., Rollins, S.A., Bianco-Caron, S., Rother, R.P., Guilmette, E.R., Burton, W.V., Albrecht, J.C., Fleckenstein, B., and Squinto, S.P. (1995) The complement control protein homolog of herpesvirus saimiri regulates serum complement by inhibiting C3 convertase activity. J Virol *69*, 3889-92.

Friedman, H.M., Cohen, G.H., Eisenberg, R.J., Seidel, C.A., and Cines, D.B. (1984) Glycoprotein C of herpes simplex virus 1 acts as a receptor for the C3b complement component on infected cells. Nature *309*, 633-5.

Friedman, H.M., Glorioso, J.C., Cohen, G.H., Hastings, J.C., Harris, S.L., and Eisenberg, R.J. (1986) Binding of complement component C3b to glycoprotein gC of herpes simplex virus type 1: mapping of gC-binding sites and demonstration of conserved C3b binding in low-passage clinical isolates. J Virol *60*, 470-5.

Friedman, H.M., Wang, L., Fishman, N.O., Lambris, J.D., Eisenberg, R.J., Cohen, G.H., and Lubinski, J. (1996) Immune evasion properties of herpes simplex virus type 1 glycoprotein gC. J Virol *70*, 4253-60.

Fries, L.F., Friedman, H.M., Cohen, G.H., Eisenberg, R.J., Hammer, C.H., and Frank, M.M. (1986) Glycoprotein C of herpes simplex virus 1 is an inhibitor of the complement cascade. J Immunol *137*, 1636-41.

Galili, U., Mandrell, R.E., Hamadeh, R.M., Shohet, S.B., and Griffiss, J.M. (1988a) Interaction between human natural anti-alpha-galactosyl immunoglobulin G and bacteria of the human flora. Infect Immun *56*, 1730-7.

Galili, U., Shohet, S.B., Kobrin, E., Stults, C.L., and Macher, B.A. (1988b) Man, apes, and Old World monkeys differ from other mammals in the expression of alpha-galactosyl epitopes on nucleated cells. J Biol Chem *263*, 17755-62.

Ganesh, V.K., Smith, S.A., Kotwal, G.J., and Murthy, K.H. (2004) Structure of vaccinia complement protein in complex with heparin and potential implications for complement regulation. Proc Natl Acad Sci U S A *101*, 8924-9.

Gerber, S.I., Belval, B.J., and Herold, B.C. (1995) Differences in the role of glycoprotein C of HSV-1 and HSV-2 in viral binding may contribute to serotype differences in cell tropism. Virology *214*, 29-39.

Ghebremariam, Y.T., Odunuga, O.O., Janse, K., and Kotwal, G.J. (2005) Humanized Recombinant Vaccinia Virus Complement Control Protein (hrVCP) with Three Amino Acid Changes, H98Y, E102K, and E120K Creating an Additional Putative Heparin Binding Site, Is 100-fold More Active Than rVCP in Blocking Both Classical and Alternative Complement Pathways. Ann N Y Acad Sci *1056*, 113-22.

Goodfellow, I.G., Evans, D.J., Blom, A.M., Kerrigan, D., Miners, J.S., Morgan, B.P., and Spiller, O.B. (2005) Inhibition of coxsackie B virus infection by soluble forms of its receptors: binding affinities, altered particle formation, and competition with cellular receptors. J Virol *79*, 12016-24.

Green, T.D., Montefiori, D.C., and Ross, T.M. (2003) Enhancement of antibodies to the human immunodeficiency virus type 1 envelope by using the molecular adjuvant C3d. J Virol *77*, 2046-55.

Harris, C.L., Spiller, O.B., and Morgan, B.P. (2000) Human and rodent decay-accelerating factors (CD55) are not species restricted in their complement-inhibiting activities. Immunology *100*, 462-70.

Harris, S.L., Frank, I., Yee, A., Cohen, G.H., Eisenberg, R.J., and Friedman, H.M. (1990) Glycoprotein C of herpes simplex virus type 1 prevents complement-mediated cell lysis and virus neutralization. J Infect Dis *162*, 331-7.

Herold, B.C., Visalli, R.J., Susmarski, N., Brandt, C.R., and Spear, P.G. (1994) Glycoprotein C-independent binding of herpes simplex virus to cells requires cell surface heparan sulphate and glycoprotein B. J Gen Virol *75 (Pt 6)*, 1211-22.

Herold, B.C., WuDunn, D., Soltys, N., and Spear, P.G. (1991) Glycoprotein C of herpes simplex virus type 1 plays a principal role in the adsorption of virus to cells and in infectivity. J Virol *65*, 1090-8.

Herzenberg, L.A. and Kantor, A.B. (1993) B-cell lineages exist in the mouse. Immunol Today *14*, 79-83; discussion 88-90.

Hileman, R.E., Fromm, J.R., Weiler, J.M., and Linhardt, R.J. (1998) Glycosaminoglycan-protein interactions: definition of consensus sites in glycosaminoglycan binding proteins. Bioessays *20*, 156-67.

Homa, F.L., Purifoy, D.J., Glorioso, J.C., and Levine, M. (1986) Molecular basis of the glycoprotein C-negative phenotypes of herpes simplex virus type 1 mutants selected with a virus-neutralizing monoclonal antibody. J Virol *58*, 281-9.

Hook, L.M., Lubinski, J.M., Jiang, M., Pangburn, M.K., and Friedman, H.M. (2006) Herpes simplex virus type 1 and 2 glycoprotein C prevents complement-mediated neutralization induced by natural immunoglobulin M antibody. J Virol *80*, 4038-46.
Huemer, H.P., Broker, M., Larcher, C., Lambris, J.D., and Dierich, M.P. (1989) The central segment of herpes simplex virus type 1 glycoprotein C (gC) is not involved in C3b binding: demonstration by using monoclonal antibodies and recombinant gC expressed in Escherichia coli. J Gen Virol *70 (Pt 6)*, 1571-8.
Huemer, H.P., Larcher, C., and Coe, N.E. (1992) Pseudorabies virus glycoprotein III derived from virions and infected cells binds to the third component of complement. Virus Res *23*, 271-80.
Huemer, H.P., Nowotny, N., Crabb, B.S., Meyer, H., and Hubert, P.H. (1995) gp13 (EHV-gC): a complement receptor induced by equine herpesviruses. Virus Res *37*, 113-26.
Huemer, H.P., Wang, Y., Garred, P., Koistinen, V., and Oppermann, S. (1993) Herpes simplex virus glycoprotein C: molecular mimicry of complement regulatory proteins by a viral protein. Immunology *79*, 639-47.
Huemer, H.P., Wechselberger, C., Bennett, A.M., Falke, D., and Harrington, L. (2003) Cloning and expression of the complement receptor glycoprotein C from Herpesvirus simiae (herpes B virus): protection from complement-mediated cell lysis. J Gen Virol *84*, 1091-100.
Hung, S.L., Peng, C., Kostavasili, I., Friedman, H.M., Lambris, J.D., Eisenberg, R.J., and Cohen, G.H. (1994) The interaction of glycoprotein C of herpes simplex virus types 1 and 2 with the alternative complement pathway. Virology *203*, 299-312.
Hung, S.L., Srinivasan, S., Friedman, H.M., Eisenberg, R.J., and Cohen, G.H. (1992) Structural basis of C3b binding by glycoprotein C of herpes simplex virus. J Virol *66*, 4013-27.
Isaacs, S.N., Kotwal, G.J., and Moss, B. (1992) Vaccinia virus complement-control protein prevents antibody-dependent complement-enhanced neutralization of infectivity and contributes to virulence. Proc Natl Acad Sci U S A *89*, 628-32.
Kapadia, S.B., Levine, B., Speck, S.H., and Virgin, H.W. 4th (2002) Critical role of complement and viral evasion of complement in acute, persistent, and latent gamma-herpesvirus infection. Immunity *17*, 143-55.
Kapadia, S.B., Molina, H., van Berkel, V., Speck, S.H., and Virgin, H.W. 4th (1999) Murine gammaherpesvirus 68 encodes a functional regulator of complement activation. J Virol *73*, 7658-70.
Kostavasili, I., Sahu, A., Friedman, H.M., Eisenberg, R.J., Cohen, G.H., and Lambris, J.D. (1997) Mechanism of complement inactivation by glycoprotein C of herpes simplex virus. J Immunol *158*, 1763-71.
Kotwal, G.J., Isaacs, S.N., McKenzie, R., Frank, M.M., and Moss, B. (1990) Inhibition of the complement cascade by the major secretory protein of vaccinia virus. Science *250*, 827-30.
Kotwal, G.J. and Moss, B. (1988) Vaccinia virus encodes a secretory polypeptide structurally related to complement control proteins. Nature *335*, 176-8.
Lewis, R.B., Matzke, D.S., Albrecht, T.B., and Pollard, R.B. (1986) Assessment of the presence of cytomegalovirus-neutralizing antibody by a plaque-reduction assay. Rev Infect Dis *8 Suppl 4*, S434-8.
Linhardt, R.J. (2003) 2003 Claude S. Hudson Award address in carbohydrate chemistry. Heparin: structure and activity. J Med Chem *46*, 2551-64.
Liszewski, M.K., Leung, M.K., Hauhart, R., Buller, R.M., Bertram, P., Wang, X., Rosengard, A.M., Kotwal, G.J., and Atkinson, J.P. (2006) Structure and regulatory profile of the monkeypox inhibitor of complement: comparison to homologs in vaccinia and variola and evidence for dimer formation. J Immunol *176*, 3725-34.
Lubinski, J., Wang, L., Mastellos, D., Sahu, A., Lambris, J.D., and Friedman, H.M. (1999) In vivo role of complement-interacting domains of herpes simplex virus type 1 glycoprotein gC. J Exp Med *190*, 1637-46.
Lubinski, J.M., Wang, L., Soulika, A.M., Burger, R., Wetsel, R.A., Colten, H., Cohen, G.H., Eisenberg, R.J., Lambris, J.D., and Friedman, H.M. (1998) Herpes simplex virus type 1 glycoprotein gC mediates immune evasion in vivo. J Virol *72*, 8257-63.
Maeda, K., Hayashi, S., Tanioka, Y., Matsumoto, Y., and Otsuka, H. (2002) Pseudorabies virus (PRV) is protected from complement attack by cellular factors and glycoprotein C (gC) Virus Res *84*, 79-87.
Mardberg, K., Trybala, E., Glorioso, J.C., and Bergstrom, T. (2001) Mutational analysis of the major heparan sulfate-binding domain of herpes simplex virus type 1 glycoprotein C. J Gen Virol *82*, 1941-50.

Mark, L., Lee, W.H., Spiller, O.B., Villoutreix, B.O., and Blom, A.M. (2006) The Kaposi's sarcoma-associated herpesvirus complement control protein (KCP) binds to heparin and cell surfaces via positively charged amino acids in CCP1-2. Mol Immunol *43*, 1665-75.

McGrath, Y., Wilkinson, G.W., Spiller, O.B., and Morgan, B.P. (1999) Development of adenovirus vectors encoding rat complement regulators for use in therapy in rodent models of inflammatory diseases. J Immunol *163*, 6834-40.

McKenzie, R., Kotwal, G.J., Moss, B., Hammer, C.H., and Frank, M.M. (1992) Regulation of complement activity by vaccinia virus complement-control protein. J Infect Dis *166*, 1245-50.

McNearney, T.A., Odell, C., Holers, V.M., Spear, P.G., and Atkinson, J.P. (1987) Herpes simplex virus glycoproteins gC-1 and gC-2 bind to the third component of complement and provide protection against complement-mediated neutralization of viral infectivity. J Exp Med *166*, 1525-35.

Morgan, B.P., Marchbank, K.J., Longhi, M.P., Harris, C.L., and Gallimore, A.M. (2005) Complement: central to innate immunity and bridging to adaptive responses. Immunol Lett *97*, 171-9.

Mullick, J., Bernet, J., Singh, A.K., Lambris, J.D., and Sahu, A. (2003). Kaposi's sarcoma-associated herpesvirus (human herpesvirus 8) open reading frame 4 protein (kaposica) is a functional homolog of complement control proteins. J Virol *77*, 3878-81.

Mullick, J., Singh, A.K., Panse, Y., Yadav, V., Bernet, J., and Sahu, A. (2005). Identification of functional domains in kaposica, the complement control protein homolog of Kaposi's sarcoma-associated herpesvirus (human herpesvirus 8). J Virol *79*, 5850-6.

Murthy, K.H., Smith, S.A., Ganesh, V.K., Judge, K.W., Mullin, N., Barlow, P.N., Ogata, C.M., and Kotwal, G.J. (2001) Crystal structure of a complement control protein that regulates both pathways of complement activation and binds heparan sulfate proteoglycans. Cell *104*, 301-11.

Olofsson, S., Bolmstedt, A., Biller, M., Mardberg, K., Leckner, J., Malmstrom, B.G., Trybala, E., and Bergstrom, T. (1999) The role of a single N-linked glycosylation site for a functional epitope of herpes simplex virus type 1 envelope glycoprotein gC. Glycobiology *9*, 73-81.

Oppermann, M., Manuelian, T., Jozsi, M., Brandt, E., Jokiranta, T.S., Heinen, S., Meri, S., Skerka, C., Gotze, O., and Zipfel, P.F. (2006) The C-terminus of complement regulator Factor H mediates target recognition: evidence for a compact conformation of the native protein. Clin Exp Immunol *144*, 342-52.

Perez-Caballero, D., Gonzalez-Rubio, C., Gallardo, M.E., Vera, M., Lopez-Trascasa, M., Rodriguez de Cordoba, S., and Sanchez-Corral, P. (2001) Clustering of missense mutations in the C-terminal region of factor H in atypical hemolytic uremic syndrome. Am J Hum Genet *68*, 478-84.

Petitou, M., Casu, B., and Lindahl, U. (2003) 1976-1983, a critical period in the history of heparin: the discovery of the antithrombin binding site. Biochimie *85*, 83-9.

Pinter, C., Beltrami, S., Stoiber, H., Negri, D.R., Titti, F., and Clivio, A. (2000) Interference with complement regulatory molecules as a possible therapeutic strategy in HIV infection. Expert Opin Investig Drugs *9*, 199-205.

Richards, A., Buddles, M.R., Donne, R.L., Kaplan, B.S., Kirk, E., Venning, M.C., Tielemans, C.L., Goodship, J.A., and Goodship, T.H. (2001) Factor H mutations in hemolytic uremic syndrome cluster in exons 18-20, a domain important for host cell recognition. Am J Hum Genet *68*, 485-90.

Rioux, P. (2001) TP-10 (AVANT Immunotherapeutics) Curr Opin Investig Drugs *2*, 364-71.

Rooijakkers, S.H. and van Strijp, J.A. (2007) Bacterial complement evasion. Mol Immunol *44*, 23-32.

Rosengard, A.M., Liu, Y., Nie, Z., and Jimenez, R. (2002) Variola virus immune evasion design: expression of a highly efficient inhibitor of human complement. Proc Natl Acad Sci U S A *99*, 8808-13.

Rundell, B.B. and Betts, R.F. (1982) Neutralization and sensitization of cytomegalovirus by IgG antibody, anti-IgG antibody, and complement. J Med Virol *10*, 109-18.

Russo, J.J., Bohenzky, R.A., Chien, M.C., Chen, J., Yan, M., Maddalena, D., Parry, J.P., Peruzzi, D., Edelman, I.S., Chang, Y., and Moore, P.S. (1996) Nucleotide sequence of the Kaposi sarcoma-associated herpesvirus (HHV8) Proc Natl Acad Sci U S A *93*, 14862-7.

Saevarsdottir, S., Vikingsdottir, T., and Valdimarsson, H. (2004) The potential role of mannan-binding lectin in the clearance of self-components including immune complexes. Scand J Immunol *60*, 23-9.

Sahu, A., Isaacs, S.N., Soulika, A.M., and Lambris, J.D. (1998) Interaction of vaccinia virus complement control protein with human complement proteins: factor I-mediated degradation of C3b to iC3b1 inactivates the alternative complement pathway. J Immunol *160*, 5596-604.

Seidel-Dugan, C., Ponce de Leon, M., Friedman, H.M., Eisenberg, R.J., and Cohen, G.H. (1990) Identification of C3b-binding regions on herpes simplex virus type 2 glycoprotein C. J Virol *64*, 1897-906.

Seidel-Dugan, C., Ponce de Leon, M., Friedman, H.M., Fries, L.F., Frank, M.M., Cohen, G.H., and Eisenberg, R.J. (1988) C3b receptor activity on transfected cells expressing glycoprotein C of herpes simplex virus types 1 and 2. J Virol *62*, 4027-36.

Sharma, A.K. and Pangburn, M.K. (1996) Identification of three physically and functionally distinct binding sites for C3b in human complement factor H by deletion mutagenesis. Proc Natl Acad Sci U S A *93*, 10996-1001.

Singh, A.K., Mullick, J., Bernet, J., and Sahu, A. (2006) Functional characterization of the complement control protein homolog of herpesvirus saimiri: ARG-118 is critical for factor I cofactor activities. J Biol Chem *281*, 23119-28.

Smith, S.A., Mullin, N.P., Parkinson, J., Shchelkunov, S.N., Totmenin, A.V., Loparev, V.N., Srisatjaluk, R., Reynolds, D.N., Keeling, K.L., Justus, D.E., Barlow, P.N., and Kotwal, G.J. (2000) Conserved surface-exposed K/R-X-K/R motifs and net positive charge on poxvirus complement control proteins serve as putative heparin binding sites and contribute to inhibition of molecular interactions with human endothelial cells: a novel mechanism for evasion of host defense. J Virol *74*, 5659-66.

Soulier, J., Grollet, L., Oksenhendler, E., Cacoub, P., Cazals-Hatem, D., Babinet, P., d'Agay, M.F., Clauvel, J.P., Raphael, M., and Degos, L. (1995). Kaposi's sarcoma-associated herpesvirus-like DNA sequences in multicentric Castleman's disease. Blood *86*, 1276-80.

Spear, G.T., Lurain, N.S., Parker, C.J., Ghassemi, M., Payne, G.H., and Saifuddin, M. (1995) Host cell-derived complement control proteins CD55 and CD59 are incorporated into the virions of two unrelated enveloped viruses. Human T cell leukemia/lymphoma virus type I (HTLV-I) and human cytomegalovirus (HCMV) J Immunol *155*, 4376-81.

Spiller, O.B., Blackbourn, D.J., Mark, L., Proctor, D.G., and Blom, A.M. (2003) Functional activity of the complement regulator encoded by Kaposi's sarcoma-associated herpesvirus. J Biol Chem *278*, 9283-9.

Spiller, O.B., Hanna, S.M., Devine, D.V., and Tufaro, F. (1997) Neutralization of cytomegalovirus virions: the role of complement. J Infect Dis *176*, 339-47.

Spiller, O.B., Mark, L., Blue, C.E., Proctor, D.G., Aitken, J.A., Blom, A.M., and Blackbourn, D.J. (2006) Dissecting the regions of virion-associated Kaposi's sarcoma-associated herpesvirus complement control protein required for complement regulation and cell binding. J Virol *80*, 4068-78.

Spiller, O.B. and Morgan, B.P. (1998) Antibody-independent activation of the classical complement pathway by cytomegalovirus-infected fibroblasts. J Infect Dis *178*, 1597-603.

Spiller, O.B., Morgan, B.P., Tufaro, F., and Devine, D.V. (1996) Altered expression of host-encoded complement regulators on human cytomegalovirus-infected cells. Eur J Immunol *26*, 1532-8.

Spiller, O.B., Robinson, M., O'Donnell, E., Milligan, S., Morgan, B.P., Davison, A.J., and Blackbourn, D.J. (2003) Complement regulation by Kaposi's sarcoma-associated herpesvirus ORF4 protein. J Virol *77*, 592-9.

Tal-Singer, R., Peng, C., Ponce De Leon, M., Abrams, W.R., Banfield, B.W., Tufaro, F., Cohen, G.H., and Eisenberg, R.J. (1995) Interaction of herpes simplex virus glycoprotein gC with mammalian cell surface molecules. J Virol *69*, 4471-83.

Tal-Singer, R., Seidel-Dugan, C., Fries, L., Huemer, H.P., Eisenberg, R.J., Cohen, G.H., and Friedman, H.M. (1991) Herpes simplex virus glycoprotein C is a receptor for complement component iC3b. J Infect Dis *164*, 750-3.

Thieblemont, N., Haeffner-Cavaillon, N., Ledur, A., L'Age-Stehr, J., Ziegler-Heitbrock, H.W., and Kazatchkine, M.D. (1993) CR1 (CD35) and CR3 (CD11b/CD18) mediate infection of human monocytes and monocytic cell lines with complement-opsonized HIV independently of CD4. Clin Exp Immunol *92*, 106-13.

Trybala, E., Bergstrom, T., Svennerholm, B., Jeansson, S., Glorioso, J.C., and Olofsson, S. (1994) Localization of a functional site on herpes simplex virus type 1 glycoprotein C involved in binding to cell surface heparan sulphate. J Gen Virol *75 (Pt 4)*, 743-52.

Vanderplasschen, A., Mathew, E., Hollinshead, M., Sim, R.B., and Smith, G.L. (1998) Extracellular enveloped vaccinia virus is resistant to complement because of incorporation of host complement control proteins into its envelope. Proc Natl Acad Sci U S A *95*, 7544-9.

Walport, M.J. (2001) Complement. First of two parts. N Engl J Med *344*, 1058-66.

Wang, F.Z., Akula, S.M., Pramod, N.P., Zeng, L., and Chandran, B. (2001) Human herpesvirus 8 envelope glycoprotein K8.1A interaction with the target cells involves heparan sulfate. J Virol *75*, 7517-27.

Watanabe, I., Ross, T.M., Tamura, S., Ichinohe, T., Ito, S., Takahashi, H., Sawa, H., Chiba, J., Kurata, T., Sata, T., and Hasegawa, H. (2003) Protection against influenza virus infection by intranasal administration of C3d-fused hemagglutinin. Vaccine *21*, 4532-8.

Wu, C.T., Levine, M., Homa, F., Highlander, S.L., and Glorioso, J.C. (1990) Characterization of the antigenic structure of herpes simplex virus type 1 glycoprotein C through DNA sequence analysis of monoclonal antibody-resistant mutants. J Virol *64*, 856-63.

Yu, H., Munoz, E.M., Edens, R.E., and Linhardt, R.J. (2005) Kinetic studies on the interactions of heparin and complement proteins using surface plasmon resonance. Biochim Biophys Acta *1726*, 168-76.

Zipfel, P.F. (2001) Complement factor H: physiology and pathophysiology. Semin Thromb Hemost *27*, 191-9.

10

C3a Receptors Signaling in Mast Cells

Asifa K. Zaidi[1] and Hydar Ali[2]

[1,2] University of Pennsylvania School of Dental Medicine, Department of Pathology, Philadelphia, PA, 19104 [1]asifaz@path.dental.upenn.edu; [2]ali@path.dental.upenn.edu

1 Introduction

Mast cells are multifunctional immune cells that play a sentinel role in innate immunity (Echtenacher, Mannel, and Hultner 1996; Marshall and Jawda, 2004; Supajatura, Ushio, Nakao, Okumura, and Ogawa 2001) but also mediate a variety of inflammatory diseases including cardiac anaphylaxis (Bani, Nistri, Mannaioni, and Masini 2006; Marone, de Crescenzo, Adt, Patella, Arbustini, and Genovese 1995) asthma (Williams and Galli 2000; Yu, Tsai, Tam, Jones, Zehnder, and Galli 2006) and rheumatoid arthritis (Maruotti, Crivellato, Cantatore, Vacca, and Ribatti 2006). Of the inflammatory diseases in which mast cells participate, their role in asthma has been studied in most detail. Asthma is a complex inflammatory disease characterized by bronchoconstriction, airway hyperresponsiveness (AHR) and inflammation. Approximately 35 million Americans suffer from asthma, and in recent years, its prevalence and severity have been increasing dramatically world-wide. It is generally accepted that the disease arises because of inappropriate immunologic responses to common environmental antigens in genetically susceptible individuals. Following antigen presentation, $CD4^+$ T cells produce T_H2 cytokines which induce B cells to synthesize IgE molecules. These IgE molecules then bind to their high affinity receptors (FcεRI) on the surface of mast cells. Subsequent cross-linking of FcεRI on mast cells by allergen results in degranulation, leukotriene generation and cytokine synthesis (Choi, Kim, Combs, Frohman, and Beaven 2002; Gonzalez-Espinosa, Odom, Olivera, Hobson, Martinez, Oliveira-Dos-Santos, Barra, Spiegel, Penninger, and Rivera 2003; Hundley, Prasad, and Beaven 2001; Sayama, Diehn, Matsuda, Lunderius, Tsai, Tam, Botstein, Brown and Galli 2002; Tkaczyk , Beaven, Brachman, Metcalfe, and Gilfillan 2003), which cause increased vascular permeability, recruit inflammatory cells to the airway and promote smooth muscle contraction (Brightling,

Bradding, Symon, Holgate, Wardlaw, and Pavord 2002; Brightling, Symon, Holgate, Wardlaw, Pavord, and Bradding 2003; Carroll, Mutavdzic, and James 2002; Panettieri 2003).

The complement system forms an important part of innate immunity against bacteria and other pathogens. As a system of 'pattern recognition molecules', foreign surface antigens and immune complexes initiate a proteolytic pathway leading to the formation of a lytic membrane attack complex. The anaphylatoxins C3a and C5a are generated as byproducts of complement activation, and interact with cell surface G protein coupled receptors (GPCRs) in target cells to mediate a variety of inflammatory responses (Bao, Osawe, Haas, and Quigg 2005; Boos, Szalai, and Barnum 2005; Kildsgaard, Hollmann, Matthews, Bian, Murad, and Wetsel 2000). Recent studies have shown that C3a levels are elevated in bronchoalveolar lavage (BAL) fluid after segmental allergen challenge in asthmatic but not healthy subjects (Castro, Schmitz-Schumann, Rother, and Kirschfink 1991; Humbles, Lu, Nilsson, Lilly, Israel, Fujiwara, Gerard, and Gerard 2000; Nakano, Morita, Kawamoto, Suda, Chida, and Nakamura 2003). Furthermore, plasma C3a level is elevated in acute exacerbations of asthma (Nakano et al. 2003) and C3a receptors are upregulated in subjects who died with asthma compared with subjects who died from other causes (Fregonese, Swan, Van Schadewijk, Dolhnikoff, Santos, Daha, Stolk, Tschernig, Sterk, Hiemstra, Rabe, and Mauad 2005). Additionally, single nucleotide polymorphism in C3 or C3a receptor (C3aR) gene increases susceptibility to asthma (Hasegawa, Tamari, Shao, Shimizu, Takahashi, Mao, Yamasaki, Kamada, Doi, Fujiwara, Miyatake, Fujita, Tamura, Matsubara, Shirakawa, and Suzuki 2004). In animal models, complement activation modulates both AHR and airway inflammation (Drouin, Corry, Kildsgaard, and Wetsel 2001; Taube, Rha, Takeda, Park, Joetham, Balhorn, Dakhama, Giclas, Holers, and Gelfand 2003). Furthermore, deletion of C3aR gene or administration of C3aR inhibitors attenuates both AHR and lung inflammation (Baelder, Fuchs, Bautsch, Zwirner, Kohl, Hoymann, Glaab, Erpenbeck, Krug, and Braun 2005; Bautsch, Hoymann, Zhang, Meier-Wiedenbach, Raschke, Ames, Sohns, Flemme, Meyer Zu Vilsendorf, Grove, Klos, and Kohl 2000; Drouin, Corry, Hollman, Kildsgaard, and Wetsel 2002; Humbles et al. 2000; Wills-Karp and Koehl 2005). Collectively, these findings demonstrate an important role for C3aR in the pathogenesis of asthma.

2 C3a Generation and Its Effect on Allergen Sensitization

There is evidence to suggest that a combination of different pathways generates C3a in the airway of individuals with asthma (Ali and Panettieri 2005). It is likely that antibody generated during sensitization interacts with allergen to activate the classical complement pathway. Additionally, airway epithelial cells and pulmonary macrophages secrete both C3 and several components of the alternate pathway of complement (factors B, H, and I and properdin) (Strunk, Eidlen, and Mason 1988; Vandermeer, Sha, Lane, and Schleimer 2004; Varsano, Kaminsky, Kaiser, and Rashkovsky 2000). Thus, activation of alternative or the lectin pathway by allergen may also lead to the generation of C3a. It is noteworthy that house dust mite protease and mast cell tryptase also directly activate the complement pathway (Haczku, Atochina, Tomer, Chen, Scanlon, Russo, Xu, Panettieri, and Beers 2001; Nakano et al. 2003; Schuh et al. 2002).

C3a receptors are expressed in both antigen-presenting cells (APCs) and activated T cells indicating that C3a may promote asthma by inducing T_H2 cytokine production and IgE synthesis (Gutzmer, Lisewski, Zwirner, Mommert, Diesel, Wittmann, Kapp, and Werfel, 2004; Kirchhoff, Weinmann, Zwirner, Begemann, Gotze, Kapp, and Werfel 2001; Soruri, Kiafard, Dettmer, Riggert, Kohl, and Zwirner 2003; Werfel, Kirchhoff, Wittmann, Begemann, Kapp, Heidenreich, Gotze, and Zwirner 2000). Drouin et al., (Drouin et al. 2002) recently showed that following allergen challenge and sensitization, antigen-specific IgE level is partially inhibited in $C3aR^{-/-}$ mice when compared to wild-type mice (Drouin et al. 2002). Although the effect of C3a in T_H2 response in the context of asthma has not been investigated in detail, Kawamoto et al (Kawamoto, Yalcindag, Laouini, Brodeur, Bryce, Lu, Humbles, Oettgen, Gerard, and Geha 2004), showed that C3a downregulates this response to epicutaneously introduced antigen in a model of skin allergy. Taube et al., (Taube et al. 2003) showed that administration of complement inhibitor in mice after sensitization but before challenge prevented the development of AHR and blocked T_H2 cytokine production as well as lung inflammation. Additionally, a small molecule antagonist of C3aR, when administered after sensitization but before challenge also significantly inhibited airway inflammation (Baelder et al. 2005). These findings suggest that although C3aR is expressed on APCs and T cells, its effect in the development of allergic asthma likely involves modification of components distal to allergen sensitization (Baelder et al. 2005; Taube et al. 2003).

2.1 Role of Mast Cells and C3aR in the Effector Phase of Asthma

Mast cells are important effector cells that orchestrate the development of AHR and inflammation (Brightling and Bradding 2005; Page, Ammit, Black, and Armour 2001; Robinson 2004; Taube, Wei, Swasey, Joetham, Zarini, Lively, Takeda, Loader, Miyahara, Kodama, Shultz, Donaldson, Hamelmann, Dakhama, and Gelfand 2004; Thangam, Venkatesha, Zaidi, Jordan-Sciutto, Goncharov, Krymskaya, Amrani, Panettieri, and Ali H 2005). In lungs of asthmatic individuals, mast cells are found in different compartments including bronchoalveolar space beneath the basement membrane, adjacent to blood vessels and scattered throughout the ASM bundles (Casolaro, Galeone, Giacummo, Sanduzzi, Melillo, and Marone 1989; Marone de Crescenzo, Adt, Patella, Arbustini, and Genovese 2005). Two subtypes of human mast cells were initially recognized based on the composition of their secretory granules. Thus, mast cell granules that contain both tryptase and chymase are designated MC_{TC} whereas those containing only tryptase are known as MC_T (Irani, Schechter, Craig, DeBlois, and Schwartz 1986). Interestingly, MC_T cells predominate in the alveolar wall and the epithelium of the lung whereas MC_{TC} cells favor bronchial smooth muscle and glandular regions (Oskeritzian, Zhao, Min, Xia, Pozez, Kiev, and Schwartz 2005). Furthermore, MC_T cell number in the respiratory epithelium increases during pollen season (Gibson, Allen, Yang, Wong, Dolovich, Denburg, and Hargreave 1993; Juliusson, Pipkorn, Karlsson, and Enerback 1992) and markedly elevated levels of MC_{TC} cells are found in bronchial smooth muscle cells of patients with asthma (Brightling et al. 2002). These findings suggest that different mast cell types may play distinct roles in the pathogenesis of asthma.

Previous studies indicated that while C3a and C5a induce mediator release in human skin mast cells, lung mast cells are unresponsive to these anaphylatoxins (el-Lati, Dahinden, and Church 1994; Fureder, Agis, Willheim, Bankl, Maier, Kishi, Muller, Czerwenka, Radaszkiewicz, Butterfield Klappacher, Sperr, Oppermann, Lechner, and Valent 1995; Schulman, Post, Henson, and Giclas 1988). Based on these findings, it was proposed that human lung mast cells do not express C3aR or C5aR. One possible reason for this discrepancy might reflect the fact that while MC_{TC} cells are the predominant cell type present in the skin they are the minority cell type found in the lung (Irani, Bradford, Kepley, Schechter, and Schwartz 1989; Xia, Kepley, Sakai, Chelliah, Irani, and Schwartz 1995). Indeed, Oskeritzian et al, (Oskeritzian et al. 2005) recently showed that MC_T cells in the lung do not express C5aR whereas MC_{TC} cells do and that this is correlated with substantial C5a-induced degranulation and LTC_4 generation in MC_{TC} cells. Although the effects of C3a on human lung MC_{TC} cells are unknown, C3aR are expressed in highly differentiated $CD34^+$-derived primary human mast cells and a newly characterized human mast cell line, LAD2 (Venkatesha, Thangam, Zaidi, and Ali 2005b; Woolhiser, Brockow, and Metcalfe 2004). Furthermore, C3a causes sustained Ca^{2+} mobilization, degranulation and chemokine production in these cells (Venkatesha et al. 2005b). In addition, we have shown that cell-cell contact between airway smooth muscle (ASM) cells and MC_{TC} cells enhance C3a-induced mast cell mediator release (Thangam et al. 2005). Elevated levels of MC_{TC} cells are found in the bronchial smooth muscle of patients with asthma and this correlates with their bronchial hyperreactivity (Brightling and Bradding 2005; Brightling et al. 2002). Furthermore, chemokines generated by ASM cells serve to recruit MC_{TC} cells into bronchial smooth muscle layers. However, unlike C3a and C5a, chemokines do not induce mast cell degranulation (Brightling, Ammit, Kaur, Black, Wardlaw, Hughes, and Bradding 2005a; Brightling, Kaur, Berger, Morgan, Wardlaw, and Bradding 2005b). These findings suggest that ASM-derived chemokines and complement components coordinately regulate mast cell recruitment and mediator release to modulate both AHR and inflammation.

3 C3aR Signaling in Mast Cells

C3aR belongs to a family of seven transmembrane domain GPCRs that couple to Gαi family of heterotrimeric G proteins. In resting condition, G proteins exist as heterotrimeric complexes consisting αβγ complex with GDP bound to the α-subunit (Gα). Receptor activation leads to a conformational change in Gα, resulting in an exchange of GTP for GDP. This interaction causes the dissociation of βγ subunit (Gβγ) from the heterotrimeric complex. Gβγ, of which there are many subtypes, plays essential roles in mediating diverse functions of GPCRs. C3a is one of most potent mast cell chemoattractants known (Hartmann, Henz, Kruger-Krasagakes, Kohl, Burger, Guhl, Haase, Lippert, and Zuberbier 1997; Nilsson, Johnell, Hammer, Tiffany,

Nilsson, Metcalfe, Siegbahn, and Murphy 1996). The ability of C3a to induce mast cell chemotaxis requires the presence of extracellular matrix protein laminin. Furthermore, pretreatment of mast cells with pertussis toxin inhibits C3a-mediated mast cell migration, indicating that Gi proteins are involved in this process.

In leukocytes, the ability of chemoattractants to induce mediator release involve the activation of distinct signaling pathways including phospholipase Cβ (PLCβ)-mediated Ca^{2+} mobilization and protein kinase C (PKC) as well as PI3 kinase, Ras-MAP kinase activation (Li, Jiang, Xie, Zhang, Smrcka, and Wu 2000). C3a induces mast cell degranulation via signaling pathways that requires PLCβ but not PI3K or extracellular signal regulated kinase (ERK) (el-Lati et al. 1994; ter Laan, Molenaar, Feltkamp-Vroom, and Pondman 1974; Thangam et al. 2005; Venkatesha, Thangam, Zaidi, and Ali 2005a). In contrast, C3a promotes cytokine gene expression in mast cells via signaling pathways that require PLCβ, PI3K as well as ERK activation (Ahamed, Venkatesha, Thangam, and Ali 2004; Ali, Ahamed, Hernandez-Munain, Baron, Krangel, and Patel 2000; Venkatesha et al. 2005a).

3.1 C3aR Desensitization

Most, if not all GPCRs, in the presence of continued stimulation undergo desensitization that dampens cellular responses. Importantly, desensitization regulates mediator release and thus prevents tissue damage (Ali, Haribabu, Richardson, and Snyderman 1997). This process involves agonist-induced receptor phosphorylation and the β-arrestin recruitment, which interacts with G protein to inhibit receptor function (Freedman and Lefkowitz, 1996). GPCRs are phosphorylated by a family of protein kinases, collectively knows as GRKs (G protein coupled receptor kinases). Of the seven known GRKs, four (GRK2, GRK3, GRK5 and GRK6) are expressed in peripheral blood leukocytes and myeloid cell lines (Fernandez et al. 2002; Lombardi, Kavelaars, Schedlowski, Bijlsma, Okihara, Van de Pol, Ochsmann, Pawlak, Schmidt, and Heijnen 1999) and presumably in mast cells. All GRKs (60-80kDa) possess a similar structural organization consisting of an amino terminal domain (185 amino acids) a catalytic domain (270 amino acids) and a carboxyl terminal domain (105 to 230 amino acids).

There are important differences in the mechanism via which GRK2/GRK3 *vs* GRK5/GRK6 are localized to the proximity of the receptor to induce receptor phosphorylation (Penn, Pronin, and Benovic 2000). GRK2 and GRK3 are found primarily in the cytoplasm and undergo translocation to the plasma membrane upon G protein activation, via their interaction with Gβγ subunit and membrane phospholipids. In contrast, GRK5 and GRK6 do not associate with Gβγ but interact with phospholipids or require lipid modification for their association with receptors. Studies with GRK overexpression and siRNA-mediated knockdown procedures indicate that different GRK family members (GRK2/GRK3 *vs* GRK5/GRK6) may modulate distinct receptor functions (Kim, Ahn, Ren, Whalen, Reiter, Wei, and Lefkowitz 2005; Langkabel, Zwirner, and Oppermann 1999; Ren, Reiter, Ahn, Kim, Chen and Lefkowitz, 2005). In this context, overexpression of GRK2, GRK3, GRK5 or GRK6 in COS cells enhances agonist-induced C3aR phosphorylation to a similar extent (Langkabel et al. 1999) but only GRK2 and GRK3 desensitizes C3a-induced G protein activation and inhibits mast cell degranulation (Ahamed, Haribabu, and Ali 2001; Langkabel et al. 1999). Furthermore, mutating the serine and threonine residues in the cytoplasmic tail of C3aR to alanine leads to more robust G protein activation, greater degranulation and LTC_4 generation when compared to wild-type receptors (Ahamed and Ali 2002; Ahamed et al. 2001). These findings suggest that C3aR phosphorylation by GRK2/GRK3 promotes desensitization but biological impact of receptor phosphorylation by GRK5 and GRK6 remains unknown.

3.2 β-arrestin as a Regulator of Down-Stream Signaling Pathway

One of the most intensely studied proteins that interact with phosphorylated GPCRs is β-arrestin. Two isoforms of β-arrestins (β-arrestin 1 and 2) are expressed in many cell types including mast cells (Miller and Lefkowitz 2001; Santini, Penn, Gagnon, Benovic, and Keen 2000). In addition to its role in receptor desensitization, β-arrestin acts as an adapter molecule to regulate diverse cellular functions. For example, β-arrestin activates small G proteins to mediate cytoskeletal reorganization in neutrophils and promote chemotaxis in T lymphocytes (Bhattacharya, Anborgh, Babwah, Dale, Dobransky, Benovic, Feldman, Verdi, Rylett, and Ferguson 2002; Fong, Premont, Richardson, Yu, Lefkowitz, and Patel 2002). β-arrestin also interacts with and activates lymphoid enhancer factor (LEF) transcription factor (Chen, Hu, Semenov, Yanagawa, Kikuchi, Lefkowitz, and Miller 2001), indicating that it might have a similar stimulatory effect on C3a-induced NF-κB activation in mast cells. However, as discussed below, recent studies indicate that β-arrestin may actually inhibit NF-κB activity.

In resting cells, most of the NF-κB is bound to a potent inhibitor IκB, thus retaining this complex in the cytoplasm (Ghosh and Karin 2002). Upon cell activation IκB is phosphorylated by IκB kinase (IKK) leading to its proteosomal degradation. NF-κB, once dissociated from IκB, rapidly translocates to the nucleus where it binds to specific promoters of the target genes. Although several IκB isoforms are known, Gao et al., (Gao, Sun, Wu, Luan, Wang, Qu, and Pei 2004) made the surprising observation that β-arrestin 2 directly interacts with IκBα to inhibit GPCR-mediated NF-κB activity. Witherow et al (Witherow, Garrison, Miller, and Lefkowitz, 2004), showed that although both β-arrestin 1 and β-arrestin 2 associate with IκBα as well as upstream kinases such as IKKα, IKKβ and NIK, only β-arrestin 1 inhibits NF-κB activity and cytokine production. Our recent studies with platelet activating factor receptor (PAFR) demonstrated that receptor phosphorylation is required for β-arrestin-mediated inhibition of NF-κB activity (Venkatesha et al. 2004). Furthermore, we have shown that overexppresion of β-arrestin enhances agonist-induced C3aR internalization and blocks chemokine chemokine CC motif ligand 2 (CCL2) generation (Fig. 1A and B). We also found that similar to the situation with PAFR (Venkatesha et al. 2004), expression of C3aR in mouse embryonic fibroblasts deficient in both β-arrestin-1 and β-arrestin-2, inhibits agonist-induced C3aR internalization but enhances NF-κB activity (unpublished data). These findings

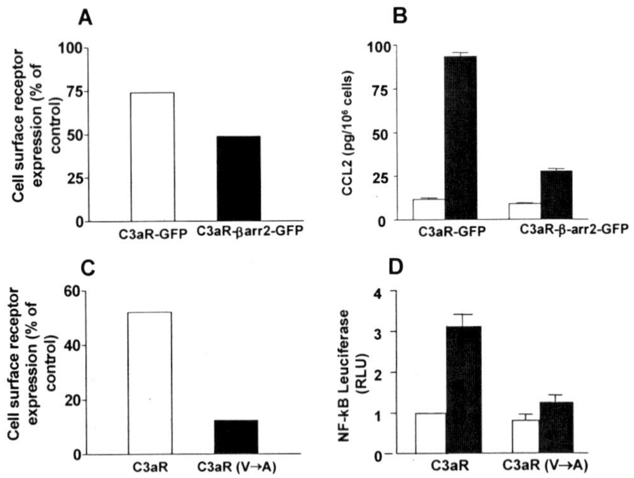

Fig. 1. Reciprocal regulation of C3aR functions in mast cells by β-arrestin and PDZ domain proteins. (A) Transient transfectants were generated in a rodent mast cell line RBL-2H3 cells coexpressing hemagglutinin (HA) tagged C3aR and GFP or C3aR and β-arrestin2-GFP (βarr2-GFP). Cells were exposed to buffer (open bar) or C3a (10 nM, closed bar) for 30 min and cell surface receptor expression was determined by flow cytometry using anti-HA antibody. The data are expressed as a percent of surface receptor expression in the absence of C3a. (B) Cells were exposed to buffer (open bar) or C3a, 10 nM (closed bar) for 6 h and chemokine CCL2 (also known as MCP-1, monocyte chemoattractant protein 1) production was determined by ELISA. Transient transfectants were generated in RBL-2H3 cells expressing C3aR or its V→A mutant with NF-κB luciferase reporter construct. (C) Cells were exposed to buffer or C3a (10 nM) for 30 min and cell surface receptor was determined. (D) Cells were exposed to buffer (open bar) or C3a, 10 nM (closed bar) for 6 h and NF-κB-luciferase activity was measured from the cell lysate.

suggest that β-arrestin inhibits C3a-induced NF-κB activity via a signaling pathway that requires receptor internalization.

3.3 Role of PDZ domain proteins on C3aR functions in mast cells

The carboxyl terminus of C3aR contains 10 serine/threonine residues and most of them are present within two distinct clusters. Cluster I (Thr463, Ser465, Thr466) is required for receptor internalization, as their mutation to alanine residues results in diminished receptor internalization (Settmacher, Rheinheimer, Hamacher, Ames, Wise, Jenkinson, Bock, Schaefer, Kohl, and Klos 2003). In contrast, mutation of phosphorylation sites within cluster II, (Ser475, Ser479, Thr480 and Thr481) with alanine enhances receptor internalization. Ahamed et al., (Ahamed et al. 2001), showed that C3a caused a robust chemokine CCL2 production in a mast cell line expressing C3aR but not ΔST-C3aR cells in which all potential phosphorylation sites were replaced with ala. These findings indicate a dual regulation of C3aR function via its phosphorylation at distinct clusters. Thus, while phosphorylation of C3aR at cluster I promote β-arrestin-mediated desensitization as well as inhibition of NF-κB activity, phosphorylation sites at cluster II provides a critical signal for chemokine gene expression via its interaction with other signaling proteins/adapter molecules.

In addition to β-arrestin, a large number of multidomain scaffolding proteins including PDZ (PSD-95/Dlg/Zo1) domain containing proteins associate with GPCRs (Becamel, Galeotti, Poncet, Jouin, Dumuis, Bockaert, and Marin 2002; Bockaert, Roussignol, Becamel, Gavarini, Joubert, Dumuis, Fagni, and Marin 2004). There are three general classes of PDZ domains; class I domains, which recognize the carboxyl terminal motif S-T-X-Φ, (where "Φ" indicates hydrophobic amino acid and X indicates any amino acid), class II domain which recognize carboxyl terminal motif Φ-X-Φ and class III domains, which recognize D/E-X-Φ as their preferred carboxyl terminal motif (Hung and Sheng 2002). There is evidence to suggest that in some systems, PDZ-domain proteins are required for NF-κB and other transcription factor activation. Thus, the plasma membrane cystic fibrosis transmembrane conductance regulator (CFTR) promotes transcriptional activation of chemokine genes in airway epithelial cells via signaling through its carboxyl terminal PDZ-interacting motif (Estell, Braunstein, Tucker, Varga, Collawn, and Schwiebert 2003). T cell receptor-mediated signaling for NF-κB activation also requires its interaction with membrane-associated PDZ-domain containing proteins that functions upstream of IκB-kinase (IKK) (Bertin, Wang, Guo, Jacobson, Poyet, Srinivasula, Merriam, DiStefano, and Alnemri, 2001; Pomerantz, Denny, and Baltimore 2002). In B cells, interleukin-5 receptor-α (IL-5Rα) interacts with PDZ domain protein to promote transcription factor activation and cytokine gene expression (Geijsen, Uings, Pals, Armstrong, McKinnon, Raaijmakers, Lammers, Koenderman and Coffer 2001). Furthermore, mutation of the terminal Val to Ala (V→A) disrupts the interaction of IL-5Rα with PDZ domain protein and prevents cytokine gene expression.

C3aR possesses a class I PDZ motif STTV at it carboxyl terminus. The sequence STT within the PDZ motif corresponds to Ser479, Thr480 and Thr481 in cluster II of C3aR's carboxyl terminus. It is noteworthy that STTV→AAAV mutant of C3aR responds to agonist for greater internalization than the wild-type receptor (Settmacher et al. 2003). Furthermore, replacing these potential phosphorylation sites with alanine blocks C3a-induced chemokine gene expression (Ahamed et al. 2001). It is therefore possible that C3aR phosphorylation (cluster II)-dependent signal for NF-κB activation is mediated via protein-protein interaction with PDZ domain proteins. Indeed, as shown in Figure 1 C and D, V→A mutant of C3aR responded to C3a for greater receptor internalization but attenuated NF-κB activity when compared to the wild-type receptor. These findings suggest that upon agonist stimulation C3aR undergoes phosphorylation at distinct clusters by different GRKs (GRK2/GKR3 *vs*

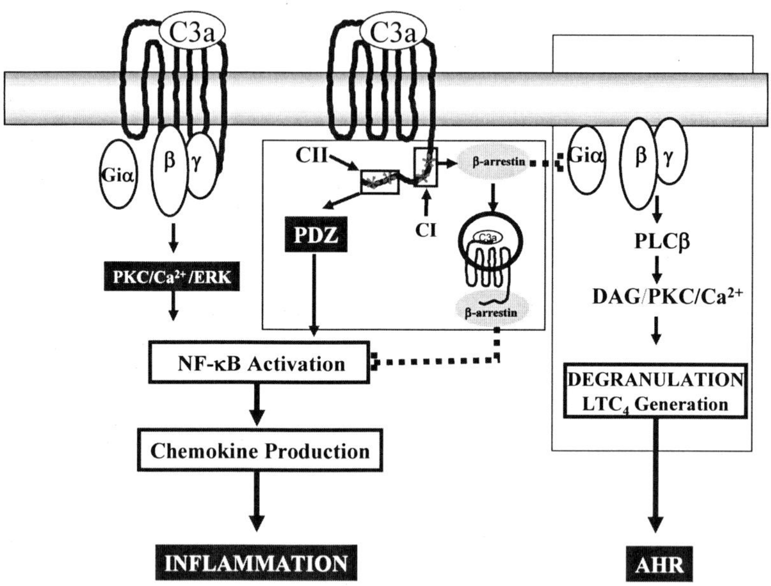

Fig. 2. Model for the possible role of mast cell C3aR signaling in the pathogenesis of asthma. C3a binds to its GPCR on mast cells to promote the release of mediators that likely causes airway hyperresponsiveness (AHR) and inflammation. C3aR phosphorylation at serine/threonine residues within its carboxyl terminus (cluster I, CI) recruits β-arrestin. This interaction desensitizes mast cell degranulation and LTC_4 generation. β-arrestin-mediated receptor internalization also inhibits NF-κB activation to block delayed chemokine gene expression. C3aR phosphorylation site at cluster II (CII) promotes recruitment of PDZ binding proteins to cause NF-κB activation and chemokine production. (Solid line represents activation and broken line represents inhibition).

GRK5/GRK6) (see Section 3.1) to recruit β-arrestin and PDZ domain proteins resulting in reciprocal regulation of chemokine gene expression (Fig. 2).

4 Conclusion

In this chapter we have reviewed the recent evidence highlighting the importance of C3aR in the pathogenesis of asthma. We have summarized the signaling pathways by which C3a induces mast cell chemotaxis, degranulation and transcription factor NF-κB activation leading to chemokine/cytokine gene expression, mediators which promote both AHR and lung inflammation. One major focus of this chapter has been to review the role of C3aR phosphorylation on its regulation. Emerging evidence suggests that agonist-induced phosphorylation of C3aR at distinct clusters by different GRKs promotes interaction with β-arrestin and PDZ domain proteins to coordinately regulate early mast cell degranulation and delayed NF-κB activation/chemokine production, mediators that induce AHR and inflammation (Fig. 2).

In addition to C3aR, C5a receptor (C5aR) also plays an important role in the effector phase of allergic asthma (Peng, Hao, Madri, Su, Elias, Stahl, Squinto, and Wang 2005; Wills-Karp and Koehl, 2005). Most importantly, it protects from the development of the T_H2 response during allergen sensitization at the level of dendritic cell/T cell interphase (Kohl, Baelder, Lewkowich, Pandey, Hawlisch, Wang, Best, Herman, Sproles, Zwirner, Whitsett, Gerard, Sfyroera, Lambris, and Wills-Karp 2006; Lambrecht, 2006). Thus, it appears that C3aR (Section 2) and C5aR play a dual role in asthma; inhibition of the T_H2 response but enhancement of the effector phase of asthma. It is quite possible that inhibition of C3a and C5a by specific receptor antagonists may block the AHR and lung inflammation but may sensitize the individual to new antigens (Kawamoto, Yalcindag, Laouini, Brodeur, Bryce, Lu, Humbles, Oettgen, Gerard, and Geha 2004; Kohl et al. 2006; Lambrecht 2006). As discussed in Section 2.1, human lung MC_{TC} mast cells express C5aR and these cells respond to C5a for robust mast cell degranulation and leukotriene generation. These findings suggests that understanding the signaling pathway via which C3a and C5a receptor functions are regulated in mast cells may provide novel approaches to develop therapeutics for the treatment of asthma.

5 Acknowledgements

This work was supported by National Institutes of Health grant RO1-HL63372.

References

Ahamed J. and Ali H. (2002) Distinct Roles of Receptor Phosphorylation, G Protein Usage, and Mitogen-activated Protein Kinase Activation on Platelet Activating Factor-induced Leukotriene C4 Generation and Chemokine Production. *J Biol Chem* **277**, 22685-91.

Ahamed J., Haribabu B. and Ali H. (2001) Cutting edge: Differential regulation of chemoattractant receptor-induced degranulation and chemokine production by receptor phosphorylation. *J Immunol* **167**, 3559-63.

Ahamed J., Venkatesha R. T., Thangam E. B. and Ali H. (2004) C3a enhances nerve growth factor-induced NFAT activation and chemokine production in a human mast cell line, HMC-1. *J Immunol* **172**, 6961-8.

Ali H., Ahamed J., Hernandez-Munain C., Baron J. L., Krangel M. S. and Patel D. D. (2000) Chemokine production by G protein-coupled receptor activation in a human mast cell line: roles of extracellular signal-regulated kinase and NFAT. *J Immunol* **165**, 7215-23.

Ali H., Haribabu B., Richardson R. M. and Snyderman R. (1997) Mechanisms of inflammation and leukocyte activation. *Med Clin North Am* **81**, 1-28.

Ali H. and Panettieri R. A., Jr. (2005) Anaphylatoxin C3a receptors in asthma. *Respir Res* **6**, 19.

Baelder R., Fuchs B., Bautsch W., Zwirner J., Kohl J., Hoymann H. G., Glaab T., Erpenbeck V., Krug N. and Braun A. (2005) Pharmacological Targeting of Anaphylatoxin Receptors during the Effector Phase of Allergic Asthma Suppresses Airway Hyperresponsiveness and Airway Inflammation. *J Immunol* **174**, 783-9.

Bani D., Nistri S., Mannaioni P. F. and Masini E. (2006) Cardiac anaphylaxis: pathophysiology and therapeutic perspectives. *Curr Allergy Asthma Rep* **6**, 14-9.

Bao L., Osawe I., Haas M. and Quigg R. J. (2005) Signaling through Up-Regulated C3a Receptor Is Key to the Development of Experimental Lupus Nephritis. *J Immunol* **175**, 1947-55.

Bautsch W., Hoymann H. G., Zhang Q., Meier-Wiedenbach I., Raschke U., Ames R. S., Sohns B., Flemme N., Meyer Zu Vilsendorf A., Grove M., Klos A. and Kohl J. (2000) Cutting edge: guinea pigs with a natural C3a-receptor defect exhibit decreased bronchoconstriction in allergic airway disease: evidence for an involvement of the C3a anaphylatoxin in the pathogenesis of asthma [In Process Citation]. *J Immunol* **165**, 5401-5.

Becamel C., Galeotti N., Poncet J., Jouin P., Dumuis A., Bockaert J. and Marin P. (2002) A proteomic approach based on peptide affinity chromatography, 2-dimensional electrophoresis and mass spectrometry to identify multiprotein complexes interacting with membrane-bound receptors. *Biol Proced Online* **4**, 94-104.

Bertin J., Wang L., Guo Y., Jacobson M. D., Poyet J. L., Srinivasula S. M., Merriam S., DiStefano P. S. and Alnemri E. S. (2001) CARD11 and CARD14 are novel caspase recruitment domain (CARD)/membrane-associated guanylate kinase (MAGUK) family members that interact with BCL10 and activate NF-kappa B. *J Biol Chem* **276**, 11877-82.

Bhattacharya M., Anborgh P. H., Babwah A. V., Dale L. B., Dobransky T., Benovic J. L., Feldman R. D., Verdi J. M., Rylett R. J. and Ferguson S. S. (2002) Beta-arrestins regulate a Ral-GDS Ral effector pathway that mediates cytoskeletal reorganization. *Nat Cell Biol* **4**, 547-55.

Bockaert J., Roussignol G., Becamel C., Gavarini S., Joubert L., Dumuis A., Fagni L. and Marin P. (2004) GPCR-interacting proteins (GIPs): nature and functions. *Biochem Soc Trans* **32**, 851-5.

Boos L., Szalai A. J. and Barnum S. R. (2005) C3a expressed in the central nervous system protects against LPS-induced shock. *Neurosci Lett* **387**, 68-71.

Brightling C. E., Ammit A. J., Kaur D., Black J. L., Wardlaw A. J., Hughes J. M. and Bradding P. (2005a) The CXCL10/CXCR3 axis mediates human lung mast cell migration to asthmatic airway smooth muscle. *Am J Respir Crit Care Med* **171**, 1103-8.

Brightling C. E. and Bradding P. (2005) The Re-emergence of the Mast Cell As a Pivotal Cell in Asthma Pathogenesis. *Curr Allergy Asthma Rep* **5**, 130-5.

Brightling C. E., Bradding P., Symon F. A., Holgate S. T., Wardlaw A. J. and Pavord I. D. (2002) Mast-cell infiltration of airway smooth muscle in asthma. *N Engl J Med* **346**, 1699-705.

Brightling C. E., Kaur D., Berger P., Morgan A. J., Wardlaw A. J. and Bradding P. (2005b) Differential expression of CCR3 and CXCR3 by human lung and bone marrow-derived mast cells: implications for tissue mast cell migration. *J Leukoc Biol.*

Brightling C. E., Symon F. A., Holgate S. T., Wardlaw A. J., Pavord I. D. and Bradding P. (2003) Interleukin-4 and -13 expression is co-localized to mast cells within the airway smooth muscle in asthma. *Clin Exp Allergy* **33**, 1711-6.

Carroll N. G., Mutavdzic S. and James A. L. (2002) Distribution and degranulation of airway mast cells in normal and asthmatic subjects. *Eur Respir J* **19**, 879-85.

Casolaro V., Galeone D., Giacummo A., Sanduzzi A., Melillo G. and Marone G. (1989) Human basophil/mast cell releasability. V. Functional comparisons of cells obtained from peripheral blood, lung parenchyma, and bronchoalveolar lavage in asthmatics. *Am Rev Respir Dis* **139**, 1375-82.

Castro F. F., Schmitz-Schumann M., Rother U. and Kirschfink M. (1991) Complement activation by house dust: reduced reactivity of serum complement in patients with bronchial asthma. *Int Arch Allergy Appl Immunol* **96**, 305-10.

Chen W., Hu L. A., Semenov M. V., Yanagawa S., Kikuchi A., Lefkowitz R. J. and Miller W. E. (2001) beta-Arrestin1 modulates lymphoid enhancer factor transcriptional activity through interaction with phosphorylated dishevelled proteins. *Proc Natl Acad Sci U S A* **98**, 14889-94.

Choi W. S., Kim Y. M., Combs C., Frohman M. A. and Beaven M. A. (2002) Phospholipases D1 and D2 regulate different phases of exocytosis in mast cells. *J Immunol* **168**, 5682-9.

Drouin S. M., Corry D. B., Hollman T. J., Kildsgaard J. and Wetsel R. A. (2002) Absence of the complement anaphylatoxin C3a receptor suppresses Th2 effector functions in a murine model of pulmonary allergy. *J Immunol* **169**, 5926-33.

Drouin S. M., Corry D. B., Kildsgaard J. and Wetsel R. A. (2001) Cutting edge: the absence of C3 demonstrates a role for complement in Th2 effector functions in a murine model of pulmonary allergy. *J Immunol* **167**, 4141-5.

Echtenacher B., Mannel D. N. and Hultner L. (1996) Critical protective role of mast cells in a model of acute septic peritonitis [see comments]. *Nature* **381**, 75-7.

el-Lati S. G., Dahinden C. A. and Church M. K. (1994) Complement peptides C3a- and C5a-induced mediator release from dissociated human skin mast cells. *J Invest Dermatol* **102**, 803-6.

Estell K., Braunstein G., Tucker T., Varga K., Collawn J. F. and Schwiebert L. M. (2003) Plasma membrane CFTR regulates RANTES expression via its C-terminal PDZ-interacting motif. *Mol Cell Biol* **23**, 594-606.

Fernandez N., Monczor F., Lemos B., Notcovich C., Baldi A., Davio C. and Shayo C. (2002) Reduction of G protein-coupled receptor kinase 2 expression in U-937 cells attenuates H2 histamine receptor desensitization and induces cell maturation. *Mol Pharmacol* **62**, 1506-14.

Fong A. M., Premont R. T., Richardson R. M., Yu Y. R., Lefkowitz R. J. and Patel D. D. (2002) Defective lymphocyte chemotaxis in beta-arrestin2- and GRK6-deficient mice. *Proc Natl Acad Sci U S A* **99**, 7478-83.

Freedman N. J. and Lefkowitz R. J. (1996) Desensitization of G protein-coupled receptors. *Recent Prog Horm Res* **51**, 319-51; discussion 352-3.

Fregonese L., Swan F. J., Van Schadewijk A., Dolhnikoff M., Santos M. A., Daha M. R., Stolk J., Tschernig T., Sterk P. J., Hiemstra P. S., Rabe K. F. and Mauad T. (2005) Expression of the anaphylatoxin receptors C3aR and C5aR is increased in fatal asthma. *J Allergy Clin Immunol* **115**, 1148-54.

Fureder W., Agis H., Willheim M., Bankl H. C., Maier U., Kishi K., Muller M. R., Czerwenka K., Radaszkiewicz T., Butterfield J. H. Klappacher G. W., Sperr W. R., Oppermann M., Lechner K. and Valent P. (1995) Differential expression of complement receptors on human basophils and mast cells. Evidence for mast cell heterogeneity and CD88/C5aR expression on skin mast cells. *J Immunol* **155**, 3152-60.

Gao H., Sun Y., Wu Y., Luan B., Wang Y., Qu B. and Pei G. (2004) Identification of beta-arrestin2 as a G protein-coupled receptor-stimulated regulator of NF-kappaB pathways. *Mol Cell* **14**, 303-17.

Geijsen N., Uings I. J., Pals C., Armstrong J., McKinnon M., Raaijmakers J. A., Lammers J. W., Koenderman L. and Coffer P. J. (2001) Cytokine-specific transcriptional regulation through an IL-5Ralpha interacting protein. *Science* **293**, 1136-8.

Ghosh S. and Karin M. (2002) Missing pieces in the NF-kappaB puzzle. *Cell* **109 Suppl**, S81-96.

Gibson P. G., Allen C. J., Yang J. P., Wong B. J., Dolovich J., Denburg J. and Hargreave F. E. (1993) Intraepithelial mast cells in allergic and nonallergic asthma. Assessment using bronchial brushings. *Am Rev Respir Dis* **148**, 80-6.

Gonzalez-Espinosa C., Odom S., Olivera A., Hobson J. P., Martinez M. E., Oliveira-Dos-Santos A., Barra L., Spiegel S., Penninger J. M. and Rivera J. (2003) Preferential signaling and induction of allergy-promoting lymphokines upon weak stimulation of the high affinity IgE receptor on mast cells. *J Exp Med* **197**, 1453-65.

Gutzmer R., Lisewski M., Zwirner J., Mommert S., Diesel C., Wittmann M., Kapp A. and Werfel T. (2004) Human monocyte-derived dendritic cells are chemoattracted to C3a after up-regulation of the C3a receptor with interferons. *Immunology* **111**, 435-43.

Haczku A., Atochina E. N., Tomer Y., Chen H., Scanlon S. T., Russo S., Xu J., Panettieri R. A., Jr. and Beers M. F. (2001) Aspergillus fumigatus-induced allergic airway inflammation alters surfactant homeostasis and lung function in BALB/c mice. *Am J Respir Cell Mol Biol* **25**, 45-50.

Hartmann K., Henz B. M., Kruger-Krasagakes S., Kohl J., Burger R., Guhl S., Haase I., Lippert U. and Zuberbier T. (1997) C3a and C5a stimulate chemotaxis of human mast cells. *Blood* **89**, 2863-70.

Hasegawa K., Tamari M., Shao C., Shimizu M., Takahashi N., Mao X. Q., Yamasaki A., Kamada F., Doi S., Fujiwara H., Miyatake A., Fujita K., Tamura G., Matsubara Y., Shirakawa T. and Suzuki Y. (2004) Variations in the C3, C3a receptor, and C5 genes affect susceptibility to bronchial asthma. *Hum Genet* **115**, 295-301.

Humbles A. A., Lu B., Nilsson C. A., Lilly C., Israel E., Fujiwara Y., Gerard N. P. and Gerard C. (2000) A role for the C3a anaphylatoxin receptor in the effector phase of asthma. *Nature* **406**, 998-1001.

Hundley T. R., Prasad A. R. and Beaven M. A. (2001) Elevated levels of cyclooxygenase-2 in antigen-stimulated mast cells is associated with minimal activation of p38 mitogen-activated protein kinase. *J Immunol* **167**, 1629-36.

Hung A. Y. and Sheng M. (2002) PDZ domains: structural modules for protein complex assembly. *J Biol Chem* **277**, 5699-702.

Irani A. A., Schechter N. M., Craig S. S., DeBlois G. and Schwartz L. B. (1986) Two types of human mast cells that have distinct neutral protease compositions. *Proc Natl Acad Sci U S A* **83**, 4464-8.

Irani A. M., Bradford T. R., Kepley C. L., Schechter N. M. and Schwartz L. B. (1989) Detection of MCT and MCTC types of human mast cells by immunohistochemistry using new monoclonal anti-tryptase and anti-chymase antibodies. *J Histochem Cytochem* **37**, 1509-15.
Juliusson S., Pipkorn U., Karlsson G. and Enerback L. (1992) Mast cells and eosinophils in the allergic mucosal response to allergen challenge: changes in distribution and signs of activation in relation to symptoms. *J Allergy Clin Immunol* **90**, 898-909.
Kawamoto S., Yalcindag A., Laouini D., Brodeur S., Bryce P., Lu B., Humbles A. A., Oettgen H., Gerard C. and Geha R. S. (2004) The anaphylatoxin C3a downregulates the Th2 response to epicutaneously introduced antigen. *J Clin Invest* **114**, 399-407.
Kildsgaard J., Hollmann T. J., Matthews K. W., Bian K., Murad F. and Wetsel R. A. (2000) Cutting edge: targeted disruption of the C3a receptor gene demonstrates a novel protective anti-inflammatory role for C3a in endotoxin-shock [In Process Citation]. *J Immunol* **165**, 5406-9.
Kim J., Ahn S., Ren X. R., Whalen E. J., Reiter E., Wei H. and Lefkowitz R. J. (2005) Functional antagonism of different G protein-coupled receptor kinases for beta-arrestin-mediated angiotensin II receptor signaling. *Proc Natl Acad Sci U S A* **102**, 1442-7.
Kirchhoff K., Weinmann O., Zwirner J., Begemann G., Gotze O., Kapp A. and Werfel T. (2001) Detection of anaphylatoxin receptors on CD83+ dendritic cells derived from human skin. *Immunology* **103**, 210-7.
Kohl J., Baelder R., Lewkowich I. P., Pandey M. K., Hawlisch H., Wang L., Best J., Herman N. S., Sproles A. A., Zwirner J., Whitsett J. A., Gerard C., Sfyroera G., Lambris J. D. and Wills-Karp M. (2006) A regulatory role for the C5a anaphylatoxin in type 2 immunity in asthma. *J Clin Invest* **116**, 783-96.
Lambrecht B. N. (2006) An unexpected role for the anaphylatoxin C5a receptor in allergic sensitization. *J Clin Invest* **116**, 628-32.
Langkabel P., Zwirner J. and Oppermann M. (1999) Ligand-induced phosphorylation of anaphylatoxin receptors C3aR and C5aR is mediated by "G protein-coupled receptor kinases. *Eur J Immunol* **29**, 3035-46.
Li Z., Jiang H., Xie W., Zhang Z., Smrcka A. V. and Wu D. (2000) Roles of PLC-beta2 and -beta3 and PI3Kgamma in chemoattractant-mediated signal transduction [see comments]. *Science* **287**, 1046-9.
Lombardi M. S., Kavelaars A., Schedlowski M., Bijlsma J. W., Okihara K. L., Van de Pol M., Ochsmann S., Pawlak C., Schmidt R. E. and Heijnen C. J. (1999) Decreased expression and activity of G-protein-coupled receptor kinases in peripheral blood mononuclear cells of patients with rheumatoid arthritis. *Faseb J* **13**, 715-25.
Marone G., de Crescenzo G., Adt M., Patella V., Arbustini E. and Genovese A. (1995) Immunological characterization and functional importance of human heart mast cells. *Immunopharmacology* **31**, 1-18.
Marone G., Triggiani M. and de Paulis A. (2005) Mast cells and basophils: friends as well as foes in bronchial asthma? *Trends Immunol* **26**, 25-31.
Marshall J. S. and Jawdat D. M. (2004) Mast cells in innate immunity. *J Allergy Clin Immunol* **114**, 21-7.
Maruotti N., Crivellato E., Cantatore F. P., Vacca A. and Ribatti D. (2006) Mast cells in rheumatoid arthritis. *Clin Rheumatol*.
Miller W. E. and Lefkowitz R. J. (2001) Expanding roles for beta-arrestins as scaffolds and adapters in GPCR signaling and trafficking. *Curr Opin Cell Biol* **13**, 139-45.
Nakano Y., Morita S., Kawamoto A., Suda T., Chida K. and Nakamura H. (2003) Elevated complement C3a in plasma from patients with severe acute asthma. *J Allergy Clin Immunol* **112**, 525-30.
Nilsson G., Johnell M., Hammer C. H., Tiffany H. L., Nilsson K., Metcalfe D. D., Siegbahn A. and Murphy P. M. (1996) C3a and C5a are chemotaxins for human mast cells and act through distinct receptors via a pertussis toxin-sensitive signal transduction pathway. *J Immunol* **157**, 1693-8.
Oskeritzian C. A., Zhao W., Min H. K., Xia H. Z., Pozez A., Kiev J. and Schwartz L. B. (2005) Surface CD88 functionally distinguishes the MCTC from the MCT type of human lung mast cell. *J Allergy Clin Immunol* **115**, 1162-8.
Page S., Ammit A. J., Black J. L. and Armour C. L. (2001) Human mast cell and airway smooth muscle cell interactions: implications for asthma. *Am J Physiol Lung Cell Mol Physiol* **281**, L1313-23.
Panettieri R. A., Jr. (2003) Airway smooth muscle: immunomodulatory cells that modulate airway remodeling? *Respir Physiol Neurobiol* **137**, 277-93.
Peng T., Hao L., Madri J. A., Su X., Elias J. A., Stahl G. L., Squinto S. and Wang Y. (2005) Role of C5 in the development of airway inflammation, airway hyperresponsiveness, and ongoing airway response. *J Clin Invest* **115**, 1590-600.

Penn R. B., Pronin A. N. and Benovic J. L. (2000) Regulation of G protein-coupled receptor kinases. *Trends Cardiovasc Med* **10**, 81-9.

Pomerantz J. L., Denny E. M. and Baltimore D. (2002) CARD11 mediates factor-specific activation of NF-kappaB by the T cell receptor complex. *Embo J* **21**, 5184-94.

Ren X. R., Reiter E., Ahn S., Kim J., Chen W. and Lefkowitz R. J. (2005) Different G protein-coupled receptor kinases govern G protein and beta-arrestin-mediated signaling of V2 vasopressin receptor. *Proc Natl Acad Sci U S A* **102**, 1448-53.

Robinson D. S. (2004) The role of the mast cell in asthma: induction of airway hyperresponsiveness by interaction with smooth muscle? *J Allergy Clin Immunol* **114**, 58-65.

Santini F., Penn R. B., Gagnon A. W., Benovic J. L. and Keen J. H. (2000) Selective recruitment of arrestin-3 to clathrin coated pits upon stimulation of G protein-coupled receptors. *J Cell Sci* **113**, 2463-70.

Sayama K., Diehn M., Matsuda K., Lunderius C., Tsai M., Tam S. Y., Botstein D., Brown P. O. and Galli S. J. (2002) Transcriptional response of human mast cells stimulated via the Fc(epsilon)RI and identification of mast cells as a source of IL-11. *BMC Immunol* **3**, 5.

Schuh J. M., Blease K., Kunkel S. L. and Hogaboam C. M. (2002) Eotaxin/CCL11 is involved in acute, but not chronic, allergic airway responses to Aspergillus fumigatus. *Am J Physiol Lung Cell Mol Physiol* **283**, L198-204.

Schulman E. S., Post T. J., Henson P. M. and Giclas P. C. (1988) Differential effects of the complement peptides, C5a and C5a des Arg on human basophil and lung mast cell histamine release. *J Clin Invest* **81**, 918-23.

Settmacher B., Rheinheimer C., Hamacher H., Ames R. S., Wise A., Jenkinson L., Bock D., Schaefer M., Kohl J. and Klos A. (2003) Structure-function studies of the C3a-receptor: C-terminal serine and threonine residues which influence receptor internalization and signaling. *Eur J Immunol* **33**, 920-7.

Soruri A., Kiafard Z., Dettmer C., Riggert J., Kohl J. and Zwirner J. (2003) IL-4 down-regulates anaphylatoxin receptors in monocytes and dendritic cells and impairs anaphylatoxin-induced migration in vivo. *J Immunol* **170**, 3306-14.

Strunk R. C., Eidlen D. M. and Mason R. J. (1988) Pulmonary alveolar type II epithelial cells synthesize and secrete proteins of the classical and alternative complement pathways. *J Clin Invest* **81**, 1419-26.

Supajatura V., Ushio H., Nakao A., Okumura K., Ra C. and Ogawa H. (2001) Protective roles of mast cells against enterobacterial infection are mediated by Toll-like receptor 4. *J Immunol* **167**, 2250-6.

Taube C., Rha Y. H., Takeda K., Park J. W., Joetham A., Balhorn A., Dakhama A., Giclas P. C., Holers V. M. and Gelfand E. W. (2003) Inhibition of complement activation decreases airway inflammation and hyperresponsiveness. *Am J Respir Crit Care Med* **168**, 1333-41.

Taube C., Wei X., Swasey C. H., Joetham A., Zarini S., Lively T., Takeda K., Loader J., Miyahara N., Kodama T., Shultz L. D., Donaldson D. D., Hamelmann E. H., Dakhama A. and Gelfand E. W. (2004) Mast cells, FcepsilonRI, and IL-13 are required for development of airway hyperresponsiveness after aerosolized allergen exposure in the absence of adjuvant. *J Immunol* **172**, 6398-406.

ter Laan B., Molenaar J. L., Feltkamp-Vroom T. M. and Pondman K. W. (1974) Interaction of human anaphylatoxin C3a with rat mast cells demonstrated by immunofluorescence. *Eur J Immunol* **4**, 393-5.

Thangam E. B., Venkatesha R. T., Zaidi A. K., Jordan-Sciutto K. L., Goncharov D. A., Krymskaya V. P., Amrani Y., Panettieri R. A., Jr. and Ali H. (2005) Airway smooth muscle cells enhance C3a-induced mast cell degranulation following cell-cell contact. *Faseb J* **19**, 798-800.

Tkaczyk C., Beaven M. A., Brachman S. M., Metcalfe D. D. and Gilfillan A. M. (2003) The PLCgamma 1-dependent pathway of Fcepsilon RI-mediated mast cell activation is regulated independently of PI-3 kinase. *J Biol Chem.*

Vandermeer J., Sha Q., Lane A. P. and Schleimer R. P. (2004) Innate immunity of the sinonasal cavity: expression of messenger RNA for complement cascade components and toll-like receptors. *Arch Otolaryngol Head Neck Surg* **130**, 1374-80.

Varsano S., Kaminsky M., Kaiser M. and Rashkovsky L. (2000) Generation of complement C3 and expression of cell membrane complement inhibitory proteins by human bronchial epithelium cell line. *Thorax* **55**, 364-9.

Venkatesha R. T., Ahamed J., Nuesch C., Zaidi A. K. and Ali H. (2004) Platelet-activating Factor-induced Chemokine Gene Expression Requires NF-{kappa}B Activation and Ca2+/Calcineurin Signaling Pathways: INHIBITION BY RECEPTOR PHOSPHORYLATION AND {beta}-ARRESTIN RECRUITMENT. *J Biol Chem* **279**, 44606-44612.

Venkatesha R. T., Berla Thangam E., Zaidi A. K. and Ali H. (2005a) Distinct regulation of C3a-induced MCP-1/CCL2 and RANTES/CCL5 production in human mast cells by extracellular signal regulated kinase and PI3 kinase. *Mol Immunol* **42**, 581-7.

Venkatesha R. T., Thangam E. B., Zaidi A. K. and Ali H. (2005b) Distinct regulation of C3a-induced MCP-1/CCL2 and RANTES/CCL5 production in human mast cells by extracellular signal regulated kinase and PI3 kinase. *Mol Immunol* **42**, 581-7.

Werfel T., Kirchhoff K., Wittmann M., Begemann G., Kapp A., Heidenreich F., Gotze O. and Zwirner J. (2000) Activated human T lymphocytes express a functional C3a receptor. *J Immunol* **165**, 6599-605.

Williams C. M. and Galli S. J. (2000) Mast cells can amplify airway reactivity and features of chronic inflammation in an asthma model in mice. *J Exp Med* **192**, 455-62.

Wills-Karp M. and Koehl J. (2005) New Insights into the Role of the Complement Pathway in Allergy and Asthma. *Curr Allergy Asthma Rep* **5**, 362-369.

Witherow D. S., Garrison T. R., Miller W. E. and Lefkowitz R. J. (2004) beta-Arrestin inhibits NF-kappaB activity by means of its interaction with the NF-kappaB inhibitor IkappaBalpha. *Proc Natl Acad Sci U S A.*

Woolhiser M. R., Brockow K. and Metcalfe D. D. (2004) Activation of human mast cells by aggregated IgG through FcgammaRI: additive effects of C3a. *Clin Immunol* **110**, 172-80.

Xia H. Z., Kepley C. L., Sakai K., Chelliah J., Irani A. M. and Schwartz L. B. (1995) Quantitation of tryptase, chymase, Fc epsilon RI alpha, and Fc epsilon RI gamma mRNAs in human mast cells and basophils by competitive reverse transcription-polymerase chain reaction. *J Immunol* **154**, 5472-80.

Yu M., Tsai M., Tam S. Y., Jones C., Zehnder J. and Galli S. J. (2006) Mast cells can promote the development of multiple features of chronic asthma in mice. *J Clin Invest* **116**, 1633-41.

11

Antimicrobial C3a –Biology, Biophysics, and Evolution

Martin Malmsten[1] and Artur Schmidtchen[2]

[1] Uppsala University, Department of Pharmacy, P.O. Box 580, SE-751 23 Uppsala, Sweden
[2] Lund University, Section of Dermatology and Venereology, Department of Clinical Sciences, Biomedical Center, Tornavägen 10, SE-221 84 Lund

1 Antimicrobial Peptides

Antimicrobial substances in blood and leukocytes were discovered over 100 years ago (see, e.g., (Skarnes and Watson 1957)). The identification of antimicrobial peptides (AMPs) in polymorphonuclear leukocytes (Zeya and Spitznagel 1963) was followed by a molecular characterization of these molecules (Selsted, Brown, DeLange, Harwig and Lehrer 1985; Selsted, Harwig, Ganz, Schilling and Lehrer 1985). The subsequent discovery of AMPs in invertebrates (Steiner, Hultmark, Engström, Bennich and Boman 1981) and cold-blooded vertebrates (Zasloff 1987), emphasized the evolutionary importance of this group of host defence molecules. Thus, cationic AMPs are widespread throughout the animal and plant kingdoms and play a fundamental role in innate immune defences. At present, over 800 different AMP peptide sequences are known (see www.bbcm.univ.trieste.it/~tossi/search.htm). Many AMPs are characterized by an amphipathic structure, where clusters of hydrophobic and cationic amino acids are spatially organized in sectors of the molecules. For example, AMPs comprise linear peptides, many of which may adopt α-helical and amphipathic conformation upon bacterial binding, peptides forming cysteine-linked antiparallel β-sheets, as well as cysteine-constrained loop structures. AMPs may also, however, be found among peptides not displaying such ordered structures as long as these are characterized by an over-representation of certain amino acids (Bulet, Stocklin and Menin 2004; Powers and Hancock 2003). Considering the role of peptide secondary structure and amphphilicity, as well as the composition of bacterial membranes, AMP function has been thought to involve direct binding to the lipid bilayer, and the interaction with bacterial membranes is a prerequisite for AMP-function. However, the modes of action of AMPs on their target bacteria are complex, and can be divided into membrane disruptive and non-membrane disruptive (see below). The fact that AMPs lack a specific molecular microbial target facilitates fast action of AMPs and minimizes resistance development. In addition, the observation that insects and mammals successfully utilize AMPs to counter microbial infections has, in combination with the growing problem of resistance to conventional antibiotics, spawned considerable interest in the development of novel AMPs for topical or systemic use, and several peptides are currently in stage III clinical trials (Zhang and Falla 2004).

2 Mode of Action of Antimicrobial Peptides

2.1 Bacterial Cell Walls

AMPs are multifunctional substances, which affect bacteria in many different ways, including inhibition of cell wall synthesis, inhibition of DNA, RNA and protein synthesis and inhibition of enzymatic activity (Brogden 2005; Lohner and Blondelle 2005; Yount, Bayer, Xiong and Yeaman 2006). However, the main mode of action of AMPs seems to be the disruption of bacterial membranes (Yount et al. 2006). The rapid bacterial breakdown in combination with the bacterial membrane being the main target is also thought to be one of the main reason for the limited bacterial resistance development observed against such peptides (Yount et al. 2006; Zasloff 2002).

The walls of bacteria are complex structures, including two lipid membranes in the case of Gram-negative bacteria (one for Gram positive bacteria), peptidoglycan envelopes for both types of bacteria, lipopolysacharides (LPS) in the case of Gram-negative bacteria, presence of proteins in the lipid membranes, as well as varying lipid compositions between different bacteria. Despite that, investigations on the interaction between AMPs and model lipid systems offer some opportunities for gaining an improved understanding on the mode of action of AMPs as well as how peptide properties such as sequence, charge and charge distribution, conformation, hydrophobicity and hydrophobicity distribution, and peptide length affect the interaction of AMPs with lipid membranes, as well as with bacteria and cell walls.

2.2 Mode of Action

From investigations on the interaction between AMPs and model lipid membranes, it has been found that antimicrobial peptides in fact display a broad range of mechanisms of action. For a number of AMPs, formation of ordered helices is observed on lipid membrane incorporation, sometimes in an oligomerized form, resulting in structurally relatively well-defined "pores". However, a number of other mechanisms have also been observed. One of these is membrane destabilization through thinning, caused by peptide adsorption in the lipid headgroup region, resulting in a lateral crowding in the headgroup region and in a structural relaxation of the phospholipid acyl chains (Lee, Hung, Chen and Huang 2005). For other peptides, yet other mechanisms have been observed, ranging from local packing defects and disordering around peptides adsorbed in the phospholipid headgroup region, to complete "membrane solubilization" and mixed micelle formation (Ohki, Marcus, Sukumaran and Arnold 1994). There are also a number of AMPs, such as the porcine cathelicidin PR-39 and human histatins which function by less well elucidated mechanisms, seemingly involving translocation through the bacterial membrane and interactions with intracellular targets. Thus, whereas PR-39 blocks bacterial DNA and protein synthesis (Boman, Agerberth and Boman 1993), histatins translocate through membranes (Den Hertog, Wong Fong Sang, Kraayenhof, Bolscher, Van't Hof, Veerman and Nieuw Amerongen 2004) and bind to a receptor in the fungal mitochondrion, whereby they may induce cell death by non-lytic ATP release, generation of reactive oxygen species and induction of G1 phase arrest (Baev, Li, Dong, Keng and Edgerton 2002; Koshlukova, Lloyd, Araujo and Edgerton 1999). Given this mode of action diversity, it seems clear that model membrane studies are important not only for the biophysical understanding of AMPs, but also for their optimization in a therapeutic context.

3 Antibacterial Activities of C3a

3.1 C3a is Antibacterial

In vertebrates, the complement system is activated by the classical, alternative, and lectin pathways, each converging at the step of C3 with release of multiple proteolytic fragments, including the anaphylatoxin C3a (Hugli 1990). As reviewed elsewhere, C3a has multiple proinflammatory functions, involving histamine release from mast cells, smooth muscle contraction, increased vascular permeability and chemoattraction against mast cells (Hugli 1990). The biological effects of C3a are regulated by the plasma protease carboxypeptidase B, which cleaves off the C-terminal arginine to generate the inactive C3a-desArg peptide (Hugli 1990). The human C3a molecule (77 aa, 9083 Da) is cationic (pI 11.3) and contains four α-helical regions In a recent report, it was demonstrated that C3a, C3a-desArg, and functional epitopes in the C3a-sequence, all function as "classical" AMPs, demonstrating a previously unknown direct antimicrobial effect of complement activation (Nordahl, Rydengard, Nyberg, Nitsche, Morgelin, Malmsten, Bjorck and Schmidtchen 2004). C3a, as well as C3a-desArg, were antibacterial (Nordahl et al. 2004) against the Gram-positive *Enterococcus faecalis* and the Gram-negative *Pseudomonas aeruginosa* and *Escherichia coli.* The holoprotein C3, on the other hand, did not exert any antibacterial effect. The minimum inhibitory concentrations (MIC) of C3a were determined and were comparable to the MICs obtained for the human cathelicidin LL-37. Transmission electron microscopy showed clear differences in the morphology between C3a-treated bacteria (**Figure 1**) and untreated control bacteria (upper left), as well as bacteria exposed to the holoprotein C3 (lower left). Thus, C3a but not C3 caused local perturbations and breaks along *P. aeruginosa* plasma membranes, and occasionally, intracellular material was found extracellularly. These findings were similar to those seen after treatment with the antimicrobial

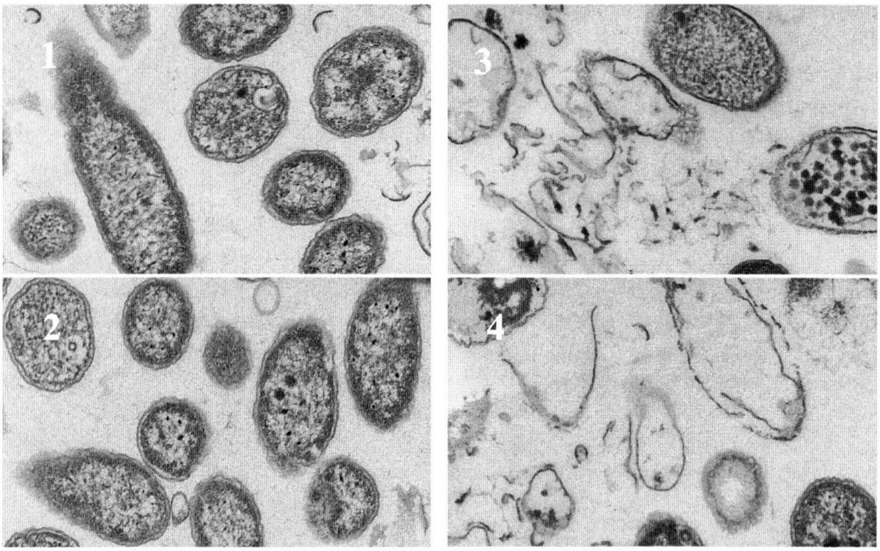

Fig. 1. Electron microscopy analysis of *Pseudomonas aeruginosa* subjected to peptides. 1; control, 2; C3, 3; C3a, 4; LL-37 (Andersson et al. 2004).

peptide LL-37 (Figure 1) (Andersson et al. 2004). In analogy, C3a caused membrane rupture and permeabilisation of liposomes, and 100% liposome leakage was noted at ~0.1 μM C3a. LL-37 yielded similar effects at ~0.5 μM (Nordahl et al. 2004). Kinetic analysis showed that ~80% of the fluorescent marker was released within 17 min for both peptides (at 0.5 μM) (Nordahl et al. 2004).

3.2 Antibacterial C3a-derived Peptides

C3a contains four helical regions, corresponding to amino acids 8-15, 17-28, 36-43 and 47-66 in human C3a. (Figure 2) (Hugli 1990) To explore whether these helical epitopes are responsible for the antibacterial effects of C3a, peptides spanning the four helices, including a peptide known to exert full anaphylatoxic activity, CNY21 (CNYITELRRQHARASHLGLAR) (Lu, Fok, Erickson and Hugli 1984), its variant devoid of the terminal R, CNY20, and the smallest peptide with any anaphylatoxic activity at all, LGL5 (LGLAR) (Caporale, Tippett, Erickson and Hugli 1980) were synthesized and tested. The experiments showed that peptides spanning the C3a-helices, including CNY21 and CNY20, indeed were antibacterial against *E. faecalis* and *P. aeruginosa*, while LGL5 showed no antimicrobial activity at all. Quantitatively, the C3a-derived peptides required a ~10 times higher concentration for efficient killing of *E. faecalis*. Furthermore, the antimicrobial activities of CNY21 and its desArg variant, CNY20 (devoid of anaphylatoxin activity), were similar (Nordahl et al. 2004). As no difference in activity was observed between the two peptides, the antimicrobial and anaphylactic functions thus seem to be separated.

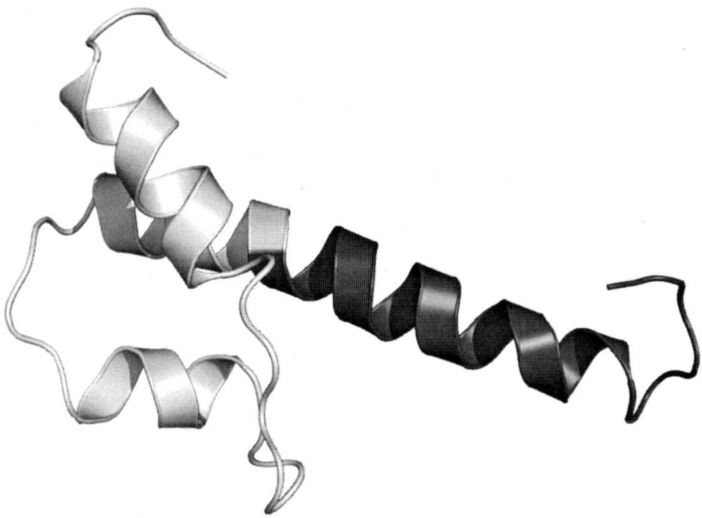

Fig. 2. Structure of C3a. The C-terminal peptide CNY21 is indicated.

3.3 C3a-peptides Function in Physiological Buffers and *In vivo*

Activities of AMP generally depend on electrolyte concentration, pH, and the presence of plasma proteins. For example, the antimicrobial activities of defensins are inhibited by the presence of physiological salt (Ganz 2001), while the cathelicidin LL-37 is inhibited by plasma (Wang, Agerberth, Lothgren, Almstedt and Johansson 1998). Other AMPs, such as azurocidin and magainins, are potentiated by acidic conditions likely to occur in biological fluids following oxidative burst response of leukocytes (Wright, Schwartz, Olson, Kosowsky and Tauber 1986). In comparison, C3a was found to be active against *P. aeruginosa* in 0.15 M NaCl at both pH 7.4 and pH 5.5 and no inhibition of bacterial killing in physiological salt was noted for CNY21 (Nordahl et al. 2004)). *P. aeruginosa* bacteria were also subjected to 10 μM C3a at physiological salt conditions in the presence of human wound fluid (20%), and found to be significantly more potent than the benchmark LL-37. Thus, C3a has potent antibacterial effects under physiological conditions. Considering that the molecule has multiple proinflammatory functions, the following strategy was employed for demonstrating an antibacterial effect *in vivo*. The CNY21 peptide and a variant thereof, CNY21 R-S (CNYITEL- SSQHASASHLGLAR), in which three arginine residues were replaced with serines, rendering the peptide totally devoid of antibacterial activity, were injected into mice infected by Gram-positive *Streptococcus pyogenes*. Compared to the non-antibacterial control peptide, treatment with CNY21 yielded significantly lower bacterial numbers in the spleen of the animals (Nordahl et al. 2004)). This, combined with the fact that both peptides contained the anaphylatoxin determinant LGLAR (Hugli 1990), demonstrated a direct antibacterial role of the C-terminal helix-forming region (Lu et al. 1984) of the C3a-molecule.

3.4 Proteolytic Generation of C3a-peptides

It has been shown that neutrophil elastase and mast cell tryptase both release C3a-like peptides (Schwartz, Kawahara, Hugli, Vik, Fearon and Austen 1983; Taylor, Crawford and Hugli 1977). Acute human wound fluid was therefore incubated with lysed neutrophils for various periods of time and the results showed that neutrophil elastase degraded C3a into several fragments. One fragment, ^{44}SLGEACKKVFLDCCNYITELRRQHA^{68}, containing a helical region of the C3a peptide, was identified by both Edman degradation and mass spectrometry. This peptide was synthesized and found to exert similar antibacterial effects as CNY21. This indicates that, apart from C3a, additional antibacterial fragments of C3 may be generated during inflammation. Taken together, the results demonstrated a new antibacterial function for C3a and its C-terminal region.

4 Antifugal Activities of C3a

4.1 Fungal Membranes

Structural differences between bacterial, fungal, and mammalian cell membranes most likely contribute to the frequently observed selectivity of AMPs towards bacteria and fungi. As previously mentioned, bacterial surfaces contain many anionic components, including lipopolysaccharides, peptidoglycans, and anionic lipids of Gram-negative bacteria, as well as teichoic and teichuronic acids of Gram-positive bacteria. Similarly, β-glucan, chitin, mannoprotein and a blend of other cell wall proteins and polysaccharides contribute to an, albeit weak, negative surface potential of fungal surfaces (Odds 1988; Pitarch, Sanchez, Nombela and Gil 2003). In contrast to eukaryotic, including fungal, membranes, which contain mostly zwitterionic lipids (e.g., phosphatidylcholine), bacterial membranes comprise various acidic phospholipids (phosphatidylglycerol, phosphatidylserine and cardiolipin), which confer a negative charge, thus facilitating AMP binding. Furthermore, the plasma membrane of eukaryotic cells contains sphingolipids and sterols, which are missing in prokaryotes (Opekarova and Tanner 2003), and which are frequently found to provide some resistance to AMP-induced membrane rupture (Epand, Ramamoorthy and Epand 2006; Glukhov, Stark, Burrows and Deber 2005; Hallock, Lee, Omnaas, Mosberg and Ramamoorthy 2002). Ergosterol is the major sterol in yeasts, whereas cholesterol is the main sterol in plasma membranes of mammalian cells (Opekarova et al. 2003). All these factors probably contribute to AMPs exhibiting different activity spectra on bacterial and fungal membranes, as well as to toxicity on eukaryotic cells. It has, however, become increasingly clear that AMP selectivity may also depend on factors such as AMP oligomerisation and preassembly (in solution and membrane)

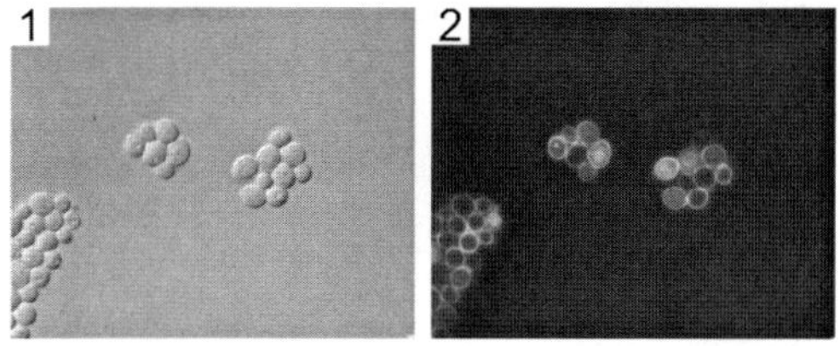

Fig. 3. Binding of TAMRA-CNY21 to *Candida albicans* ATCC 90028. Panel 2 shows red fluorescence of *Candida* cells stained with TAMRA-conjugated CNY21. The corresponding Nomarski image is shown in panel 1 (Sonesson et al. 2006)

(for references see Sonesson, Ringstad, Andersson Nordahl, Malmsten, Morgelin and Schmidtchen 2006)

4.2 C3a Kills *Candida*

Candida is a dimorphous fungi colonizing the mucous membranes of the mouth and vagina in a saprophytic manner. However, *Candida* is also known to be involved in several diseases such as cutaneous infections, atopic eczema, oroesophageal candidiasis, candida vaginosis, and septicaemia (Odds 1988; Savolainen, Lammintausta, Kalimo and Viander 1993). In animal models it has been shown that *Candida* activates the complement system and C3 fragments are deposited at the cell surface of Candida (Kozel 1996; Kozel, Weinhold and Lupan 1996; Sohnle and Kirkpatrick 1976). In humans, generation of C3a has been implicated in the pathogenesis of skin diseases, such as atopic dermatitis (Kawamoto, Yalcindag, Laouini, Brodeur, Bryce, Lu, Humbles, Oettgen, Gerard and Geha 2004; Werfel, Kirchhoff, Wittmann, Begemann, Kapp, Heidenreich, Gotze and Zwirner 2000). In a recent report, it was shown that the anaphylatoxin peptide C3a exerts antifungal activity (Sonesson et al. 2006). Notably, C3a was active against *Candida* and exerted a stronger inhibitory effect than the benchmark LL-37, at least under the low-salt conditions investigated. For example, in solution phase assays (viable count), the concentrations of C3a and LL37 required to kill 50% of the microorganisms were determined to be 5.0 μM and 11.5 μM, respectively. It was noted that among the C3a helical regions tested, CNY21 displayed a significant antifungal effect.

Given its antifungal effect, it is not surprising that CNY21 binds to *Candida*, as shown using fluorescently labelled CNY21 (TAMRA-CNY21; Figure 3) (Sonesson et al. 2006) To further analyze the effects of CNY21 on the yeast cells, *Candida* cells were incubated with this peptide, and analyzed by scanning electron microscopy. The human cathelicidin LL-37 was included for comparison and a CNY21-derived peptide containing the anaphylatoxin determinant LGLAR, but totally devoid of antibacterial activity (Nordahl et al. 2004), CNY21 R-S (CNYITELSSQHASASHLGLAR), was used as negative control. Compared to the control cells, CNY21-treated *Candida* cells displayed extracellular material and membrane perturbations (Figure 4).

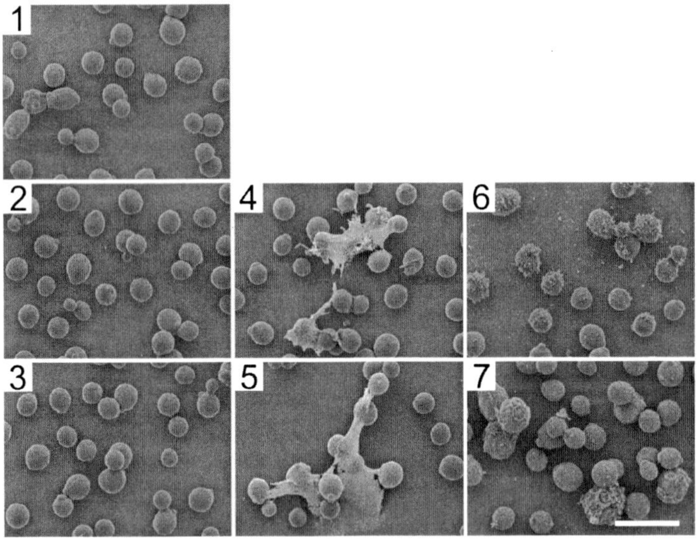

Fig. 4. Treatment of *Candida* with C3a-derived peptides and LL-37. *Candida* cell suspensions were incubated with the peptides and visualized by scanning electron microscopy. LL37 and CNY21R-S served as positive and negative controls, respectively. Panel 1 shows *Candida* cells alone, Panels 2, 4, and 6 show effects of 50 μM of CNY21R-S, CNY21 and LL-37, respectively and panels 3, 5, and 7 show the corresponding peptides at 100 μM. Scale bar represents 10 μm (Sonesson et al. 2006).

In analogy to the situation with bacteria, *Candida* was found to generate similar C3a-like peptides in human serum (Sonesson et al. 2006), thus presenting a proof of the concept that antimicrobial peptides may indeed be generated in response to *Candida* infection. Clearly, considering the multifunctionality of C3a, as well as the high redundancy of innate immune defenses, the physiological *in vivo* significance of the direct antifungal activity of C3a needs to be determined.

4.3 Structural Requirements for Antifungal C3a-peptides

In order to determine the structural requirements of CNY21 and its actions against *Candida*, the net charge, hydrophobicity, and helical propensity of the peptide was modified. Replacement of the two histidine residues by lysine (CNY21H-K) increased the antifungal activity, possibly as a result of the increased net charge of the peptide. Increasing the hydrophobicity by replacing the two histidines with leucines (CNY21 H-L), on the other hand, had little effect on the antimicrobial action of the peptide. Interestingly, introduction of helix breaking proline residues (CNY21H-P) significantly reduced the activity of the peptide, suggesting that a helical conformation may be important for CNY21 action on *Candida* membranes. In analogy to the abolishment of the antimicrobial activity of CNY21 through replacement of the central three arginines by the polar but uncharged residue serine completely abolished also the antifungal activity of CNY21. Taken together, these results indicate that a positive net charge, hydrophobicity, and a partial helical conformation is required for the antifungal and membrane-breaking action of the C-terminal part of C3a, as represented by CNY21 (Sonesson et al. 2006).

5 Evolution of Antimicrobial C3a

Phylogenetic analysis of representative C3a peptides from invertebrates as well as vertebrates indicates that C3a has evolved via multiple changes in the protogene, a finding consistent with previous analyses of the evolution of the complement system (Nonaka and Yoshizaki 2004; Zhu, Thangamani, Ho and Ding 2005). On the other hand, C4a and C5a form separate clades, C5a being most distant from C3a. This pattern suggests that C5a and C4a likely evolved from C3a and that C5a is the paralog to C4a and C3a (Zhu et al. 2005).

5.1 Structural Considerations

Given this phylogenetic relation, it was of interest to examine this molecular family closer, both from a structural and a functional perspective (**Figure 5**). Several common structural features exist for the corresponding C3a, C4a, and C5a sequences of various organisms, crucial for the integrity and stability of the molecules (**Figure 2 and 5**). A most notable feature is the existence of six disulfide-bonded cysteines, which are conserved not only in C3a, but also in C4a and C5a. The three disulfide bonds stabilize the conformation of the internal "core" portion of the molecules, represented by residues 22-57 in the human C3a sequence. Apart from these cysteines, the four glycines (positions 13, 26, 46, 74), phenylalanine (53), as well as lysines and arginines (21, 64, 77), constitute additional conserved features, suggesting their importance for the structural stability as well as function of C3a (Pasupuleti, Walse, Nordahl, Morgelin, Malmsten and Schmidtchen 2006).

5.2 Structural Modeling of Anaphylatoxin Peptides

Computational modeling, utilizing available structural data on human C3a (Huber, Scholze, Paques and Deisenhofer 1980; Janssen, Huizinga, Raaijmakers, Roos, Daha, Nilsson-Ekdahl, Nilsson and Gros 2005) , as well as C5a peptides (Williamson and Madison 1990; Zhang, Boyar, Toth, Wennogle and Gonnella 1997), was employed to provide further structural information. Considering the recent identification of C3, and a putative anaphylactic peptide in the arthropod *C. rotundicauda* (Zhu et al. 2005), these two molecules were compared. As seen in Figure 5, the similarity at the 3-D level between human C3a and the predicted *C. rotundicauda* C3a peptide is apparent, albeit an extensive overall sequence discrepancy (26% sequence identity and 38% similarity). Both peptides share a striking similarity in the four helical regions and in the two first loops located before the cysteines at positions 22 and 36 and positions 17 and 31 in the human and *C. rotundicauda* peptides, respectively. Analogously to human C3a, the *C. rotundicauda* peptide contains a prominent cationic and

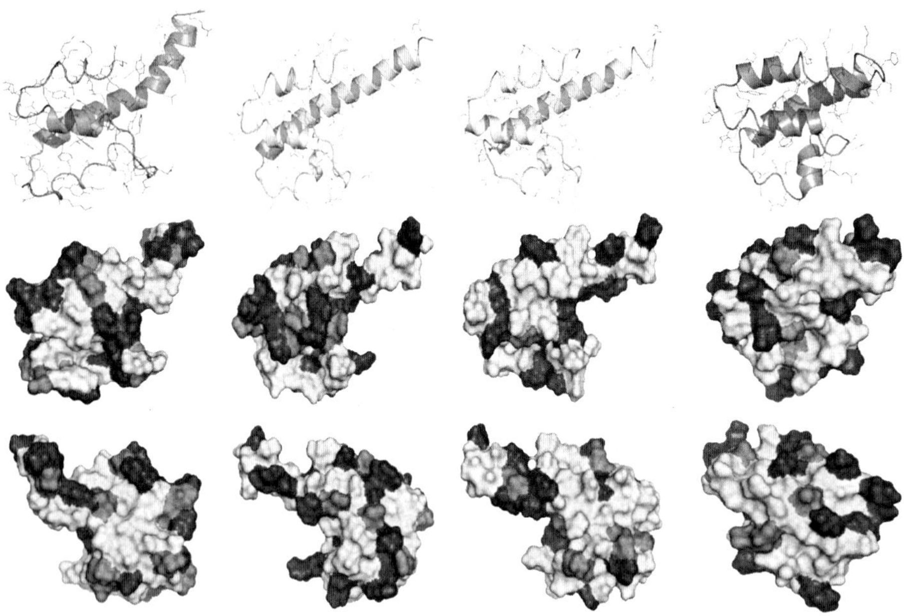

Fig. 5. Molecular models of anaphylatoxin peptides (Pasupuleti et al. 2006).

amphipathic helical C-terminus, predicted to extend a few residues further than in human C3a. As stated above, the C-terminal part of human C3a being flexible in solution, strongly conforms to an α-helical conformation in anisotropic environments (Hugli 1990).

In order to explore the structure-function relationships of C3a epitopes further, overlapping peptide sequences comprising 20mers were synthesized and screened for antimicrobial activities against *P. aeruginosa*, a ubiquitous pathogen found among both vertebrates as well as invertebrates. The experiments showed that peptides derived from the C-terminal regions of *C. rotundicauda*, *Ciona*, and *Branchiostoma*, as well as from the vertebrates *Homo, Sus, Mus, Rattus,* and *Guinea*, were antimicrobial (Pasupuleti et al. 2006), whereas peptides originating from *Cobra, Paralicchtys, Onchorhyncus*, *Eptatretus, Xenopus*, and *Gallus* did not show any activity against bacteria (Pasupuleti et al. 2006). Properties common for most AMPs include minimum levels of cationicity, amphipathicity, and hydrophobicity (Yount et al. 2006). Therefore, it was interesting to note that C-terminal regions of the two former groups comprised cationic peptides, whereas the latter group comprised negatively charged peptides (Pasupuleti et al. 2006). These results corresponded with the phylogenetic analysis, which showed that these animals belong to different subclades and with single nodes of divergence (Pasupuleti et al. 2006). The global analysis of biophysical parameters showed that in general, peptides displaying

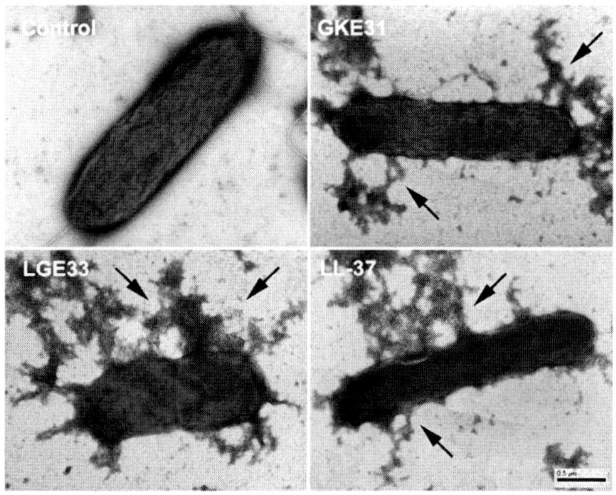

Fig. 6. Electron microscopy analysis of *Pseudomonas aeruginosa* subjected to the C3a-derived peptide GKE31 (*C. rotundicauda*), LGE33 (*Homo*), and LL-37 (Pasupuleti et al. 2006).

antimicrobial activity had a net charge of +2-+4 and ~20-40% hydrophobic amino acids. The degree of amphipathicity, as judged by the relative hydrophobic moment (μHrel), ranged ~0.2-0.4, values comparable with those observed in many helical AMPs (Zelezetsky and Tossi 2006b). Considering these findings, it is notable that disproportionate alteration of charge appear to characterize the evolution of β-defensins (Maxwell, Morrison and Dorin 2003; Tennessen 2005). Analogous relationships were recently reported to apply to the evolution of primate cathelicidin, showing positive selection affecting charge while keeping hydrophobicity and amphipathicity fairly constant (Zelezetsky, Pontillo, Puzzi, Antcheva, Segat, Pacor, Crovella and Tossi 2006a).

5.3 Structural and Functional Congruence of C-termini of C3a

In analogy to the discussion above, the *C. rotundicauda* peptide GKE31 (GKETCMAAFLGCCNEKHLYLLKNIEKEGRGR), spanning the whole C-terminal part of *C. rotundicauda* C3a, exerted similar antibacterial effects as the human homolog LGE33 (LGEACKKVFLDCCNYITELRRQHARASHLGLAR) against Gram-negative *P. aeruginosa* and *E. coli,* as well as Gram-positive *Bacillus subtilis* (Pasupuleti et al. 2006). Using electron microscopy, clear differences in the morphology of peptide-treated bacteria in comparison with the control were demonstrated. The peptides both caused breaks along *P. aeruginosa* plasma membranes, and occasionally, intracellular material was found extracellularly. These findings were similar to those seen after treatment with the benchmark peptide LL-37 (Figure 6).

CD spectroscopy was used to study the structure and the organization of the GKE31 and LGE33 peptides in solution and upon interaction with *E. coli* LPS

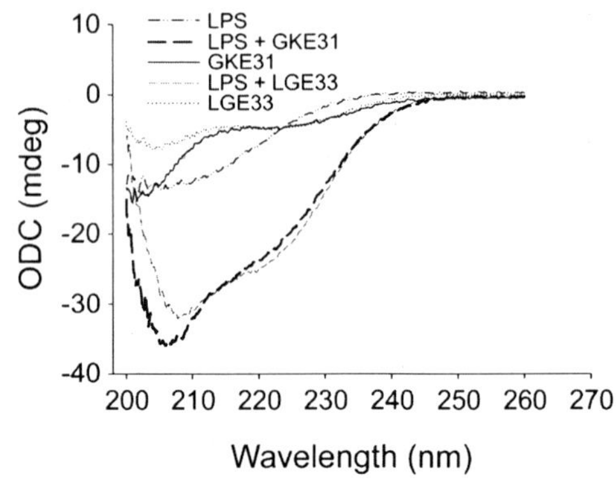

Fig. 7. CD-spectroscopy analysis of GKE31 and LGE33 in absence and presence of LPS (Pasupuleti et al. 2006).

(Figure7). While neither GKE31 nor LGE33 adopted an ordered conformation in aqueous solution, a significant and almost identical structural change, largely indicating helix induction, was seen in the presence of *E. coli* LPS. The similar profile of both peptides indicates that despite a marked difference in primary sequence, the peptides structure in a similar way in presence of negatively charged LPS-rich bacterial membranes. Both peptides also induced leaking of liposomes, thus establishing their membrane breaking activities (Pasupuleti et al. 2006). In relation to the latter, kinetic analysis showed that ~80% of the maximum fluorescence was reached within ~200 seconds for both peptides (at 1 μM) (Pasupuleti et al. 2006). The results therefore indicate that the GKE31 and LGE33 peptides indeed function like most helical AMPs such as LL-37 (Yount et al. 2006; Zelezetsky et al. 2006b), by interactions with LPS and, most likely, peptidoglycan at bacterial surfaces, leading to induction of an α-helical conformation, which in turn facilitates membrane interactions, membrane destabilization and finally, bacterial killing. The fact that the two C3a-derived peptides are separated by as much as over a half billion years of evolutionary distance demonstrates the remarkable structural and functional conservation of this C-terminal peptide region.

5.4 Comparison of C3a Molecules of *C. rotundicauda* and *C. intestinalis*

The C3a molecule of *Ciona* is functionally active; it exerts chemotactic effects (Pinto, Chinnici, Kimura, Melillo, Marino, Spruce, De Santis, Parrinello and Lambris 2003), and has a C-terminal antibacterial part (Pasupuleti et al. 2006). However, in contrast to the other C3a molecules, C3a of *Ciona* lacks one disulfide bridge (Pinto et al. 2003). Despite this, and a sequence identity of only 17% (37% similarity) with human C3a, molecular modeling and conformational analysis showed that *Ciona* C3a adopts a predicted conformation similar to other C3a molecules (Figure 8).

Interestingly, the first missing cysteine in *Ciona* C3a is replaced by a glycine residue (Gly 18), while the second cysteine is replaced by a glutamate residue

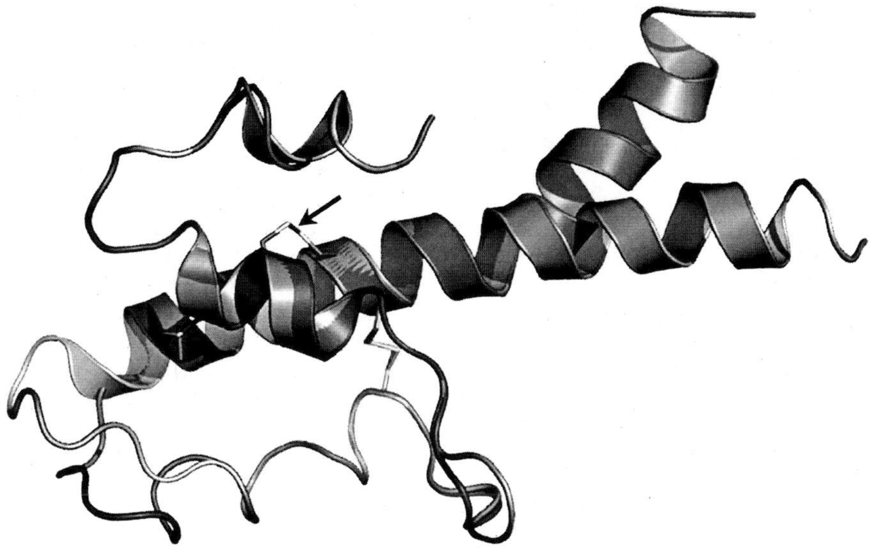

Fig. 8. Comparison of *C. rotundicauda* and *Ciona* C3a.

(Glu57). Hypothetically, these changes could enable the second helix to approach to, and interact with, the C-terminal helix by formation of a salt bridge between Glu 57 and Lys 22, thus compensating for the loss of a disulfide bond (Pasupuleti et al. 2006). These interactions, together with main structural constraints imposed by the remaining two disulfide pairs preserve the overall topology of *Ciona* C3a (Figure 8). Furthermore, it is notable that residues forming the inner "core" throughout the anaphylatoxin family (Gly 21, Ile/Val 39, and Phe 54) are all conserved in *Ciona* C3a. Taken together, these structural considerations, combined with the functional data, further underscore the evolutionary conservation of C3a.

5.5 Structure and Activities of C4a and C5a

Previous phylogenetic analyses has indicated that C5a evolved separately from the family of C3a as well as C4a molecules (see (Pasupuleti et al. 2006) and references herein). To address whether this also reflected a functional difference, the antibacterial activities of human C3a, C4a, and C5a, as well as their corresponding peptides from the C-termini, were compared. C3a and C4a both exerted antibacterial effects, whereas C5a was inactive (Pasupuleti et al. 2006)). Corroborating results were obtained with the corresponding C-terminal peptides (Pasupuleti et al. 2006). Thus, while the C3a and C4a-derived peptides (CNY21 and CQF21, respectively) were antimicrobial, the C5a-peptide (CVV20) had no activity against bactera. As seen in the 3-D models, C5a display a significantly different structure when compared with both C4a and C3a, lacking the typical C-terminal and antimicrobial protruding peptide. This observation, paired with the separate evolutionary development of C5a, indicate that this molecule has evolved separately in higher organisms into a purely chemotactic and highly spasmogenic molecule.

6 Biophysical Studies of C3a-Derived Peptides

Although the detailed mechanism of action of antimicrobial peptides most likely depends on a complex interplay between these peptides and the various components in the bacterial walls, some information was aimed for by investigations of the effect of peptide hydrophobicity and charge on peptide interaction with model lipid bilayers. Of interest in the present context, phospholipid liposomes, bilayers and monolayers were employed in the case of C3a-derived CNY21, together with a method combination of fluorescence spectroscopy, circular dichroism, ellipsometry, z-potential and photon correlation spectroscopy measurements (Ringstad, Andersson Nordahl, Schmidtchen and Malmsten 2007). For both zwitterionic and anionic liposomes, the membrane-disruptive potency for CNY21-variants increased with increasing net positive charge and mean hydrophobicity, and was completely lost on elimination of all peptide positive charges. Analogous effects of elimination of the peptide positive net charge in particular were found regarding bacteria killing for both *P. aeruginosa* and *B. subtilis*. The peptides, characterized by moderate helix content both in buffer and when attached to the liposomes, displayed high adsorption for the net positively charged peptide variants, while adsorption was non-measurable for the uncharged peptide (CNY21 R-S). That electrostatically driven adsorption represents the main driving force for membrane disruption in lipid systems was demonstrated also by a drastic reduction in both liposome leakage and peptide adsorption with increasing ionic strength, and that this salt inactivation can be partly avoided by increasing the peptide hydrophobicity. This increased electrolyte resistance translates also to a higher antibacterial effect for the hydrophobically modified (CNY21 H-L) variant at high salt concentration. Overall, our findings demonstrate the importance of the peptide adsorption and resulting peptide interfacial density for membrane-disruptive effects of these peptides.

The nature of the defect formation caused by CNY21 was subsequently investigated in follow-up investigations, using electrocemical methods, solid state NMR, and high resulution AFM, respectively (to be published). First, from the use of impedance spectroscopy and electroactive ions of different valency and size, a few additional pieces of information were obtained. In particular, the effect of increasing the peptide hydrophobicity by replacing the histidin residues of CNY21 with L resulted in semi-quantitatively potency enhancement for liposomes (i.e., bilayers) as for supported monolayers. This clearly shows that transmembrane configurations of CNY21 are not involved in the phospholipid membrane disruption of this peptide, as previously suggested also by the lack of extensive helix induction in the peptide when binding to phospholipid membranes. This also suggests peptide binding in the surface region of the phospholipid membranes, in line with results obtained from ellipsometry, z-potential measurements, and solid state NMR. Moreover, both solid state NMR and

AMF demonstrated that the presence of CNY21 results in a disordering of the phospholipid membrane.

While the biophysical investigations on C3a-derived peptides has so far been largely limited to CNY21 variants, the importance of lipid membrane disruption in the action of also other C3a-derived peptides has been demonstrated. Specifically, liposome studies comparing the C3-derived human peptide LGE31 with GKE31, the latter derived from *Carcinoscorpius rotundicauda*, showed both peptides to cause lipid membrane disruption, most likely contributing to their antibacterial mode of action (Shai 2002).

Interestingly, both peptides also displayed a pronounced helix transition together with LPS. Thus, while neither GKE31 nor LGE33 adopted an ordered conformation in aqueous solution (or with phospholipid liposomes), a significant and almost identical structural change, largely indicating an induction of helicity, taking place in the presence of *E. coli* LPS. This suggests that despite a marked difference in primary sequence, the peptides may structure in a similar way in presence of negatively charged LPS-rich bacterial membranes. It also demonstrates the limitations of phospholipid membrane models, and suggests that more complex interactions, involving both lipid membranes and carbohydrate compartments of bacteria walls, are involved in the antimicrobial action of these peptides.

7 Outlook

In conclusion, the combination of phylogenetic, structural, biophysical and biological analyses points at a preservation of structures crucial for antimicrobial activity of C3a in invertebrates as well as vertebrate lineages. In contrast, significant changes in cationicity are observed in cathelicidins and defensins even in closely related species (such as primates) (Semple, Rolfe and Dorin 2003; Zelezetsky et al. 2006a). Furthermore, gene duplications and subsequent variations has lead to further generation of different AMPs (Tennessen 2005). For example, the C-terminal domains of cathelicidins may range from a dozen to over eighty residues forming cysteine-bridged hairpins (bovine bactenecin), tryptophan-rich (bovine indolicidin), proline-rich (porcine PR-39), or α-helical molecules (human LL-37) (Lehrer et al. 2002). In this context, it is interesting to note that a unifying structural motif (γ-core) was recently revealed in many diverse AMPs, such as defensins, protegrins and chemokines, thus indicating the existence of previously undisclosed higher level structural motifs governing antimicrobial function (Yount et al. 2006). In invertebrates and vertebrates, C3a has maintained strikingly similar features governing antimicrobial activity, in spite of a significant primary sequence variation. Thus, in addition to Nature's "innovative" generation of AMPs, the herein reviewed findings on C3a illustrate an alternative concept where molecules are subjected to strong and precise selection forces aiming at maintaining structurally intact innate defense system.

Can this knowledge be utilized in the development of novel, improved AMPs based on the structure of the C-terminal region of C3a? In a present study (Pasupuleti,

Malmsten, Schmidtchen, in manuscript), the peptide CNY20, CNYITELRRQHARASHLGLA, was utilized as a framework to study the effects on antimicrobial activity and spectrum using peptides with alterations affecting predicted helical propensity, net charge, and hydrophobic moment. Among the results obtained for some 150 peptide variants investigated, it was noted that the effects of the CNY20 variant peptides against various Gram-positive bacteria as well as fungi correlated strongly to net charge, and amphipathic peptides with net charge of +6-7 exerted significantly higher antimicrobial activities than the parent peptide. Furthermore, a central, and evolutionary conserved, histdidine residue (H11), which induces a break in the hydrophobic surface of CNY20, was associated with absence of hemolysis in a group of optimized peptides. Thus, based on a sequence template for CNY20, and taking both structural as well as evolutionary relationships into account, it indeed seems possible to design non-hemolytic and otherwise low-toxic peptides with enhanced antimicrobial effects for therapeutic use.

References

Andersson, E., Rydengard, V., Sonesson, A., Morgelin, M., Bjorck, L., and Schmidtchen, A. (2004) Antimicrobial activities of heparin-binding peptides. Eur J Biochem 271: 1219-1226.

Baev, D., Li, X.S., Dong, J., Keng, P., and Edgerton, M. (2002) Human salivary histatin 5 causes disordered volume regulation and cell cycle arrest in Candida albicans. Infect Immun 70: 4777-4784.

Boman, H.G., Agerberth, B., and Boman, A. (1993) Mechanisms of action on Escherichia coli of cecropin P1 and PR-39, two antibacterial peptides from pig intestine. Infect Immun 61: 2978-2984.

Brogden, K.A. (2005) Antimicrobial peptides: pore formers or metabolic inhibitors in bacteria? Nat Rev Microbiol 3: 238-250.

Bulet, P., Stocklin, R., and Menin, L. (2004) Anti-microbial peptides: from invertebrates to vertebrates. Immunol Rev 198: 169-184.

Caporale, L.H., Tippett, P.S., Erickson, B.W., and Hugli, T.E. (1980) The active site of C3a anaphylatoxin. J Biol Chem 255: 10758-10763.

Den Hertog, A.L., Wong Fong Sang, H.W., Kraayenhof, R., Bolscher, J.G., Van't Hof, W., Veerman, E.C., and Nieuw Amerongen, A.V. (2004) Interactions of histatin 5 and histatin 5-derived peptides with liposome membranes: surface effects, translocation and permeabilization. Biochem J 379: 665-672.

Epand, R.F., Ramamoorthy, A., and Epand, R.M. (2006) Membrane lipid composition and the interaction of pardaxin: the role of cholesterol. Protein Pept Lett 13: 1-5.

Ganz, T. (2001) Antimicrobial proteins and peptides in host defense. Semin Respir Infect 16: 4-10.

Glukhov, E., Stark, M., Burrows, L.L., and Deber, C.M. (2005) Basis for selectivity of cationic antimicrobial peptides for bacterial versus mammalian membranes. J Biol Chem 280: 33960-33967.

Hallock, K.J., Lee, D.K., Omnaas, J., Mosberg, H.I., and Ramamoorthy, A. (2002) Membrane composition determines pardaxin's mechanism of lipid bilayer disruption. Biophys J 83: 1004-1013.

Huber, R., Scholze, H., Paques, E.P., and Deisenhofer, J. (1980) Crystal structure analysis and molecular model of human C3a anaphylatoxin. Hoppe Seylers Z Physiol Chem 361: 1389-1399.

Hugli, T.E. (1990) Structure and function of C3a anaphylatoxin. Curr Top Microbiol Immunol 153: 181-208.

Janssen, B.J., Huizinga, E.G., Raaijmakers, H.C., Roos, A., Daha, M.R., Nilsson-Ekdahl, K., Nilsson, B., and Gros, P. (2005) Structures of complement component C3 provide insights into the function and evolution of immunity. Nature 437: 505-511.

Kawamoto, S., Yalcindag, A., Laouini, D., Brodeur, S., Bryce, P., Lu, B., Humbles, A.A., Oettgen, H., Gerard, C., and Geha, R.S. (2004) The anaphylatoxin C3a downregulates the Th2 response to epicutaneously introduced antigen. J Clin Invest 114: 399-407.

Koshlukova, S.E., Lloyd, T.L., Araujo, M.W., and Edgerton, M. (1999) Salivary histatin 5 induces non-lytic release of ATP from Candida albicans leading to cell death. J Biol Chem 274: 18872-18879.

Kozel, T.R. (1996) Activation of the complement system by pathogenic fungi. Clin Microbiol Rev 9: 34-46.
Kozel, T.R., Weinhold, L.C., and Lupan, D.M. (1996) Distinct characteristics of initiation of the classical and alternative complement pathways by Candida albicans. Infect Immun 64: 3360-3368.
Lee, M.T., Hung, W.C., Chen, F.Y., and Huang, H.W. (2005) Many-body effect of antimicrobial peptides: on the correlation between lipid's spontaneous curvature and pore formation. Biophys J 89: 4006-4016.
Lehrer, R.I., and Ganz, T. (2002) Cathelicidins: a family of endogenous antimicrobial peptides. Curr Opin Hematol 9: 18-22.
Lohner, K., and Blondelle, S.E. (2005) Molecular mechanisms of membrane perturbation by antimicrobial peptides and the use of biophysical studies in the design of novel peptide antibiotics. Comb Chem High Throughput Screen 8: 241-256.
Lu, Z.X., Fok, K.F., Erickson, B.W., and Hugli, T.E. (1984) Conformational analysis of COOH-terminal segments of human C3a. Evidence of ordered conformation in an active 21-residue peptide. J Biol Chem 259: 7367-7370.
Maxwell, A.I., Morrison, G.M., and Dorin, J.R. (2003) Rapid sequence divergence in mammalian beta-defensins by adaptive evolution. Mol Immunol 40: 413-421.
Nonaka, M., and Yoshizaki, F. (2004) Evolution of the complement system. Mol Immunol 40: 897-902.
Nordahl, E.A., Rydengard, V., Nyberg, P., Nitsche, D.P., Morgelin, M., Malmsten, M., Bjorck, L., and Schmidtchen, A. (2004) Activation of the complement system generates antibacterial peptides. Proc Natl Acad Sci U S A 101: 16879-16884.
Odds, F.C. (1988) *Candida and candidosis.* London ; Philadelphia: Baillière Tindall.
Ohki, S., Marcus, E., Sukumaran, D.K., and Arnold, K. (1994) Interaction of melittin with lipid membranes. Biochim Biophys Acta 1194: 223-232.
Opekarova, M., and Tanner, W. (2003) Specific lipid requirements of membrane proteins--a putative bottleneck in heterologous expression. Biochim Biophys Acta 1610: 11-22.
Pasupuleti, M., Walse, B., Nordahl, E.A., Morgelin, M., Malmsten, M., and Schmidtchen, A. (2006) Preservation of antimicrobial properties of complement peptide C3a - from invertebrates to humans. J Biol Chem.
Pinto, M.R., Chinnici, C.M., Kimura, Y., Melillo, D., Marino, R., Spruce, L.A., De Santis, R., Parrinello, N., and Lambris, J.D. (2003) CiC3-1a-mediated chemotaxis in the deuterostome invertebrate Ciona intestinalis (Urochordata). J Immunol 171: 5521-5528.
Pitarch, A., Sanchez, M., Nombela, C., and Gil, C. (2003) Analysis of the Candida albicans proteome. II. Protein information technology on the Net (update 2002). J Chromatogr B Analyt Technol Biomed Life Sci 787: 129-148.
Powers, J.P., and Hancock, R.E. (2003) The relationship between peptide structure and antibacterial activity. Peptides 24: 1681-1691.
Ringstad, L., Andersson Nordahl, E., Schmidtchen, A., and Malmsten, M. (2007) Composition Effect on Peptide Interaction with Lipids and Bacteria: Variants of C3a Peptide CNY21. Biophys J 92: 87-98.
Savolainen, J., Lammintausta, K., Kalimo, K., and Viander, M. (1993) Candida albicans and atopic dermatitis. Clin Exp Allergy 23: 332-339.
Schwartz, L.B., Kawahara, M.S., Hugli, T.E., Vik, D., Fearon, D.T., and Austen, K.F. (1983) Generation of C3a anaphylatoxin from human C3 by human mast cell tryptase. J Immunol 130: 1891-1895.
Selsted, M.E., Brown, D.M., DeLange, R.J., Harwig, S.S., and Lehrer, R.I. (1985) Primary structures of six antimicrobial peptides of rabbit peritoneal neutrophils. J Biol Chem 260: 4579-4584.
Selsted, M.E., Harwig, S.S., Ganz, T., Schilling, J.W., and Lehrer, R.I. (1985) Primary structures of three human neutrophil defensins. J Clin Invest 76: 1436-1439.
Semple, C.A., Rolfe, M., and Dorin, J.R. (2003) Duplication and selection in the evolution of primate beta-defensin genes. Genome Biol 4: R31.
Shai, Y. (2002) Mode of action of membrane active antimicrobial peptides. Biopolymers 66: 236-248.
Skarnes, R.C., and Watson, D.W. (1957) Antimicrobial factors of normal tissues and fluids. Bacteriol Rev 21: 273-294.
Sohnle, P.G., and Kirkpatrick, C.H. (1976) Deposition of complement in the lesions of experimental cutaneous candidiasis in guinea pigs. J Cutan Pathol 3: 232-238.
Sonesson, A., Ringstad, L., Andersson Nordahl, E., Malmsten, M., Morgelin, M., and Schmidtchen, A. (2006) Antifungal activity of C3a and C3a-derived peptides against Candida. Biochim Biophys Acta.
Steiner, H., Hultmark, D., Engström, A., Bennich, H., and Boman, H.G. (1981) Sequence and specificity of two antibacterial proteins involved in insect immunity. Nature 292: 246-248.

Sunyer, J.O., Boshra, H., and Li, J. (2005) Evolution of anaphylatoxins, their diversity and novel roles in innate immunity: insights from the study of fish complement. Vet Immunol Immunopathol 108: 77-89.
Taylor, J.C., Crawford, I.P., and Hugli, T.E. (1977) Limited degradation of the third component (C3) of human complement by human leukocyte elastase (HLE): partial characterization of C3 fragments. Biochemistry 16: 3390-3396.
Tennessen, J.A. (2005) Molecular evolution of animal antimicrobial peptides: widespread moderate positive selection. J Evol Biol 18: 1387-1394.
Wang, Y., Agerberth, B., Lothgren, A., Almstedt, A., and Johansson, J. (1998) Apolipoprotein A-I binds and inhibits the human antibacterial/cytotoxic peptide LL-37. J Biol Chem 273: 33115-33118.
Werfel, T., Kirchhoff, K., Wittmann, M., Begemann, G., Kapp, A., Heidenreich, F., Gotze, O., and Zwirner, J. (2000) Activated human T lymphocytes express a functional C3a receptor. J Immunol 165: 6599-6605.
Williamson, M.P., and Madison, V.S. (1990) Three-dimensional structure of porcine C5adesArg from 1H nuclear magnetic resonance data. Biochemistry 29: 2895-2905.
Wright, J., Schwartz, J.H., Olson, R., Kosowsky, J.M., and Tauber, A.I. (1986) Proton secretion by the sodium/hydrogen ion antiporter in the human neutrophil. J Clin Invest 77: 782-788.
Yount, N.Y., Bayer, A.S., Xiong, Y.Q., and Yeaman, M.R. (2006) Advances in antimicrobial peptide immunobiology. Biopolymers.
Zasloff, M. (1987) Magainins, a class of antimicrobial peptides from Xenopus skin: isolation, characterization of two active forms, and partial cDNA sequence of a precursor. Proc Natl Acad Sci U S A 84: 5449-5453.
Zasloff, M. (2002) Antimicrobial peptides of multicellular organisms. Nature 415: 389-395.
Zelezetsky, I., Pontillo, A., Puzzi, L., Antcheva, N., Segat, L., Pacor, S., Crovella, S., and Tossi, A. (2006a) Evolution of the primate cathelicidin - correlation between structural variations and antimicrobial activity. J Biol Chem.
Zelezetsky, I., and Tossi, A. (2006b) Alpha-helical antimicrobial peptides-Using a sequence template to guide structure-activity relationship studies. Biochim Biophys Acta.
Zeya, H.I., and Spitznagel, J.K. (1963) Antibacterial and Enzymic Basic Proteins from Leukocyte Lysosomes: Separation and Identification. Science 142: 1085-1087.
Zhang, L., and Falla, T.J. (2004) Cationic antimicrobial peptides - an update. Expert Opin Investig Drugs 13: 97-106.
Zhang, X., Boyar, W., Toth, M.J., Wennogle, L., and Gonnella, N.C. (1997) Structural definition of the C5a C terminus by two-dimensional nuclear magnetic resonance spectroscopy. Proteins 28: 261-267.
Zhu, Y., Thangamani, S., Ho, B., and Ding, J.L. (2005) The ancient origin of the complement system. Embo J 24: 382-394.

12

C5L2 – an Anti-inflammatory Molecule or a Receptor for Acylation Stimulating Protein (C3a-desArg)?

Kay Johswich[1] and Andreas Klos[2]

[1] Medical School Hannover, Department of Medical Microbiology, 30625 Hannover, Germany, johswich.kay@mh-hannover.de
[2] Medical School Hannover, Department of Medical Microbiology, 30625 Hannover, Germany, klos.andreas@mh-hannover.de

1 Introduction

Sometimes, two seemingly unrelated research areas suddenly overlap to the surprise of the scientists involved. This has occurred to the researchers investigating in the complement field the anaphylatoxins C5a, C5a-desArg, and C3a and their receptors: the C5a-receptor (C5aR, CD88), the C5a-receptor like 2 (C5L2), and the C3a-receptor (C3aR). Anaphylatoxins and their receptors play a crucial role in the inflammation reaction against foreign material in the body (for review see (Ember, Jagels, and Hugli 1998)). Only seconds after the generation of C3a, serum carboxypeptidase removes the C-terminal arginine that is indispensable for the binding of C3a to the C3aR. Thus, among complement researchers, C3a-desArg is mainly regarded as a non-functional complement degradation product which serves as a valuable measure of complement activation (Hartmann, Lubbers, Casaretto, Bautsch, Klos, and Kohl 1993). However, C3a-desArg (and C3a) might play a role in bone marrow cell engraftment (Ratajczak, Reca, Wysoczynski, Kucia, Baran, Allendorf, Ratajczak, and Ross 2004). Additionally, C5L2 has been mainly considered as a non-signaling decoy receptor for C5a and C5a-desArg and thus as an antagonist of the C5aR (Okinaga, Slattery, Humbles, Zsengeller, Morteau, Kinrade, Brodbeck, Krause, Choe, Gerard, and Gerard 2003). Most people involved in this research field thought that they were focusing on inflammation and immunity.

On the other hand scientists involved in research on lipid and carbohydrate metabolism and related diseases must have been equally surprised when purified acylation stimulating protein (ASP), a potent anabolic activator of triglyceride synthesis and of glucose uptake in cultured adipocytes and skin fibroblasts, turned out to be identical to C3a-desArg (Cianflone, Sniderman, Walsh, Vu, Gagnon, and Rodriguez 1989b). Moreover, recently it has been claimed that C5L2 acts additionally as the functional receptor for C3a and ASP/C3a-desArg leading to triglyceride synthesis and glucose uptake (Kalant, Cain, Maslowska, Sniderman, Cianflone, and Monk 2003; Kalant, Maclaren, Cui, Samanta, Monk, Laporte, and Cianflone 2005). Based on these observations, downstream signaling of ASP such as the activation of the phospholipase C pathway has been attributed to C5L2. Investigations on patients have been started to associate mutations of this receptor with reduced ASP-stimulated triglyceride synthesis and glucose transport (Marcil, Vu, Cui, Dastani, Engert, Gaudet, Castro-Cabezas, Sniderman, Genest, and Cianflone 2006), and increased plasma levels of triglycerides, cholesterol, low-density lipoprotein (LDL) cholesterol, apolipoprotein B and ASP (Maslowska, Legakis, Assadi, and Cianflone 2006).

However, very recently it has been shown definitively that C5L2 does not bind C3a or ASP/C3a-desArg (Johswich, Martin, Thalmann, Rheinheimer, Monk, and Klos 2006) and thus, that it is unlikely that C5L2 is directly involved in lipid or carbohydrate metabolism. These results might lead to some confusion in both fields. This review provides an overview about the different lines of research on C5L2 and ASP/C3a-desArg, reevaluates former results and address critically the following questions: What is the evidence for the involvement of the complement system and in particular of C3 in glucose and lipid metabolism? How strong is the evidence for the identity of C3a-desArg with ASP? What are data which indicate that C5L2 might be the receptor for ASP/C3a-desArg and what are the facts demonstrating that this is not the case? Moreover, is there still a potential role for C3a-desArg or C5L2 in carbohydrate/lipid metabolism? On the other hand, what is the role of C5L2 as a specific receptor for C5a and C5a-desArg? What is the evidence that C5L2 acts as an anti-inflammatory mediator of C5a-induced inflammation?

2 The C5L2 receptor

2.1 C5L2 as a Member of the Anaphylatoxin Receptor Family

Anaphylatoxins exert their functions within the complement system by binding to their specific receptors of the anaphylatoxin receptor family consisting of the closely related members C3aR, C5aR and C5L2. They belong to the superfamily of rhodopsin-like G-protein coupled receptors with seven transmembrane domains. These three receptors are highly homologous, especially in the transmembrane regions. The anaphylatoxin receptors show a high similarity to the chemotactic fMLP-receptor and to the chemokine receptors CXCR1 and CXCR2 recognizing IL-8. This reflects not only structural but also functional relations, as all these receptors are involved in cell migration towards inflammatory mediators. All three anaphylatoxin receptors show a broad expression pattern which is not restricted to the hemopoietic cell lineage.

The first Anaphyatoxin receptor, cloned 1991, was C5aR (Boulay, Mery, Tardif, Brouchon, and Vignais 1991; Gerard and Gerard 1991). Sequence analysis revealed a seven transmembrane domain receptor with a mass of about 42 kDa, consisting of 350 amino acids of which asparagine 5 is glycosylated and tyrosines 11 and 14 are sulfated (Farzan, Schnitzler, Vasilieva, Leung, Kuhn, Gerard, Gerard, and Choe 2001). The single copy gene for human C5aR is located on chromosome 19q13.4 together with other genes for chemotactic receptors (fMLP receptor, FPRH1, FPRH2)

(Bao, Gerard, Eddy, Jr., Shows, and Gerard 1992). Unlike most other G-protein coupled receptors, usually consisting of a single exon, it is made up of two exons separated by a 9 kb intron located directly behind the start codon. While most G-protein coupled receptors share a high interspecies similarity (~85-98%), the gene for C5aR is quite variable in different species. The overall similarity among human, mouse, rat, rabbit and bovine C5aR is only about 68% (Gerard, Bao, Orozco, Pearson, Kunz, and Gerard 1992; Perret, Raspe, Vassart, and Parmentier 1992). C5aR binds C5a with a K_d of 1 nM and, to a lesser extent, also its desarginated form. The affinity of C5aR on human monocytes for C5a-desArg is roughly 10-100 fold lower than for C5a (Marder, Chenoweth, Goldstein, and Perez 1985). However, this study did not take in account another C5a-desArg binding receptor. In recent reports, the K_d of C5a-desArg binding to C5aR was estimated at 412-660 nM (Okinaga et al. 2003). Binding of C5a to C5aR involves two different sites of C5aR. One site is the aspartate-rich, negatively charged N-terminus and the other site is a moiety built up of the transmembrane segments which interact with the agonistic C-terminus of C5a (Siciliano, Rollins, DeMartino, Konteatis, Malkowitz, Van Riper, Bondy, Rosen, and Springer 1994). Upon ligand binding, C5aR initiates intracellular signaling by activating pertussis-toxin sensitive $Gi\alpha_2$, $Gi\alpha_3$ (Rollins, Siciliano, Kobayashi, Cianciarulo, Bonilla-Argudo, Collier, and Springer 1991), or pertussis-toxin insensitive $G\alpha_{16}$ expressed in hemopoietic cells (Amatruda, III, Gerard, Gerard, and Simon 1993). Interaction of C5a with C5aR leads to activation of several signaling pathways including phophatidylinositol-bisphosphate-3-kinase (Perianayagam, Balakrishnan, King, Pereira, and Jaber 2002), phospholipase D (Mullmann, Siegel, Egan, and Billah 1990) as well as protein kinase C and mitogen activated protein-(MAP)-kinases (Buhl, Avdi, Worthen, and Johnson 1994; la Sala, Gadina, and Kelsall 2005). As C5aR is involved in chemotaxis, oxidative burst and other inflammatory mechanisms it was initially thought to play a role predominantly on myeloid cells. Actually, C5aR is highly expressed in granulocytes, monocytes/macrophages, mast cells and dendritic cells (Chenoweth and Goodman 1983; Chenoweth and Hugli 1978; Gerard, Hodges, Drazen, Weller, and Gerard 1989). More recently C5aR was found to be expressed in most organs and many different cell types like endothelial cells (Laudes, Chu, Huber-Lang, Guo, Riedemann, Sarma, Mahdi, Murphy, Speyer, Lu, Lambris, Zetoune, and Ward 2002), neurons (Farkas, Baranyi, Takahashi, Fukuda, Liposits, Yamamoto, and Okada 1998; Stahel, Frei, Eugster, Fontana, Hummel, Wetsel, Ames, and Barnum 1997), astrocytes, microglia (Gasque, Singhrao, Neal, Gotze, and Morgan 1997), and other organs including kidney, lung, liver, spleen, intestine and heart (Fayyazi, Scheel, Werfel, Schweyer, Oppermann, Gotze, Radzun, and Zwirner 2000; Wetsel 1995). Using C5aR knockout mice or C5a/C5aR-inhibitors it has been shown that C5a is a critical player in various diseases such as sepsis (Guo, Riedemann, and Ward 2004), Arthus reaction (Kohl and Gessner 1999), Alzheimer's disease (Mukherjee and Pasinetti 2000), asthma (Kohl, Baelder, Lewkowich, Pandey, Hawlisch, Wang, Best, Herman, Sproles, Zwirner, Whitsett, Gerard, Sfyroera, Lambris, and Wills-Karp 2006) and in various models of ischemia-reperfusion injury (Arumugam, Shiels, Woodruff, Granger, and Taylor 2004).

In contrast to C5a, the binding of C3a to cells was thought until recently to occur by nonspecific interaction of the positive charged ligand with the negative charged cell surface (Fukuoka and Hugli 1990; Gervasoni, Jr., Conrad, Hugli, Schwartz, and Ruddy 1986). The specific receptor for C3a, C3aR, was cloned in 1996 independently by two different groups (Ames, Li, Sarau, Nuthulaganti, Foley, Ellis, Zeng, Su, Jurewicz, Hertzberg, Bergsma, and Kumar 1996; Crass, Raffetseder, Martin, Grove, Klos, Kohl, and Bautsch 1996). Sequence analysis revealed that it is identical to the orphan receptor AZ3B (Roglic, Prossnitz, Cavanagh, Pan, Zou, and Ye 1996). C3aR is a 54 kDa protein consisting of 482 amino acids with glycosylations at asparagine 9 and 195. A remarkable feature of this receptor is its extremely large second extracellular loop making up roughly a third of the whole protein and being crucial for ligand binding. A key residue for ligand/receptor interaction is the sulfated tyrosine 174 within this loop (Gao, Choe, Bota, Wright, Gerard, and Gerard 2003). The human gene for C3aR is located on chromosome 12p13 and comprises two exons separated by a 6 kb intron residing 11 nucleotides upstream of the start codon (Paral, Sohns, Crass, Grove, Kohl, Klos, and Bautsch 1998). C3aR binds C3a with high affinity (K_d: ~1nM), whereas C3a-desArg does not act as a ligand for C3aR (Crass et al. 1996; Wilken, Gotze, Werfel, and Zwirner 1999). Like C5aR, C3aR transduces signaling via pertussistoxin-sensitive Giα (Norgauer, Dobos, Kownatzki, Dahinden, Burger, Kupper, and Gierschik 1993) or Gα16 in hemopoietic cells (Crass et al. 1996). It is also thought to interact with pertussis-toxin insensitive $G\alpha_{12}$ and $G\alpha_{13}$ in endothelial cells (Schraufstatter, Trieu, Sikora, Sriramarao, and DiScipio 2002). In contrast to C5aR, no activation of phophatidylinositol-bisphosphate-3-kinase is observed after activation of C3aR. Thus, C3a-dependent Ca^{2+} influx occurs only from the extracellular medium, whereas C5a/C5aR-interaction leads also to mobilization from intracellular calcium stores (Norgauer et al. 1993). C3aR activates protein kinase C via phospholipase C (Langkabel, Zwirner, and Oppermann 1999). In astrocytes, stimulation of either C3aR or C5aR leads to activation of MAP-kinases p42 and p44 (Sayah, Jauneau, Patte, Tonon, Vaudry, and Fontaine 2003).

The third member of the anaphylatoxin receptor family, C5L2, was discovered in 2000 by Ohno et al. (Ohno, Hirata, Enomoto, Araki, Ishimaru, and Takahashi 2000). C5L2 is a 37 kDa protein consisting of 337 amino acids. Asparagine 3 serves as a potential glycosylation site. The gene for C5L2 is located on chromosome 19q13.4, directly neighboring C5aR. C5L2 shares 58 % sequence identity with C5aR and 55 % with C3aR in the transmembrane regions (Lee, George, Cheng, Nguyen, Liu, Brown, Lynch, and O'Dowd 2001). The gene comprises two exons interspersed by an 3,6 kb intron upstream of the start codon, thereby resembling the genetic architecture of C5aR and C3aR. Soon after its discovery C5L2 was found to bind C5a with a high affinity (K_d: 2,5 nM) and additionally C5a-desArg (K_d: 12 nM), resulting in the opinion that C5L2 is the second C5a binding chemoattractant receptor. Notably, C5a-desArg binds with a 20-30 fold higher affinity to C5L2 than to C5aR (Cain and Monk 2002; Okinaga et al. 2003). However, C5aR and C5L2 show significant differences with respect to their velocity of ligand binding as the onrate of C5aR is about 100 fold faster than that of C5L2. In the last years there was a controversy regarding the affinity of C5L2 for C3a, C3a-desArg, C4a and C4a-desArg (Cain et al. 2002; Kalant et al. 2003; Okinaga et al. 2003).

To date, no direct signalling functions of C5L2 have been observed. Activation of either C5aR or C3aR usually leads to calcium fluxes. In contrast, no mobilization of intracellular Ca^{2+} was observed in C5L2 transfected cells after administration of C5a, C5a-desArg, C3a or C4a (Cain et al. 2002; Okinaga et al. 2003). Additionally, no calcium fluxes were observed in endogenously C5L2 expressing cell lines (Johswich et al. 2006) or in neutrophils from $C5aR^{-/-}$ mice after stimulation with C5a (Hopken, Lu, Gerard, and Gerard 1996). This behavior emerges from differences in the DRY motif which is important for the interaction of the seven transmembrane receptors with their corresponding G-proteins. The DRY motif is located in the third transmembrane domain and is highly conserved through G-protein coupled receptors such as chemokine receptors, fMLP-receptor, C5aR and C3aR. The DRY motif is DRF in C5aR and DRC in C3a, but in C5L2 it is DLC. The central arginine residue plays a key role in coordinating the transmembrane domains 3 and 6 and mutagenesis of this residue in C5aR leads to loss of function (Kolakowski, Jr., Lu, Gerard, and Gerard 1995). When DLC in C5L2 is restored to DRC, C5L2 couples weakly to intracellular Ca^{2+} fluxes in 293 cells coexpressing $G\alpha_{16}$ (Okinaga et al. 2003). Another moiety differing in C5L2 is the very short third intracellular loop that lacks a conserved basic region (KTLK in C5aR or KTFR in C3aR, respectively) as well as serine/threonine residues that are potential protein kinase C phosphorylation sites. Mutation in C5aR of these serine/threonine residues to alanine leads to uncoupling of the receptor from intracellular signaling (Bock, Martin, Gartner, Rheinheimer, Raffetseder, Arseniev, Barker, Monk, Bautsch, Kohl, and Klos 1997). Taken together, these findings strongly suggest that C5L2 is uncoupled from G-proteins. Additionally, no activation of the MAP kinase pathway by binding of C5a to C5L2 was observed in stably transfected pre B L1.2 cells (Okinaga et al. 2003). Furthermore, no degranulation of C5L2 transfected RBL cells occurred after stimulation with C5a, C5a-desArg, or C3a (Cain et al. 2002). However, preloading of the tyrosine kinase coupled IgE-receptors (FcεRI) with DNP-specific IgE resulted in a small (~1,3 fold) increase in β-hexosaminidase release when C5a, C5a-desArg, C3a or C4a were administered prior to crosslinking the IgE-receptors with the specific antigen (Cain et al. 2002). Ultimately, C5a binding to C5L2 in bone marrow cells derived from $C5aR^{-/-}$ mice does not lead to changes in the mRNA expression pattern in these cells (Okinaga et al. 2003). A central feature of G-protein coupled receptors is their internalization after ligand binding. Whereas C5aR and C3aR are internalized rapidly after binding of their specific ligands (Bock et al. 1997; Settmacher, Bock, Saad, Gartner, Rheinheimer, Kohl, Bautsch, and Klos 1999), two independent studies found no internalization of C5L2 upon stimulation with C5a, C5a-desArg, C3a or C4a (Cain et al. 2002; Okinaga et al. 2003). A prerequisite for receptor internalization is the phosphorylation of C-terminal serine or threonin residues by G-protein coupled receptor kinases (GRKs), protein kinase A or protein kinase C, followed by their association with arrestins to the receptor. C5L2 in transfected mouse L1.2 cells elicited only a very basal level of phosphorylation after C5a stimulation, although possessing a serine/threonine rich C-terminus (Okinaga et al. 2003). However, in another study using transfected HEK293 cells, C5L2 was found to be robustly phosphorylated after stimulation with 20 μM C3a-desArg. Furthermore, a cotransfected β-arrestin-GFP fusion protein translocated from the plasma membrane to endocytic vesicles in these cells after stimulation with C5a, C3a or C3a-desArg, thereby monitoring functionality of C5L2 (Kalant et al. 2005).

Like C5aR and C3aR, C5L2 is expressed in various tissues of myeloid and non-myeloid origin. Surface expression of C5L2 was detected in lung, liver, heart, kidney (Gao, Neff, Guo, Speyer, Sarma, Tomlins, Man, Riedemann, Hoesel, Younkin, Zetoune, and Ward 2005), in adipose tissue and in skin fibroblasts (Kalant et al. 2005), in neutrophils (Huber-Lang, Sarma, Rittirsch, Schreiber, Weiss, Flierl, Younkin, Schneider, Suger-Wiedeck, Gebhard, McClintock, Neff, Zetoune, Bruckner, Guo, Monk, and Ward 2005) and in immature, but not in mature dendritic cells (Ohno et al. 2000). Additionally, C5L2 transcripts were detected in the brain (Gavrilyuk, Kalinin, Hilbush, Middlecamp, McGuire, Pelligrino, Weinberg, and Feinstein 2005; Lee et al. 2001), in placenta, ovary, testis, spleen and colon (Ohno et al. 2000). Notably, comparative southern blots revealed congruent expression of C5L2 and C5aR in the same tissues (Okinaga et al. 2003), suggestive for a coexpression of both receptors on the cell surface and possibly a functional interplay.

2.2 Postulated functions of C5L2

The following main functions have been proposed for C5L2:

1) C5L2 serves as a non-signaling scavenger receptor for C5a and C5a-desArg as its rapidly generated, still biologically active derivative. C5L2 controls C5a/C5a-desArg mediated inflammation limiting it to the site of complement activation. Elimination of these anaphylatoxins helps to sustain their chemotactic gradient.

2) C5L2 is the functional receptor for C3a and ASP/C3a-desArg being responsible for changes in the lipid and carbohydrate metabolism.

3) C5L2 might activate other cell responses after binding of its ligands by so far unidentified intracellular signaling pathways.

3 C5L2 and Lipid and Carbohydrate Metabolism

3.1 The Adipsin / ASP Model

The adipose tissue plays a crucial role in energy homeostasis and insulin sensitivity. Adipocytes produce a number of hormones which effect energy intake and expenditure. Thereby, ASP, leptin, and adiponectin are considered as such mediators and keyplayers within the carbohydrate and lipid metabolism [for a detailed review on lipid metabolism see (Havel 2004)]. ASP is regarded as an anabolic determinant of energy homeostasis and insulin action. Moreover, a malfunction of the ASP receptor or ASP-dependent signaling is considered as a potential cause of familial hyperapobetalipoproteinemia (hyperapoB) which leads to arteriosclerosis and coronary artery disease. In hyperapoB patients, an overproduction of hepatic apolipoprotein B-100 (apoB-100) occurs. This is postulated to be responsible for the observed postprandial hyperlipidemia with increased triglyceride-rich remnants from processed chylomicrons, increased plasma fatty acid levels, insulin resistence, as well as elevated levels of very low density protein (VLDL) and reduced levels of high density lipoprotein (HDL). Furthermore, ASP might be generally involved in fat storage, obesity, and their medical consequences for public health. During the last 4 years C5L2 was considered as the functional ASP receptor. Therefore, ASP and C5L2 are discussed as drug targets. The interest on this topic is reflected by ~120 MEDLINE listed publications on ASP which appeared during the last two decades and ~20 publications on C5L2, in part overlapping, at the present time.

3.2 Historical Development of the Adipsin/ASP Model

In 1987 a partially purified protein fraction has been described which stimulated fatty acid uptake and esterification by human adipocytes and cultured skin fibroblast (Cianflone, Kwiterovich, Walsh, Forse, Rodriguez, and Sniderman 1987). In the presence of lipoprotein-deficient serum the *de novo* sythesis of triglycerides was substantially smaller in cultured skin fibroblast of hyperapoB patients as compared to cells from healthy donors. These findings suggested that a serum protein stimulates triglyceride synthesis and that hyperapoB fibroblasts are less responsive to it (Cianflone, Rodriguez, Walsh, Vu, and Sniderman 1988). In 1989 this protein with a molecular weight of 14,000 and a pI of 9.0 was purified to (assumed) homogeneity, and named acylation stimulating protein (ASP). The stimulation of triacylglycerol synthesis determined in an [^{14}C]oleate incorporation assay was greater with ASP than with insulin (Cianflone et al. 1989b). A competitive ELISA using a polyclonal antiserum against purified ASP was applied to demonstrate that a sustained increase in ASP occurred after an oral fat but not a glucose load in normal subjects (Cianflone, Vu, Walsh, Baldo, and Sniderman 1989a). ASP-dependent triglyceride synthesis was markedly reduced in cultured skin fibroblast from hyperapoB patients but not from healthy donors or patients suffering from familial hypercholesterolemia or hypertriglyceridemia without hyperapoB (Cianflone, Maslowska, and Sniderman 1990). In ^{125}I-ASP binding studies, a K_d of ~1 μM was determined for the binding of ASP to skin fibroblast of healthy donors and signaling and decrease triacylglycerol synthesis. The decreased uptake of fatty acids released by lipoproteinlipase from chylomicrons in adipose tissue might lead to their increased transport by the general circulation and uptake in the liver, and as a secondary effect to increased hepatic VLDL production and other imbalances of the lipid content in serum in this metabolic disease. This hypothesis is in detail described in (Sniderman, Cianflone, Summers, Fielding, and Frayn 1997).

It was shown that ASP stimulates transport of glucose in human skin fibroblasts and that the glucose transporter GLUT 1 is translocated from intracellular stores to the plasma membrane (Germinario, Sniderman, Manuel, Lefebvre, Baldo, and Cianflone 1993). Glucose is also involved in triglycerid synthesis since it is required for the generation of the backbone of glycerol-3-phosphate to which the free fatty acid is linked. In 1993 Baldo et. al demonstrated by amino acid terminal sequencing that the purified ASP was identical to C3a-desArg (Baldo, Sniderman, St Luce,

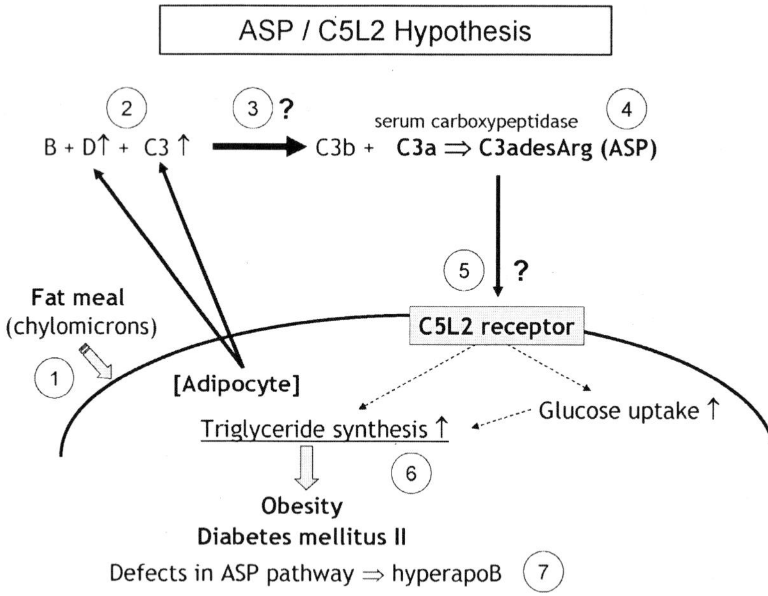

Fig. 1. The ASP / C5L2 Hypothesis. (1) After fat ingestion chylomicrons activate adipocytes and skin fibroblasts. (2) In response, complement factor C3 and D are locally produced. (3) An unknown trigger activates C3 degradation, most likely by the alternative pathway: C3 is cleaved into C3b and the anaphylatoxin C3a. (4) By serum carboxypeptidase, C3a is immediately processed to C3a-desArg/ASP which can no longer bind to the C3a receptor. (5) Within an autocrine loop, ASP binds to the C5L2 receptor on adipocytes. (6) ASP increases triglyceride synthesis and glucose uptake via C5L2. Thus, ASP influences lipid metabolism and participates in obesity or diabetes mellitus. (7) If the ASP/C5L2 signaling is disturbed. e.g. in case of a decreased expression of the ASP receptor, triglyceride synthesis in adipocytes is reduced and lipids remain in the blood stream. They are transported to the liver where they finally lead (indirectly) to metabolic diseases such as hyperapoB with an increased risk for arteriosklerose and cardiovascular diseases.

Avramoglu, Maslowska, Hoang, Monge, Bell, Mulay, and Cianflone 1993). C3a/C3a-desArg generated *in vitro* from purified complement C3, B, and D (adipsin) increased triacylglycerol synthesis in human skin fibroblast (increase of ~200%). A maximal biological effect was reached in the range of 50 - 200 ng/mL (~5-20 nM). It is known that differentiating murine 3T3 adipocytes express complement factors (Cook, Min, Johnson, Chaplinsky, Flier, Hunt, and Spiegelman 1987). It was postulated, that C3, in conjunction with factor B and D are produced and released by adipocytes after lipid ingestion. According to this hypothesis, the alternative pathway is then activated by an unidentified trigger. C3 is cleaved into C3b and C3a which is immediately further processed by serum carboxypeptidase N to the more stable autocrine hormone ASP/C3a-desArg (see model in Fig. 1). Mature human adipocytes contain mRNA coding for the three complement factors. Cultured differentiating adipocytes produce

ASP proportional to the degree of accumulated triacylglycerol. ASP is more potent in differentiated cells. ASP induces an intrinsic change in the activity of triacylglycerol synthesizing enzymes (Cianflone, Roncari, Maslowska, Baldo, Forden, and Sniderman 1994). The maximal increase in triacylglycerol synthesis was reached here at 100 - 200 ng/ml (~10-20 nM) purified ASP.

In ASP-dependent signal transduction an increase in diacylglycerol and a translocation and activation of PKC occurs (Baldo, Sniderman, St Luce, Zhang, and Cianflone 1995). Inhibitors of PKC diminish the effects of ASP on triglyceride synthesis. In a more detailed analysis on 3T3 preadipocytes (Maslowska et al. 2006) it was demonstrated that the intracellular signaling involves within minutes to hours sequential activation of phosphatidylinositol 3-kinase, and phospholipase C, with downstream activation of protein kinase C, Akt phosphorylation, MAP/extracellular signal-regulated kinase (ERK) 1 / 2 and calcium-dependent phospholipase A2. Pertussis toxin had no effect on triglyceride sysnthesis. The signaling pathways used by ASP overlapped only partially with that used by insulin.

From the nineties on, a variety of studies on patients were peformed. One of them showed that obese female subjects had a higher ASP level in plasma (116 vs 53 nM) which dropped progressively during a prolonged fast along with in increase in free fatty acids (Cianflone, Kalant, Marliss, Gougeon, and Sniderman 1995). Furthermore, ASP measured in fasting samples was higher in patients with coronary artery disease (Cianflone, Zhang, Genest, Jr., and Sniderman 1997).

In differentiated human adipocytes, insulin increased ASP production up to twofold. In contrast, addition of chylomicrons obtained from postprandial plasma increased ASP and C3 levels more than 100fold (Maslowska, Scantlebury, Germinario, and Cianflone 1997). It could also be demonstrated that ASP regulates glucose transport in a rat muscle cell-line. ASP increased the V_{max} for glucose transport due to a translocation of the glucose transporters GLU1, GLUT3, and GLUT4 to the plasma membrane (Tao, Cianflone, Sniderman, Colby-Germinario, and Germinario 1997).

In 1997, recombinant ASP (from *E. coli*) highly purified by HPLC was characterized and compared with purified plasma ASP and *in vitro* from factors C3, B and D generated ASP (Murray, Parker, Kirchgessner, Tran, Zhang, Westerlund, and Cianflone 1997). The functional responses obtained with recombinant and purified ASP on human adipocytes and cultured skin fibroblasts were similar. The ASP effects were here observed at 100fold higher concentrations as earlier described, i.e. in the micromolar range. Competition binding assays on 3T3 fibroblasts or human differentiated adipocytes competing up to 500 nM ^{125}I-labelled ASP with nonlabelled ASP indicated an IC_{50} of ~2.7 μM or 420 nM, respectively, for recombinant ASP. Moreover, it was shown that 0.7 μM human ASP can stimulate triacylglycerol synthesis of omental primary cells from Cynomologous and African Green monkey. As an additional control for specificity, immunoprecipitation of recombinant purified or *in vitro* complement generated ASP with a monoclonal anti-C3a/ASP antibody drastically reduced triglyceride synthesis. In 1999 synthetic peptides related to C3a/C3a-desArg were used as well as enzymatically or chemically modified forms of ASP (Murray, Kohl, and Cianflone 1999a). Here, 1 nM ^{125}I-labbeled ASP was applied as tracer. The IC_{50} values on human skin fibroblasts were 44 nM for ASP, and 120 nM for the peptide P117 (C3a63-76; LRRQAWRASALGLAR with three underlined deviations from the native C3a-sequence), an activator of the C3aR. Two other peptides with smaller partial C3a-sequences were much less effective competitors. As described by others, the C-terminal Arg of C3a does not play any role in ASP binding and activity. Modifications which altered the Cys-bound core of C3a/C3a-desArg had little effect of specific binding of ASP to fibroblasts. The removal of the N-terminal nine amino acids did not influence ASP binding. However, modifications of Lys residues of which several are located in the N-terminal part of the C3a core structure as well as alterations of 2 C-terminal His residues decreased binding affinity substantially. Similar to ASP, micromolar concentrations of P117 could stimulate triglyceride synthesis in human skin fibroblast and Swiss 3T3 cells. However, the maximal stimulation reached by increasing concentrations of this peptide was smaller than those obtained with ASP. These data pointed to a specific receptor for ASP which is different from the C3aR.

From 1999 on, C3 deficient mice, unable to produce ASP, were investigated. These mice (in particular males) have delayed postprandial triglyceride clearance and an altered postprandial and adipose tissue metabolism (Murray, Sniderman, and Cianflone 1999b; Murray, Sniderman, Havel, and Cianflone 1999c). Their energy intake was increased and feed efficiency reduced. The amount of postprandial detected nonesterified fatty acids was also higher in mice lacking C3 and ASP. Intraperitonal administration of ASP accelerated *in vivo* triglyceride clearance and reduced plasma glucose (Murray, Sniderman, and Cianflone 1999a).

Intriguingly, Kajita and Hugli had also instillated C3a and C3a-desArg in the peritoneal cavity of rodents (Kajita and Hugli 1991). After introduction of micromolar levels of human C3a or C3a-desArg, both peptides stimulated chymase release with similar activity and C3a degradation by this proteinase in rats. However, these authors had concluded that "the mechanism of rat mast cell activation by C3a (C3ades Arg) was nonspecific [...] as C3a is a highly cationic molecule."

Wetsel and his colleagues could not confirm Murray's data (Wetsel, Kildsgaard, Zsigmond, Liao, and Chan 1999). In their study, no significant differences were found in the triglyceride, cholesterol, or free fatty acid concentrations in the plasma of fasted normal and ASP-deficient mice. In addition, plasma lipoprotein content including apoB-100 and apoB-48 or HDL cholesterol was not significantly different. When challenged with an oral fat load, ASP deficient mice showed no impaired ability to clear triglycerides and free fatty acids from their circulation. The authors called for a reevaluation of the hyperapoB hypothesis. In response it was proposed that the observed contradicting results in the C3-knockout mouse were a consequence of the background strain (Cianflone, Xia, and Chen 2003).

Another divergent result was published by Ylitalo and colleagues who did not find any differences in plasma ASP levels comparing hypertriglyceridemic Finnish patients suffering from familial combined hyperipidemia and their normotriglyceridemic relatives (Ylitalo, Pajukanta, Meri, Cantor, Mero-Matikainen, Vakkilainen, Nuotio, and Taskinen 2001). Additionally, no increase in plasma ASP after a fatty meal in patients or control subjects was observable.

3.3 C5L2 as a Receptor for ASP?

In 2003 a new era of ASP research began with the assumed identification of C5L2 as a receptor for C3a/C3a-desArg/ASP (Cain et al. 2002; Kalant et al. 2003). This interpretation was based on competitive binding studies on RBL cells stably transfected with human C5L2, supported by FACS analysis on transfected HEK293 cells. Additionally, RT-PCR was used to demonstrate the presence of mRNA coding for C5L2 on the same cells which showed specific ASP binding. ^{125}I-C5a as tracer could be competed with increasing concentrations of non-labeled C5a with an IC_{50} of ~20 nM, and C5a-desArg (IC_{50} of ~400 nM), but not with C3a (IC_{50} of ~25 μM), or ASP/C3a-desArg on C5L2-expressing RBL-cells. However, applying ^{125}I-C3a as tracer, non-labeled C3a, C3a-desArg/ASP, C5a, and even C4a and C4a-desArg were able to compete with increasing concentrations with IC_{50} between ~160 and ~530 nM. This suggested that C5L2 is not only a specific receptor for C5a/C5a-desArg, but also a receptor for C3a/C3a-desArg and C4a/C4a-desArg - with partially distinct binding-sites. FACS analysis of transfected HEK cells using FLUOS labeled ASP/C3a-desArg and C3a showed weak but dose-dependent binding of both ligands. For human skin fibroblasts an IC_{50} of 50-70 nM was determined for C3a binding. In 2005 C5L2 was described as a functional receptor for ASP (Kalant et al. 2005). Stimulation of HEK cells stably transfected with human C5L2 led to a dose-dependend increase in triglyceride synthesis of ~200% compared to non-transfected cells (IC_{50} ~2μM). A smaller difference was seen in a similar experiment for glucose uptake. Only when ASP and C5L2 were simultaneously present, an increase in diacylglycerol acyltransferase activity could be detected. Transfection of antisense oligonucleotides to human C5L2 diminished C5L2 expression and triglyceride synthesis in human skin fibroblasts and in 3T3 cells. By immunoprecipitation rapid phosphorylation of C5L2 was observed after incubation with ASP. Finally, within 40 min after incubation with ASP fluorescence-tagged β-arrestin was redistributed in the cytosol of transfected cells suggesting translocation to the membrane and association with C5L2.

However, other groups could not observe any C5L2 internalization (Cain et al. 2002; Okinaga et al. 2003). Additionally, it was claimed that there was no interaction of C5L2 with C3a in transfected HEK293 or preB-L1.2 cells (Okinaga et al. 2003).

Assuming that a variability in the gene of the putative ASP receptor might play a role in hyperapoB, DNA sequencing of the C5L2 coding region and restriction fragment polymorphism was performed as a broad screening (Marcil et al. 2006). A C5L2 variant (S323I) was found in one family. Comparing the homozygote and his 8 heterozygote relatives with other family members, the abnormal allele seemed to be associated with increased plasma triglyceride, cholesterol, LDL cholesterol, apoB and ASP. Heterozygote members of this family had a 50% reduction in ASP-stimulated triglyceride synthesis and glucose transport.

However, using ^{125}I-labeled ligand binding assays and flow cytometry with fluorescently labeled ligands we have shown recently that C3a and C3a-desArg are no ligands for C5L2 (Johswich et al. 2006). Receptor-independent "specific" binding of ^{125}I-C3a to plastic or filter membranes could be identified as the possible cause for former divergent results (see below).

3.4 Critical Evaluation of Experimental Findings

Science is a quest for the truth and within the huge amount of scientific findings within a field, there will happen - almost unavoidable - also a few mistakes or misinterpretations. The misinterpretation of C5L2 as the specific functional receptor for ASP is probably one example. The ASP research field is dominated by one group, reflected by an impressive output of publications. The downside to this situation is that confirmatory results obtained by different approaches or alternative techniques are rare and thus, it is more difficult to distinguish facts from the potential artefacts that could mislead us.

3.4.1 Quality of Purified ASP, Recombinant ASP and Synthetic Peptides as Stimuli, and Biologic Activity of ASP

ASP is a cleavage product of C3. Therefore, all early quantifications of ASP in serum by a polyclonal antiserum [such as (Cianflone et al. 1989a; Cianflone et al. 1990) had quite possibly determined the concentration of the more abundant precursor molecule C3 instead of C3a-desArg/ASP. Additionally, as only partially purified ASP had been used as antigen, this polyclonal antiserum might also have cross-reacted to some degree with co-purified contaminants. However, from ~1993 on, C3a/C3a-desArg specific assays were applied. In a 1994 publication it was mentioned that "the purified material at this (former) time yielded a single band on SDS-polyacrylamid gel electrophoresis which later proved to be misled when analyzed by amino acid sequencing and ion spray mass spectroscopy" (Cianflone et al. 1994). This indicates that data from former studies must be considered with caution. However, in 1997 recombinant ASP was used, and the results in functional assays were comparable to those obtained with ASP purified from serum (Murray et al. 1997). Regrettably, in subsequent studies the purified and not the recombinant ASP seem to have been preferentially used. Additionally, peptides which are homologous to C3a-desArg/ASP showed functional activity. Therefore one can concluded, that ASP and not contaminants are responsible for the observed increase in triglyceride synthesis (Murray et al. 1999a).

There seem to have been problems with the determination of the biologic activity of ASP over time: Due to an "improved purification procedure" ASP with an "increased biologic activity" was obtained. The maximal response in triacylglycerol synthesis was reached applying 50 - 200 ng/ml (~5-20 nM) purified ASP (Baldo et al. 1993; Cianflone et al. 1994). However, when purified and recombinant C3a were later compared, ASP effects were observed in the micromolar range (Murray et al. 1997), the same concentrations which had also been used in experiments with C5L2.

Because of this relatively high concentrations (1 and 10 μM), contamination is nevertheless still of some concern when purified ASP is applied in functional assays such as in the publications on functionality of C5L2 (Kalant et al. 2005). In comparison, C5a (or C3a) are ~1,000-fold more active when interacting with the C5a-receptor (or the C3a-receptor, respectively). Thus, small contaminations with C5a or similar substances could mislead us in these experiments. The use of recombinant ASP in key experiments could diminish this risk.

3.4.2 Binding Data (Calculated K_d) - and Recent Results Indicating that C5L2 is not the Receptor for ASP

When recombinant ASP from *E. coli* was purified the typical problems dealing with a highly cationic peptide had to be overcome. As one example, all tubes and plastic tips for pipettes must be siliconized (Murray et al. 1997). Expression cloning of C3aR by our group was based on ^{125}I-C3a binding to transfected HEK cells. A critical problem was the binding of ^{125}I-C3a to cell culture dishes - even in the presence of BSA or other proteins. The solution had been the pretreatment of the plastic surface with cationic protamine sulfate (Crass et al. 1996). This difficulty must be always considered when handling C3a/C3a-desArg since it might lead for example to the loss of the peptide and to false negative results. Moreover, due to their charge C3a and C3a-desArg bind also to cell surfaces. This problem is specific for C3a/C3a-desArg - it does not occur for closely related but less cationic C5a. Several techniques are available in binding studies to separate free from cell-bound ^{125}I-C3a. Centrifugation of the cells through a sucrose cushion was not affected by this feature of C3a/C3a-desArg (Ames et al. 1996; Klos, Bank, Gietz, Bautsch, Kohl, Burg, and Kretzschmar 1992; Martin, Bock, Arseniev, Tornetta, Ames, Bautsch, Kohl, Ganser, and Klos 1997). However, most laboratories have switched to less tedious filter plate assays in microtitre plates or similar techniques.

When we addressed the controversial discussion about the role of C5L2 in competitive ^{125}I-C3a binding studies, our filter plate assay led to unexpected results in controls. Even in the absence of any cells we observed specific binding with a high affinity ($IC_{50} \approx 25$ nM). Pretreatment of the filter plates with protamine sulfate or poly-L-lysine prevented that phenomenon: Using pretreated plates, only C3aR expressing HEK and RBL cells showed specific ^{125}I-C3a binding, whereas ^{125}I-C3a-desArg did not bind at all. C5L2 expressing cells were clearly negative for either ^{125}I-C3a or ^{125}I-C3a-desArg binding in the optimized assay. C5L2 expression was confirmed by ^{125}I-C5a binding. Moreover, fluorescently labeled C3a, C3a-desArg and C5a (50 nM in the absence or presence of a 20 fold excess of non-labelled ligands) were used to analyze binding by flow cytometry as a second independent method. Both methods showed that C5L2 is not a receptor for C3a or C3a-desArg/ASP. Thus, C5L2 is unlikely to be directly involved in lipid metabolism (Johswich et al. 2006).

Noteworthy, in their review Kildsgaard and his colleagues, as well as Fukuoka and Hugli in their manuscript (Fukuoka et al. 1990) had already postulated such an effect for the highly cationic peptide C3a/C3a-desArg, although for cell surfaces and not for filter plates (Kildsgaard, Zsigmond, Chan, and Wetsel 1999). Therefore, the interpretation of all former ^{125}I-ASP binding studies must be critically re-evaluated.
Moreover, there is a striking divergence between the determined high affinity of ~20 nM of ASP to its assumed receptor (Baldo et al. 1993; Cianflone et al. 1994; Murray et al. 1997) and the concentrations needed to obtain functional responses which are in the range of 1 to 10 μM (Kalant et al. 2005). Theoretically, one should expect that at

ASP concentrations 10-fold above the K_d (i.e. ~200-400 nM) almost all receptors are saturated leading to robust functional responses.

3.4.3 Receptor Dependent or Independent Functional Responses and Intracellular Signal Transduction ?

Functional dose-dependent responses can be clearly observed after stimulation with ASP purified from human serum, recombinant ASP or synthetic peptides homologous to the C-terminus of ASP. Furthermore, as an additional proof of specificity, pre-incubation with antibodies directed against ASP can block increased triglyceride synthesis (Murray et al. 1999a; Murray et al. 1997). If C5L2 is not the receptor for ASP, is there another receptor which is mediating these responses, or are the effects due to a receptor independent interaction?

A study by Murray clearly shows that the binding of ASP to its assumed receptor must be quite different from the specific binding of C3a to C3aR (Murray et al. 1999a). Unfortunately, the experiments competing the binding of ^{125}I-ASP /C3a-desArg with the synthetic peptide P117 must again be considered with care since P117 is also cationic. Strikingly, exchanges of positive charged Lys or Histidine decreased the binding of the positively charged ASP dramatically, whereas structural modifications by an alteration of the Cys-bound core had little effect. The synthetic peptide caused a partial stimulation of triglyceride synthesis on human skin fibroblasts and Swiss 3T3 cells. However, unfortunately no unrelated control peptide with similar physico-chemical properties was applied.

Noteworthy, in micromolar concentrations human ASP enhances triacylglycerol biosynthesis in oilseed rape (Weselake, Kazala, Cianflone, Boehr, Middleton, Rennie, Laroche, and Recnik 2000). It seems unlikely that a homologous ASP receptor should exist in this plant. There is no correlate to the complement system in this type of organisms. This observation might also suggest a receptor-independent activation of cells by cationic ASP.

3.4.4 Determinations and Concentrations of ASP (or C3) in Human Serum and Observations in Humans and Animals

As mentioned above, from ~1993 ASP/C3a-desArg specific assays were used for quantification. The ASP concentration found then in the peripheral blood was in the range of 50 - 100 nM. For comparison, C3, which is mainly synthesized in the liver, is in range of 1 - 1.5 mg/ml (~5-8 μM). Thus, to obtain the concentration of ASP necessary to achieve high functional responses in cell culture (1 - 10 μM), 20 to 100% of C3 in serum would have to be converted. In the periphery this might happen in sepsis or other dramatic disturbances of the complement system. Therefore, such a drastic activation of the complement system with all its consequences after a fatty meal would have to be restricted locally in lipid tissue. Likewise, this high amount of ASP would have to be eliminated in lipid tissue immediately before it reaches the peripheral veins again. Otherwise, ASP/C3a-desArg concentrations determined in peripheral blood should be much higher than observed. Although the ASP hypothesis is intriguing, there is still missing evidence for important parts of it: a) the proof of such high local ASP concentrations in lipid tissue; b) evidence for an upregulation of regulatory complement proteins such as CD59 which would be necessary to protect the adipocytes from their destruction by a locally activated complement system; c) the identification of a trigger for complement activation in lipid tissue. A postprandial upregulation of the synthesis of the precursor C3 and other complement factors alone is not sufficient to start the cascade.

Besides the generation of ASP/C3a-desArg, the majority of effector functions of the complement system are also blocked in C3 (ASP) knock-out mice. Therefore, these animals are frequently used to study the effect of a hampered immune system. Thus, if carbohydrate or lipid metabolism is different in C3 deficient mice, this could also be caused by the absence of any other downstream product of complement activation, such as C3b, C5a or the membrane attack complex. In a specific pathogen free (SPF) environment, C3/ASP deficient mice do not have any excess illness. However, it can not be excluded that their immune systems must constantly fight against microorganisms of the normal flora overcoming barriers that are normally controlled by a functional complement system. This might be an energy consuming process independent of the absence of ASP. It is also impossible to exclude this explanation for the differences in oxygen uptake or body weight observed in these knock-out mice (Xia, Sniderman, and Cianflone 2002; Xia, Stanhope, Digitale, Simion, Chen, Havel, and Cianflone 2004). The complement system seems to participate in the pathogenesis of arteriosclerosis. Thus, as discussed in more detail in a review by Kildsgaard and colleagues, observed changes in the level of complement factors or complement activation in serum of patients suffering from arteriosclerotic related diseases might be caused by artherosclerotic plaque formation (Kildsgaard et al. 1999).

The divergent results in different mice background strains of C3 (ASP) knockout mice (Murray et al. 1999b; Murray et al. 1999c; Wetsel et al. 1999) as well as the divergent results in patients and healthy human beings by a Finish study (Ylitalo et al. 2001) might point to high genetic variability in respect to ASP and lipid metabolism.

Taken together, the data clearly show that C5L2 is not the receptor for C3a or C3a-desArg/ASP. Previous binding data obtained with ^{125}I-ASP must be considered with caution because non-receptor binding to plastic surfaces/filter plates of this peptide can simulate competitive specific binding in the absence of any cells. ASP causes an increase in lipid metabolism and glucose uptake in cell culture. It remains unclear whether these effects on the lipid metabolism are mediated by a specific receptor for ASP or whether they only depend on charge dependent interaction of the cationic peptide with the cell surface. As a consequence, knockout mice specifically lacking ASP binding are not available. Observations on C3 knockout mice suggest that complement components downstream of this factor can modify lipid metabolism. However, ASP is just one of several candidates. ASP would have to be locally present in micromolar concentrations to account for the postulated effects. Increased C3, B, and D synthesis in lipid tissue could participate in the regulation, but it is not sufficient to explain the necessary high degree of complement activation, its limitation to the lipid tissue and the protection of cells from the deleterious effects of a highly activated complement system. There is an association of elevated levels of

ASP/C3a-desArg with fat consumption and arteriosclerosis related diseases in men. Yet, the reason for this association remains unclear. It can not be excluded, that ASP acts physiologically as an autocrine hormone in lipid tissue. However, presently there is not enough evidence to proof this hypothesis.

4 C5L2 as an Anti-inflammatory Molecule

C5a is a highly potent inflammatory mediator, therefore its biological activity has to be tightly controlled to avoid detrimental effects on healthy tissue surrounding a site of complement activation. The first step to limit the pro-inflammatory potential of C5a is its rapid degradation to C5a-desArg by serum carboxypeptidase (Huey, Bloor, Kawahara, and Hugli 1983). C5a-desArg still weakly binds to C5aR and possesses about 10% of the potency of C5a. It accumulates at sites of inflammation due to ongoing generation of C5a followed by its degradation. The ability of C5L2 to bind both, C5a and C5a-desArg together with its incapability of G-protein mediated intracellular signaling is highly suggestive for a scavenger or decoy receptor for C5a and C5a-desArg. Additionally, C5L2 binds C5a with nearly the same affinity and C5a-desArg even with a 20- to 50-fold higher affinity compared to the C5aR (Cain et al. 2002; Okinaga et al. 2003). As C5L2 and C5aR are usually coexpressed on cells, it is likely that there is a competition of both receptors for these ligands potentially leading to an attenuation of C5aR dependent effects by C5L2. Indeed, it was shown *in vitro* that the production of pro-inflammatory IL-6 is enhanced when C5L2 is blocked by anti-C5L2-antibodies in rat neutrophils stimulated with C5a plus LPS. Furthermore, serum levels of IL-6 are enhanced in anti-C5L2-antibody treated rats *in vivo* in a cecum ligation and puncture (CLP) model of sepsis (Gao et al. 2005). Another study using C5L2 knockout mice revealed a more severe course of disease in these animals compared to wild type mice in a pulmonary immune complex injury with about 4 fold higher concentrations of TNF-alpha and IL-6 in the lung. Additionally, elevated numbers of inflammatory cells, predominantly neutrophils, infiltrated into the lung tissue. Moreover, bone marrow cells from C5L2 knockout mice show enhanced chemotaxis towards a C5a gradient (Gerard, Lu, Liu, Craig, Fujiwara, Okinaga, and Gerard 2005). These *in vivo* and *in vitro* data demonstrate that C5L2 has protective effects against inflammatory diseases. If this is only due to a competition of both receptors for C5a and C5a-desArg, regulation of C5L2 should alter C5aR dependent effects.

C5L2 was found to be upregulated *in vitro* by the anti-inflammatory neutrotransmitter noradrenaline in rat astrocytes. Additionally, *in vivo* depletion of noradrenaline producing neurons led to downregulation of C5L2 in the frontal cortex of rats. Knockdown of C5L2 by RNA interference in rat astrocytes led to a stronger increase of nitric oxide synthase type 2 and higher levels of nitrite after stimulation with LPS (Gavrilyuk et al. 2005). These experiments demonstrated a possible involvement of C5L2 in inflammatory disorders of the brain. There was also an immediate focus on neutrophils in different studies. C5L2 was reported to be regulated in neutrophils of rats in a CLP model of sepsis. Surface expression of C5L2 peaked 24 hours after cecal injury of the animals and dropped afterwards. C5aR expression was exact inversed with a minimum surface expression at 24 hours after CLP. Additionally, C5L2 expression was shown to be increased 12 hours after CLP in lung, liver and heart but not in kidney of rats (Gao et al. 2005). In contrast, another study demonstrated downregulation of C5L2 on neutrophils of rats after CLP during a 36 hours time course, unless the animals were treated with anti-C5a-antibodies prior to surgery. In the same study, C5L2 was found to be downregulated on neutrophils from blood of human sepsis patients within 4 hours. Interestingly, C5L2 expression levels on neutrophils correlated with the survival rate of sepsis patients: C5L2 was not completely downregulated in patients surviving, but in patients developing a multi organ failure almost no surface expression of C5L2 was detectable (Huber-Lang et al. 2005). This indicates a protective role of C5L2 in systemic inflammatory disorders such as sepsis.

It is unclear what happens on molecular basis after binding of C5a or C5a-desArg to C5L2. No ligand-dependent internalization of C5L2 upon stimulation with C5a, C5a-desArg or C3a could be observed in direct assays (Cain et al. 2002; Okinaga et al. 2003). However, internalization without intracellular signaling is an important feature of decoy receptors for chemokines. By now three of these silent receptors with promiscuous ligand binding patterns are identified: D6, DARC and CCX-CKR. Like C5L2, they lack signal transduction capacity and are uncoupled from G-proteins due to mutations in the highly conserved DRYLAIVHA sequence, also known as the DRY-motif (Luo, Chaudhuri, Johnson, Neote, Zbrzezna, He, and Pogo 1997; Nibbs, Wylie, Pragnell, and Graham 1997; Nibbs, Wylie, Yang, Landau, and Graham 1997; Townson and Nibbs 2002). The decoy receptors are efficiently internalized, thereby acting as a sink for numerous inflammatory chemokines. In the case of D6 and CCX-CRK, internalization targets the bound ligands to the endosome, leading to their degradation (Comerford, Milasta, Morrow, Milligan, and Nibbs 2006; Galliera, Jala, Trent, Bonecchi, Signorelli, Lefkowitz, Mantovani, Locati, and Haribabu 2004). D6 internalizes constitutively and independent of ligand binding (Galliera et al. 2004), resembling a conveyer for inflammatory chemokines right into the cells endosome. DARC seems not only to act as a sink for chemokines when expressed on erythrocytes but is also discussed to facilitate chemokine transcytosis through the endothelium of postcapillary venules (Chaudhuri, Nielsen, Elkjaer, Zbrzezna, Fang, and Pogo 1997; Middleton, Neil, Wintle, Clark-Lewis, Moore, Lam, Auer, Hub, and Rot 1997). However, it remains questionable to what extent the function of C5L2 is related to that of the chemokine decoy receptors, because simply binding C5a and keeping it at the cell surface does not provide a system by which C5a or C5a-desArg can be efficiently eliminated. Our knowledge about C5L2 is still limited due to the fact that to date many of the functional investigations have been performed in transfected cells. Future analysis concerning the functionality of C5L2 that focus more on primary and endogenously expressing cells will hopefully reveal a mechanism for C5L2's mode of action. Additionally, animal models will shed more light on the role of C5L2 during inflammatory diseases.

A new publication from spring 2007 by Chen N-J et al. (doi:10.1038/nature05559) on C5L2 knockout mice suggests that C5L2 might be a positive modulator of effects mediated by the two other anaphylatoxin receptors.

References

Amatruda, T. T., III, Gerard, N. P., Gerard, C., and Simon, M. I. (1993). "Specific interactions of chemoattractant factor receptors with G- proteins." *J.Biol.Chem.* 268:10139-10144.

Ames, R. S., Li, Y., Sarau, H. M., Nuthulaganti, P., Foley, J. J., Ellis, C., Zeng, Z., Su, K., Jurewicz, A. J., Hertzberg, R. P., Bergsma, D. J., and Kumar, C. (1996). "Molecular cloning and characterization of the human anaphylatoxin C3a receptor." *J.Biol.Chem.* 271:20231-20234.

Arumugam, T. V., Shiels, I. A., Woodruff, T. M., Granger, D. N., and Taylor, S. M. (2004). "The role of the complement system in ischemia-reperfusion injury." *Shock.* 21:401-409.

Baldo, A., Sniderman, A. D., St Luce, S., Avramoglu, R. K., Maslowska, M., Hoang, B., Monge, J. C., Bell, A., Mulay, S., and Cianflone, K. (1993). "The adipsin-acylation stimulating protein system and regulation of intracellular triglyceride synthesis." *J.Clin.Invest.* 92:1543-1547.

Baldo, A., Sniderman, A. D., St Luce, S., Zhang, X. J., and Cianflone, K. (1995). "Signal transduction pathway of acylation stimulating protein: involvement of protein kinase C." *J.Lipid Res.* 36:1415-1426.

Bao, L., Gerard, N. P., Eddy, R. L., Jr., Shows, T. B., and Gerard, C. (1992). "Mapping of genes for the human C5a receptor (C5AR), human FMLP receptor (FPR), and two FMLP receptor homologue orphan receptors (FPRH1, FPRH2) to chromosome 19." *Genomics.* 13:437-440.

Bock, D., Martin, U., Gartner, S., Rheinheimer, C., Raffetseder, U., Arseniev, L., Barker, M. D., Monk, P. N., Bautsch, W., Kohl, J., and Klos, A. (1997). "The C terminus of the human C5a receptor (CD88) is required for normal ligand-dependent receptor internalization." *Eur.J.Immunol.* 27:1522-1529.

Boulay, F., Mery, L., Tardif, M., Brouchon, L., and Vignais, P. (1991). "Expression cloning of a receptor for C5a anaphylatoxin on differentiated HL-60 cells." *Biochemistry.* 30:2993-2999.

Buhl, A. M., Avdi, N., Worthen, G. S., and Johnson, G. L. (1994). "Mapping of the C5a receptor signal transduction network in human neutrophils." *Proc.Natl.Acad.Sci.U.S.A.* 91:9190-9194.

Cain, S. A. and Monk, P. N. (2002). "The orphan receptor C5L2 has high affinity binding sites for complement fragments C5a and C5a des-Arg(74)." *J.Biol.Chem.* 277:7165-7169.

Chaudhuri, A., Nielsen, S., Elkjaer, M. L., Zbrzezna, V., Fang, F., and Pogo, A. O. (1997). "Detection of Duffy antigen in the plasma membranes and caveolae of vascular endothelial and epithelial cells of nonerythroid organs." *Blood.* 89:701-712.

Chenoweth, D. E. and Goodman, M. G. (1983). "The C5a receptor of neutrophils and macrophages." *Agents Actions Suppl.* 12:252-273.

Chenoweth, D. E. and Hugli, T. E. (1978). "Demonstration of specific C5a receptor on intact human polymorphonuclear leukocytes." *Proc.Natl.Acad.Sci.U.S.A.* 75:3943-3947.

Cianflone, K., Kalant, D., Marliss, E. B., Gougeon, R., and Sniderman, A. D. (1995). "Response of plasma ASP to a prolonged fast." *Int.J.Obes.Relat Metab Disord.* 19:604-609.

Cianflone, K., Kwiterovich, P. O., Walsh, M., Forse, A., Rodriguez, M. A., and Sniderman, A. D. (1987). "Stimulation of fatty acid uptake and triglyceride synthesis in human cultured skin fibroblasts and adipocytes by a serum protein." *Biochem.Biophys.Res.Commun.* 144:94-100.

Cianflone, K., Rodriguez, M. A., Walsh, M., Vu, H., and Sniderman, A. D. (1988). "The effect of a plasma protein fraction on lipid synthesis in cultured skin fibroblasts from normals and patients with hyperapobetalipoproteinemia." *Clin.Invest Med.* 11:99-107.

Cianflone, K., Roncari, D. A., Maslowska, M., Baldo, A., Forden, J., and Sniderman, A. D. (1994). "Adipsin/acylation stimulating protein system in human adipocytes: regulation of triacylglycerol synthesis." *Biochemistry.* 33:9489-9495.

Cianflone, K., Vu, H., Walsh, M., Baldo, A., and Sniderman, A. (1989a). "Metabolic response of Acylation Stimulating Protein to an oral fat load." *J.Lipid Res.* 30:1727-1733.

Cianflone, K., Xia, Z., and Chen, L. Y. (2003). "Critical review of acylation-stimulating protein physiology in humans and rodents." *Biochim.Biophys.Acta.* 1609:127-143.

Cianflone, K., Zhang, X. J., Genest, J., Jr., and Sniderman, A. (1997). "Plasma acylation-stimulating protein in coronary artery disease." *Arterioscler.Thromb.Vasc.Biol.* 17:1239-1244.

Cianflone, K. M., Maslowska, M. H., and Sniderman, A. D. (1990). "Impaired response of fibroblasts from patients with hyperapobetalipoproteinemia to acylation-stimulating protein." *J.Clin.Invest.* 85:722-730.

Cianflone, K. M., Sniderman, A. D., Walsh, M. J., Vu, H. T., Gagnon, J., and Rodriguez, M. A. (1989b). "Purification and characterization of acylation stimulating protein." *J.Biol.Chem.* 264:426-430.

Comerford, I., Milasta, S., Morrow, V., Milligan, G., and Nibbs, R. (2006). "The chemokine receptor CCX-CKR mediates effective scavenging of CCL19 in vitro." *Eur.J.Immunol.* 36:1904-1916.

Cook, K. S., Min, H. Y., Johnson, D., Chaplinsky, R. J., Flier, J. S., Hunt, C. R., and Spiegelman, B. M. (1987). "Adipsin: a circulating serine protease homolog secreted by adipose tissue and sciatic nerve." *Science*. 237:402-405.

Crass, T., Raffetseder, U., Martin, U., Grove, M., Klos, A., Kohl, J., and Bautsch, W. (1996). "Expression cloning of the human C3a anaphylatoxin receptor (C3aR) from differentiated U-937 cells." *Eur.J.Immunol.* 26:1944-1950.

Farkas, I., Baranyi, L., Takahashi, M., Fukuda, A., Liposits, Z., Yamamoto, T., and Okada, H. (1998). "A neuronal C5a receptor and an associated apoptotic signal transduction pathway." *J.Physiol (Lond)*. 507 (Pt 3):679-687.

Farzan, M., Schnitzler, C. E., Vasilieva, N., Leung, D., Kuhn, J., Gerard, C., Gerard, N. P., and Choe, H. (2001). "Sulfated tyrosines contribute to the formation of the C5a docking site of the human C5a anaphylatoxin receptor." *J.Exp.Med.* 193:1059-1066.

Fayyazi, A., Scheel, O., Werfel, T., Schweyer, S., Oppermann, M., Gotze, O., Radzun, H. J., and Zwirner, J. (2000). "The C5a receptor is expressed in normal renal proximal tubular but not in normal pulmonary or hepatic epithelial cells." *Immunology*. 99:38-45.

Fukuoka, Y. and Hugli, T. E. (1990). "Anaphylatoxin binding and degradation by rat peritoneal mast cells. Mechanisms of degranulation and control." *J.Immunol.* 145:1851-1858.

Galliera, E., Jala, V. R., Trent, J. O., Bonecchi, R., Signorelli, P., Lefkowitz, R. J., Mantovani, A., Locati, M., and Haribabu, B. (2004). "beta-Arrestin-dependent constitutive internalization of the human chemokine decoy receptor D6." *J.Biol.Chem.* 279:25590-25597.

Gao, H., Neff, T. A., Guo, R. F., Speyer, C. L., Sarma, J. V., Tomlins, S., Man, Y., Riedemann, N. C., Hoesel, L. M., Younkin, E., Zetoune, F. S., and Ward, P. A. (2005). "Evidence for a functional role of the second C5a receptor C5L2." *FASEB J.* 19:1003-1005.

Gao, J., Choe, H., Bota, D., Wright, P. L., Gerard, C., and Gerard, N. P. (2003). "Sulfation of tyrosine 174 in the human C3a receptor is essential for binding of C3a anaphylatoxin." *J.Biol.Chem.* 278:37902-37908.

Gasque, P., Singhrao, S. K., Neal, J. W., Gotze, O., and Morgan, B. P. (1997). "Expression of the receptor for complement C5a (CD88) is up-regulated on reactive astrocytes, microglia, and endothelial cells in the inflamed human central nervous system." *Am.J.Pathol.* 150:31-41.

Gavrilyuk, V., Kalinin, S., Hilbush, B. S., Middlecamp, A., McGuire, S., Pelligrino, D., Weinberg, G., and Feinstein, D. L. (2005). "Identification of complement 5a-like receptor (C5L2) from astrocytes: characterization of anti-inflammatory properties." *J.Neurochem.* 92:1140-1149.

Gerard, C., Bao, L., Orozco, O., Pearson, M., Kunz, D., and Gerard, N. P. (1992). "Structural diversity in the extracellular faces of peptidergic G- protein-coupled receptors. Molecular cloning of the mouse C5a anaphylatoxin receptor." *J.Immunol.* 149:2600-2606.

Gerard, N. P. and Gerard, C. (1991). "The chemotactic receptor for human C5a anaphylatoxin." *Nature*. 349:614-617.

Gerard, N. P., Hodges, M. K., Drazen, J. M., Weller, P. F., and Gerard, C. (1989). "Characterization of a receptor for C5a anaphylatoxin on human eosinophils." *J.Biol.Chem.* 264:1760-1766.

Gerard, N. P., Lu, B., Liu, P., Craig, S., Fujiwara, Y., Okinaga, S., and Gerard, C. (2005). "An anti-inflammatory function for the complement anaphylatoxin C5a binding protein, C5L2." *J.Biol.Chem.*

Germinario, R., Sniderman, A. D., Manuel, S., Lefebvre, S. P., Baldo, A., and Cianflone, K. (1993). "Coordinate regulation of triacylglycerol synthesis and glucose transport by acylation-stimulating protein." *Metabolism*. 42:574-580.

Gervasoni, J. E., Jr., Conrad, D. H., Hugli, T. E., Schwartz, L. B., and Ruddy, S. (1986). "Degradation of human anaphylatoxin C3a by rat peritoneal mast cells: a role for the secretory granule enzyme chymase and heparin proteoglycan." *J.Immunol.* 136:285-292.

Guo, R. F., Riedemann, N. C., and Ward, P. A. (2004). "Role of C5a-C5aR interaction in sepsis." *Shock*. 21:1-7.

Hartmann, H., Lubbers, B., Casaretto, M., Bautsch, W., Klos, A., and Kohl, J. (1993). "Rapid quantification of C3a and C5a using a combination of chromatographic and immunoassay procedures." *J.Immunol.Methods*. 166:35-44.

Havel, P. J. (2004). "Update on adipocyte hormones: regulation of energy balance and carbohydrate/lipid metabolism." *Diabetes*. 53 Suppl 1:S143-S151.

Hopken, U. E., Lu, B., Gerard, N. P., and Gerard, C. (1996). "The C5a chemoattractant receptor mediates mucosal defence to infection." *Nature*. 383:86-89.

Huber-Lang, M., Sarma, J. V., Rittirsch, D., Schreiber, H., Weiss, M., Flierl, M., Younkin, E., Schneider, M., Suger-Wiedeck, H., Gebhard, F., McClintock, S. D., Neff, T., Zetoune, F., Bruckner, U., Guo, R. F., Monk, P. N., and Ward, P. A. (2005). "Changes in the novel orphan, C5a receptor (C5L2), during experimental sepsis and sepsis in humans." *J.Immunol.* 174:1104-1110.

Huey, R., Bloor, C. M., Kawahara, M. S., and Hugli, T. E. (1983). "Potentiation of the anaphylatoxins in vivo using an inhibitor of serum carboxypeptidase N (SCPN). I. Lethality and pathologic effects on pulmonary tissue." *Am.J.Pathol.* 112:48-60.

Johswich, K., Martin, M., Thalmann, J., Rheinheimer, C., Monk, P. N., and Klos, A. (2006). "Ligand specificity of the anaphylatoxin C5L2 receptor and its regulation on myeloid and epithelial cell-lines." *J.Biol.Chem.*

Kajita, T. and Hugli, T. E. (1991). "Evidence for in vivo degradation of C3a anaphylatoxin by mast cell chymase. I. Nonspecific activation of rat peritoneal mast cells by C3ades Arg." *Am.J.Pathol.* 138: 1359-1369.

Kalant, D., Cain, S. A., Maslowska, M., Sniderman, A. D., Cianflone, K., and Monk, P. N. (2003). "The chemoattractant receptor-like protein, C5L2, binds C3a desArg77/acylation stimulating protein." *J.Biol.Chem.* 278(13):11123-11129.

Kalant, D., Maclaren, R., Cui, W., Samanta, R., Monk, P. N., Laporte, S. A., and Cianflone, K. (2005). "C5L2 Is a Functional Receptor for Acylation-stimulating Protein." *J.Biol.Chem.* 280:23936-23944.

Kildsgaard, J., Zsigmond, E., Chan, L., and Wetsel, R. A. (1999). "A critical evaluation of the putative role of C3adesArg (ASP) in lipid metabolism and hyperapobetalipoproteinemia." *Mol.Immunol.* 36: 869-876.

Klos, A., Bank, S., Gietz, C., Bautsch, W., Kohl, J., Burg, M., and Kretzschmar, T. (1992). "C3a receptor on dibutyryl-cAMP-differentiated U937 cells and human neutrophils: the human C3a receptor characterized by functional responses and 125I-C3a binding." *Biochemistry.* 31:11274-11282.

Kohl, J., Baelder, R., Lewkowich, I. P., Pandey, M. K., Hawlisch, H., Wang, L., Best, J., Herman, N. S., Sproles, A. A., Zwirner, J., Whitsett, J. A., Gerard, C., Sfyroera, G., Lambris, J. D., and Wills-Karp, M. (2006). "A regulatory role for the C5a anaphylatoxin in type 2 immunity in asthma." *J.Clin.Invest.* 116:783-796.

Kohl, J. and Gessner, J. E. (1999). "On the role of complement and Fc gamma-receptors in the Arthus reaction." *Mol.Immunol.* 36:893-903.

Kolakowski, L. F., Jr., Lu, B., Gerard, C., and Gerard, N. P. (1995). "Probing the "message:address" sites for chemoattractant binding to the C5a receptor. Mutagenesis of hydrophilic and proline residues within the transmembrane segments." *J.Biol.Chem.* 270:18077-18082.

la Sala, A., Gadina, M., and Kelsall, B. L. (2005). "G(i)-protein-dependent inhibition of IL-12 production is mediated by activation of the phosphatidylinositol 3-kinase-protein 3 kinase B/Akt pathway and JNK." *J.Immunol.* 175:2994-2999.

Langkabel, P., Zwirner, J., and Oppermann, M. (1999). "Ligand-induced phosphorylation of anaphylatoxin receptors C3aR and C5aR is mediated by "G protein-coupled receptor kinases." *Eur.J.Immunol.* 29:3035-3046.

Laudes, I. J., Chu, J. C., Huber-Lang, M., Guo, R. F., Riedemann, N. C., Sarma, J. V., Mahdi, F., Murphy, H. S., Speyer, C., Lu, K. T., Lambris, J. D., Zetoune, F. S., and Ward, P. A. (2002). "Expression and function of C5a receptor in mouse microvascular endothelial cells." *J.Immunol.* 169:5962-5970.

Lee, D. K., George, S. R., Cheng, R., Nguyen, T., Liu, Y., Brown, M., Lynch, K. R., and O'Dowd, B. F. (2001). "Identification of four novel human G protein-coupled receptors expressed in the brain." *Brain Res.Mol.Brain Res.* 86:13-22.

Luo, H., Chaudhuri, A., Johnson, K. R., Neote, K., Zbrzezna, V., He, Y., and Pogo, A. O. (1997). "Cloning, characterization, and mapping of a murine promiscuous chemokine receptor gene: homolog of the human Duffy gene." *Genome Res.* 7:932-941.

Marcil, M., Vu, H., Cui, W., Dastani, Z., Engert, J. C., Gaudet, D., Castro-Cabezas, M., Sniderman, A. D., Genest, Jr J., and Cianflone, K. (2006). "Identification of a Novel C5L2 Variant (S323I) in a French Canadian Family With Familial Combined Hyperlipemia." *Arterioscler.Thromb.Vasc.Biol.*

Marder, S. R., Chenoweth, D. E., Goldstein, I. M., and Perez, H. D. (1985). "Chemotactic responses of human peripheral blood monocytes to the complement-derived peptides C5a and C5a des Arg." *J.Immunol.* 134:3325-3331.

Martin, U., Bock, D., Arseniev, L., Tornetta, M. A., Ames, R. S., Bautsch, W., Kohl, J., Ganser, A., and Klos, A. (1997). "The human C3a receptor is expressed on neutrophils and monocytes, but not on B or T lymphocytes." *J.Exp.Med.* 186:199-207.

Maslowska, M., Legakis, H., Assadi, F., and Cianflone, K. (2006). "Targeting the signaling pathway of acylation stimulating protein." *J.Lipid Res.* 47:643-652.
Maslowska, M., Scantlebury, T., Germinario, R., and Cianflone, K. (1997). "Acute in vitro production of acylation stimulating protein in differentiated human adipocytes." *J.Lipid Res.* 38:1-11.
Middleton, J., Neil, S., Wintle, J., Clark-Lewis, I., Moore, H., Lam, C., Auer, M., Hub, E., and Rot, A. (1997). "Transcytosis and surface presentation of IL-8 by venular endothelial cells." *Cell.* 91:385-395.
Mukherjee, P. and Pasinetti, G. M. (2000). "The role of complement anaphylatoxin C5a in neurodegeneration: implications in Alzheimer's disease." *J.Neuroimmunol.* 105:124-130.
Mullmann, T. J., Siegel, M. I., Egan, R. W., and Billah, M. M. (1990). "Complement C5a activation of phospholipase D in human neutrophils. A major route to the production of phosphatidates and diglycerides." *J.Immunol.* 144:1901-1908.
Murray, I., Kohl, J., and Cianflone, K. (1999a). "Acylation-stimulating protein (ASP): structure-function determinants of cell surface binding and triacylglycerol synthetic activity." *Biochem.J.* 342 (Pt 1): 41-48.
Murray, I., Parker, R. A., Kirchgessner, T. G., Tran, J., Zhang, Z. J., Westerlund, J., and Cianflone, K. (1997). "Functional bioactive recombinant acylation stimulating protein is distinct from C3a anaphylatoxin." *J.Lipid Res.* 38:2492-2501.
Murray, I., Sniderman, A. D., and Cianflone, K. (1999a). "Enhanced triglyceride clearance with intraperitoneal human acylation stimulating protein in C57BL/6 mice." *Am.J.Physiol.* 277:E474-E480.
Murray, I., Sniderman, A. D., and Cianflone, K. (1999b). "Mice lacking acylation stimulating protein (ASP) have delayed postprandial triglyceride clearance." *J.Lipid Res.* 40:1671-1676.
Murray, I., Sniderman, A. D., Havel, P. J., and Cianflone, K. (1999c). "Acylation stimulating protein (ASP) deficiency alters postprandial and adipose tissue metabolism in male mice." *J.Biol.Chem.* 274: 36219-36225.
Nibbs, R. J., Wylie, S. M., Pragnell, I. B., and Graham, G. J. (1997). "Cloning and characterization of a novel murine beta chemokine receptor, D6. Comparison to three other related macrophage inflammatory protein-1alpha receptors, CCR-1, CCR-3, and CCR-5." *J.Biol.Chem.* 272:12495-12504.
Nibbs, R. J., Wylie, S. M., Yang, J., Landau, N. R., and Graham, G. J. (1997). "Cloning and characterization of a novel promiscuous human beta-chemokine receptor D6." *J.Biol.Chem.* 272:32078-32083.
Norgauer, J., Dobos, G., Kownatzki, E., Dahinden, C., Burger, R., Kupper, R., and Gierschik, P. (1993). "Complement fragment C3a stimulates Ca2+ influx in neutrophils via a pertussis-toxin-sensitive G protein." *Eur.J.Biochem.* 217:289-294.
Ohno, M., Hirata, T., Enomoto, M., Araki, T., Ishimaru, H., and Takahashi, T. A. (2000). "A putative chemoattractant receptor, C5L2, is expressed in granulocyte and immature dendritic cells, but not in mature dendritic cells." *Mol.Immunol.* 37:407-412.
Okinaga, S., Slattery, D., Humbles, A., Zsengeller, Z., Morteau, O., Kinrade, M. B., Brodbeck, R. M., Krause, J. E., Choe, H. R., Gerard, N. P., and Gerard, C. (2003). "C5L2, a nonsignaling C5A binding protein." *Biochemistry.* 42:9406-9415.
Paral, D., Sohns, B., Crass, T., Grove, M., Kohl, J., Klos, A., and Bautsch, W. (1998). "Genomic organization of the human C3a receptor." *Eur.J.Immunol.* 28:2417-2423.
Perianayagam, M. C., Balakrishnan, V. S., King, A. J., Pereira, B. J., and Jaber, B. L. (2002). "C5a delays apoptosis of human neutrophils by a phosphatidylinositol 3- kinase-signaling pathway." *Kidney Int.* 61:456-463.
Perret, J. J., Raspe, E., Vassart, G., and Parmentier, M. (1992). "Cloning and functional expression of the canine anaphylatoxin C5a receptor. Evidence for high interspecies variability." *Biochem.J.* 288 (Pt 3):911-917.
Ratajczak, M. Z., Reca, R., Wysoczynski, M., Kucia, M., Baran, J. T., Allendorf, D. J., Ratajczak, J., and Ross, G. D. (2004). "Transplantation studies in C3-deficient animals reveal a novel role of the third complement component (C3) in engraftment of bone marrow cells." *Leukemia.* 18:1482-1490.
Roglic, A., Prossnitz, E. R., Cavanagh, S. L., Pan, Z., Zou, A., and Ye, R. D. (1996). "cDNA cloning of a novel G protein-coupled receptor with a large extracellular loop structure." *Biochim.Biophys.Acta.* 1305:39-43.
Rollins, T. E., Siciliano, S., Kobayashi, S., Cianciarulo, D. N., Bonilla-Argudo, V., Collier, K., and Springer, M. S. (1991). "Purification of the active C5a receptor from human polymorphonuclear leukocytes as a receptor-Gi complex." *Proc.Natl.Acad.Sci.U.S.A.* 88:971-975.

Sayah, S., Jauneau, A. C., Patte, C., Tonon, M. C., Vaudry, H., and Fontaine, M. (2003). "Two different transduction pathways are activated by C3a and C5a anaphylatoxins on astrocytes." *Brain Res.Mol.Brain Res.* 112:53-60.

Schraufstatter, I. U., Trieu, K., Sikora, L., Sriramarao, P., and DiScipio, R. (2002). "Complement c3a and c5a induce different signal transduction cascades in endothelial cells." *J.Immunol.* 169:2102-2110.

Settmacher, B., Bock, D., Saad, H., Gartner, S., Rheinheimer, C., Kohl, J., Bautsch, W., and Klos, A. (1999). "Modulation of C3a activity: internalization of the human C3a receptor and its inhibition by C5a." *J.Immunol.* 162:7409-7416.

Siciliano, S. J., Rollins, T. E., DeMartino, J., Konteatis, Z., Malkowitz, L., Van Riper, G., Bondy, S., Rosen, H., and Springer, M. S. (1994). "Two-site binding of C5a by its receptor: an alternative binding paradigm for G protein-coupled receptors." *Proc.Natl.Acad.Sci.U.S.A.* 91:1214-1218.

Sniderman, A. D., Cianflone, K., Summers, L., Fielding, B., and Frayn, K. (1997). "The acylation-stimulating protein pathway and regulation of postprandial metabolism." *Proc.Nutr.Soc.* 56:703-712.

Stahel, P. F., Frei, K., Eugster, H. P., Fontana, A., Hummel, K. M., Wetsel, R. A., Ames, R. S., and Barnum, S. R. (1997). "TNF-alpha-mediated expression of the receptor for anaphylatoxin C5a on neurons in experimental Listeria meningoencephalitis." *J.Immunol.* 159:861-869.

Tao, Y., Cianflone, K., Sniderman, A. D., Colby-Germinario, S. P., and Germinario, R. J. (1997). "Acylation-stimulating protein (ASP) regulates glucose transport in the rat L6 muscle cell line." *Biochim.Biophys.Acta.* 1344:221-229.

Townson, J. R. and Nibbs, R. J. (2002). "Characterization of mouse CCX-CKR, a receptor for the lymphocyte-attracting chemokines TECK/mCCL25, SLC/mCCL21 and MIP-3beta/mCCL19: comparison to human CCX-CKR." *Eur.J.Immunol.* 32:1230-1241.

Weselake, R. J., Kazala, E. C., Cianflone, K., Boehr, D. D., Middleton, C. K., Rennie, C. D., Laroche, A., and Recnik, I. (2000). "Human acylation stimulating protein enhances triacylglycerol biosynthesis in plant microsomes." *FEBS Lett.* 481:189-192.

Wetsel, R. A. (1995). "Expression of the complement C5a anaphylatoxin receptor (C5aR) on non- myeloid cells." *Immunol.Lett.* 44:183-187.

Wetsel, R. A., Kildsgaard, J., Zsigmond, E., Liao, W., and Chan, L. (1999). "Genetic deficiency of acylation stimulating protein (ASP(C3ades-Arg)) does not cause hyperapobetalipoproteinemia in mice." *J.Biol.Chem.* 274:19429-19433.

Wilken, H. C., Gotze, O., Werfel, T., and Zwirner, J. (1999). "C3a(desArg) does not bind to and signal through the human C3a receptor." *Immunol.Lett.* 67:141-145.

Xia, Z., Sniderman, A. D., and Cianflone, K. (2002). "Acylation-stimulating protein (ASP) deficiency induces obesity resistance and increased energy expenditure in ob/ob mice." *J.Biol.Chem.* 277: 45874-45879.

Xia, Z., Stanhope, K. L., Digitale, E., Simion, O. M., Chen, L., Havel, P., and Cianflone, K. (2004). "Acylation-stimulating protein (ASP)/complement C3adesArg deficiency results in increased energy expenditure in mice." *J.Biol.Chem.* 279:4051-4057.

Ylitalo, K., Pajukanta, P., Meri, S., Cantor, R. M., Mero-Matikainen, N., Vakkilainen, J., Nuotio, I., and Taskinen, M. R. (2001). "Serum C3 but not plasma acylation-stimulating protein is elevated in Finnish patients with familial combined hyperlipidemia." *Arterioscler.Thromb.Vasc.Biol.* 21:838-843.

13

The Exosporium of *B.cereus* Contains a Binding Site for gC1qR/p33: Implication in Spore Attachment and/or Entry*

Berhane Ghebrehiwet[1], Lee Tantral[1], Mathew A. Titmus[1], Barbara J. Panessa-Warren[2], George T. Tortora[3], Stanislaus S. Wong[4], John B. Warren[5].

Stony Brook University, Departments of [1] Medicine, [4] Chemistry, and [3] Clinical Microbiology Laboratory, NY.11794, berhane.ghebrehiwet@stonybrook.edu
[2] Department of Materials Science Bldg.480, Brookhaven National Laboratory, Upton, NY 11973
[5] Instrumentation Division Bldg.535B, Brookhaven National Laboratory, Upton, NY 11973.

Abstract. *B. cereus,* is a member of a genus of aerobic, gram-positive, spore-forming rod-like bacilli, which includes the deadly, *B. anthracis*. Preliminary experiments have shown that gC1qR binds to *B.cereus* spores that have been attached to microtiter plates. The present studies were therefore undertaken, to examine if cell surface gC1qR plays a role in *B.cereus* spore attachment and/or entry. Monolayers of human colon carcinoma (Caco-2) and lung cells were grown to confluency on 6 mm coverslips in shell vials with gentle swirling in a shaker incubator. Then, 2 μl of a suspension of strain SB460 *B.cereus* spores ($3x10^8$/ml, in sterile water), were added and incubated (1-4 h; 36^0 C) in the presence or absence of anti-gC1qR mAb-carbon nanoloops. Examination of these cells by EM revealed that: (1) When *B. cereus* endospores contacted the apical Caco-2 cell surface, or lung cells, gClqR was simultaneously detectable, indicating upregulation of the molecule. (2) In areas showing spore contact with the cell surface, gClqR expression was often adjacent to the spores in association with microvilli (Caco-2 cells) or cytoskeletal projections (lung cells). (3) Furthermore, the exosporia of the activated and germinating spores were often decorated with mAb-nanoloops. These observations were further corroborated by experiments in which *B.cereus* spores were readily taken up by monocytes and neutrophils, and this uptake was partially inhibited by mAb 60.11, which recognizes the C1q binding site on gC1qR. Taken together, the data suggest a role, for gC1qR at least in the initial stages of spore attachment and/or entry.

1 Introduction

Bacillus cereus is an opportunistic pathogen involved in several types of food poisoning and tissue destruction (Davey and Tauber 1987; Callegan, Engel, Hill, and O'Callaghan 1994; Beecher, Pulido, Barney and Wong, 1995; Helgason, Økstad, Caugant, Johansen, Fouet, Mock, Hegna, and Koltsø, 2000; Day, Smith, Gregg,

Turnbull, Head, Ives, and Ho 1981). It belongs to the genus of spore-forming aerobic bacteria, which includes *B. thuringiensis,* a bacterium which produces intracellular protein crystals toxic to a number of insect larvae —commonly used in many parts of the world as a biological pesticide in the protection of crops (Helgason, Økstad, Caugant, Johansen, Fouet, Mock, Hegna, and Koltsø, 2000); and *B.anthracis,* the causative agent of anthrax and a potential bio-warfare agent (Helgason, Økstad, Caugant, Johansen, Fouet, Mock, Hegna, and Koltsø, 2000; Whitney, Beatty, Taylor, Weyant, Sobel, Arduino, and Ashford). Traditionally, *B. cereus* has been regarded more as an opportunistic nuisance found in hospitals and in milk products causing mild gastroenteritis, rather than something more sinister. Toxin formation as well as formation of heat-resistant spores, which survive pasteurization, has been associated with food poisoning since many strains produce enterotoxins and emetic toxins that cause gastroenteritis, diarrhea, and vomiting (Davey and Tauber 1987; Callegan, Engel, Hill, and O'Callaghan 1994; Beecher, Pulido, Barney and Wong, 1995; Helgason, Økstad, Caugant, Johansen, Fouet, Mock, Hegna, and Koltsø, 2000; Day, Smith, Gregg, Turnbull, Head, Ives, and Ho 1981; Whitney, Beatty, Taylor, Weyant, Sobel, Arduino, and Ashford; Gould and Dring, 1974). However, a plethora of experimental evidence has been rapidly accumulating showing that *B.cereus* infection is more serious than it had been previously thought. *B.cereus* is one of the most important causes of post-traumatic and metastatic bacterial endophthalmitis causing an average of 39% of trauma associated cases of endophthalmitis (Davey and Tauber 1987; Callegan, Engel, Hill, and O'Callaghan 1994; Beecher, Pulido, Barney and Wong, 1995) and is the most frequent cause of metastatic bacterial endophthalmitis in drug abusers (Davey and Tauber 1987; Callegan, Engel, Hill, and O'Callaghan 1994; Beecher, Pulido, Barney and Wong, 1995). Indeed, *B.cereus* is considered to be one of the most destructive organisms to affect the eye causing almost certain blindness even when aggressive and appropriate anti-microbial therapy is administered before the loss of visual acuity (Davey and Tauber 1987; Callegan, Engel, Hill, and O'Callaghan 1994; Beecher, Pulido, Barney and Wong, 1995). This is because highly potent toxins and virulence factors continue to damage the tissue and promote inflammation (Davey and Tauber 1987). Among these virulence factors are, cereolysin O, a thiol-activated cytolysin, which recognizes the cholesterol component of cell membranes prior to lysis (Shaney, Bernheimer, Grushoff, and Kim, 1974); cereolysin AB, another cytolytic unit unrelated to cereolysin O (Gilmore, Cruz-Rodz, Leimeister-Wachter, kreft and Goebel, 1989); hemolysin BL, a toxin with hemolytic, dermo-necrotic, and emetic activities (Beecher, Pulido, Barney and Wong, 1995); and collagenase, capable of degrading the collagen constituent of the vitreous and retinal architecture (Soderling and Pauino, 1981; Harrington, 1996).

Formed in response to starvation, bacterial spores such as those of *B. cereus* and its relative *B.anthracis* are among the most resilient cell types known (Harrington, 1996). Although they remain dormant for decades in a nutrient-deprived environment enduring a wide range of environmental stresses, they retain the capacity to germinate and give rise to bacterial colonies once the nutrient conditions return to a favorable state (Helgason, Økstad, Caugant, Johansen, Fouet, Mock, Hegna, and Koltsø, 2000). Fundamental to spore resistance and germination is the coat, a tough proteinaceous, multilayered shield that provides mechanical integrity and excludes large toxic molecules, while allowing small nutrient molecules to access germination receptors beneath the coat (Driks 2002; Chada, Sanstad, Wang, and Driks 2003). For some spore species, such as *B. anthracis*, *B. cereus* and Clostridial species, an outer responsive membrane called the exosporium comprised of protein, carbohydrate, and lipid has been shown to function in spore attachment but is still poorly understood (Chada, Sanstad, Wang, and Driks 2003; Beaman, Pankratz, and Gerhardt 1971; Matz, Beaman, and Gerhardt 1970; Todd, Moir, Johnson, and Moir 2003; Panessa-Warren, Tortora, and Warren, 1997; Panessa-Warren, Tortora, and Warren,1994). Because the exosporium is the outermost layer of the spore, it is the first that makes initial contact with the host cells and tissues (16-18). Therefore, in addition to being the protective layer encasing the spore conferring unique adherence and hydrophobic properties, the exosporium may express molecules that play a significant role in spore attachment and pathogenicity (Driks 2002; Chada, Sanstad, Wang, and Driks 2003; Chada, Sanstad, Wang, and Driks 2003; Beaman, Pankratz, and Gerhardt 1971; Matz, Beaman, and Gerhardt 1970; Todd, Moir, Johnson, and Moir 2003). The present studies were therefore undertaken to investigate whether the exosporium expresses molecule(s) that participate in spore-host cell [surface] attachment/colonization, and if the multifunctional cell surface protein gC1qR, which has affinity for a number of bacterial outer surface antigens serves as one of the spore attachment site(s) (Ghebrehiwet, Peerschke, Willis, Hong, and Reid, 1994; Ghebrehiwet, Lim, Kumar, Feng and Peerschke, 2001).

2 Materials and Methods

2.1 Chemicals and reagents

The following reagents and chemicals were purchased or obtained from the sources indicated. Alkaline phosphatase (AP)-streptavidin, and *p*-Nitrophenyl Phosphate (pNPP) (Pierce Co,); single walled carbon nanotubes (SWNTs) (Carbolex Inc., Lexington, KY).

2.2 Expression of recombinant gC1qR/p33:

The strategy for the construction of a plasmid containing the full-length gC1qR cDNA was described in detail in our earlier publication (Ghebrehiwet, Peerschke, Willis, Hong, and Reid, 1994). Briefly, the cDNAs were subcloned downstream of the glutathione-S-transferase (GST) gene in the expression plasmid, pGex-2T (Pharmacia Biotech Inc.), transformed into *E.coli* BL-21 (DE3) and protein expression induced by 0.5 mM IPTG (isopropyl β-D-1 thiogalacto pyranoside). The proteins were expressed as GST fusion products with the GST at the N-terminus of the gC1qR. The fusion products were purified on glutathione-Sepharose 4B column. The gC1qR fusion products were then cleaved by thrombin and the GST-free gC1qR purified on fast protein liquid chromatography (FPLC, Pharmacia) using a Mono-Q ion exchange column. After verification by ELISA and Western blotting using monoclonal antibody to gC1qR (Ghebrehiwet, Zhang, Lim, Eggleton, Leigh, Reid, and Peerschke, 1996), the single peak containing the gC1qR was pooled, concentrated to 1-2 mg/ml, and stored at -80^0 C. When necessary, 5 mM pro-phe-arg chloromethyl ketone (PPACK), a specific and rapid thrombin inhibitor, was added to eliminate the possibility of the presence of a trace amount of the enzyme in the preparation.

2.3 Monoclonal antibodies

The production, characterization and purification of various monoclonal antibodies to the various domains and epitopes of gC1qR have been described in detail in our earlier publications (Ghebrehiwet, Zhang, Lim, Eggleton, Leigh, Reid, and Peerschke, 1996). The monoclonal antibodies 60.11 and 74.5.2 used in the present studies have been used extensively in various functional studies (Ghebrehiwet, Lim, Kumar, Feng and Peerschke, 2001). They are both IgG1 kappa isotype and recognize the C1q binding site (residues 74-96) and kininogen binding site (residues 204-218) of the molecule.

2.4 Binding studies

To determine the interaction between gC1qR and *B.cereus* spores, a standard ELISA was used. Briefly, 100 µl (1×10^7/ ml) of *B. cereus* SB460 [originally isolated from a blood culture of an intravenous drug user (Panessa-Warren, Tortora, Wong, Ghebrehiwet, and Warren, 2003)] spores were first attached to duplicate microtiter plate wells by incubating at either 37^0 C or 60° C for 2 hr. After blocking with 1% BSA, and washing with Tris-buffered saline, pH 7.5, the spores were incubated with biotinylated gC1qR/p33 in TBS alone or containing either 10 mM Ca^{2+} or 10 mM EDTA. The bound protein was then detected by sequential reaction with alkaline phosphatase (AP)-streptavidin, and p-Nitrophenyl phosphate (pNPP) and read at 410 nm in a Dynatech, MR 700 ELISA reader.

2.5 Phagocytosis assay

B. cereus spores were preincubated with or without gC1qR (60 min, 37^0 C) before addition to citrated blood, and incubated for 60 min. Alternatively, mAb 60.11 was added to the blood and prior to addition of spores and incubation (60 min, 37° C). In each of these experiments, the blood was thin smeared onto glass slides and stained with Dif-quick. Visualization of the internalized spores was achieved by spore-specific staining with malachite green (7.6%, in distilled H_2O) and counterstained with 0.5% safranin (2.5 g Safranin O, first dissolved in 100 ml of 95% ethanol and brought to 1L with distilled H_2O) as described previously (Schaeffer and Fulton, 1933). Phagocytosis was then evaluated by counting under a light microscope, neutrophils or monocytes that ingested $\geq$3 spores per cell.

2.6 Preparation of carbon nanoloops

Single walled carbon nanotubes (SWNTs) (Carbolex Inc., Lexington, KY) were chemically cleaned in acid, cut and formed into 20-40 nm loops (Panessa-Warren, Wong, Ghebrehiwet, Tortora, and Warren, 2002). The nanoloops were then functionalized with purified recombinant gClqR (gC1qR-nanoloops), a non-immune mouse IgG1k, (IgG1k-nanoloops) or a monoclonal antibody to gC1qR (mAb 60.11, or 74.5.2 mAb-nanoloops). By atomic force microscopy and transmission electron microscopy these functionalized carbon nanoloops measured between 28.2 ± 4.2 and 36.7 ± 10.2 nm mean diameter. The preliminary trials exposing the carbon nanoloops to bacteria and tissue culture cells produced no alterations in normal cell behavior and physiology. The nanoloops provided good visualization for TEM and FESEM without any signs of bioaccumulation within cells, or cytotoxicity over the 4.5 hr incubation period.

The rationale for using antibody functionalized carbon nanoloops is to identify where and when gClqR was expressed on the apical surface of human colon and lung cells. The carbon nanoloops provided a platform for attaching a high density of antibody molecules, and served as a carrier to deliver this excess of antibody to the cell surface using in vitro culture. Unlike more conventional gold labeling, this newer, albeit not much tested, technique does not produce background noise, or cause adverse reactions with the bacteria or human cells in culture. Preliminary exploratory studies showed that there was neither accumulation nor intracellular localization of the nanoloops, nor evidence of necrosis or nuclear damage in control (non-functionalized or IgG and gClqR functionalized nanoloops) or mAb-functionalized carbon nanoloops incubated with cells. Furthermore, cells and bacteria showed no reactivity or binding to the loop itself. However, the ability to carry multiple antibody molecules at one time to the cell surface, insured the attachment and localization of the nanoloop, even at the earliest time periods of gClqR expression.

2.7 Preparation of cells for microscopic analyses

For *in vitro* experiments, mAb-nanoloops, IgG1*k*-nanoloops, and naked-nanoloops were incubated (36^0 C) in shell vials with Caco-2 (human colon carcinoma) cells or NCI-292 human epithelial lung cells grown to confluency as monolayers on glass coverslips (Diagnostic Hybrids, Athens, OH). Continual rotary mixing during cell incubation to ensure that the nanoloops did not clump, but were well dispersed with the bacterial spores and evenly distributed over the cell monolayer. Cells were incubated for 20 min (rotary mixing) with control or mAb-nanoloops, followed by spore inoculation with 2 µl of 3 X 10^8 spore/ml deionized sterile water, and incubated for 30,60,90, and 150- 270 min. Cells incubated with naked-nanoloops, and cells with inoculated spores but no nanoloops were also incubated under the same conditions. Following incubation, the media were removed, the surface of the monolayers washed 2X with PBS and the cells fixed in 2.8% glutaraldehyde in 0.1M cacodylate buffer with 0.1% $CaCl_2$ and 10% sucrose overnight. The saved media and PBS washes were spun at 3400 rpm for 30 min and the pellets screened for bacteria, cells, nanoloops, and any contaminants. For microscopy, the cell monolayers were post fixed in osmium tetroxide, dehydrated in acetone, and either critical point dried for FESEM, or embedded in epoxy resin, polymerized, and thin sectioned for TEM. For each time period studied 4 coverslips with culture cells were prepared for electron microscopy. By FESEM, the coverslips were surveyed using five random fields imaged at the same magnifications — 500X, 2000X, 16,000X, 25,000X, 40,000X and 60,000X — at the same tilt angle and accelerating voltage. Microvilli, cell surfaces, and the background were screened for the presence of nanoloops, as well as contaminating cell damage and compared to controls. With TEM samples, control and experimental tissues were thin-sectioned and imaged both stained (uranyl acetate and lead citrate) and unstained to verify the presence of nanoloops.

2.8 Microscopic analyses

2.8.1 Light microscopy

Blood smears on glass slides were reacted with *B.cereus* spores as described above and stained by a previously detailed method for spore visualization using malachite green (7.6%) and counterstained with safranin (0.25%) (Schaeffer and Fulton, 1933). Slides were examined and photographed (100 ASA Kodak color film) under oil using an Olympus BH2 photomicroscope (Plainview, NY) with a polarizing filter at 1025X.

2.8.2 Electron Microscopy

Glutaraldehyde (2.8% in 0.1M cacodylate buffer, pH 7.2, with added 10% sucrose) fixed Caco-2 and lung monolayers were washed in 0.1 M cacodylate buffer with 10% sucrose, osmicated, and prepared for microscopy with half of the coverslip processed for TEM, and the other half of the cleaved coverslip prepared for FESEM. Dehydration (dropwise) with acetone and infiltration with epoxy resin was done slowly (2-3 weeks) to prevent the bacterial spores from rupturing. Coverslips were placed on cleaned glass slides and a BEEM capsule filled with epoxy placed over each coverslip and polymerized at 56°C for 2 days. The samples were then frozen in liquid nitrogen and the BEEM capsules broken away from the glass slide, leaving the tissue culture cells in the plastic of the BEEM capsule. This layer of cells was then removed from the BEEM capsule and flat embedded to provide longitudinal cell sections showing the apical and basal cell surfaces. For TEM, thin sections were stained with uranyl acetate and lead citrate and imaged at 80 KV using a Philips 300.

For FESEM, critical point dried monolayers on glass coverslips were attached with silver conducting paint to brass stubs, metal coated with 3-5 nm Pt, and imaged in a JEOL6500F field emission scanning electron microscope at 5-15 KV at 30° tilt. To prevent sample bias, 5 fields were randomly chosen on each sample, and 15 or more images taken at each site at the same magnification. To ensure the correct interpretation of microscopic data, TEM and FESEM images from the same sample were compared to verify nanoloop localizations, cell morphology, and bacterial interactions.

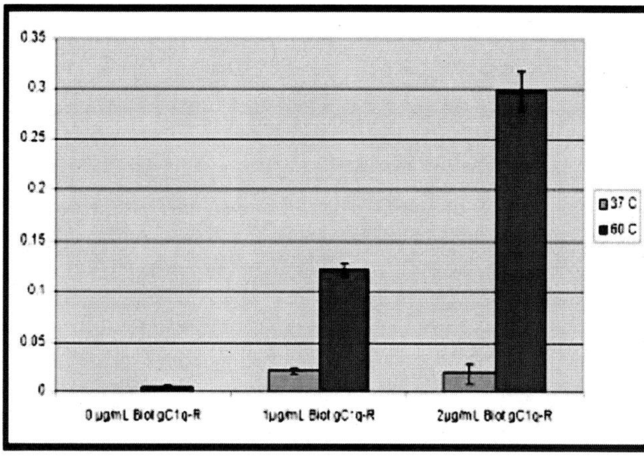

Fig. 1. The binding of gC1qR/p33 to *B. cereus* is temperature dependent. To determine the effect of temperature variation, a standard ELISA was used. Spores at a concentration of $1x10^7$ /ml were attached to microtiter plates by incubation at either, 37^oC and 60^oC (2 hr). After blocking with 1%BSA, and washing, the spores were incubated with biotinylated gC1qR/p33, detected by sequential reaction with alkaline phosphatase (AP)-streptavidin, and *p*-Nitrophenyl phosphate (pNPP) and read at 410 nm in an ELISA reader. Each data point is a mean of 3 separate experiments run in duplicates.

3 Results

3.1 The binding of gC1qR to B. cereus is temperature and calcium dependent

To confirm preliminary observations and to determine the effect of temperature variation on the interaction between gC1qR and spores, a standard solid phase ELISA was used. Spores at a concentration of $1x10^7$ were attached to microtiter plates as described in *Methods*, incubated with biotinylated gC1qR/p33 and developed. The results (Fig. 1, n=3) not only show that spores attach much more (~10) readily at higher temperatures (60^0 C $\geq 37^0$ C), but also that gC1qR/p33 binding to *B. cereus* spores is dose-dependent. Furthermore, although good binding of spores was observed in TBS alone, the addition of exogenous calcium consistently showed a significantly enhanced binding in the presence of additional Ca^2, (TBS, pH 7.5, containing 10 mM Ca^{2+}) implying that gC1qR binding to *B.cereus* is may be Ca^{2+} dependent (Fig. 2).

3.2 EDTA abrogates gC1qR binding to B. cereus

To verify the importance of endogenous or exogenous Ca^{2+}, the experiments in Fig. 2 were repeated except that the added gC1qR was in TBS buffer that contained 0-10 mM EDTA instead of Ca^{2+}. As expected results obtained from this experiment showed a dose-dependent decrease in gC1qR binding (n=3, data not shown).

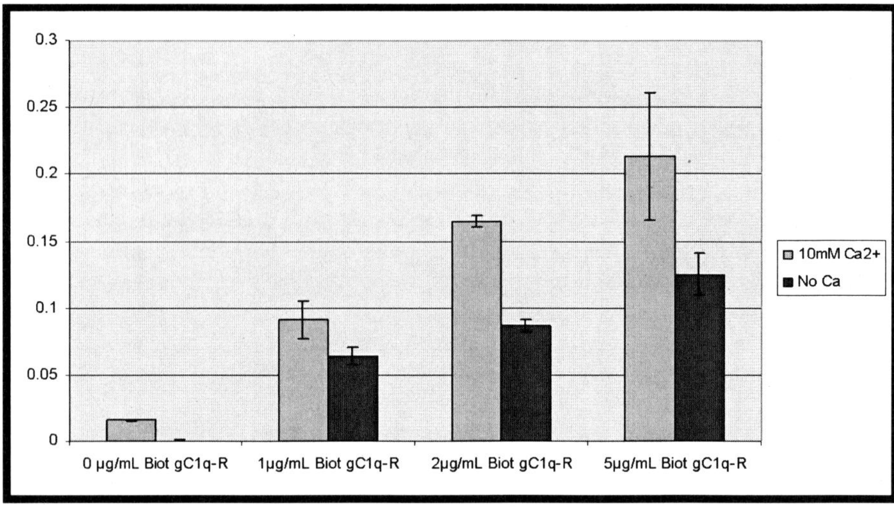

Fig. 2. Exogenous Ca^{2+} enhances gC1qR/p33 binding to *B. cereus*. Microtiter plates were first coated with *B. cereus* spores as in Fig.1 by incubation at $60^{O}C$. Varying concentrations of gC1qR/p33 in Tris buffered saline (TBS), pH 7.5, containing 10 mM Ca^{2+} were then added to the spores and developed as described. Each data point is a mean of 4 separate experiments run in duplicates.

However, complete inhibition was not achieved even with the highest dose of EDTA (10 mM). This is probably due to excessive amount of endogenous calcium, which is released by the spores during activation and attachment (Panessa-Warren, Tortora, and Warren, 1997).

3.3 Neutrophils and monocytes readily take up B. cereus spores

To get insight into the biologic relevance of the interaction between gC1qR and *B.cereus* spores, we chose a phagocytosis model in which the ability of monocytes or neutrophils to ingest *B. cereus* spores was examined. To this end spores were incubated with citrated blood for 30 min at 37° C and after incubation, the blood was thin-smeared onto glass slides, stained and evaluated by counting the cells that ingested ≥3 spores under a light microscope. As shown in Fig. 3, *B.cereus* spores are readily engulfed and taken up by neutrophils and also by monocytes (not shown). To examine whether this ingestion is mediated by gC1qR, experiments were performed in which mAb 60.11 or isotype- and species-matched control antibody was added to the blood together with the spores. After incubation (30 min, 37° C), the degree and quality of ingestion was evaluated by counting the number of cells that ingested spores and the number of ingested spores under a light microscope as described above. The results from these studies (n=3; data not shown) showed a decrease in both in the number of spores ingested and the number of monocytes/neutrophils that were engaged in phagocytosis as compared with controls.

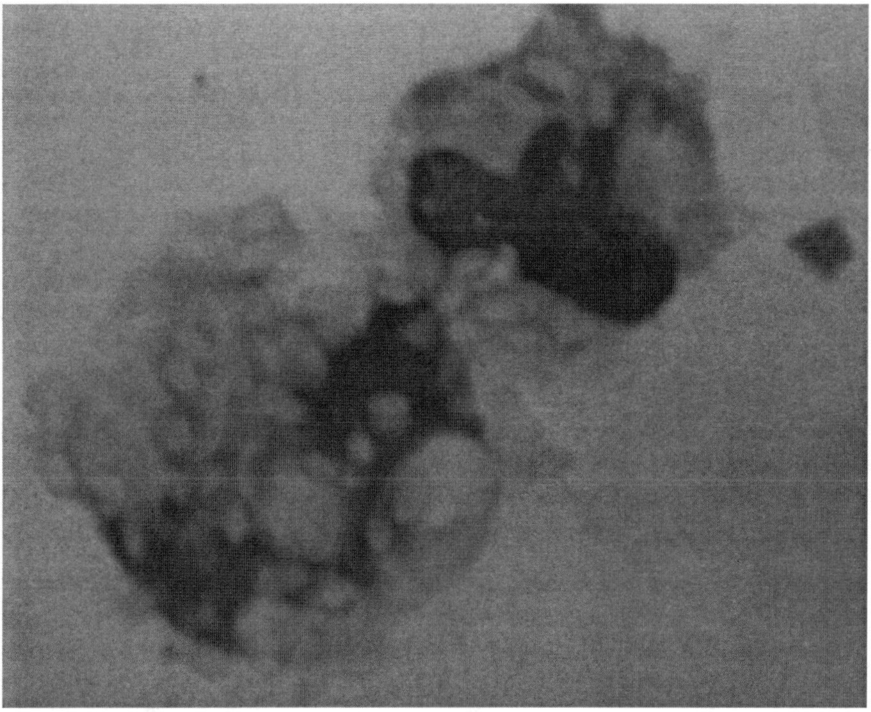

Fig. 3. Neutrophils readily take up *B. cereus* spores. After incubation of spores with whole blood, smears of blood were made on glass slides and stained by a previously detailed method for spore visualization using malachite green (7.6%) and counterstained with safranin (0.25%) (23). Slides were examined and photographed (100 ASA Kodak color film) under oil using an Olympus BH2 photomicroscope with a polarizing filter at 1025X. Shown is the image of neutrophils that have ingested an robust number of spores (>10) seen as translucent elliptical spheres.

3.4 Electron microscopic analyses

Certain nanotube-fabricated loops (24-40 nm diameter) were used to convey multiple monoclonal antibodies to the specimen surface. Control experiments using plain carbon nanoloops showed no cellular toxicity to tissue culture cells or bacteria. Using these immunocarriers more antibody could be brought to the sample surface and visualized without loosing antibody reactivity. When Caco-2 or lung cells were incubated with naked-nanoloops, no carbon nanoloops were found attached to the apical surface of the cells, the microvilli or cytoskeletal projections in the case of lung cells. Cells appeared robust with turgid microvilli (Fig. 4A), and maintained cell junctions in both TEM (Fig. 4B), and FESEM images (Fig.4A), indicating that over the time period of 30 min to 4.5 hours, the nanoloops did not cause cell toxicity, bioaccumulation, or produce spurious binding. For all samples analyzed no binding of the nanoloops was observed when IgG1k-nanoloops were incubated with cell

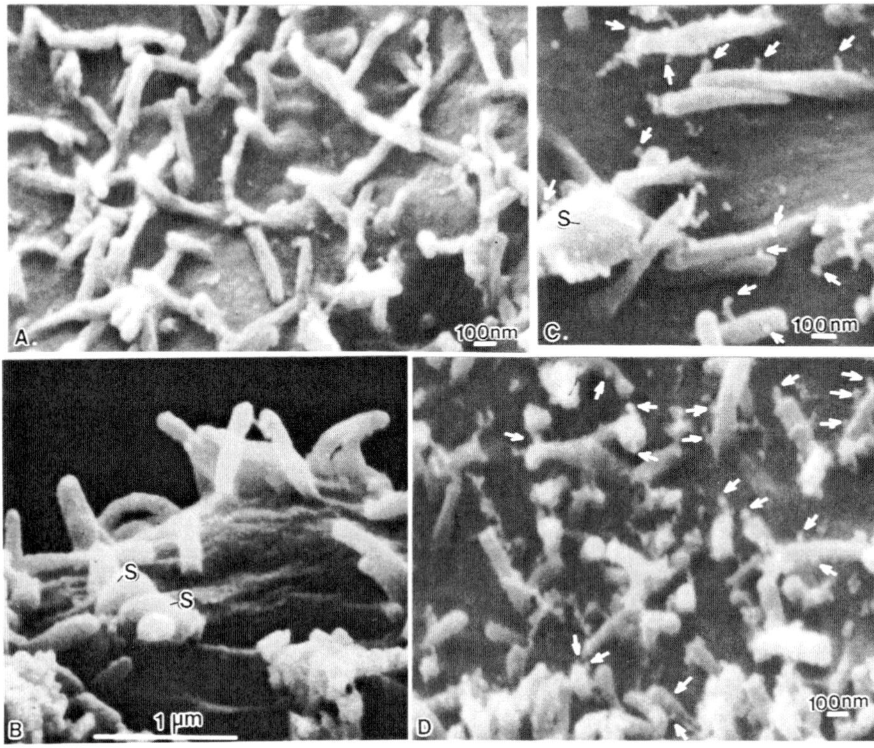

Fig. 4. FESEM images of Caco-2 cell microvilli incubated for 2.5 hr with mAb-nanoloops in the presence and absence of *B. cereus* spores. (A) Following incubation with mAb -74.5.2 functionalized nanoloops, the microvillous surface showed no marker uptake in the absence of spores. (B) Caco-2 microvilli incubated with *B. cereus* spores (s) and control IgG1k-nanoloops showed no marker uptake, even on microvilli adjacent to attached endospores. (C) However, numerous mAb-nanoloops (arrows), attached to microvilli and the exosporia of *B. cereus* spores. (D) At lower magnification, the extensive localization of gC1qR/p33 with mAb -nanoloops (arrows) is apparent on the microvilli adjacent to *B. cereus* spores.

monolayers and *B.cereus* spores. Fig. 4B clearly shows spores (S) on the Caco-2 cell surface adjacent to turgid, clean microvilli, with no binding of nanoloops on the cell or microvilli surface. Nanoloops functionalized with purified recombinant gClqR also failed to bind to the cells or microvilli (not shown). However mAb-functionalized nanoloops readily attached to the surface of microvilli following spore contact with the cell surface (Fig. 4C, D). Increased mAb-binding to gClqR was found overtime as more spores attached to the apical surface (Fig. 4D arrows) with some microvilli showing many bound nanoloops (arrows).

By TEM, cells did not show any mAb-nanoloop binding at 30-60 min incubation (Fig. 5A). Some spores (S) were seen passing towards the Caco-2 apical surface. Some of the spores (S) were still dehydrated (blackened spore), while others were

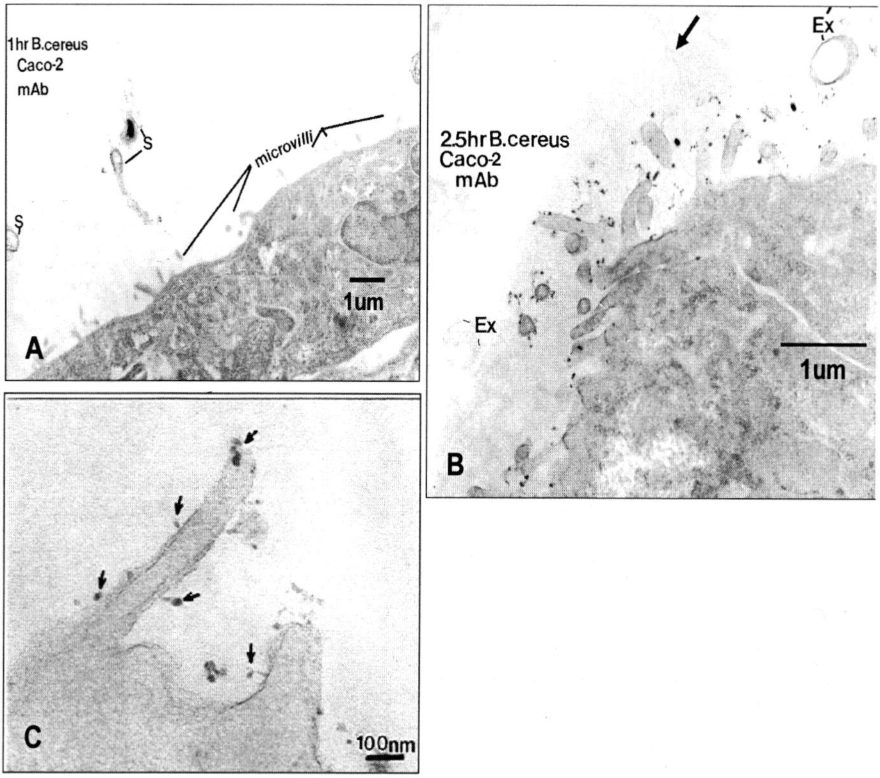

Fig. 5. FESEM images showed that the attachment of *B. cereus* spores to Caco-2 cells induced surface expression of gClqR. The composite images show that: (A) No detectable cell surface gC1qR is expressed after 1 h incubation of Caco-2 cells with spores. (B) After 2.5 h incubation, significant mAb-nanoloop binding indicates sites of gC1qR localization (arrows), which was predominantly localized on the surface of microvilli and occasionally on the surface of the exosporium (Ex). (C). A higher magnification of the microvilli reveals clusters of mAb-nanoloop binding (arrows) along the base and tip of the microvilli outer membranes, revealing sites of gC1qR expression.

germinating as they approached the host cell apical surface. Surface microvilli showed no mAb-nanoloop binding indicating that no gClqR was being expressed on the cell surface as the spores approached. However at 2.5 hr following the introduction of mAb-nanoloops and *B.cereus* endospores, numerous sites of mAb binding were seen on the microvilli of the host cells under direct bacterial attack (Fig. 5B). Some empty exosporia (Ex) in contact with microvilli were seen. Numerous clusters of nanoloops decorated the microvilli adjacent to attached spores. At higher magnification individual microvilli revealed clusters of carbon nanotube loops (Fig. 5C).

A similar gC1qR distribution pattern following *B.cereus* inoculation of Caco-2 cells, was seen when lung cells revealed attached *B.cereus* endospores with the adjacent cytoskeletal projections heavily labeled with mAb-functionalized nanoloops (Fig. 6A), localizing sites of gClqR expression. Following analysis of more than

500-field of tissue culture cells, no mAb-nanoloop binding was seen on adjacent cells, or on the same cell in areas lacking bound spores. Intact spores could be seen in surface invaginations (Fig. 6B), with heavy labeling of mAb-nanoloops (arrows) on the cytoskeletal projections and immediate apical surface surrounding the spore. From FESEM and TEM analyses of colon or lung cells incubated with *B.cereus*, gC1qR expression was only observed at cell surface sites with *B.cereus* spores, suggesting that spore attachment or the presence on the host cell surface did not cause gC1qR expression on all adjoining cells of the monolayer.

4 Discussion

The data presented in this report collectively suggest that *B. cereus* spores express on their surface a yet unidentified surface molecule(s) through which they bind gC1qR. The binding of microtiter-bound *B.cereus* spores to purified gC1qR was found to be calcium-dependent, which may be provided by the spore itself during the process of activation (Panessa-Warren, Tortora, and Warren, 1997).

Additional data in support of the involvement of gC1qR in the attachment of *B.cereus* to cell surfaces is provided by the following electron microscopic observations: (i) Spore attachment induces upregulation of gC1qR on the cell surface of Caco-2 cells, and NCI 292 lung cells. This upregulation appears after 1 hr exposure and peaks at about 2.5 hr. (ii) gC1qR expression is robust at or near sites of spore attachment at least when Caco-2 cells and lung cells are used as target cells for attachment, and this expression is predominantly localized on the surface of microvilli (Caco-2, Fig 4 and 5) or apical cytoskeletal projections (lung cells, Fig.6). (iii) Concomitant with the appearance of gC1qR on the microvilli, the spore exosporium itself is often decorated with gC1qR. (iv) *B. cereus* spores normally attach and colonize the surface of Caco-2 cells. However, when mAb-nanoloops block host cell surface gClqR, the *B.cereus* protoplasts were found to enter the host cell cytoplasm, rather than colonize the host cell surface thus altering the invasion paradigm (Panessa-Warren, Wong, Ghebrehiwet, Tortora, and Warren, 2002). Spore protoplast entry, which occurs only when the target cells are preincubated with blocking antibody, is initiated following attachment of the spore to Caco-2 microvilli and simultaneous shedding of the outermost coating, or exosporium (not shown). (v) In the absence of antibody blockade, the spores germinate, colonizing the host cell surface, but fail to get inside the cell. A similar finding with *E.coli* sequestered within bladder cells (Anderson, Palermo, Schilling Roth, Heuser, and Hultgren, 2003) whereby the bacteria could emerge to produce subsequent infections, may be analogous to the storage of *B.cereus* protoplasts within the colon cells. This process may allow the spore DNA-containing protoplasts to survive for a prolonged period of time undamaged even by the hostile intracellular environment until they are released.

Both Caco-2 and NCI 292 cells allowed *B. cereus* spore attachment on their cell surfaces regardless of gClqR blockade with mAb-nanoloops. Yet, the presence of activated endospores on the surface of both Caco-2 and human lung cells produced copious overexpression of gClqR only in those areas with spore attachment but not on cells without any bacterial spores. On Caco-2 cells, gClqR surface expression was

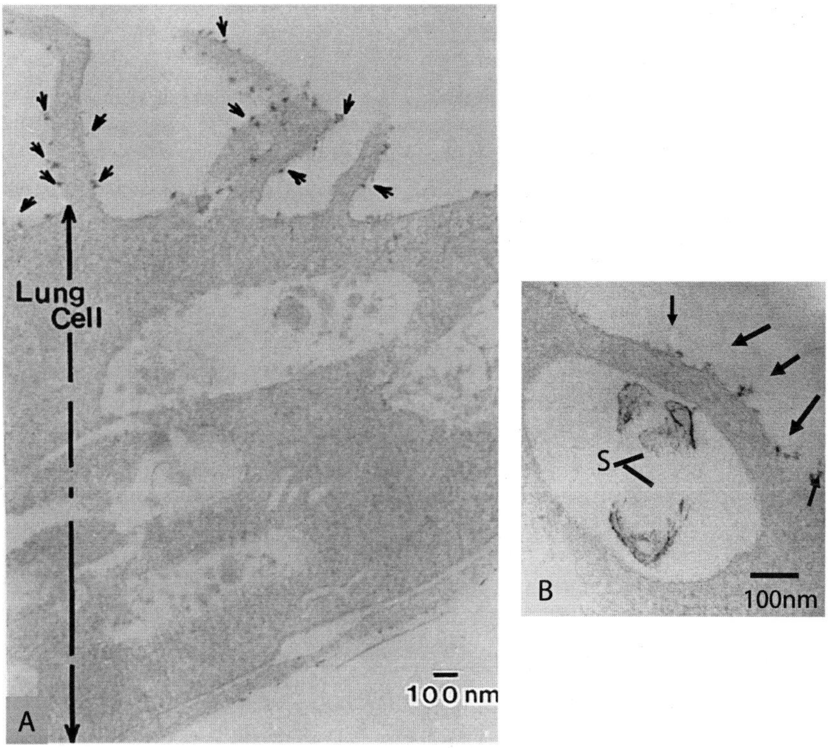

Fig. 6. Surface expression of gC1qR on NCI-292 human epithelial lung cells. The figure shows TEM images of human lung epithelial cells after incubation with *B. cereus* spores. Following incubation with *B. cereus* endospores, lung cells that had attached spores formed large cytoskeletal projections, which revealed profuse surface binding of mAb-functionalized nanoloops [(arrows (A)]. This profuse expression of gClqR was only found focally where a spore had attached. Adjacent areas of the same cell did not show mAb-nanoloop binding, or the cytoskeletal projections. (B) Shows a higher magnification of an apical area of the lung cell with an incorporated intact *B. cereus* spore with exosporium and spore coats within a vacuole. Just above the vacuole containing the spore on the apical lung surface, extensive mAb binding can be seen in the plasmalemma indicating the presence of gClqR.

localized to the microvilli within a 6-10 μm radius of attached spores. However, neighboring cells and the adjacent areas to the responding microvilli showed no gClqR surface expression. Lung cells on the other hand, produced cytoskeletal extensions where *B. cereus* endospores were present. The cytoskeletal extensions adjacent to the attached spores showed extensive mAb-nanoloop binding indicating the presence of gC1qR, whereas the surrounding cells lacking endospores showed no detectable surface gClqR expression.

The possible involvement of gC1qR in the process of spore attachment is further corroborated by phagocytosis experiments. When spores were exposed to whole blood, there was a rapid and robust uptake by monocytes and neutrophils, but the uptake of spores was significantly reduced (not shown) when mAb-60-11 was added to the blood prior to addition of the spores. The presence of a putative gC1qR binding site on the *B.cereus* exosporium suggests that spores are able to interact directly with cell surface gC1qR through a molecule on their surface that mimics one of the multiple ligands of gC1qR including C1q and kininogen (Ghebrehiwet, Lim, Kumar, Feng and Peerschke, 2001). Identification and characterization of this molecule(s) will therefore be of biologic significance with therapeutic potential.

The spore-forming family of Bacillus, which includes *B.cereus, B. thuringiensis* and *B. anthracis*, possess a loose balloon-like structure called exosporium (Todd, Moir, Johnson, and Moir, 2003; Panessa-Warren, Tortora, Wong, Ghebrehiwet, Warren, 2003). However, it is not known with certainty whether different strains of the same species including the SB460 isolate used in our studies possess an identical structure or a modified version of the exosporium. The particular adherence (Bowen, Fenton, Lovitt, and Wright, 2002) and hydrophobic properties (Koshikawa, Yamazaki, Yoshimi., Ogawa, Yamada, Watanabe and Torii, 1989) conferred by the exosporium suggest that this structure may be of significance to spore attachment and pathogenicity (Todd, Moir, Johnson, and Moir, 2003). The exosporium of *B.cereus* contains 43-52% protein, 15-18% lipid, and 23% carbohydrates of dry weight, respectively (Beaman, Pnkratz, and Gerhardt, 1971). It is not surprising therefore that identification and characterization of molecules of the exosporium has become of primordial significance in recent years. For example, a 205 kDa glycoprotein multimer (70 KDa monomer) has been identified and partially characterized from the spore of *B.thuringiensis* (Garcia-Patrone and Tandecarz) More recently, a major structural protein of *Bacillus anthracis* exosporium has been identified and molecularly cloned (Sylvestre, Couture-Tosi and Mock, 2002). The gene encodes a protein of 382 amino acid residues, the central part of which contains a region of GXX triplet repeats, similar to those found in the triple-helix region of mammalian collagen. Interestingly, this collagen-like surface glycoprotein, designated BclA (***Bacillu**s* **c**ollagen-**l**ike protein of ***a**nthracis*) was detected only in spores and sporulating cells but not in vegetative cells suggesting that it is a spore-specific glycoprotein (Sylvestre, Couture-Tosi and Mock, 2002). BclA is a major exosporium protein and is a structural constituent of the filamentous appendages present on its outer layer. Because the exosporium is exposed on the surface and is presumably the first structure of the pathogen to interact with the host cells, it is postulated to play a significant role in the infection process possibly through the interaction of spore proteins such as BclA, with cell surface proteins, of which gC1qR may be a possible candidate. This hypothesis is particularly attractive since gC1qR, is a ubiquitously distributed cellular protein with affinity for a number of viral and bacterial antigens including protein A of *Staphylococcus aureus* (Nguyen, Ghebrehiwet, Peerschke, 2000) and internalin B of *L. monocytogenes* (Braun, Ghebrehiwet, and Cossart), 2000). It has been shown that the genome of *Bacillus anthracis* is virtually identical to that of *B.cereus* with the exception of their respective toxin production ((Helgason, Økstad, Caugant, Johansen, Fouet, Mock, Hegna, and Koltsø, 2000; Todd, Moir, Johnson, and Moir, 2003). Therefore, although results obtained from the use of a surrogate bacterium, regardless of how closely related, may not necessarily reflect the function of the desired target bacterium (Finlay and Cossart, 1997), understanding the mechanism of *B. cereus* spore attachment and entry may nonetheless, serve as a proxy for understanding the mechanism by which spores of *B. anthracis* exploit the mammalian host cell functions to their advantage (Finlay and Cossart, 1997). Identifying a molecule(s) on the *B. cereus* exosporium, which allows interaction with cell surface proteins or receptors, may therefore predict the existence of a molecularly and functionally related molecule on *B. anthracis* or vice versa. In support of this hypothesis is the recently published report (Rety, Salamitou, Garcia-Verdugo, Hulmes, Le Hagarat, Chaby and Lewit-Bentley, 2005), which showed that the crystal structure of BclA, the *B. anthracis* spore surface protein, shares remarkable similarity to the C1q/TNF family of mammalian proteins. It will therefore be interesting to see whether *B. cereus* spores express a BclA-like molecule that has the ability to bind gC1qR. Experiments are therefore presently underway not only to address the veracity of this postulate but also to determine how the interaction between the spores and the host cell surface contributes to the pathogenicity of *B. cereus*.

5 Acknowledgements

*This work has been authored by Brookhaven Science Associates, LLC under contract No. DE-AC02- 98CH10886 with the US Department of Energy. The United States Government retains, and the publisher, by accepting the article for publication, acknowledges, a worldwide license to publish or reproduce the published form of this manuscript, or allow others to do so, for the United States Government purposes.
This work was also supported in part by grants R01-AI 060866 from the National Institutes of Allergy and Infectious Diseases (B.G.) and a generous gift from Larry and Sheila Dalzell (to B.G.).

6 Abbreviations

gC1qR, 33 kDa cellular protein which binds to the globular heads of C1q; MBL; FESEM, field emission scanning electron microscopy; TEM, transmission electron microscopy.

References

Anderson, G.G., Palermo, J. J., Schilling, J.D., Roth, R., Heuser, J., and Hultgren, S.J. (2003) Intracellular bacterial biofilm-like pods in urinary tract infections. Science 301, 105-107.

Beaman, T.C., Pankratz, H.S., and Gerhardt, P. (1971) Paracrystalline sheets reaggregated from solubilized exosporium of *Bacillus cereus*. J. Bacteriol 107, 320-324.

Beecher, D.J., Pulido, J.S., Barney, N.P., and Wong, A.C.L. (1995) Extracellular virulence factors in *Bacillus cereus* endophthalmitis: Methods and implication of involvement of hemolysin BL. Infect Immun 63, 632-639.

Bowen, W.R., Fenton, A.S., Lovitt, R.W., and Wright, C.J. (2002) The measurement of *Bacillus mycoides* spore adhesion using atomic force microscopy, simple counting methods, and spinning disc technique. Biotechnol Bioeng 79, 170-179.

Braun, L., Ghebrehiwet, B., and Cossart, P. (2000) gC1q-R/p32, a C1q-binding protein, is a receptor for the InlB invasion protein of *Listeria monocytogenes*. EMBO J 19, 1458-1466.

Callegan, M.C., Engel, L.S., Hill, J.M., and O'Callaghan, R.J. (1994) Corneal virulence of *Staphylococcus aureus*: Roles of alpha-toxin and protein A in corneal pathology. Infect Immun 62, 2478-2482.

Chada, V.G., Sanstad, E.A., Wang, R., and Driks. A. (2003) Morphogenesis of bacillus spore surfaces. J. Bacteriol 185, 6255-6261.

Davey, R.T., and Tauber, W.B. (1987) Posttraumatic endophthalmitis: the emerging role of *Bacillus cereus* infection. Rev Infect Dis. 9 110-123.

Driks, A., (2002) Maximum shields: the armor plating of the bacterial spore. Trends in Microbiol 10, 251-254.

Finlay, B.B., and Cossart, P. (1997). Exploitation of mammalian host cell functions by bacterial pathogens. Science 276, 718-725.

Garcia-Patrone, M., and Tandecarz, J.S. (1995) A glycoprotein multimer from *Bacillus thuringiensis* sporangia: dissociation into subunits and sugar composition. Mol. Cell. Biochem 145, 29-37.

Ghebrehiwet, B., Lim, W., Peerschke, E.I.B., Willis, A.C., Hong, Y., and Reid, K.B.M. (1994) Isolation, cDNA cloning and overexpression of a 33 kDa cell surface glycoprotein which binds to the globular "heads" of C1q (gC1q-R*)*. J Exp Med 179, 1809-1821.

Ghebrehiwet, B., Lim, B-L., Kumar, R., Feng, X., and Peerschke, E.I.B. (2001) gC1q-R/p33: a member of a new class of multifunctional and multicompartmental cellular proteins (MMCPs) is involved in inflammation and infection. Immunol Rev 180, 65-77.

Ghebrehiwet, B., Lu, P.D., Zhang, W., Lim, B-L., Eggleton, P., Leigh, L.E.A., Reid, K.B.M. and Peerschke, E.I.B. (1996). Identification of functional domains on gC1q-R, a cell surface protein, which binds to the globular "heads" of C1q, using monoclonal antibodies and synthetic peptides. Hybridoma 15, 333-342.

Gilmore, M.S. , Cruz-Rodz, A.L., Leimeister-Wachter, M., Kreft, J., and Goebel, W.A. (1989) *Bacillus cereus* cytolytic determinant, cereolysin AB, which comprises the phospholipase C and sphingomyelinase genes: Nucleotide sequence and genetic linkage. J. Bacteriol 17, 744-753.

Gould, G.W., and Dring, J.G. (1974) Mechanisms of spore heat resistance. *In Adv Microbial Physiol* (Rose A.H. and D.W. Tempest, eds) vol 11:pp137 Academic press, London.

Harrington, D.J. (1996) Bacterial collagenases and collagen-degrading enzymes and their potential role in human disease. Infect Immun 64, 1885-1891.

Helgason, E., Økstad, O.A. Caugant, D.A., Johansen, H.A., Fouet, A., Mock, M. Hegna, I., Koltsø, A-B. (2000) *Bacillus anthracis, Bacillus cereus,* and *Bacillus thuringiensis*—one species on the basis of genetic evidence. Appl Environm Microbiol 66, 2627-26230.

Koshikawa, T., Yamazaki, M., Yoshimi, M., Ogawa, S., Yamada, A., Watabe, K., and Torii, M. (1989) Surface hydrophobicity of spores of *Bacillus* species. J. Gen. Microbiol 135, 2717-2722.

Matz, L.L., Beaman, T.C., and Gerhardt, P. (1970) Chemical composition of the exosporium from spores of *B. cereus*. *J. Bacteriol* **101**(1): 196-201.

Nguyen, T., Ghebrehiwet, B., and Peerschke, E.I.B. (2000) *Staphylococcus aureus* protein A recognizes platelet gC1q-R/p33: A novel mechanism for staphylococcal interactions with platelets. Infect Immun 84, 2061-2068.

O'Day, D.M., Smith, R.S., Gregg, C.R., Turnbull, P.C.B., Head, W.S., Ives, J.A., and Ho, P.C. (1981). The problem of Bacillus species with special emphasis on the virulence of *Bacillus cereus.* Ophthalmology 88, 833-838.

Panessa-Warren, B., Tortora, G., and Warren, J. (1997). Exosporial membrane plasticity of *Clostridium sporogenes* and *Clostridium difficile*. Tissue and Cell 29, 449-461.

Panessa-Warren, B., Tortora, G., and Warren, J. (1994) Electron Microscopy of *C. sporogenes* endospore attachment and germination. Scanning 16, 227-240.

Panessa-Warren, B., Tortora, G., Wong, S.S., Ghebrehiwet, B., and Warren, J. (2003) Bacillus and Clostridium Spore Attachment/Entry of Human Colon Cells. Microscopy & Microanalysis 9 (suppl. 2) 1378-1379: Cambridge University Press, NY. 2003.

Panessa-Warren, B., Wong, S.S., Ghebrehiwet, B., Tortora, G.T., and Warren, J. (2002) Carbon Nanotube Membrane Probes: Immuno-labelling by LM, TEM & FESEM, *Microscopy and Microanalysis*, Voelkl, E., Piston, D., Gauvin, R., Lockley, A., Bailey, G., & McKernan, S. (Eds.) Cambridge University Press, New York, N.Y. **8** (suppl. 2), 726-727CD.

Rety, S., Salamitou, S., Garcia-Verdugo, I., Hulmes, D.J.S., Le Hagarat, F., Chaby, R. , and Lewit-Bentley, A. (2005) The crystal structure of the Bacillus anthracis spore surface protein BclA shows remarkable similarity to mammalian proteins. J Biol Chem 280, 43073-43078.

Schaeffer, A., and Fulton, D. (1933) A simplified method for staining spores. Science 77, 194-195.

Shany, S., Bernheimer, A.W., Grushoff, P.S., and Kim, K.W. (1974) Evidence for membrane cholesterol as the common binding site for cereolysin, strpetolysin O and saponin. Molec Cell Biochem 3, 179-186.

Soderling, E., and Pauino, K.U. (1981) Conditions of production and properties of the collagenolytic enzymes by two Bacillus strains from dental plaque. J Periodontal Res. 16, 513-523.

Sylvestre, P., Couture-Tosi, E., and Mock, M. (2002) A collagen-like surface glycoprotein is a structural component of the *Bacillus anthracis* exosporium. Molec Microbiol 45, 169-178.

Todd, S.J., Moir, A.J.G., Johnson, M.J., and Moir, A. (2003) Genes of *Bacillus cereus and Bacillus anthracis* encoding proteins of the exosporium. J. Bacteriol 185 (11): 3373-3378.

Whitney, E.A.S., Beatty, M.E., Taylor Jr, T.H., Weyant, R., Sobel, J., Arduino, M.J., and Ashford, D.A. (2003) Inactivation of *Bacillus anthracis* spores. Emerging Infect Dis 9, 623-627.

14

Immunity in Borreliosis with Special Emphasis on the Role of Complement

Kristina Nilsson Ekdahl[1], Anna J. Henningsson[2], Kerstin Sandholm[3], Pia Forsberg[4], Jan Ernerudh[4], and Christina Ekerfelt[4]

[1] University of Uppsala, Sweden, Department of Oncology, Radiology and Clinical Immunology, Kristina.Nilsson_Ekdahl@klinimm.uu.se
[2] Ryhov County Hospital, Sweden, Department of Infectious Diseases Anna.Henningsson.Jonsson@lj.se
[3] University of Kalmar, Sweden, Department of Chemistry and Biomedical Sciences kerstin.sandholm@hik.se
[4] Linköping University, Sweden, Department of Molecular and Clinical Medicine pia.forsberg@imk.liu.se, Jan.Ernerudh@lio.se, christina.ekerfelt@imk.liu.se

1 Lyme Borreliosis

Lyme borreliosis is a complex inflammatory disorder caused by the spirochete *Borrelia burgdorferi sensu lato (s. l.)*, which is transmitted to humans and animals by infected *Ixodes* ticks (Burgdorfer, Barbour, Hayes, Benach, Grunwaldt, and Davis 1982; Benach, Bosler, Hanrahan, Coleman, Habicht, Bast, Cameron, Ziegler, Barbour, Burgdorfer, Edelman, and Kaslow 1983; Steere, Grodzicki, Kornblatt, Craft, Barbour, Burgdorfer, Schmid, Johnson, and Malawista 1983; Johnson, Schmid, Hyde, Steigerwalt, and Brenner 1984). *B. burgdorferi s. l.* can be subdivided into at least 10 species of which *B. burgdorferi sensu stricto, B. garinii* and *B. afzelii* are pathogenic to humans (Wang, vanDam, Schwartz, and Dankert 1999). Recently, a novel human pathogenic species, A14S, was isolated, of which the name *B. spielmani* has been suggested (Wang, vanDam, and Dankert 1999; Richter, Schlee, Algower, and Matuschka 2004). Only *B. burgdorferi sensu stricto* has been identified in the U.S. (Wang et al. 1999), whereas all four human pathogenic species are found in Europe (Wang et al. 1999; Ornstein, Berglund, Bergström, Norrby, and Barbour 2002).

B. burgdorferi s. l. infection is the most common vector-borne disease in temperate areas of the Northern Hemisphere. The incidence varies considerably in different geographic locations. In the U. S an overall incidence of six per 100 000 inhabitants has been reported, the highest 105 per 100 000 reported from Lyme Connecticut (Anon 1997; Pena, and Strickland 1999). In Europe the incidence has been assessed as up to 155 per 100 000 inhabitants, with marked regional variability, even within the same geographic area (Stanek, Satz, Strle, and Wilske 1993; Berglund,

Eitrem, Ornstein, Lindberg, Ringer, Elmrud, Carlsson, Runehagen, Svanborg, and Norrby 1995; Huppertz, Bohme, Standaert, Karch, and Plotkin 1999; Strle 1999; Korenberg, Gorban, Kovalevskii, Frizen, and Karanov 2001).The disease was named during the seventies, when an extraordinary high number of children in Lyme, Connecticut, U. S. developed an arthritis, resembling juvenile rheumatoid arthritis. The etiological agent was isolated from *Ixodes* ticks for the first time in 1982 by Willy Burgdorfer (Burgdorfer et al. 1982), and the spirochete was named *B. burgdorferi* after him and according to its relationship to *B. recurrentis*, the infectious agent of relapsing fever.

B. burgdorferi s.l. has been called "the great imitator" because of the complexity of the clinical features of human borreliosis (Pachner 1988) and the diverse consequences of infection (Duffy 1990). The onset of the disease was originally reported to usually be heralded by the appearance of a characteristic skin lesion, erythema migrans (EM), several days to weeks after a tick bite, but not all patients manifest this unique marker (Steere, Malawista, Hardin, Ruddy, Askenase, and Andiman 1977), and recent reports have estimated that only 35-59% or even fewer cases, present with an EM rash (Stricker, and Phillips 2003; Johnson, and Stricker 2004). Unless treatment is initiated early, the disease may easily disseminate, resulting in mainly neurological, joint or skin manifestations, but other, rare manifestations, e.g. cardiac, as well as non-specific symptoms have also been reported (Steere et al. 1977; Stanek and Strle 2003). The different subspecies seem to have specific tissue preferences, and are thus also primarily associated with different manifestations; *B. burgdorferi sensu stricto* with arthritis, *B. garinii* with neurological manifestations and *B. afzelii* with the skin manifestation acrodermatitis chronicum atrophicans but each species may cause all symptoms (Balmelli and Piffaretti 1995). A clinical consensus divides Lyme borreliosis into an acute (localized) EM and a disseminated stage, with the latter divided into early and late phases (Garcia-Monco and Benach 1995).

The clinical outcome following infection with borrelia differs between individuals. There are strong indications that borrelia infection can be asymptomatic in certain individuals (Berglund et al. 1995; Ekerfelt, Forsberg, Svenvik, Roberg, Bergström, and Ernerudh 1999; Ekerfelt, Masreliez, Svenvik, Ernerudh, Roberg, and Forsberg 2001). Borrelia infection can also cause clinical symptoms with a benign subacute course, which following adequate antibiotic treatment resolves within six months. In some individuals, however, infection with borrelia despite antibiotic treatment causes symptoms persisting more than six months, here called chronic borreliosis (Kaiser 1994; Nocton, and Steere 1995; Oschmann, Dorndorf, Hornig, Schafer, Wellensiek, and Pflughaupt 1998; Steere 2001). A chronic disease course may be rather common as reported frequencies range between 25-53% in patients who develop disseminated infection (Asch and Bujac 1994; Berglund, Stjernberg, Ornstein, Tykesson-Joelsson, and Walter 2002; Vrethem, Hellblom, Widlund, Ahl, Danielsson, Ernerudh, and Forsberg 2002).

The mechanisms behind chronic borreliosis are not known and are a subject of controversy (Stricker, Lautin, and Burrascano 2005). Current hypotheses include i) long-time persistence of the spirochete despite antibiotic treatment (Preac-Mursic, Weber, Pfister, Wilske, Gross, Bauman, and Prokop 1989; Frey, Jaulhac, Piemont,

Marcellin, Boohs, Vautravers, Jesel, Kuntz, Monteil, and Sibilia 1998; Oksi, Marjamaki, Nikoskelainen, and Viljanen 1999), possibly explained by so called cyst form or "stress-starvation form" of borrelia, that may lie dormant in the human host (Kersten, Poitschek, Rauch, and Aberer 1995; Brorson, and Brorson 1998; Alban, Johnson, and Nelson 2000), ii) symptoms caused by an aberrant, insufficient or excessive, immune response, due to host-specific differences leading to chronic inflammation, possibly including post infectious autoimmunity (Sigal 1997; Roessner, Trivedi, Gaur, Howard, Aversa, Cooper, Sigal, and Budd 1998; Hemmer, Gran, Zhao, Marques, Pascal, Tzou, Kondo, Cortese, Bielekova, Straus, McFarland, Houghten, Simon, Pinilla, and Martin 1999), iii) symptoms emanating from irreversible tissue injury caused by an aberrant initial inflammation, that may or may not be associated with general symptoms, e.g. fatigue, which so far, is of unknown etiology, iiii) pathogen-specific genomic differences between subspecies of *B. burgdorferi* which can influence the clinical course as demonstrated by Wang and co-workers (Wang, Ojaimi, Wu, Saksenberg, Iyer, Liveris, McClain, Wormser, and Schwartz 2002). Irrespective of the mechanism for development of long-lasting symptoms, the establishment of an optimal immune response against borrelia is of crucial importance for resolution of the disease.

2 Immune Responses in Lyme Borreliosis

2.1 Innate Immunity

Inflammatory innate immune responses are critical in the control of early, disseminated infection (Weis, McCracken, Ma, Fairbairn, Roper, Morrison, Weis, Zachary, Doerge, and Teuscher 1999). The outer surface of *B. burgdorferi s.l.*, which contains an unusual high number of lipoproteins (Noppa 1998), presumably plays an important role in the host-parasite interactions. Spirochetal lipoproteins are recognized by Toll-like receptor (TLR)-2 and the ligation leads to activation of monocytes/ macrophages, neutrophils, lymphocytes, endothelial cells and fibroblasts (Lien, Sellati, Yoshimura, Flo, Rawadi, Finberg, Carroll, Espevik, Ingalls, Radolf, and Golenbock 1999), with subsequent secretion of inflammatory cytokines (Hirschfeld, Kirschning, Schwandner, Wesche, Weis, Wooten, and Weis 1999). In addition to sensing of the presence of microbes and inducing inflammation, individual TLRs induce immune responses that are tailored to a given microbial infection, including innate as well as adaptive responses (Barton and Medzhitov 2003). Outer surface protein (Osp)A and OspB, two abundant lipoproteins located at the surface of *B. burgdorferi s. l.* (Barbour 1984), were the first borrelial lipoproteins described to be highly immunogenic (Ma and Weis 1993) and provided the basis for the development of a Lyme disease vaccine (Steere, Sikand, Meurice, Parenti, Fikrig, Schoen, Nowakowski, Schmid, Laukamp, Buscarino, and Krause 1998), which later was withdrawn due to suspected side effects. The inflammatory activities attributed to OspA and OspB include an ability to directly induce NF-κB nuclear translocation, resulting in cytokine secretion, adhesion molecule expression and generation of nitric oxide and super oxide (Ma, Seiler, Tai, Yang, Woods, and Weis 1994; Norgard, Arndt, Akins, Curetty, Harrich, and Radolf 1996; Wooten, Modur, McIntyre, and Weis 1996; Morrison, Weis, and Weis 1997). Thus, studies so far strongly suggest a crucial role for TLR-2 in controlling *B. burgdorferi s. l.* infection. Experimental studies on TLR-2-deficient mice suggest that TLR-2 is required for innate, but not adaptive immunity to borrelia but are contradictory regarding the effect of TLR-2 deficiency on disease severity (Wooten, Ma, Yoder, Brown, Weis, Zachary, Kirschning, and Weis 2002; Wang, Ma, Buyuk, McClain, Weis, and Schwarz 2004).

2.2 Importance of Innate Cytokine Responses

We have previously reported significantly higher serum levels of the innate pro-inflammatory cytokine tumour necrosis factor-α (TNF-α) in cerebrospinal fluid (CSF), accompanied by an increase of the anti-inflammatory transforming growth factor (TGF)-β1, in the early disease course of patients with sub acute neuroborreliosis, compared with patients who later developed chronic neuroborreliosis (Widhe, Grusell, Ekerfelt, Vrethem, Forsberg, and Ernerudh 2002). This finding supports a beneficial role of a strong pro-inflammatory response in the target organ early in borrelia infection and in addition a role for TGF-β1 in controlling the systemic immune response, minimising host damage.

The delicate balance between eradication of the microbe and protection of the host from immune mediated tissue damage, mirrored by pro- and anti-inflammatory cytokines respectively, was illustrated in an experimental study. Mice deficient of the anti-inflammatory cytokine interleukin (IL-10) had tenfold lower spirochete burden but developed more severe arthritis compared with wild-type mice (Brown, Zachary, Teuscher, and Weis 1999). We recently made a similar observation in patients with a history of borrelia infection. Asymptomatic borrelia seropositive individuals showed higher numbers of TNF-α secreting dendritic cells in response to live borrelia spirochetes than did individuals with subacute borreliosis. The asymptomatic individuals also showed a tendency for increased IL-10 compared with both subacute and chronic borreliosis (Sjöwall, Carlsson, Vaarala, Bergström, Ernerudh, Forsberg, and Ekerfelt 2005). There was, however, no difference in TNF-α secretion compared with chronic borreliosis. Interestingly, the latter group seemed to be divided into two populations regarding TNF-α levels. One showed very low numbers of TNF-α secreting dendritic cells, indicating impairment of TNF-α responses, which may be needed for optimal eradication. The other population showed similar high levels of TNF-α as the asymptomatic individuals which combined with the tendency to lowered IL-10 secretion instead may indicate a defect in the down-regulation of the inflammatory response. Although this observation on differences within the chronic group is weak, development of chronic borreliosis may well be due to dysregulation of immune mechanisms at different levels in different individuals.

3 Adaptive Immunity

3.1 T-cell Mediated Responses

Studies on borrelia-specific T-cell responses in patients with borreliosis have unambiguously shown predominating Th1-like responses, most pronounced within the target organ (Forsberg, Ernerudh, Ekerfelt, Roberg, Vrethem, and Bergstrom 1995; Wang, Fredrikson, Sun, and Link 1995; Oksi, Savolainen, Pene, Bousquet, Laippala, and Viljanen 1996; Ekerfelt, Ernerudh, Bunikis, Vrethem, Aagesen, Roberg, Bergström, and Forsberg 1997; Yin, Braun, Neure, Wu, Eggens, Krause, Kamradt, and Sieper 1997; Ekerfelt, Ernerudh, Forsberg, and Bergström 1998; Gross, Steere, and Huber 1998; Widhe, Ekerfelt, Forsberg, Bergstrom, and Ernerudh 1998; Grusell, Widhe, and Ekerfelt 2002; Widhe, Jarefors, Ekerfelt, Vrethem, Bergström, Forsberg, and Ernerudh 2004). These findings strongly suggest a role for Th1-mediated effector responses in the eradication of borrelia, i. e. phagocytosis and cytotoxicity, which is supported by infiltrates of macrophages and T cells in EM lesions (Muellegger, McHugh, Ruthazer, Binder, Kerl, and Steere 2000).

The importance of strong Th1-responses early in borrelia infection is supported by several experimental studies on borrelia infection in mice. Mice that fail to eradicate the spirochetes, and develop clinical signs of infection, show weak Th1-responses initially in the infection, followed by lack of down-regulation, but instead a gradual increase in the Th1 response. In contrast, resistant mice, which eradicate the spirochetes without showing any clinical signs of infection, initially show strong Th1 responses, that subsequently are down regulated by the induction of Th2-responses (Kang, Barthold, Persing, and Bockenstedt 1997). The significance of a Th1-like response in eradication of borrelia is further supported by studies where experimental co-infection with borrelia and *Anaplasma phagocytophilum* led to significantly lower Th1 responses against borrelia, resulting in significantly higher borrelia spirochete burden and aggravated clinical disease, compared with mice infected with borrelia only (Zeidner, Dolan, Massung, Piesman, and Fish 2000; Thomas, Anguita, Barthold, and Fikrig 2001). Interestingly, in a study on the effect of exposure to *A. phagocytophilum* on the cytokine patterns in patients with EM, we found significantly lower numbers of mononuclear blood cells secreting IL-12 in response to *in vitro* stimulation with borrelial lipoproteins in patients with erythema migrans previously exposed to *A. phagocytophilum* compared with erythema migrans patients seronegative for the same agent (Jarefors, Karlsson, Forsberg, Eliasson, Ernerudh, and Ekerfelt 2005). This finding, that needs to be further investigated, indicates that a previous infection with *A. phagocytophilum* may affect the clinical outcome of borreliosis also in humans.

In human borreliosis, similarily to the mouse model, we found an initial interferon(IFN)-γ response (Th1) followed by a subsequent up regulation of IL-4 (Th2), associated with non-chronic manifestations of human borreliosis, whereas a persistent up regulation of IFN-γ was associated with chronic manifestations (Widhe et al. 2004). Furthermore, washed whole blood cells, including erythrocytes and all leukocyte populations, from asymptomatic borrelia seropositive individuals showed increased secretion of the Th1-inducing cytokine IL-12 when exposed to live borrelia spirochetes, as compared to patients with a history of subacute or chronic borreliosis, further indicating the importance of a strong initial Th1-response for optimal eradication of borrelia (Sjöwall et al. 2005). A strong enough initial Th1 response will also induce a proper antagonistic Th2 response, which turns the Th1-inflammation off when the infection is cleared (Kang et al. 1997), avoiding chronic inflammation. This is supported by our findings of persistent up regulation of IFN-γ in chronic borreliosis (Widhe et al. 2004).

The nature of the IFN-γ mediated inflammation is, however, not known. We have previously reported phenotypes indicating cytolytic properties of borrelia specific IFN-γ secreting cells in chronic neuroborreliosis (Ekerfelt, Jarefors, Tynngård, Hedlund, Sander, Bergström, Forsberg, and Ernerudh 2003), suggesting that cytotoxic cells mediate tissue injury. The significance of cytotoxic cells in the Th1-response was further supported in a recent study, where we found that patients with a history of chronic Lyme borreliosis displayed a decreased up-regulation of the IL-12-receptor β2-chain (the signalling part of the IL-12 receptor) on CD8+ cells in response to borrelia antigen and also a lower number of borrelia-induced IFN-γ-secreting cells compared to asymptomatic borrelia-exposed individuals (Jarefors, Janefjord, Forsberg, Jenmalm, and Ekerfelt 2006). This is intriguing since enrichment of cytotoxic cells in the central nervous system has been reported in studies of multiple sclerosis (MS) (Babbe, Roers, Waisman, Lassmann, Goebels, Hohlfeld, Friese, Schroder, Deckert, Schmidt, Ravid, and Rajewsky 2000; Jacobsen, Cepok, Quak, Happel, Gaber, Ziegler, Schock, Oertel, Sommer, and Hemmer 2002). Chronic neuroborreliosis shares many features with MS (Garcia-Monco, and Benach 1995; Oksi, Kalimo, Marttila, Marjamaki, Sonninen, Nikoskelainen, and Viljanen 1996), including autoimmune responses (Baig, Olsson, Höjeberg, and Link 1986; Weigelt, Schneider, and Lange 1992).

3.2 B-cell Mediated Responses

The humoral response in Lyme borreliosis has since long been proposed to be essential for clearance of the spirochete (Barthold, deSouza, and Feng 1996; Seiler, and Weis 1996). An epitope spreading of the antibody response, leading to production of antibodies against different components of borrelia is especially seen in disseminated infection (Fung, McHugh, Leong, and Steere 1994; Kalish, Leong, and Steere 1995; Akin, McHugh, Flavell, Fikrig, and Steere 1999), supporting a continuous activation of B-cells. Interestingly, IgG antibodies to OspA and OspB, which are found in the late-stage of disease, have been shown to correlate with severe Lyme arthritis (Akin et al. 1999). This is in line with previous findings that treatment-resistant chronic Lyme arthritis is associated with antibody reactivity to OspA and OspB in combination with the HLA class II genotype HLA-DR4 (Kalish, Leong, and Steere 1993), in addition indicating a T-cell dependent secretion of antibodies.

Borrelia spirochetes have been shown to be killed extra-cellularly by antibodies plus complement via the classical pathway, as well as by phagocytes through apparently non-oxidative means (Montgomery and Malawista 1994). Considering the latter, neutrophils, which previously have been shown to play a critical role in spirochete clearance in mice (Barthold, and deSouza 1995), have been suggested to be very important in controlling numbers of Ig-opsonized borrelia spirochetes in TLR-2-deficient mice (Singh and Girshick 2006).

The importance of complement activation in the clearance of borrelia is supported by our previous finding of a predominance of IgG1 and IgG3, which both are complement activating isotypes, among borrelia-specific IgG antibodies in sera from Lyme borreliosis patients (Widhe et al. 1998). This can also be considered as a read-out of the borrelia-induced Th1-response, since switching to opsonizing and complement-activating isotypes is part of the Th1-mediated effector response. The significance of the classical pathway of complement activation in the control of *B. burgdorferi s. l.* infection is further underlined by studies on immune evasion of borrelia, where the down regulation of several *B. burgdorferi s. l.* genes leads to insufficient killing of the spirochetes by complement and antibodies (Kraiczy, Skerka, Kirschfink, Zipfel, and Brade 2002; Liang, Jacobs, Bowers, and Philipp 2002; Liang, Nelson, and Fikrig 2002).

4 The Complement System

4.1 Activation, main biological effects, and control of complement

The complement system has a primary function in host defence and clears the body of foreign cells and organisms, either by direct lysis or by recruitment of leukocytes which promote phagocytosis. It consists of approximately 30 plasma and cellular proteins (receptors and regulators). The central complement reaction is the cleavage of C3 into C3b and C3a. Two multi-molecular enzyme complexes, the C3-convertases, which are assembled by three different recognition and activation pathways, promote this cleavage. The alternative pathway (AP) may be triggered directly by foreign surfaces, which do not provide adequate down-regulation of the convertase, C3bBb. Formation of the classical pathway (CP) convertase, C4bC2a, is triggered either by the formation of antigen-antibody complexes or by the binding of certain plasma proteins, mannan-binding lectin (MBL) or ficolins, to carbohydrates, in particular on the surface of microorganisms.

The nascent C3b molecule has the specific property to bind to carbohydrates and proteins, on the surface of e.g. microorganisms, via free hydroxyl or amino groups, resulting in covalent ester and amide bonds, respectively. The AP also serves as a major amplification loop, so an initial weak stimulus, mediated by any of the pathways, may be markedly enhanced (Harboe, Ulvund, Vien, Fung, and Mollnes 2004). The activation pathways converge into a common pathway forming the membrane attack complex (sC5b-9) that elicits cell lysis by insertion into the lipid bilayer of cell membranes. The anaphylatoxins (C3a and C5a) activate and recruit phagocytes, while target-bound C3 fragments facilitate binding to and activation of the recruited cells.

In vivo, activation of the complement system is under control of numerous soluble and membrane-bound regulators. Most of the regulators are members of the "regulators of complement activation" (RCA) superfamily, which are encoded by a gene cluster on chromosome 1 and mainly regulate the two types of C3-convertases. In addition, CD59 is a regulator of the sC5b-9 complex at the C8 level and the C1 inhibitor (C1INH) regulates the recognition complexes C1 and MBL-MASP (MBL associated serine proteases). The plasma proteins factor H (FH) and C4b-binding protein (C4BP), and the membrane proteins CR1, MCP (CD46) and DAF (CD55) all belong to this family and act either by directly dissociating the convertase or as cofactors to factor I to enable its digestion of particle bound C3b or C4b. Both FH and C4BP mediate protection to host cells by binding to heparan sulfate on the cellular surfaces (Blom 2002; Zipfel, Skerka, Hellwage, Jokiranta, Meri, Brade, Kraiczy, Noris, and Remuzzi 2002).

4.2 Control of complement activation by *B. burgdorferi*

These potent complement regulatory systems, in particular those aimed to control the C3-convertases, are being imitated or exploited by various microorganisms in order to survive host complement. E.g. variola and vaccinia viruses encode endogenous RCA proteins designated smallpox inhibitor of complement enzymes (SPICE) and vaccinia complement-control protein (VCP), respectively. Both proteins are functional homologues of mammalian FH and C4BP and cause inactivation of C3b and C4b (Dunlop, Oehlberg, Reid, Avci, and Rosengard 2003). In contrast, some bacteria, e.g. *Streptococcus pyogenes* (group A streptococcus) and *B. burgdorferi*, acquire resistance against complement by expressing surface proteins that bind human RCAs. In the case of *S. pyogenes* it is members of the M protein family that bind C4BP with high affinity (Lindahl, Sjöbring, and Johnsson 2000), while the complement restistance of *B. burgdorferi* is dependent on the expression of different types of bacterial proteins which bind FH and FH-like protein-1 (FHL-1) e.g. (Alitalo, Meri, Lankinen, Seppälä, Ladhdenne, Hefty, Akins, and Meri 2002; Hartmann, Corvey, Skerka, Kirschfink, Karas, Brade, Miller, Stevenson, Wallich, Zipfel, and Kraiczy 2006). All pathogenic *Borrelia* genospecies have been shown to activate the CP and/or the AP of complement *in vitro* (Kraiczy et al. 2002), and complement-resistant strains of *Borrelia* isolates can inactivate human complement *in vitro* (Brade, Kleber, and Acker 1992; Breitner-Ruddock, Wurzner, Schulze, and Brade 1997; van Dam, Oei, Jaspars, Fijen, Wilske, Spanjaard, and Dankert 1997; Kraiczy, Hunfeld, Breitner-Ruddock, Wurzner, Acker, and Brade 2000; Alitalo, Meri, Ramo, Jokiranta, Heikkila, Seppala, Oksi, Viljanen, and Meri 2001).

The borrelia proteins capable of binding host inhibitory FH and FHL-1 include a family of lipoproteins, e.g. OspE, ErpA and ErpP (Hellwage, Meri, Heikkila, Alitalo, Panelius, Lahdenne, Seppala, and Meri 2001; Alitalo et al. 2002; Berglund et al. 2002; Alitalo, Meri, Comstedt, Jeffery, Tornberg, Strandin, Lankinen, Bergstrom, Cinco, Vuppala, Akins, and Meri 2005) as well as the group of BbCRASP proteins (Kraiczy, Skerka, Brade, and Zipfel 2001; Cordes, Kraiczy, Roversi, Simon, Brade, Jahraus, Wallis, Goodstadt, Ponting, Skerka, Zipfel, Wallich, and Lea 2006; Hartmann et al. 2006; Rossmann, Kitiratschky, Hofmann, Kraiczy, Simon, and Wallich 2006). The FH and FHL-1 bound to the bacteria then enable the inactivation of C3b through factor I-mediated cleavage (Kraiczy, Hellwage, Skerka, Becker, Kirschfink, Simon, Brade, Zipfel, and Wallich 2004).

In addition, a CD59-like complement inhibitor with affinity for C9 has been identified in complement-resistant strains of *B. burgdorferi* (Pausa, Pellis, Cinco, Giulianini, Presani, Perticarari, Murgia, and Tedesco 2003).

Another link to complement has been suggested in that Complement receptor 3 (CR3, CD11b/CD18) has been shown to bind to *B. burgdorferi* OspA and OspB in a manner that is independent of the physiological ligand iC3b (Garcia, Murgia, and Cinco 2005).

4.3 Complement activation *in vivo* in borreliosis

Different genospecies of *B. burgdorferi* are classified as complement-resistant, complement-sensitive, and intermediate complement-sensitive. Most isolates belonging to the genospecies *B. afzelii* are complement-resistant while particularly *B. garinii* isolates were rapidly killed by complement. In general, isolates of the genospecies *B. burgdorferi s.s* are intermediate complement-sensitive (Kraiczy, Skerka, Kirschfink, Zipfel, and Brade 2001). Earlier studies demonstrate complement activation by *B. burgdorferi s.l.* in mice (Bockenstedt, Barthold, Deponte, Marcantonio, and Kantor 1993; Lawrenz, Wooten, Zachary, Drouin, Weis, Wetsel, and Norris 2003), but until now published data on complement activation patterns from human *in vivo* studies have been lacking.

In a recent study we therefore monitored the concentration and activation of complement in plasma and CSF from patients who were admitted to hospital with symptoms of suspected neuroborreliosis. Out of these patients 8% were eventually diagnosed with neuroborreliosis and another 16% with borreliosis without central nervous system (CNS) manifestations. The most significant observation was that we detected high levels of C1q and C3a in CSF in patients with neuroborreliosis. There was no correlation between the levels of C3a found in plasma and CSF suggesting intrathecal and not systemic complement activation in these patients. Since a positive correlation was found between the levels of C3a and C1q we propose that complement activation may occur via the CP (Henningsson, Ernerudh, Sandholm, Carlsson, Granlund, Jansson, Nyman, Forsberg, and Nilsson Ekdahl 2006).

We believe that complement is synthesized locally within the CNS in neuroborreliosis and that intrathecal complement activation may have a detrimental role in this disease as has been described in neurodegenerative disorders such as Alzheimer's. It has been reported that astrocytes, microglia, neurons, and oligodendrocytes all produce complement proteins, in particular after stimulation with pro-inflammatory cytokines. This subject has been extensively studied by Gasque and collaborators (Francis, Van Beek, Canova, Neal, and Gasque 2003).

5 Concluding remarks

Inflammatory innate immune responses are critical in the control of early infection. Borrelia infection in humans is initially characterized by a strong production of the innate proinflammatory cytokine TNF-α in CSF, accompanied by an increase of the anti-inflammatory cytokine TGF-β1. Later during the course of infection both in mouse models and in patients there is an initial Th1 response, characterized by IFN-γ generation followed by a later Th2 response monitored as IL-4 generation. In sera from Lyme borreliosis patients, the borrelia specific antibodies are predominantely of the IgG1 and IgG3 subclasses, underscoring the link to Th1 response. Borrelia spirochetes are known to activate complement *in vitro*, but are protected from lysis by binding the complement inhibitory molecule FH to its surface. We have found evidence for inthratechal complement synthesis and activation *in vivo* in samples from a material of patients with Lyme neuroborreliosis. In conclusion, we suggest that the complement system plays a role in the host's immune response in borreliosis and propose that measurement of complement-related parameters in CSF should be considered in the pathogenic and diagnostic evaluation of Lyme borreliosis.

6 Acknowledgements

We acknowledge economic support from the Health Research Council in the South East of Sweden (FORSS), the Swedish Research Council, the University of Kalmar, the University Hospital of Linköping, Futurum - the Academy of Healthcare, County Council, Jönköping, and the Family Olinder-Nielsen's Foundation.

References

Akin, E., McHugh, G. L., Flavell, R. A., Fikrig, E. and Steere, A. C. (1999). The immunoglobulin (IgG) antibody response to OspA and OspB correlates with severe and prolonged Lyme arthritis and the IgG response to P35 correlates with mild and brief arthritis. Infect Immun 67: 173-181.

Alban, P. S., Johnson, P. W. and Nelson, D. R. (2000). Serum-starvation-induced changes in protein synthesis and morphology of Borrelia burgdorferi. Microbiology 146: 119-127.

Alitalo, A., Meri, T., Comstedt, P., Jeffery, L., Tornberg, J., Strandin, T., Lankinen, H., Bergstrom, S., Cinco, M., Vuppala, S. R., Akins, D. R. and Meri, S. (2005). Expression of complement factor H binding immunoevasion proteins in Borrelia garinii isolated from patients with neuroborreliosis. Eur J Immunol 35(10): 3043-3053.

Alitalo, A., Meri, T., Lankinen, H., Seppälä, I., Ladhdenne, P., Hefty, P. S., Akins, D. and Meri, S. (2002). Complement inhibitor factor H binding to Lyme disease spirochetes is mediated by inducible expression of multiple plasmid-encoded Outer Surface Protein E paralogs. J Immunol 169: 3847-3853.

Alitalo, A., Meri, T., Ramo, L., Jokiranta, T. S., Heikkila, T., Seppala, I. J., Oksi, J., Viljanen, M. and Meri, S. (2001). Complement evasion by Borrelia burgdorferi: serum-resistant strains promote C3b inactivation. Infect Immun 69(6): 3685-91.

Anon (1997). Case definitions for infectious conditions under public health surveillance: Lyme disease (revisited 9/96). MMWR Morb Mortal Wkly Rep 46: 20-21.

Asch, E. S. and Bujac, M. (1994). Lyme disease: an infectious and postinfectious syndrome. J Rheumatol 21: 454-461.

Babbe, H., Roers, A., Waisman, A., Lassmann, H., Goebels, N., Hohlfeld, R., Friese, M., Schroder, R., Deckert, M., Schmidt, S., Ravid, R. and Rajewsky, K. (2000). Clonal expansions of CD8(+) T cells dominate the T cell infiltrate in active multiple sclerosis lesions as shown by micromanipulation and single cell polymerase chain reaction. J Exp Med 192(3): 393-404.

Baig, S., Olsson, T., Höjeberg, B. and Link, H. (1986). Cells secreting antibodies to myelin basic protein in cerebrospinal fluid of patients with Lyme neuroborreliosis. Neurology 41: 581-587.

Balmelli, T. and Piffaretti, J. C. (1995). Association between different clinical manifestations of Lyme disease and different species of Borrelia burgdorferi sensu lato. Res Microbiol 146(4): 329-340.

Barbour, A. G. (1984). Immunochemical analysis of Lyme disease spirochetes. Yale Journal of Biological Medicine 57: 581-586.

Barthold, S. W. and deSouza, M. (1995). Exacerbation of Lyme arthritis in beige mice. J Infect Dis 172: 772-784.

Barthold, S. W., deSouza, M. and Feng, S. (1996). Serum-mediated resolution in Lyme-arthritis in mice. Lab Invest 74: 57-67.

Barton, G. M. and Medzhitov, R. (2003). Toll-like receptor signaling pathways. Science 300: 1524-1525.

Benach, J. L., Bosler, E. M., Hanrahan, J. P., Coleman, J. L., Habicht, G. S., Bast, T. F., Cameron, D. J., Ziegler, J. L., Barbour, A. G., Burgdorfer, W., Edelman, E. and Kaslow, R. A. (1983). Spirochetes isolated from the blood of two patients with Lyme disease. N Engl J Med 308: 740-742.

Berglund, J., Eitrem, R., Ornstein, K., Lindberg, A., Ringer, A., Elmrud, H., Carlsson, M., Runehagen, A., Svanborg, C. and Norrby, R. (1995). An epidemiologic study of Lyme disease in southern Sweden. N Engl J Med 333(20): 1319-1327.

Berglund, J., Stjernberg, L., Ornstein, K., Tykesson-Joelsson, K. and Walter, H. (2002). 5-y Follow-up study of patients with neuroborreliosis. Scand J Infect Dis 34(6): 421-425.

Blom, A. M. (2002). Structural and functional studies of complement inhibitor C4b-binding protein. Biochem Soc Trans 30(Pt 6): 978-982.

Bockenstedt, L. K., Barthold, S., Deponte, K., Marcantonio, N. and Kantor, F. S. (1993). Borrelia burgdorferi infection and immunity in mice deficient in the fifth component of complement. Infect Immun 61(5): 2104-2107.

Brade, V., Kleber, I. and Acker, G. (1992). Differences of two Borrelia burgdorferi strains in complement activation and serum resistance. Immunobiology 185(5): 453-465.

Breitner-Ruddock, S., Wurzner, R., Schulze, J. and Brade, V. (1997). Heterogeneity in the complement-dependent bacteriolysis within the species of Borrelia burgdorferi. Med Microbiol Immunol (Berl) 185(4): 253-260.

Brorson, O. and Brorson, S. H. (1998). A rapid method for generating cystic forms of Borrelia burgdorferi, and their reversal to mobile spirochetes. Apmis 106(12): 1131-1141.

Brown, J. P., Zachary, J. F., Teuscher, C. and Weis, J. J. (1999). Dual role of interleukin-10 in murine Lyme disease: regulation of arthritis severity and host defence. Infect Immun 67: 5142-5150.

Burgdorfer, W., Barbour, A. G., Hayes, S. F., Benach, J. L., Grunwaldt, E. and Davis, J. P. (1982). Lyme disease - a tick-borne spirochetosis? Science 216: 1317-1319.

Cordes, F. S., Kraiczy, P., Roversi, P., Simon, M. M., Brade, V., Jahraus, O., Wallis, R., Goodstadt, L., Ponting, C. P., Skerka, C., Zipfel, P. F., Wallich, R. and Lea, S. M. (2006). Structure-function mapping of BbCRASP-1, the key complement factor H and FHL-1 binding protein of Borrelia burgdorferi. Int J Med Microbiol 296 Suppl 40: 177-184.

Duffy, J. (1990). Lyme disease. Ann Allergy 65(1): 1-13.

Dunlop, L. R., Oehlberg, K. A., Reid, J. J., Avci, D. and Rosengard, A. M. (2003). Variola virus immune evasion proteins. Microbes Infect 5(11): 1049-1056.

Ekerfelt, C., Ernerudh, J., Bunikis, J., Vrethem, M., Aagesen, J., Roberg, M., Bergström, S. and Forsberg, P. (1997). Compartmentalization of antigen specific cytokine responses to the central nervous system in CNS borreliosis: Secretion of IFN-γ predominates over IL-4 secretion in response to outer surface proteins of Lyme disease *Borrelia spirochetes*. J Neuroimmunol 79: 155-162.

Ekerfelt, C., Ernerudh, J., Forsberg, P. and Bergström, S. (1998). Augmented intrathecal secretion of interferon-g in response to *Borrelia garinii* in neuroborreliosis. J Neuroimmunol 89: 177-181.

Ekerfelt, C., Forsberg, P., Svenvik, M., Roberg, M., Bergström, S. and Ernerudh, J. (1999). Asymptomatic *Borrelia* seropositive individuals display the same incidence of*Borrelia* specific interferon-gamma (IFN-γ)- secreting cells in blood as patients with clinical *Borrelia* infection. Clin Exp Immunol 115: 498-502.

Ekerfelt, C., Jarefors, S., Tynngård, N., Hedlund, M., Sander, B., Bergström, S., Forsberg, P. and Ernerudh, J. (2003). Phenotypes indicating cytolytic properties of Borrelia-specific interferon-g secreting cells in chronic Lyme neuroborreliosis. J Neuroimmunol 145: 115-126.

Ekerfelt, C., Masreliez, C., Svenvik, M., Ernerudh, J., Roberg, M. and Forsberg, P. (2001). Antibodies and T-cell reactivity to Borrelia burgdorferi in an asymptomatic population: a study of healthy blood donors in an inland town district in the south-east of Sweden. Scand J Infect Dis 33(11): 806-808.

Forsberg, P., Ernerudh, J., Ekerfelt, C., Roberg, M., Vrethem, M. and Bergstrom, S. (1995). The outer surface proteins of Lyme disease borrelia spirochetes stimulate T cells to secrete interferon-gamma (IFN-gamma): diagnostic and pathogenic implications. Clin Exp Immunol 101(3): 453-460.

Francis, K., Van Beek, J., Canova, C., Neal, J. W. and Gasque, P. (2003). Innate immunity and brain inflammation: the key role of complement. Expert Rev Mol Med 2003: 1-19.

Frey, M., Jaulhac, B., Piemont, Y., Marcellin, L., Boohs, P. M., Vautravers, P., Jesel, M., Kuntz, J. L., Monteil, H. and Sibilia, J. (1998). Detection of Borrelia burgdorferi DNA in muscle of patients with chronic myalgia related to Lyme disease. Am J Med 104(6): 591-594.

Fung, B. P., McHugh, G. L., Leong, J. M. and Steere, A. C. (1994). Humoral immune response to outer surface protein C of Borrelia burgdorferi in Lyme disease: role of the immunoglobulin M response in the serodiagnosis of early infection. Infect Immun 62(8): 3213-3221.

Garcia, R. C., Murgia, R. and Cinco, M. (2005). Complement receptor 3 binds the Borrelia burgdorferi outer surface proteins OspA and OspB in an iC3b-independent manner. Infect Immun 73(9): 6138-6142.

Garcia-Monco, J. C. and Benach, J. L. (1995). Lyme neuroborreliosis. Ann Neurol 37(6): 691-702.

Gross, D. M., Steere, A. C. and Huber, B. T. (1998). T helper 1 response is dominant and localized to the synovial fluid in patients with Lyme arthritis. J Immunol 160(2): 1022-1028.

Grusell, M., Widhe, M. and Ekerfelt, C. (2002). Increased expression of the Th1-inducing cytokines interleukin-12 and interleukin-18 in cerebrospinal fluid but not in sera from patients with Lyme neuroborreliosis. Journal of Neuroimmunology 131: 173-178.

Harboe, M., Ulvund, G., Vien, L., Fung, M. and Mollnes, T. E. (2004). The quantitative role of alternative pathway amplification in classical pathway induced terminal complement activation. Clin Exp Immunol 138(3): 439-446.

Hartmann, K., Corvey, C., Skerka, C., Kirschfink, M., Karas, M., Brade, V., Miller, J. C., Stevenson, B., Wallich, R., Zipfel, P. F. and Kraiczy, P. (2006). Functional characterization of BbCRASP-2, a distinct outer membrane protein of Borrelia burgdorferi that binds host complement regulators factor H and FHL-1. Mol Microbiol 61(5): 1220-1236.

Hellwage, J., Meri, T., Heikkila, T., Alitalo, A., Panelius, J., Lahdenne, P., Seppala, I. J. and Meri, S. (2001). The complement regulator factor H binds to the surface protein OspE of Borrelia burgdorferi. J Biol Chem 276(11): 8427-8435.

Hemmer, B., Gran, B., Zhao, Y., Marques, A., Pascal, J., Tzou, A., Kondo, T., Cortese, I., Bielekova, B., Straus, S. E., McFarland, H. F., Houghten, R., Simon, R., Pinilla, C. and Martin, R. (1999). Identification of candidate T-cell epitopes and molecular mimics in chronic Lyme disease. Nat Med 5(12): 1375-1382

Henningsson, A. J., Ernerudh, J., Sandholm, K., Carlsson, S.-A., Granlund, H., Jansson, C., Nyman, D., Forsberg, P. and Nilsson Ekdahl, K. (2006). Complement activation in Lyme neuroborreliosis – increased levels of C1q and C3a in cerebrospinal fluid indicate complement activation in the CNS. J Neuroimmunol in press.

Hirschfeld, M., Kirschning, C. J., Schwandner, R., Wesche, H., Weis, J. H., Wooten, R. M. and Weis, J. J. (1999). Cutting edge: inflammatory signaling by Borrelia burgdorferi lipoproteins is mediated by toll-like receptor 2. J Immunol 163(5): 2382-1286.

Huppertz, H. I., Bohme, M., Standaert, S. M., Karch, H. and Plotkin, S. A. (1999). Incidence of Lyme borreliosis in the Wurtzburg region of Germany. Eur J Clin Microbiol Infect Dis 18: 697-703.

Jacobsen, M., Cepok, S., Quak, E., Happel, M., Gaber, R., Ziegler, A., Schock, S., Oertel, W. H., Sommer, N. and Hemmer, B. (2002). Oligoclonal expansion of memory CD8+ T cells in cerebrospinal fluid from multiple sclerosis patients. Brain 125(Pt 3): 538-550.

Jarefors, S., Janefjord, C. K., Forsberg, P., Jenmalm, M. C. and Ekerfelt, C. (2006). Decreased up-regulation of the interleukin-12Rb2-chain and interferon-g secretion and increasde number of forkhead box P3-expressing cells in patients with a history of chronic Lyme borreliosis compared with asymptomatic individuals. Clin Exp Immunol In press.

Jarefors, S., Karlsson, M., Forsberg, P., Eliasson, I., Ernerudh, J. and Ekerfelt, C. (2005). Reduced numbers of interleukin-12 secreting cells in patients with Lyme borreliosis previously exposed to Anaplasma phagocytophilum. Clin Exp Immunol 143: 322-328.

Johnson, L. and Stricker, R. B. (2004). Treatment of Lyme disease: a medicolegal assessment. Expert Rev Anti Infect Ther 2: 533-557.

Johnson, R. C., Schmid, G. P., Hyde, F. W., Steigerwalt, A. G. and Brenner, D. J. (1984). *Borrelia burgdorferi* sp. nov.: etiologic agent of Lyme disease. Int J Syst Bacteriol 34: 496-497.

Kaiser, R. (1994). Variable CSF findings in early and late Lyme neuroborreliosis: a follow-up study in 47 patients. J Neurology 242: 26-36.
Kalish, R. A., Leong, J. M. and Steere, A. C. (1993). Association of treatment-resistant chronic Lyme arthritis with HLA-DR4 and antibody reactivity to OspA and OspB of Borrelia burgdorferi. Infect Immun 61(7): 2774-2779.
Kalish, R. A., Leong, J. M. and Steere, A. C. (1995). Early and late antibody responses to full-length and truncated constructs of outer surface protein A of Borrelia burgdorferi in Lyme disease. Infect Immun 63(6): 2228-2235.
Kang, I., Barthold, S. W., Persing, D. H. and Bockenstedt, L. K. (1997). T-helper-cell cytokines in the early evolution of murine Lyme arthritis. Infect Immun 65(8): 3107-3111.
Kersten, A., Poitschek, C., Rauch, S. and Aberer, E. (1995). Effects of penicilline, ceftriaxone, and doxycycline on the morphology of Borrelia burgdorferi. Agents Chemother 39: 1127-1133.
Korenberg, E. I., Gorban, L. Y., Kovalevskii, Y. V., Frizen, V. I. and Karanov, A. S. (2001). Risk for human tick-borne encephalitis, borreliosis and double infection in the pre-Ural region of Russia. Emerg Infect Dis 7: 459-462.
Kraiczy, P., Hellwage, J., Skerka, C., Becker, H., Kirschfink, M., Simon, M. M., Brade, V., Zipfel, P. F. and Wallich, R. (2004). Complement resistance of Borrelia burgdorferi correlates with the expression of BbCRASP-1, a novel linear plasmid-encoded surface protein that interacts with human factor H and FHL-1 and is unrelated to Erp proteins. J Biol Chem 279(4): 2421-2429.
Kraiczy, P., Hunfeld, K. P., Breitner-Ruddock, S., Wurzner, R., Acker, G. and Brade, V. (2000). Comparison of two laboratory methods for the determination of serum resistance in Borrelia burgdorferi isolates. Immunobiology 201(3-4): 406-419.
Kraiczy, P., Skerka, C., Brade, V. and Zipfel, P. F. (2001). Further characterization of complement regulator-acquiring surface proteins of Borrelia burgdorferi. Infect Immun 69(12): 7800-7809.
Kraiczy, P., Skerka, C., Kirschfink, M., Zipfel, P. F. and Brade, V. (2001). Mechanism of complement resistance of pathogenic Borrelia burgdorferi isolates. Int Immunopharmacol 1(3): 393-401.
Kraiczy, P., Skerka, C., Kirschfink, M., Zipfel, P. F. and Brade, V. (2002). Immune evasion of Borrelia burgdorferi: insufficient killing of the pathogens by complement and antibody. Int J Med Microbiol 291 Suppl 33: 141-146.
Lawrenz, M. B., Wooten, R. M., Zachary, J. F., Drouin, S. M., Weis, J. J., Wetsel, R. A. and Norris, S. J. (2003). Effect of complement component C3 deficiency on experimental Lyme borreliosis in mice. Infect Immun 71(8): 4432-4440.
Liang, F. T., Jacobs, M. B., Bowers, L. C. and Philipp, M. T. (2002). An immune evasion mechanism for spirochetal persistance in the host. J Exp Med 195: 415-422.
Liang, F. T., Nelson, F. K. and Fikrig, E. (2002). Molecular adaption of Borrelia in the murine host. J Exp Med 196: 275-280.
Lien, E., Sellati, T. J., Yoshimura, A., Flo, T. H., Rawadi, G., Finberg, R. W., Carroll, J. D., Espevik, T., Ingalls, R. R., Radolf, J. D. and Golenbock, D. T. (1999). Toll-like receptor 2 functions as a pattern recognition receptor for diverse bacterial products. J Biol Chem 274(47): 33419-33425.
Lindahl, G., Sjöbring, U. and Johnsson, E. (2000). Human complement regulators: a major target for pathogenic microorganisms. Curr Opin Immunol 12(1): 44-51.
Ma, Y., Seiler, K. P., Tai, K. F., Yang, L., Woods, M. and Weis, J. J. (1994). Outer surface lipoproteins of Borrelia burgdorferi stimulate nitric oxide production by the cytokine-inducible pathway. Infect Immun 62(9): 3663-3671.
Ma, Y. and Weis, J. J. (1993). Borrelia burgdorferi outer surface lipoproteins OspA and OspB possess B-cell mitogenic and cytokine-stimulatory properties. Infect Immun 61: 3843-3853.
Montgomery, R. R. and Malawista, S. E. (1994). Borrelia burgdorferi and the macrophage: routine annihilation but occasional haven? Parasitol Today 10(4): 154-157.
Morrison, T. B., Weis, J. H. and Weis, J. J. (1997). Borrelia burgdorferi outer surface protein A (OspA) activates and primes human neutrophils. J Immunol 158(10): 4838-4845.
Muellegger, R. R., McHugh, G., Ruthazer, R., Binder, B., Kerl, H. and Steere, A. C. (2000). Differential expression of cytokine mRNA in skin specimens from patients with erythema migrans or acrodermatitis chronica atrophicans. J Invest Dermatol 115: 1115-1123.
Nocton, J. J. and Steere, A. C. (1995). Lyme disease. Adv intern med 40: 69-115.
Noppa, L. (1998). Membrane proteins and periplasmic flagella of *Borrelia* spirochetes. (PhD thesis) Umeå University, Umeå, Sweden

Norgard, M. V., Arndt, L. L., Akins, D. R., Curetty, L. L., Harrich, D. A. and Radolf, J. D. (1996). Activation of human monocytic cells by Treponema pallidum and Borrelia burgdorferi lipoproteins and synthetic lipopeptides proceeds via a pathway distinct from that of lipopolysaccharide but involves the transcriptional activator NF-kappa B. Infect Immun 64(9): 3845-3852.

Oksi, J., Kalimo, H., Marttila, R. J., Marjamaki, M., Sonninen, P., Nikoskelainen, J. and Viljanen, M. K. (1996). Inflammatory brain changes in Lyme borreliosis. A report on three patients and review of literature. Brain 119 (Pt 6): 2143-2154.

Oksi, J., Marjamaki, M., Nikoskelainen, J. and Viljanen, M. K. (1999). Borrelia burgdorferi detected by culture and PCR in clinical relapse of disseminated Lyme borreliosis. Ann Med 31: 225-232.

Oksi, J., Savolainen, J., Pene, J., Bousquet, J., Laippala, P. and Viljanen, M. K. (1996). Decreased interleukin-4 and increased gamma interferon production by peripheral blood mononuclear cells of patients with Lyme borreliosis. Infect Immun 64(9): 3620-3623.

Ornstein, K., Berglund, J., Bergström, S., Norrby, R. and Barbour, A. G. (2002). Three major Lyme Borrelia genospecies (Borrelia burgdorferi sensu stricto, B. afzelii and B. gariniii) identified by PCR in cerebrospinal fluid from patients with neuroborreliosis in Sweden. Scand J Infect Dis 34: 341-346.

Oschmann, P., Dorndorf, W., Hornig, C., Schafer, C., Wellensiek, H. J. and Pflughaupt, K. W. (1998). Stages and syndromes of neuroborreliosis. J Neurol 245(5): 262-272.

Pachner, A. R. (1988). Borrelia burgdorferi in the nervous system: the new "great imitator". Ann N Y Acad Sci 539: 56-64.

Pausa, M., Pellis, V., Cinco, M., Giulianini, P. G., Presani, G., Perticarari, S., Murgia, R. and Tedesco, F. (2003). Serum-resistant strains of Borrelia burgdorferi evade complement-mediated killing by expressing a CD59-like complement inhibitory molecule. J Immunol 170(6): 3214-3222.

Pena, C. A. and Strickland, G. T. (1999). Incidence rates of Lyme disease in Maryland: 1993 through 1996. Md Med J 48: 68-73.

Preac-Mursic, V., Weber, K., Pfister, H. W., Wilske, B., Gross, B., Bauman, A. and Prokop, B. (1989). Survival of Borrelia burgdorferi in antibiotically treated patients with Lyme borreliosis. Infection 17: 355-359.

Richter, D., Schlee, D. B., Algower, R. and Matuschka, F. R. (2004). Relationships of a novel Lyme disease spirochete, Borrelia spielmani sp nov., with its hosts in Central Europe. Appl Environ Micrrobiol 70: 6414-6419.

Roessner, K., Trivedi, H., Gaur, L., Howard, D., Aversa, J., Cooper, S. M., Sigal, L. H. and Budd, R. C. (1998). Biased T-cell antigen receptor repertoire in Lyme arthritis. Infect Immun 66(3): 1092-1099.

Rossmann, E., Kitiratschky, V., Hofmann, H., Kraiczy, P., Simon, M. M. and Wallich, R. (2006). BbCRASP-1 of the Lyme disease spirochetes is expressed in humans and induces antibody responses restricted to non-denatured structural determinants. Infect Immun in press.

Seiler, K. P. and Weis, J. J. (1996). Immunity to Lyme disease: protection, pathology and persistence. Curr Opin Immunol 8(4): 503-509.

Sigal, L. H. (1997). Immunologic mechanisms in Lyme neuroborreliosis: the potential role of autoimmunity and molecular mimicry. Semin Neurol 17(1): 63-68.

Singh, S. K. and Girshick, H. J. (2006). Toll-like receptors in Borrelia burgdorferi-induced inflammation. Clin Microbiol Infect 12: 705-717.

Sjöwall, J., Carlsson, A., Vaarala, O., Bergström, S., Ernerudh, J., Forsberg, P. and Ekerfelt, C. (2005). Innate immune responses in Lyme borreliosis: enhanced tumour necrosis factor-a and interleukin-12 in asymptomatic individuals in response to live spirochetes. Clin Exp Immunol 141: 89-98.

Stanek, G., Satz, N., Strle, F. and Wilske, B. (1993). Epidemiology of Lyme borreliosis. In:. K. Weber and W. Burgdorfer (Eds.), *Aspects of Lyme borreliosis.* Springer-Verlag, Berlin, pp 358-370.

Stanek, G. and Strle, F. (2003). Lyme borreliosis. Lancet 362(9396): 1639-47.

Steere, A. C. (2001). Lyme disease. N Engl J Med 345: 115-125.

Steere, A. C., Grodzicki, R. L., Kornblatt, A. N., Craft, J. E., Barbour, A. G., Burgdorfer, W., Schmid, G. P., Johnson, E. and Malawista, S. E. (1983). The spirochetal etiology of Lyme disease. N Engl J Med 308: 733-740.

Steere, A. C., Malawista, S. E., Hardin, J. A., Ruddy, S., Askenase, P. W. and Andiman, W. A. (1977). Erythema chronicum migrans and Lyme arthritis: the enlarging clinical spectrum. Ann Intern Med 86: 685-698.

Steere, A. C., Sikand, V. K., Meurice, F., Parenti, D. L., Fikrig, E., Schoen, R. T., Nowakowski, J., Schmid, C. H., Laukamp, S., Buscarino, C. and Krause, D. S. (1998). Vaccination against Lyme disease with recombinant Borrelia burgdorferi outer-surface lipoprotein A with adjuvant. Lyme Disease Vaccine Study Group. N Engl J Med 1998: 209-215.
Stricker, R. B., Lautin, A. and Burrascano, J. J. (2005). Lyme disease: point/counterpoint. Expert Rev Anti Infect Ther 3(2): 155-165.
Stricker, R. B. and Phillips, S. E. (2003). Lyme disease without erythema migrans: cause for concern? Am J Med 115: 72.
Strle, F. (1999). Lyme borreliosis in Slovenia. Zentralbl Bakteriol 289: 634-652.
Thomas, V., Anguita, J., Barthold, S. W. and Fikrig, E. (2001). Coinfection with Borrelia burgdorferi and the agent of human granulocytic ehrlichiosis alters murine immune responses, pathogen burden, and severity of Lyme arthritis. Infect Immun 69(5): 3359-3371.
van Dam, A. P., Oei, A., Jaspars, R., Fijen, C., Wilske, B., Spanjaard, L. and Dankert, J. (1997). Complement-mediated serum sensitivity among spirochetes that cause Lyme disease. Infect Immun 65(4): 1228-36.
Vrethem, M., Hellblom, L., Widlund, M., Ahl, M., Danielsson, O., Ernerudh, J. and Forsberg, P. (2002). Chronic symptoms are common in patients with neuroborreliosis -- a questionnaire follow-up study. Acta Neurol Scand 106(4): 205-208.
Wang, G., Ma, Y., Buyuk, A., McClain, S., Weis, J. J. and Schwarz, I. (2004). Impaired host defence to infection and Toll-like receptor 2-independent killing of Borrelia burgdorferi clinical isolates in TLR2-deficient C3H/HeJ mice. FEMS Microbiol Lett 231: 219-225.
Wang, G., Ojaimi, C., Wu, H., Saksenberg, V., Iyer, R., Liveris, D., McClain, S. A., Wormser, G. P. and Schwartz, I. (2002). Disease severity in a murine model of lyme borreliosis is associated with the genotype of the infecting Borrelia burgdorferi sensu stricto strain. J Infect Dis 186(6): 782-791.
Wang, G., vanDam, A. P. and Dankert, J. (1999). Phenotypic and genetic characterization of a novel Borrelia burgdorferi sensu lato isolate from a patient with Lyme borreliosis. J Clin Microbiol 37: 3025-3028.
Wang, G., vanDam, A. P., Schwartz, I. and Dankert, J. (1999). Molecular typing of Borrelia burgdorferi sensu lato: taxonomic, epidemiological, and clinical implications. Clin Microbiol Rev 12: 633-653.
Wang, W. Z., Fredrikson, S., Sun, J. B. and Link, H. (1995). Lyme neuroborreliosis: evidence for persistent up-regulation of Borrelia burgdorferi-reactive cells secreting interferon-gamma. Scand J Immunol 42(6): 694-700.
Weigelt, W., Schneider, T. and Lange, R. (1992). Sequence homology between spirochaete flagellin and human myelin basic protein. Immunol Today 13(7): 279-280.
Weis, J. J., McCracken, B. A., Ma, Y., Fairbairn, D., Roper, R. J., Morrison, T. B., Weis, J. H., Zachary, J. F., Doerge, R. W. and Teuscher, C. (1999). Identification of quantitative trait loci governing arthritis severity and humoral responses in the murine model of Lyme disease. J Immunol 162: 948-956.
Widhe, M., Ekerfelt, C., Forsberg, P., Bergstrom, S. and Ernerudh, J. (1998). IgG subclasses in Lyme borreliosis: a study of specific IgG subclass distribution in an interferon-gamma-predominated disease. Scand J Immunol 47(6): 575-581.
Widhe, M., Jarefors, S., Ekerfelt, C., Vrethem, M., Bergström, S., Forsberg, P. and Ernerudh, J. (2004). Borrelia specific IFN-γ and IL-4 secretion in blood and CSF during the course of human Lyme borreliosis: relation to clinical outcome. J Infect Dis 189: 1881-1891.
Widhe, M. E. A., Grusell, M., Ekerfelt, C., Vrethem, M., Forsberg, P. and Ernerudh, J. (2002). Cytokines in Lyme borreliosis: Lack of early TNFα and TGFβ1 responses are associated with chronic neuroborreliosis. Immunology 107: 46-55.
Wooten, R. M., Ma, Y., Yoder, R. A., Brown, J. P., Weis, J. H., Zachary, J. F., Kirschning, C. J. and Weis, J. J. (2002). Toll-like receptor 2 is required for innate, but not acquired, host defense to Borreliaburgdorferi. J Immunol 168: 348-355.
Wooten, R. M., Modur, V. R., McIntyre, T. M. and Weis, J. J. (1996). Borrelia burgdorferi outer membrane protein A induces nuclear translocation of nuclear factor-kappa B and inflammatory activation in human endothelial cells. J Immunol 157(10): 4584-4590.
Yin, Z., Braun, J., Neure, L., Wu, P., Eggens, U., Krause, A., Kamradt, T. and Sieper, J. (1997). T cell cytokine pattern in the joints of patients with Lyme arthritis and its regulation by cytokines and anticytokines. Arthritis Rheum 40(1): 69-79.

Zeidner, N. S., Dolan, M. C., Massung, R., Piesman, J. and Fish, D. (2000). Coinfection with Borrelia burgdorferi and the agent of human granulocytic ehrlichiosis suppresses IL-2 and IFN gamma production and promotes an IL-4 response in C3H/HeJ mice. Parasite Immunol 22(11): 581-588.

Zipfel, P. F., Skerka, C., Hellwage, J., Jokiranta, S. T., Meri, S., Brade, V., Kraiczy, P., Noris, M. and Remuzzi, G. (2002). Factor H family proteins: on complement, microbes and human diseases. Biochem Soc Trans 30(Pt 6): 971-978.

Murine CR1/2 Targeted Antigenized Single-Chain Antibody Fragments Induce Transient Low Affinity Antibodies and Negatively Influence an Ongoing Immune Response

József Prechl[1], Eszter Molnár[2], Zsuzsanna Szekeres[2], Andrea Isaák[2], Krisztián Papp[2], Péter Balogh[3], Anna Erdei[1,2,*]

[1] Research Group of the Hungarian Academy of Sciences at the Department of Immunology, Eötvös Loránd University, Budapest
[2] Department of Immunology, Eötvös Loránd University, Pázmány P.s.1/C, 1117 Budapest, Hungary
[3] Department of Immunology and Biotechnology, University of Pécs, Faculty of Medicine Szigeti út 12, 7643, Pécs, Hungary

Abstract. We have generated a single-chain antibody which recognizes murine CR1/2 and carries a genetically fused influenza hemagglutinin derived peptide. Theoretically such a construct is able to crosslink the B cell antigen receptor and CR1/2 on peptide specific B cells. The construct was able to reach its CR1/2 positive target cells, yet intraperitoneal delivery of the construct elicited an IgM response only slightly exceeding that induced by the free peptide. Providing T cell help by the injection of peptide specific lymphocytes did not alter the response in essence, that is anti-peptide IgG was not detectable even after booster immunizations. When used as a booster vaccine following injection of the peptide in adjuvant, the construct even inhibited the development of IgG1 and IgG3 anti-peptide antibodies.

These data indicate that although targeting of antigen to CR1/2 on B cells can enhance transient proliferation or differentiation of antigen specific B cells it cannot induce strong, long-lasting humoral immune responses. Furthermore, CR1/2 targeting constructs may negatively influence an ongoing immune reaction.

1 Introduction

The role of the complement system in the initiation of the adaptive immune responses was the focus of several studies in recent years. Type 2 complement receptors (CR2) were shown to be required for the generation of a humoral immune response induced by limiting amounts of antigen (Molina *et al.*, 1996;Ahearn *et al.*, 1996;Croix *et al.*, 1996;Haas *et al.*, 2002); though they were found to be disposable when adjuvants were given with the antigen. The affinity of the antibodies, induced in animals deficient in CR1/2, was found to be altered as compared to wild type mice (Wu *et al.*, 2000;Chen *et al.*, 2000;Applequist *et al.*, 2000). While IgM mediated feedback enhancement of humoral responses depends on CR1/2, IgE and IgG2a can promote responses to antigen even in CR1/2 deficient animals (Applequist, 2000).

Because of the suspected adjuvant role of complement activation fragments deposited on antigens, a number of experiments addressed the potency of C3b or C3d to improve immunogenicity. Activation products of C3 were coupled to the antigen by genetic engineering (Dempsey *et al.*, 1996), via the physiological thioester link (Villiers *et al.*, 1999), and by photochemical reactions (Lou and Kohler, 1998). The naked DNA vaccination route of such constructs has also been tested (Ross *et al.*, 2000;Ross *et al.*, 2001) successfully. These studies reinforced the notion that C3 fragments act as molecular adjuvants. The results of experiments using C3d, the preferred ligand of CR2, together with the data obtained with knock-out animals, pointed to the primary involvement of complement receptors on B lymphocytes in the efficient induction of humoral immunity (Fearon and Carroll, 2000). Since CR2 forms a coreceptor complex together with CD19 and CD81, which positively influences B cell activation, it was suggested that the adjuvant effect was exerted by lowering the thresholds of B cell activation.

Although C3d binds mainly to CR2, it is also able to bind to integrins such as CR3 and CR4 with lower affinity and enhance phagocytosis (Gaither *et al.*, 1987). Therefore the results obtained by immunizations with antigen-C3d complexes cannot be interpreted as the outcome of sheer CR1/2 targeting. To overcome this difficulty, we have generated constructs that target antigen to CR1/2 by a single-chain antibody fragment. The model antigen in our system is an influenza hemagglutinin derived peptide. Earlier we have shown that the hemagglutinin 317-342 peptide could be cleaved off during intracellular processing and associated with MHCII molecules to be presented to T cells (Prechl *et al.*, 1999). In this paper we describe the construction of an antigenized scFv that contains an additional, immunodominant Balb/c T cell epitope (HA306-316), and the results of immunizations with this CR1/2 targeted hemagglutinin peptide-scFv construct.

2 Materials and methods

2.1 Generation of hH2-7G6scFv

The sequence coding for amino acids 306-342 in the hemagglutinin of influenza virus Japan/57 (H2) strain was amplified by RT-PCR after isolation of RNA from virions using Tri-Reagent (Sigma-Aldrich, Hungary) as described (Prechl *et al.*, 1999). The PCR introduced NcoI and RcaI sites, which were used for subcloning the fragment from pBluescript to pET11b-7G6scFv vector. Upon insertion the peptide sequence MGPKYVKSEKLVLATGLRNVPQIESRGLFGAIAGFIEGGGSGGGV is fused to the N terminal end of the scFv antibody; thus the peptide fused form of 7G6scFv we call hH2-7G6scFv. Primers used for the reaction were: flu306Nco 5'-CCATGGGCCCCAAATATGTAAAATCGG-3'; flu3Bsp 5'-GCTCATGACTCCGCCACCTGAGCCTCCCCCTTCAATAAAACC-3'. Recombinant proteins – 7G6scFv, hH2-7G6scFv - were produced as described (Prechl *et al.*, 1999).

2.2 Immunizations

6-8 weeks old Balb/c mice (Charles River Laboratories) in groups of 4-6 were injected intraperitoneally with equimolar amounts of antigen: 2 μg peptide/mouse or 20 μg hH2-7G6scFv/mouse, in 200 μl PBS. Booster immunization was carried out 21 days after the primary injection, using the same antigen doses and formulation. Blood was drawn from the orbital sinus after the indicated time periods following immunizations.

T cells, enriched for HA306-329 specificity, were generated by injecting 10μg of the peptide mixed with complete Freund adjuvant (Sigma-Aldrich, Hungary) into the hind limbs and the base of the tail. Draining lymphnodes were removed 7 days later and passed through a wire mash. B cells were eliminated by negative panning on Petri dishes coated with rabbit anti-mouse IgM. The resulting population of cells was >90% CD5 positive by cytofluorimetry. To allow homing to the lymphoid organs $2x10^5$/mouse of these cells were injected intravenously 24 hours before the immunizations.

For testing the boosting capacity of hH2-7G6scFv, mice were injected subcutaneously with 100 μg HA306-329 peptide emulsified in complete Freund's adjuvant (Sigma-Aldrich, Hungary). 14 days later animals were given an intraperitoneal injection of PBS, free peptide (10 μg/mouse) or hH2-7G6scFv (100 μg/mouse). Blood was collected on day 28.

2.3 Flow cytometry and immunohistochemistry

Spleens were removed 2 hours after intraperitoneal injection of 7G6scFv, and halves of the spleens were saved for immunohistochemistry, the other half was mechanically disrupted using a sterile syringe piston. After the lysis of erythrocytes isolated murine splenocytes were incubated sequentially with or without 7G6scFv, biotinylated anti-myc mAb 9E10 and anti-biotin FITC (Sigma-Aldrich, Hungary). Cells were analysed on a FACSCalibur flow cytometer (BD Biosciences, Mountain View, CA), 10,000 lymphocyte events were collected.

Frozen sections of 10-μm thickness were cut and air dried. Five representative sections were collected and processed together. Following acetone fixation, the sections were rehydrated and the endogenous peroxidase activity was quenched with PBS containing 0.1% phenylhydrazine-HCl. After washing, the sections were saturated with 10% BSA, then incubated with biotinylated 9E10 mAb followed by incubating the sections with streptavidin-HRP conjugate. The reaction was visualized using 3-amino-9-ethylcarbazole in a 0.1 M sodium acetate buffer containing 0.03% H_2O_2 substrate.

2.4 ELISA

96 well plates were coated with 50μl of 5 μg/ml HA306-329 peptide in PBS at 4°C overnight. After three washes in PBS-0.05%Tween20, 50 μl of serum diluted in PBS-Tween was added and incubated at 37°C for 1 hour. HRPO conjugated goat anti-mouse Ig with the indicated isotype specificity (Southern Biotechnologies, USA) was added after another round of washes, and incubated for yet an hour. TMB (Sigma-Aldrich, Hungary) was used as a detection reagent.

Kruskal-Wallis ANOVA (Sigmastat) was used for the statistical analysis of the results.

3 Results

3.1 Peritoneally injected 7G6scFv binds to CR1/2 positive cells

As a first step to characterize in vivo effects of the CR1/2 specific 7G6scFv we tested biodistribution of the molecule following intraperitoneal injection using flow cytometry and immunohistochemistry. Splenic B cells were all positive for the presence of myc-tag, indicating that 7G6scFv was stable in the circulation and reached its targets (Fig 1A). Addition of 7G6scFv ex vivo to the lymphocytes further increased the fluorescence signal, indicating that although all B cells carried the antibody, on particular cells not all CR1/2 receptors were saturated. Likewise blood B cells, and to a lesser extent monocytes and granulocytes were found to carry 7G6scFv on their surface (data not shown). Immunohistochemistry revealed 7G6scFv positive B cell areas of splenic follicles and marginal zones (Fig.1B). Thus, following intraperitoneal injection 7G6scFv can reach all CR1/2 positive target cells.

3.2 CR1/2 targeted constructs induce low-titer anti-peptide response

An antigenized scFv construct, which we have described earlier, contained only a subdominant T cell epitope (Prechl *et al.*, 1999). In pilot experiments using that construct we aimed to determine the optimal route of administration and dosage in order to obtain antibodies reacting with the peptide fused to the scFv. However, we could not generate detectable amounts of such antibodies. Therefore we decided to modify the antigen and include an additional T cell epitope (HA306-317) which is dominant in Balb/c mice (Rajnavolgyi *et al.*, 1997). The capacity of this construct, designated hH2-7G6scFv, to induce anti-peptide antibodies was tested by injecting it intraperitoneally and determining the anti-peptide humoral response. As compared to preimmune serum, the constructs elicited antibodies that reacted with HA306-329 (Fig.2.). Targeting the peptide to CR1/2 enhanced the antibody response, however, the difference was only detectable at low dilutions and was transient in nature, as it decreased after the booster immunization (Fig.2.C). Moreover, the isotype of the peptide specific antibodies was restricted to IgM.

injected	PBS	7G6scFv	7G6scFv
1ry Ab	none	none	7G6scFv
2ry Ab	9E10	9E10	9E10

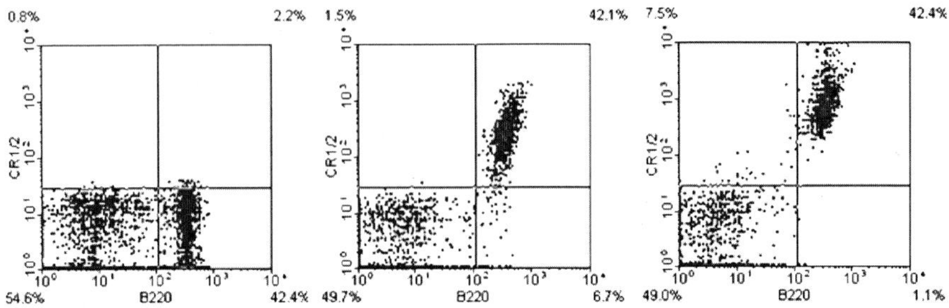

A

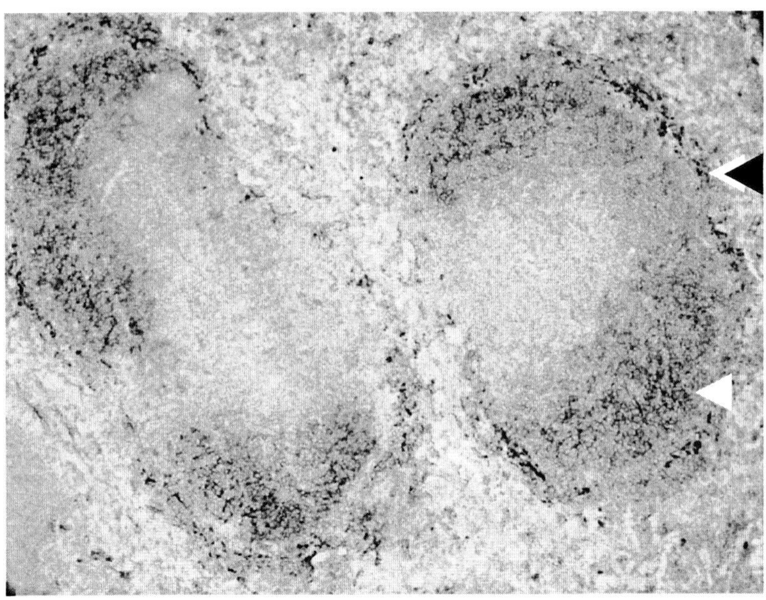

B

Fig. 1. Distribution of 7G6scFv after intraperitoneal injection. Either PBS or 100 μg of the construct was injected intraperitoneally and the mice were sacrificed 2 hours later. (A) Splenocytes were labelled as indicated with the anti-myc tag antibody 9E10 or with 7G6scFv and then anti-myc tag antibody. (B) Frozen sections from the spleen were stained with anti-myc tag antibody to detect bound 7G6scFv molecules. Arrowheads indicate follicular(◀) and marginal zone(◀◀) areas.

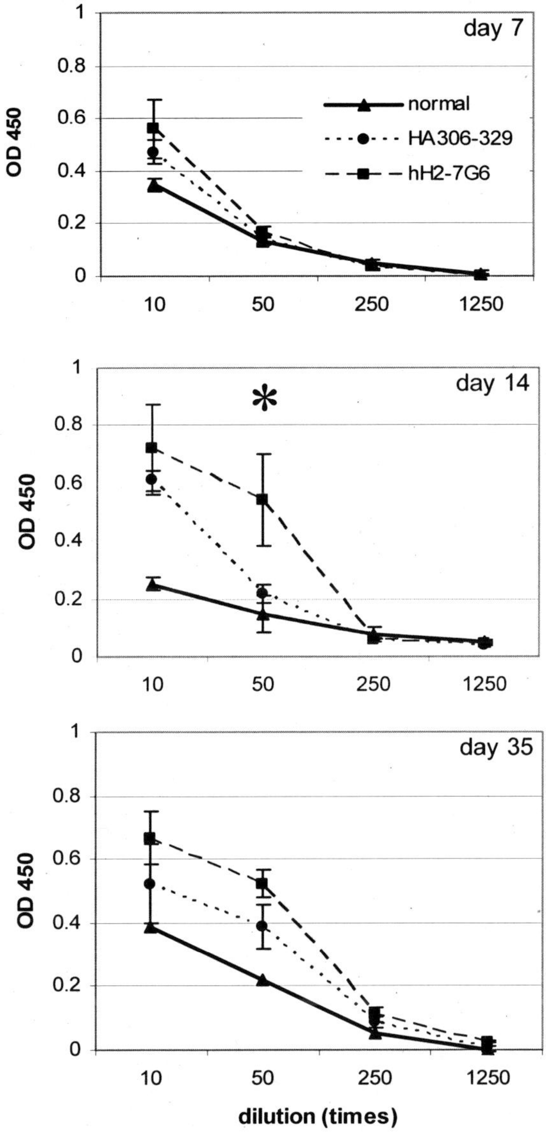

Fig. 2. The hH2-7G6scFv construct induces antibodies against HA306-329.
Total Ig response against HA306-329 was measured by ELISA 7 days and 14 days after the immunization, and 14 days after the booster immunization(day 35), from sera of mice immunized with equimolar amounts of hH2-7G6scFv or HA306-329 and compared with nonimmune serum (normal). Shown are the mean absorbance values, error bars represent standard deviation. (* $p<0.05$, ANOVA, Kruskal-Wallis test)

3.3 Provision of T cell help cannot qualitatively improve the response

We hypothesized that the inability of the constructs to induce class switch and affinity maturation was due to lack of adequate T cell help. To investigate this possibility, we transferred syngeneic peptide specific T lymphocytes 24 hours before the primary immunization. The quality of the response was not altered by this pretreatment: only low titer IgM antibodies were detected and no anti-peptide IgG was found in the sera. While there was a significant difference between the response of animals to hH2-7G6scFv and PBS or HA306-329 when 100-fold diluted sera after the primary immunization were tested, this difference was not improved or disappeared after the secondary immunizations (Fig. 3.).

3.4 CR1/2 targeted booster immunization negatively influences isotype switch

Since hH2-7G6scFv was not able to induce a robust immune response as a primary immunizing agent, we next tested its ability to boost an immune response. 14 days after injecting animals with free hemagglutinin peptide in complete Freund adjuvant, we injected PBS, free peptide or hH2-7G6scFv intraperitoneally. Peptide specific antibody responses are shown in figure 4. Interestingly, boosting with hH2-7G6scFv even lowered IgG1 and IgG3 response against the peptide, with a modest increase in IgM and no change in IgG2a, as compared to the control groups.

4 Discussion

We have generated scFv molecules which recognize murine CR1/2 and carry an influenza virus derived peptide. The CR1/2 specific construct, 7G6scFv reaches and binds to CR1/2 positive cells in the circulation and the spleen upon intraperitoneal administration. This construct is capable of inducing antibodies that recognize the viral peptide. This fact indicates that the peptide in the construct is available for surface IgM on B cells. Thus, the constructs can potentially coligate sIgM with CR1/2 on peptide specific B cells. One would therefore expect an enhanced response on the side of B cells as compared to the response induced by a free peptide. Indeed, we observed that sera from hH2-7G6 injected mice reacted more strongly than sera of peptide immunized animals with HA306-329. The titer of the sera was, however, very low. The likely explanation is the presence of higher amounts of low affinity IgM antibodies which might be the result of enhanced proliferation or differentiation of peptide specific B cells.

The question whether B cells can activate naïve T cells is still controversial. Antigen presentation by resting B cells is rather tolerogenic, resulting in anergy of the T cell. Since our CR1/2 specific constructs can be taken up by any B cell without activating the cell via the BCR, such a mechanism might be responsible for the ineffectiveness of the constructs to induce IgG responses. Therefore we transferred peptide specific effector T cells into the mice receiving immunizations. If lack of T cells was to blame for the absence of IgG antibodies, this treatment could have

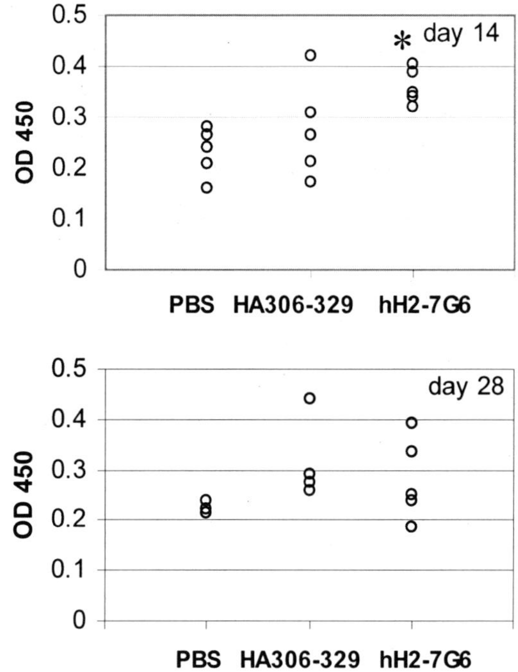

Fig. 3. IgM response against HA306-329 in individual mice after transfer of peptide specific T cells. Data points represent anti-HA306-329 IgM values of 100-fold diluted sera in individual mice from the indicated immunization group 14 days after the primary (day 14) or the booster (day 28) immunization. Immunization with hH2-7G6scFv had a significant effect (* $p<0.05$) on day 14 as compared to the free peptide.

overcome the problem. In spite of this, no changes were observed after T cell transfer. We hence conclude that a different/another factor is required to help the B cells to go through affinity maturation and switch antibody isotype. Dendritic cells have been implicated to deliver stimuli to B cells during the formation of germinal centres (Wykes *et al.*, 1998). CR1/2 targeted constructs might miss this event as they directly reach B cells and follicular dendritic cells upon administration. An alternative explanation could be the ineffectiveness of antigen transfer to FDC. It was recently

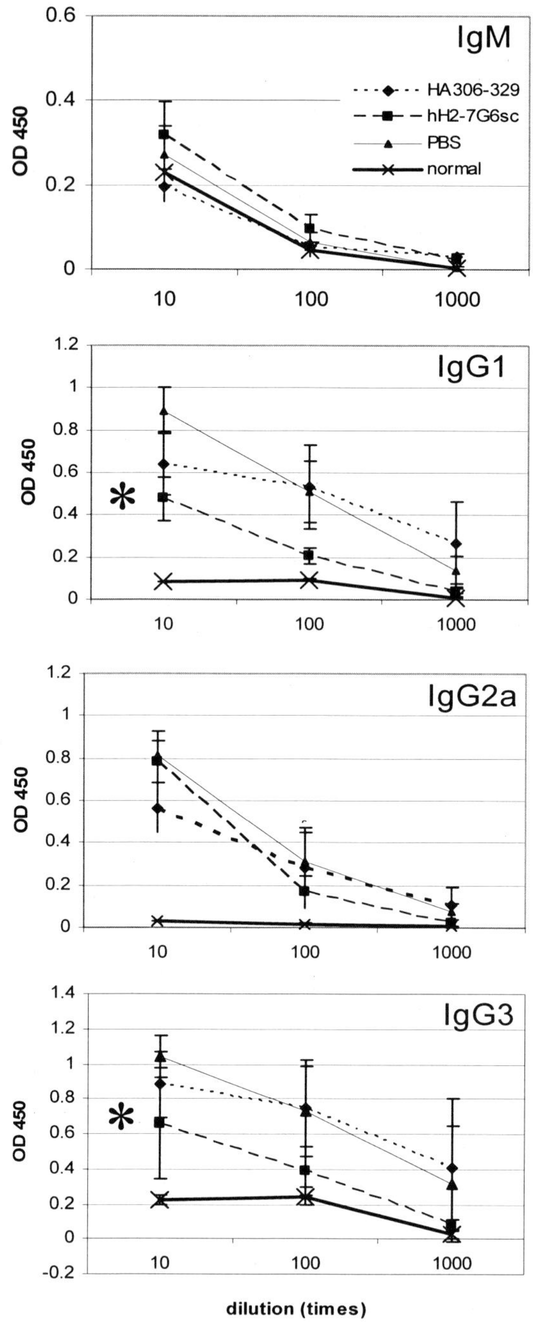

Fig. 4. CR1/2 targeted booster immunization negatively influences isotype switching. Mice were injected with a mixture of peptide and CFA, two weeks later they received a booster immunization with the indicated substances. Anti-peptide responses of different isotypes are shown. Note that IgM values are plotted on a different scale to help visualize differences. Treatments had a significant effect on IgG1 and IgG3 titers (* $p<0.05$).

shown that the monoclonal antibody 7G6 – a whole Ig – after an initial localization to the marginal zone was transferred to FDC by MZ B cells (Whipple *et al.*, 2004). This transfer reaction involved proteolysis of the CR1/2 molecule on the donor cells and binding of the complex to either FcγRII or CR1/2 on the acceptor FDC. Our monovalent scFv constructs which lack Fc are therefore not suitable agents for such a transfer reaction. Earlier we have used whole monoclonal antibody 7G6 chemically conjugated to ovalbumin – a complex amenable for a transfer reaction – and found it to be highly immunogenic, indeed (Baiu *et al.*, 1999).

Unexpectedly, CR1/2 targeting did not even work as a booster immunization, but rather lowered the peptide specific IgG titers. This data again indicates that targeting CR1/2 bearing cells may not be adequate for inducing humoral immunity. B cells need survival signals while interacting with FDC, and CR1/2 has been shown to mediate such signals (Qin *et al.*, 1998). Covering CR1/2 both on B cells and FDC may interrupt these interactions and thereby inhibit isotype switching. Another possibility is that BCR-CR1/2 coligation pushes B cells towards proliferation and IgM production instead of entering a germinal center reaction. Increased affinity of antibodies generated in CR1/2 deficient mice supports the letter hypothesis (Chen *et al.*, 2000).

In the light of these data we also need to reinterpret results with antigen-C3d conjugates (Dempsey *et al.*, 1996;Ross *et al.*, 2000). Increased immunogenicity of those constructs must at least partly rely on factors other than CR1/2, perhaps on CR3. The fact that more than one copy of C3d needs to be fused to an antigen in order to boost immunogenicity – in contrast to C3b from which one is sufficient – may imply a need for multiple CR1/2 crosslinkage, but also may indicate that avidity for other receptors must be increased. Enhanced uptake of antigen-C3d by myeloid cells bearing CR3, then relay of the antigen to B cells and T cells with CR1/2 on their surface may be the way of action of these constructs. Indeed, a recent report showed that C3d can function as a molecular adjuvant in the absence of CR1/2 expression (Haas *et al.*, 2004). Experiments using various targeted antigens in knockout mice lacking the relevant receptors will provide the answers to these questions and help the rational design of receptor targeted vaccines.

5 Acknowledgements

We wish to thank Árpád Mikessy for his invaluable help with immunizations and animal care, and Zsuzsa Greff for technical assistance.
This work was supported in part by grants OTKA T047151, OTKA TS044711, NKFP 1A/040/2004.

6 Abbreviations

scFv, single-chain antibody fragment; BCR, B cell antigen receptor; CR1/2, complement receptor type 1 and 2

References

Ahearn, J.M., Fischer, M.B., Croix, D., Goerg, S., Ma, M., Xia, J., Zhou, X., Howard, R.G., Rothstein, T.L., Carroll, M.C., 1996. Disruption of the Cr2 locus results in a reduction in B-1a cells and in an impaired B cell response to T-dependent antigen. Immunity. 4, 251-262.

Applequist, S.E., Dahlstrom, J., Jiang, N., Molina, H., Heyman, B., 2000. Antibody production in mice deficient for complement receptors 1 and 2 can be induced by IgG/Ag and IgE/Ag, but not IgM/Ag complexes. J. Immunol. 165, 2398-2403.

Baiu, D.C., Prechl, J., Tchorbanov, A., Molina, H.D., Erdei, A., Sulica, A., Capel, P.J., Hazenbos, W.L., 1999. Modulation of the humoral immune response by antibody-mediated antigen targeting to complement receptors and Fc receptors. J. Immunol. 162, 3125-3130.

Chen, Z., Koralov, S.B., Gendelman, M., Carroll, M.C., Kelsoe, G., 2000. Humoral immune responses in Cr2-/- mice: enhanced affinity maturation but impaired antibody persistence. J. Immunol. 164, 4522-4532.

Croix, D.A., Ahearn, J.M., Rosengard, A.M., Han, S., Kelsoe, G., Ma, M., Carroll, M.C., 1996. Antibody response to a T-dependent antigen requires B cell expression of complement receptors. J. Exp. Med. 183, 1857-1864.

Dempsey, P.W., Allison, M.E., Akkaraju, S., Goodnow, C.C., Fearon, D.T., 1996. C3d of complement as a molecular adjuvant: bridging innate and acquired immunity. Science %19;271, 348-350.

Fearon, D.T., Carroll, M.C., 2000. Regulation of B lymphocyte responses to foreign and self-antigens by the CD19/CD21 complex. Annu. Rev. Immunol. 18, 393-422.

Gaither, T.A., Vargas, I., Inada, S., Frank, M.M., 1987. The complement fragment C3d facilitates phagocytosis by monocytes. Immunology 62, 405-411.

Haas, K.M., Hasegawa, M., Steeber, D.A., Poe, J.C., Zabel, M.D., Bock, C.B., Karp, D.R., Briles, D.E., Weis, J.H., Tedder, T.F., 2002. Complement receptors CD21/35 link innate and protective immunity during Streptococcus pneumoniae infection by regulating IgG3 antibody responses. Immunity. 17, 713-723.

Haas, K.M., Toapanta, F.R., Oliver, J.A., Poe, J.C., Weis, J.H., Karp, D.R., Bower, J.F., Ross, T.M., Tedder, T.F., 2004. Cutting edge: C3d functions as a molecular adjuvant in the absence of CD21/35 expression. J. Immunol. 172, 5833-5837.

Lou, D., Kohler, H., 1998. Enhanced molecular mimicry of CEA using photoaffinity crosslinked C3d peptide. Nat. Biotechnol. 16, 458-462.

Molina, H., Holers, V.M., Li, B., Fung, Y., Mariathasan, S., Goellner, J., Strauss-Schoenberger, J., Karr, R.W., Chaplin, D.D., 1996. Markedly impaired humoral immune response in mice deficient in complement receptors 1 and 2. Proc. Natl. Acad. Sci. U. S. A 93, 3357-3361.

Prechl, J., Tchorbanov, A., Horvath, A., Baiu, D.C., Hazenbos, W., Rajnavolgyi, E., Kurucz, I., Capel, P.J., Erdei, A., 1999. Targeting of influenza epitopes to murine CR1/CR2 using single-chain antibodies. Immunopharmacology 42, 159-165.

Qin, D., Wu, J., Carroll, M.C., Burton, G.F., Szakal, A.K., Tew, J.G., 1998. Evidence for an important interaction between a complement-derived CD21 ligand on follicular dendritic cells and CD21 on B cells in the initiation of IgG responses. J. Immunol. 161, 4549-4554.

Rajnavolgyi, E., Horvath, A., Gogolak, P., Toth, G.K., Fazekas, G., Fridkin, M., Pecht, I., 1997. Characterizing immunodominant and protective influenza hemagglutinin epitopes by functional activity and relative binding to major histocompatibility complex class II sites. Eur. J. Immunol. 27, 3105-3114.

Ross, T.M., Xu, Y., Bright, R.A., Robinson, H.L., 2000. C3d enhancement of antibodies to hemagglutinin accelerates protection against influenza virus challenge. Nat. Immunol. 1, 127-131.

Ross, T.M., Xu, Y., Green, T.D., Montefiori, D.C., Robinson, H.L., 2001. Enhanced avidity maturation of antibody to human immunodeficiency virus envelope: DNA vaccination with gp120-C3d fusion proteins. AIDS Res. Hum. Retroviruses 17, 829-835.

Villiers, M.B., Villiers, C.L., Laharie, A.M., Marche, P.N., 1999. Amplification of the antibody response by C3b complexed to antigen through an ester link. J. Immunol. 162, 3647-3652.

Whipple, E.C., Shanahan, R.S., Ditto, A.H., Taylor, R.P., Lindorfer, M.A., 2004. Analyses of the in vivo trafficking of stoichiometric doses of an anti-complement receptor 1/2 monoclonal antibody infused intravenously in mice. J. Immunol. 173, 2297-2306.

Wu, X., Jiang, N., Fang, Y.F., Xu, C., Mao, D., Singh, J., Fu, Y.X., Molina, H., 2000. Impaired affinity maturation in Cr2-/- mice is rescued by adjuvants without improvement in germinal center development. J. Immunol. 165, 3119-3127.

Wykes, M., Pombo, A., Jenkins, C., MacPherson, G.G., 1998. Dendritic cells interact directly with naive B lymphocytes to transfer antigen and initiate class switching in a primary T-dependent response. J. Immunol. 161, 1313-1319.

16

The Third Complement Component as Modulator of Platelet Production

Marcin Wysoczynski[1], Janina Ratajczak[1], Ryan Reca[1], Magda Kucia[1] and Mariusz Z. Ratajczak[1]

[1] James Graham Brown Cancer Center University of Louisville, Stem Cell Biology Program, Louisville, KY, mzrata01@louisville.edu

1 Introduction

We have demonstrated that the complement (C) system is activated/cleaved in bone marrow (BM) after irradiation, exposure to cytostatics or granulocyte colony stimulating factor (G-CSF) infusion (Reca R, Mastellos D, Majka M, Marquez L, Ratajczak J, Franchini S, Glodek A, Honczarenko M, Spruce LA, Janowska-Wieczorek A, Lambris JD, Ratajczak MZ 2003 ; Ratajczak MZ, Reca R, Wysoczynski M, Kucia M, Baran JT, Allendorf DJ, Ratajczak J, Ross GD 2004; Ratajczak J, Reca R, Kucia M, Majka M, Allendorf DJ, Baran JT, Janowska-Wieczorek A, Wetsel RA, Ross GD, Ratajczak MZ 2004; Allendorf DJ, Ostroff G, Baran JT, Dyke CW, Ratajczak MZ, Ross GD 2003). To explain this C3 is present at a relatively high concentration in serum (1mg/ml) that flows through the BM. In addition we have shown that human BM stromal fibroblasts secrete C3 (Reca R et al. 2003). It is also well known that BM macrophages synthesize and secrete all complement components. Thus, C3 is a physiological constituent of the BM environment (Volanakis JE and Frank MM 1998). We also recently reported that the complement (C) system and third complement protein (C3) cleavage fragments C3a and ${}_{desArg}$C3a enhance the responsiveness of hematopoietic stem/progenitor cells (HSPC) to an alpha-chemokine stromal-derived factor-1 (SDF-1) gradient, suggesting that C3 cleavage fragments modulate the trafficking of HSPC (Reca R et al. 2003; Ratajczak MZ et al. 2004).

It is known that C3a binds to $G_{\alpha i}$-protein coupled seven transmembrane receptor (C3aR) and we reported recently that C3aR is expressed on human and murine HSPC (Reca R et al. 2003; Crass T, Raffetseder U, Martin U, Grove M, Klos A, Kohl J, Bautsch W 1996; Tornetta MA, Foley JJ, Sarau HM and Ames RS 1997). In contrast the receptor for ${}_{desArg}$C3a has not been identified yet. The fact that HSPC express both

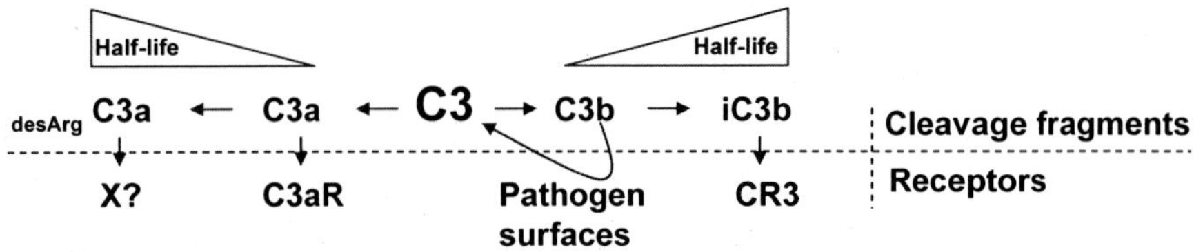

Fig. 1. Cleavage of C3 into biologically active fragments. Cleavage/activation of C3 in BM is initiated by a C3-convertase to generate fluid-phase C3a and stromal cell-bound C3b. Both C3a and C3b have a short half-life. Fluid C3a is rapidly degraded to $_{des\text{-}Arg}$C3a and bound C3b is proteolyzed to iC3b. While C3a mainly activates the C3aR, $_{des\text{-}Arg}$C3a binds to another unidentified receptor X? iC3b tethers HSPC by interacting with CR3.

functional SDF-1 binding receptor CXCR4 and C3a cleavage fragments binding receptors support the concept that the C system may modulate the responsiveness of HSPC to chemokines.

Since C is activated in several clinical conditions associated with high platelet counts (e.g., chronic inflammation, hemorrhage), we hypothesized that C3a and $_{des\text{-}Arg}$C3a are involved in stress/inflammation-related thrombocytosis. Supporting this are our observations that C3-deficient mice, which have normal platelet counts in steady state conditions (Ratajczak MZ et al. 2004), display delayed recovery of platelets after sublethal irradiation or hematopoietic transplants (Ratajczak MZ et al. 2004). These findings stimulated our interest in determining whether C3 cleavage fragments could play a role in stress-related thrombocytosis. Others already demonstrated a pivotal role of SDF-1-CXCR4 axis in this phenomenon (Wang JF, Liu ZY, Groopman JE 1998; Hamada T, Mohle R, Hesselgesser J, Hoxie J, Nachman RL, Moore MA, Rafii S 1998 ; Kowalska MA, Ratajczak J, Hoxie J, Brass LF, Gewirtz A, Poncz M, Ratajczak MZ 1999; Riviere C, Subra F, Cohen-Solal K, Cordette-Lagarde V, Letestu R, Auclair C, Vainchenker W, Louache F 1999 ; Hodohara K, Fujii N, Yamamoto N, Kaushansky K 2000; Majka M, Janowska-Wieczorek A, Ratajczak J, Kowalska MA, Vilaire G, Pan ZK, Honczarenko M, Marquez LA, Poncz M, Ratajczak MZ 2000) .

2 The complement system as an underappreciated regulator of hematopoiesis

As shown in Figure 1 two groups of C3 cleavage fragments are distinguished: fluid phase (C3a, $_{des\text{-}Arg}$C3a) and cell- or extracellular matrix-bound (C3b, iC3b) fragments (Volanakis et al. 1998; Walport MJ 2001; Walport MJ 2001). C3a and C3b are the first cleavage products of C3 and each has a short half-life in plasma. C3a is processed by serum carboxypeptidase N (SCPN) to $_{desArg}$C3a (long half-life), and C3b is cleaved into iC3b (long half-life) by factor I. Accordingly, while C3a activates the classical C3aR, $_{des\text{-}Arg}$C3a selectively activates a still unknown receptor (Receptor X?) (Volanakis et al. 1998; Walport MJ 2001; Walport MJ 2001). We noticed that HSPC express abundant amounts of C3aR (Reca R et al. 2003) and it has been reported that they also express CR3 in abundance (Coombe DR, Watt SM and Parish CR 1994; Teixido J, Hemler ME, Greenberger JS and Anklesaria P 1992). Thus, since C3 is cleaved/activated to C3a and $_{des\text{-}Arg}$C3a (liquid phase) and iC3b (solid phase) in BM during conditioning for transplantation or mobilization (Ratajczak MZ et al. 2004; Ratajczak J et al. 2004), all of these cleavage fragments may directly interact with e.g., C3aR$^+$ or CR3$^+$ HSPC – which are important for modulating the SDF-1 dependent trafficking of these cells.

3 Megakaryopoiesis – implications for new regulators of platelet production

Megakaryopoiesis is regulated by several factors that affect the proliferation and differentiation of megakaryopoietic cells (Long MW and Hoffman R 1999; Gewirtz AM 1995). The fact that mice with a "knock-out" of thrombopoietin (TPO) and thrombopoietin receptor (c-mpl) show a 90% reduction in both the number of megakaryocytes (Megs) in BM and circulating platelets indicates that TPO is a crucial regulator of megakaryopoiesis (Gurney AL, Carver-Moore K, de Sauvage FJ, Moore MW 1994; Gurney AL and de Sauvage FJ 1996 ; Bunting S, Widmer R, Lipari T, Rangell L, Steinmetz H, Carver-Moore K, Moore MW, Keller GA, de Sauvage FJ 1997). Nevertheless, the fact that some Megs and platelets are still present in these animals even in the total absence of TPO suggests that some other cytokines can compensate for TPO deficiency and the cytokines that signal through gp130 such as interleukin-6 (IL-6), IL-11, leukemia inhibitory factor, cilliary neurotropic factor and oncostatin M have been suggested to be such factors (Sasaki H, Hirabayashic Y, Ishibashid T, Inoueb T, Matsudaa M, Kaia S, Ikutaa S, Yokoyamad K, Yokotac T, Maruyamad Y and Matsuyamaa S 1995 ; Burstein SA, Mei RL, Henthorn J, Friese P, Turner K 1992 ; Neben TY, Loebelenz J, Hayes L, McCarthy K, Stoudemire J, Schaub R, Goldman SJ 1993).

However, using a serum-free liquid culture system we observed that of all the gp-130 signaling cytokines only IL-6 protected CD34$^+$ cells and Megs from undergoing apoptosis, and only IL-6 when used in combination with TPO, stimulated the growth of colony-forming unit-Megs (CFU-Meg) in semisolid serum-free medium (Majka M, Ratajczak J, Villaire G, Kubiczek K, Marquez LA, Janowska-Wieczorek A, Ratajczak MZ 2002). Furthermore, others have reported that IL-6 and IL-11 do not induce platelet production in thrombocytopenic, TPO-deficient (Tpo$^{-/-}$) and TPO-receptor-deficient (Mpl$^{-/-}$) mice (Gainsford T, Nandurkar H, Metcalf D, Robb L, Begley CG, Alexander WS 2000). Thus, it is unlikely that gp-130 signaling cytokines are significantly involved in platelet production in these animals (Gurney AL et al. 1996; Bunting et al. 1997; Sasaki H et al. 1995; Burnstein SA et al. 1992; Neben TY et al. 1993; Majka M et al. 2002; Gainsford T et al. 2000).

Lately, the α-chemokine SDF-1 has been proposed as a new regulator of Megs maturation and platelet formation (Majka M et al. 2000; Wang JF, Liu ZY, Groopman JE 1998; Hamada T, Mohle R, Hesselgesser J, Hoxie J, Nachman RL, Moore MA,

Rafii S 1998 ; Kowalska MA, Ratajczak J, Hoxie J, Brass LF, Gewirtz A, Poncz M, Ratajczak MZ 1999; Riviere C, Subra F, Cohen-Solal K, Cordette-Lagarde V, Letestu R, Auclair C, Vainchenker W, Louache F 1999 ; Hodohara K, Fujii N, Yamamoto N, Kaushansky K 2000). Accordingly a recently published study clearly demonstrated that SDF-1, together with fibroblast growth factor-4 (FGF-4), restored thrombopoiesis in $Tpo^{-/-}$ and $Mpl^{-/-}$ mice by directing trans-localization of megakaryocytic progenitors positive for CXCR4, the receptor for SDF-1, from the endostial to the vascular niche, thereby promoting survival, maturation to megakaryocytes and platelet production (Avecilla ST, Hattori K, Heissig B, Tejada R, Liao F, Shido K, Jin DK, Dias S, Zhang F, Hartman TE, Hackett NR, Crystal RG, Witte L, Hicklin DJ, Bohlen P, Eaton D, Lyden D, de Sauvage F, Rafii S 2004). These findings stimulated our interest in determining whether C3 cleavage fragments could play a role in stress-related thrombocytosis.

Since we noticed that C3aR is expressed on normal human Megs, it suggested a role for C3 cleavage fragments in regulating normal human megakaryopoiesis. Highly purified human Megs were found to respond to C3a stimulation by calcium flux (Reca R et al. 2003). Furthermore, we found that C3a but not ${}_{desArg}C3a$ induced F-actin polymerization in human Megs. The fact that ${}_{desArg}C3a$ did not induce F-actin polymerization and SB290157 inhibited C3a induced F-actin polymerization (not shown) indicates that the C3aR is responsible for the polymerization of F-actin (Wysoczynski M et al. 2007). More important these observations collectively support our concept that C3 cleavage fragments may modulate migration and adhesion of Megs.

4 Megakaryopoiesis is regulated differently at the osteoblastic and endothelial niche

The concept of developmental allocation of Megs from the bone-niche to the endothelial niche has been proposed by S. Rafii et al. This allocation is tightly regulated by an SDF-1 gradient and we envision that during stressed situations (e.g., chronic inflamation, hemorrhage) (Avecilla ST et al. 2004), C3 cleavage fragments are important co-regulators of this process. Accordingly, we proposes that C3 is cleaved/activated in BM sinusoids and as a result of this C3 cleavage fragments (C3a and ${}_{desArg}C3a$) modulate/enhance the responsiveness of maturing Megs to an SDF-1 gradient (Wysoczynski M et al. 2007). We envision that Megs are chemoattracted by SDF-1 expressed by marrow endothelial cells and re-allocate from the bone-marrow niche to the endothelial niche where proplatelets are formed and platelets shed from the Meg surface. At the molecular level we envision that C3 cleavage fragments sensitize responsiveness of Megs to SDF-1 by enhancing similarly as in other types of cells its inclusion into membrane lipid rafts (Ratajczak M.Z., Zuba-Surma E, Kucia M, Reca R, Wojakowski W, Ratajczak J. 2006). CXCR4 if included into membrane lipid rafts regulates Megs migration more efficiently as well as adhesion to endothelium and secretion of metalloproteinase -9 (MMP-9) and vascular endothelial growth factor (VEGF) (Wysoczynski M et al. 2007). All of these processes are crucial

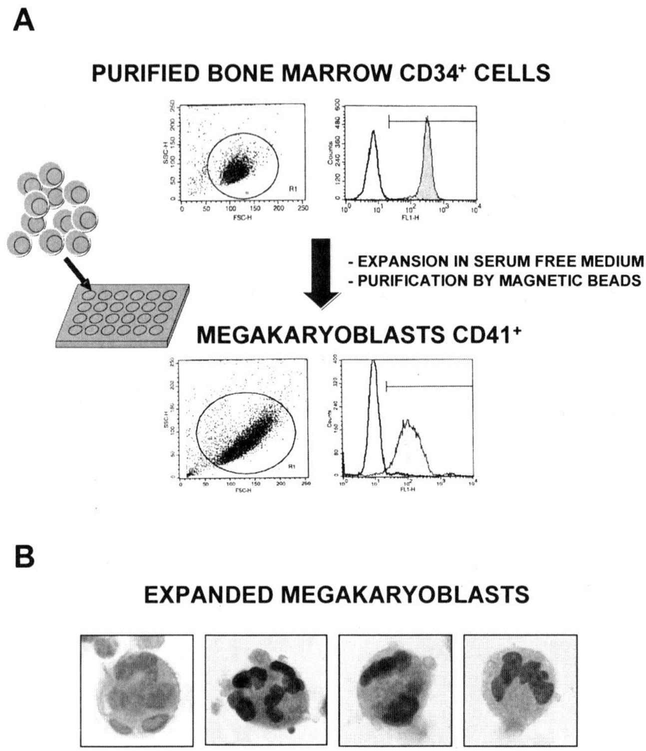

Fig. 2. Expansion system for human megakaryoblasts. Purified by immunomagnetic beads CD34$^+$ cells are cultured in liquid serum-free cultures for 11 days in a presence of TpO or TpO + IL-3. **Panel A** – upper part - purity of isolated CD34$^+$ cells for expansion, lower part - purity of expanded megakaryoblasts (CD41$^+$ cells). **Panel B** – Giemsa staining of expanded ex vivo Megs.

for the i) interaction of Megs with endothelium, ii) proplatelet formation and iii) platelet shedding which are enhanced by C3 cleavage fragments during reactive thrombocytosis.

5 Extensive bleeding in C3-deficient mice leads to a lower platelet count and in normal mice causes activation of C3

Given that excessive bleeding leads to reactive thrombocytosis (Schafer AI 2001; Christenson JT 1999), and C becomes activated in several stress-related situations

(Goldfarb RD and Parrillo JE 2005) our aim was to determine whether C3 cleavage fragments might be implicated in post-bleeding thrombocytosis. To address this we first investigated the kinetics of platelet counts after excessive bleeding (20% of blood vol.) in wild type (wt) and C3-deficient mice and, in parallel, measured C activation by detecting the level of C3 cleavage fragment ($_{desArg}$C3a) in the peripheral blood of bled animals.

We noticed that both wt and C3-deficient mice display a biphasic increase in peripheral blood counts at days 1-2 and 6-10 (Wysoczynski M et al. 2007). While the first elevation in platelet counts may reflect the supply of platelets by Megs located in the BM endothelial niche, the second elevation is likely related to generation of new Megs from a pool of CFU-Meg that are attracted to this niche from the osteoblastic niche. Although C3-deficient mice have normal platelet counts in normal steady state conditions, they generated fewer platelets after excessive bleeding as compared to wt littermates.

6 Serum-free expansion systems for human megakaryoblasts

To study an influence of C3 cleavage fragments on Megs we employed a serum-free expansion model. The serum-free liquid culture system that we employ in our laboratory to expand human megakaryoblasts *ex vivo* is depicted in Figure 2. The cultures are initiated with bone marrow- or cord blood-derived CD34$^+$ cells purified by MiniMacs magnetic beads (Figure 2).

In this expansion system highly purified CD34$^+$ cells are placed in Iscove's DMEM supplemented with 25% of artificial serum (delipidated albumin, iron-saturated transferrin, liposomes and insulin) and stimulated with TpO or TpO + IL-3. If TpO alone is employed as a stimulator pure megakaryoblasts grow in these cultures (~100% of expanded cells express the CD41 antigen, $\alpha_{IIb}\beta_3$ integrin). The number of cells generated in the presence of TpO only, however, is relatively low and consequently, to increase the expansion efficiency of megakaryocytic cells, we use a combination of TpO + interleukin-3 (IL-3).

7 C3a and $_{desArg}$C3a do not affect proliferation and survival of Megs

Next, we investigated whether C3a and $_{desArg}$C3a could influence the proliferation/survival of normal human Megs. Highly purified CD34$^+$ cells were cultured in a serum-free liquid cultures or serum-free methylcellulose cloning medium with various doses of C3a or $_{desArg}$C3a (0.1–10 μg/mL) and stimulated to grow CFU-Meg colonies with sub-optimal (10 ng/ml) and optimal (100 ng/ml) doses

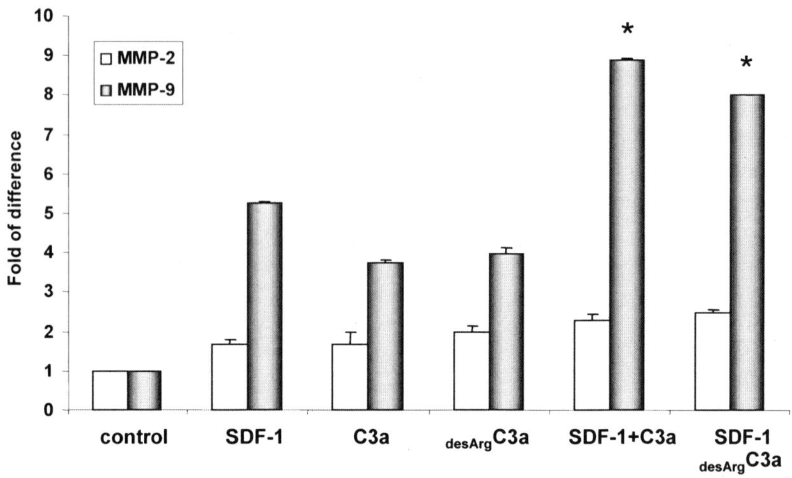

Fig. 3. C3a and $_{desArg}$C3a enhance expression of MMP-9 by Megs. Real-time analysis of MMP-2 and MMP-9 expression in human Megs stimulated by SDF-1 (300 ng/ml), C3a (1 µg/mL), $_{desArg}$C3a (1 µg/mL), SDF-1 + C3a or SDF-1 + $_{desArg}$C3a. The data represent the average of four independent experiments. Data are mean ± SD. * $P<0.00001$ vs. SDF-1 alone or control. ** $P<0.00001$ vs. control.

of TPO alone or TPO + SDF-1. In another set of experiments we incubated Megs for 24 h in serum-free medium in the presence or absence of C3a or $_{desArg}$C3a (0.1–10 µg/mL) and estimated cell viability/Annexin V binding after this period of time. In all of these experiments, however, no effect of C3a or $_{desArg}$C3a on the proliferation/survival of Megs was observed (Reca R et al. 2003).

8 C3a and $_{desArg}$C3a increase responsiveness of Megs to an SDF-1 gradient and enhance SDF-1-dependent migration and adhesion

Although C3a and $_{desArg}$C3a alone did not chemoattract Megs in a transwell system, we observed that they did increase their response to a low, "threshold" dose of SDF-1. In the presence of C3a or $_{desArg}$C3a the chemotactic response of Megs to a threshold dose of SDF-1 was comparable to that at a high dose of this chemokine (Reca R et al. 2003; Wysoczynski M et al. 2007).

Next, since SDF-1 enhances adhesion of Megs to fibrinogen, we focused on the effect of C3 cleavage fragments on SDF-1-dependent adhesion of Megs to fibrinogen (Majka M et al. 2000) and found that C3a and $_{desArg}$C3a significantly enhanced SDF-1 dependent adhesion of Megs. At the same time adhesion of Megs was also slightly enhanced in the presence of C3a alone, but not $_{desArg}$C3a. This C3a-induced adhesion of Megs correlated with the fact that C3a employed alone was able to enhance F-actin polymerization in these cells (Wysoczynski M et al. 2007).

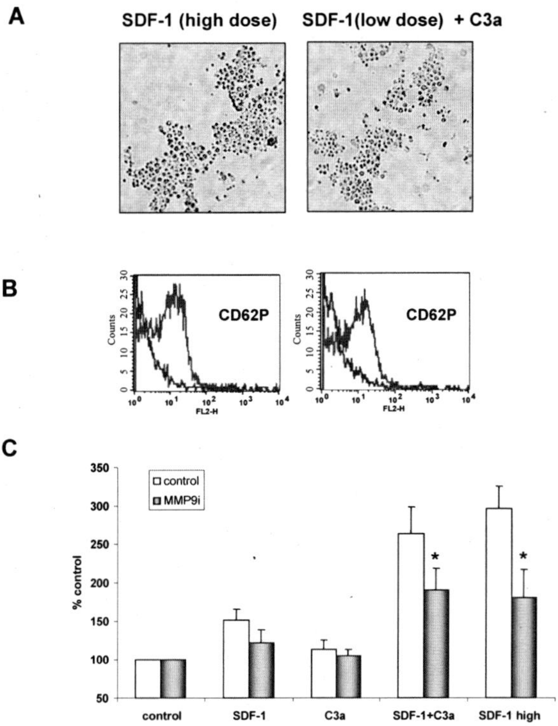

Fig. 4. C3 cleavage fragments enhance SDF-1-dependent platelet production in vitro. Panel A – platelets produced in the presence of SDF-1 (300 ng/ml) or SDF-1 (10 ng/ml) + C3a (1mg/ml). **Panel B** – platelets produced in our model were evaluated in a functional assay. Stimulation by Thrombin 1U/ml for 5' induces expression of CD62P (red line). **Panel C** - Platelet production in a trans-endothelial migration assay by normal human Megs (control) and Megs that were preincubated with MMP-9 inhibitor (MMP9i) (5-phenylo-1,10-phenanthroline monohydrate, 5mM for 30 minutes before assay). Platelets were identified and analyzed by FACS. Data are pooled from quadruplicate samples from two independent experiments. $^*P<0.0001$.

As migration of Megs in the BM environment is dependent on the secretion of MMP-9 (Majka M et al. 2000; Lane WJ, Dias S, Hattori K, Heissig B, Choy M, Rabbany SY, Wood J, Moore MA, Rafii S 2000) and VEGF-mediated interaction with endothelium in the BM endothelial niche (Hamada T et al. 1998), we examined the influence of C3a and $_{desArg}$C3a on MMP-9 and VEGF expression in these cells. We found that C3 cleavage fragments alone only slightly up regulated the expression of MMP-9 in normal megakaryocytic cells; however, when C3a or $_{desArg}$C3a were used together with SDF-1, expression of MMP-9 was up regulated in a synergistic manner (Figure 3). Furthermore, C3a and $_{desArg}$C3a increased the production of VEGF by human Megs (Wysoczynski M et al. 2007).

9 C3a and $_{desArg}$C3a enhance SDF-1-dependent platelet production

It is widely accepted that Megs produce platelets by extending proplatelet protrusions between endothelial cells (Hamada T et al. 1998) and it had been recently reported that SDF-1 enhances platelet formation by Megs that are in contact with endothelium (Hamada T et al. 1998). To address whether this process could be affected by C3a and $_{desArg}$C3a we evaluated platelet production in a transwell-migration assay (Lane WJ et al. 2000). In this assay transwell membranes were covered by confluent IL-1β activated HUVEC and megakaryocytic cells derived from day 12 expansion cultures were placed in the upper chambers; C3a, $_{desArg}$C3a and SDF-1 together or alone were placed in the lower chambers. After 48 h the number of platelets present in the lower chambers was evaluated by FACS after staining with CD41 Abs. Furthermore, in control experiments platelets were evaluated both morphologically by microscopy and functionally by evaluating the increase in expression of CD62P after stimulation with thrombin. We found that C3a and $_{desArg}$C3a, if combined with SDF-1, significantly enhanced platelet formation by human Megs (Figure 4 panel A). Since production of platelets by Megs is MMP-9 dependent (Hamada T et al. 1998), we pre-incubated in control experiments human Megs before the assay with MMP-9 inhibitor (5-phneyl-1,10-phenanthroline monohydrate, 5mM for 30 minutes) and as expected this treatment significantly inhibited platelet formation after stimulation by SDF-1 or SDF-1 + C3a (Figure 4 panel B).

10 Molecular analysis of signaling pathways activated by C3a and $_{desArg}$C3a in normal megakaryoblasts

Having demonstrated that both C3a and $_{desArg}$C3a enhance SDF-1 dependent migration and adhesion of megakaryocytic cells, we turned our attention to signaling pathways that could regulate these processes. First, we observed that, like SDF-1 (Majka M et al. 2000), C3a and $_{desArg}$C3a induced phosphorylation of MAPK p42/44 in normal human Megs (Wysoczynski M et al. 2007).

Further, given our finding that C3a, like SDF-1, activates F-actin polymerization in Megs, we sought to identify which small GTP-ases from the Rho family, Rho, Rac or cdc42, which are known to mediate F-actin reorganization, are involved in this process. We found that C3a but not $_{desArg}$C3a activated Rho (Figure 5 panel A), which corresponds with our observation that C3a but not $_{desArg}$C3a induces F-actin polymerization in Megs.

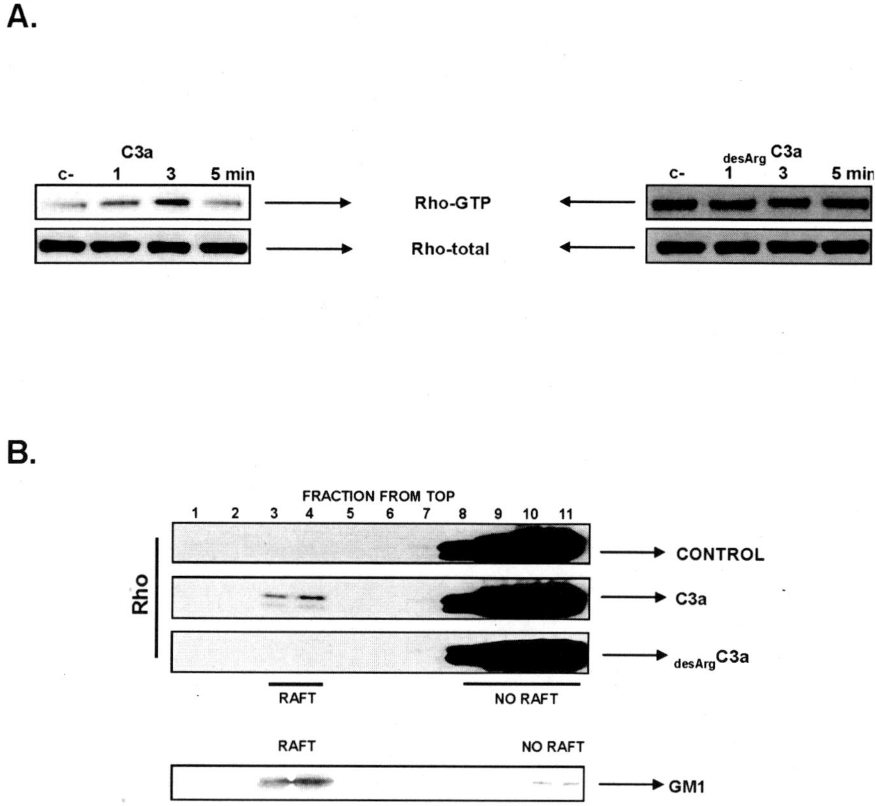

Fig. 5. Western blot analysis of C3a and $_{desArg}$C3a-dependent activation of Rho. Panel A – C3a but not $_{desArg}$C3a activates Rho GTP-ase. The detection of the GTP bound (active) form of Rho in human Megs not stimulated (c), or stimulated by C3a or $_{desArg}$C3a. Experiments were performed three times with similar results. Representative blots are shown. **Panel B** - Analysis of the co-localization of Rho in various fractions of cell membranes enriched in lipid rafts (fractions 3–5) and depleted of lipid rafts (fractions 9–11). Meg cell line MO7E cells were stimulated by C3a (1 μg/mL) or $_{desArg}$C3a (1 μg/mL) or not stimulated (control). Rho was detected in these membrane fractions by Western blot along with ganglioside M1 (GM1), a marker of lipid rafts. Experiments were performed three times with similar results. A representative blot is shown.

In our previous investigations on hematopoietic stem/progenitor cells, we reported that C3a and $_{desArg}$C3a may enhance incorporation of CXCR4 into membrane lipid rafts (Ratajczak MZ et al. 2004; Wysoczynski M, Reca R, Ratajczak J, Kucia M, Shirvaikar N, Honczarenko M, Mills M, Wanzeck J, Janowska-Wieczorek A, Ratajczak MZ 2005) thereby allowing this receptor to form an optimal association with downstream signaling proteins. To determine whether a similar mechanism plays a role in megakaryocytic lineage cells, we evaluated the incorporation of CXCR4 into membrane lipid rafts in the human Meg line MO7E stimulated with C3a or $_{desArg}$C3a.

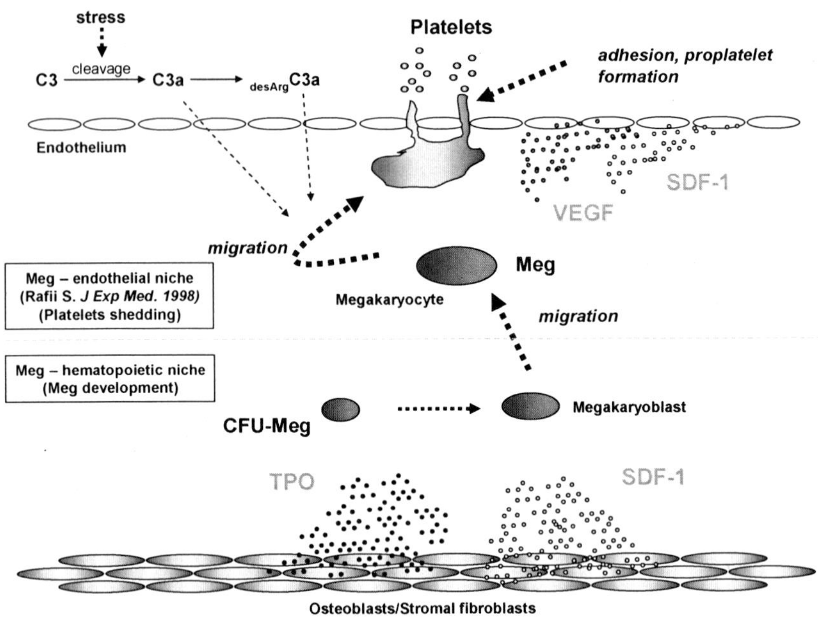

Fig. 6. C3a and $_{desArg}$C3a modulate SDF-1-dependent attraction of Megs from bone niche to endothelial niche and platelet production. This reallocation is tightly regulated by an SDF-1 gradient and we envision that during stressed situations (e.g. chronic inflammation, hemorrhage), C3 cleavage fragments are important co-regulators of this process.

As expected, Western blot analysis revealed that both molecules enhanced incorporation of CXCR4 into membrane lipid rafts (Wysoczynski M et al. 2007). Furthermore, Western blot analysis performed on MO7E cells revealed that C3a but not $_{desArg}$C3a both activates and enhances the incorporation of Rho into membrane lipid rafts (Figure 5 panel B). Thus C3a-mediated activation) and incorporation of both CXCR4 and Rho into membrane lipid rafts correlated with an increase in F-actin polymerization in Megs stimulated by C3a alone or C3a + SDF-1 (Figure 3).

11 Platelet formation is lipid raft dependent

C3 cleavage fragments enhance incorporation of CXCR4 into membrane lipid rafts allowing optimal SDF-1 signaling by this receptor (Nguyen DH and Taub D 2002; Christenson JT 1999). Thus, we determined whether perturbation of lipid raft formation (by depletion of membrane cholesterol) affects SDF-1-dependent platelet formation (Nguyen DH et al. 2002; Christenson JT 1999). When human Megs were preincubated for 1 h in 5mM MβCD, which depletes cholesterol from cell membranes (Ratajczak MZ et al. 2004; Nguyen DH et al. 2002), we found that perturbation of lipid raft formation by MβCD significantly inhibited platelet formation dependent on both SDF-1 + C3a and SDF-1 + $_{desArg}$C3a (Wysoczynski M et al. 2007).

12 Conclusions

In conclusion, both C3a and $_{des\text{-}Arg}$C3a clearly have a role in modulating the responsiveness of Megs to an SDF-1 gradient. The crosstalk between the C3aR and CXCR4-G-protein coupled receptors we identified here with respect to Megs plays an important and previously unrecognized role in stress and inflammation-dependent platelet formation. We hypothesize that a crucial role is played here by the reallocation of Megs from the osteoblastic to endothelial niche (Figure 6). This reallocation is tightly regulated by an SDF-1 gradient and we envision that during stressed situations (e.g., chronic inflammation, hemorrhage), C3 cleavage fragments are important co-regulators of this process. Accordingly, we propose that C3 is cleaved/activated in BM and as result of this C3 cleavage fragments (C3a and $_{desArg}$C3a) modulate/enhance the responsiveness of maturing Megs to an SDF-1 gradient. As shown, Megs are chemoattracted by SDF-1 expressed by marrow endothelial cells and re-allocate from the bone-marrow niche to the endothelial niche where proplatelets are formed and platelets shed from the Meg surface (Avecilla ST et al. 2004). At the molecular level we envision that C3 cleavage fragments sensitize responsiveness of Megs to SDF-1 by enhancing its inclusion into membrane lipid rafts. CXCR4 if included into membrane lipid rafts, regulates Megs migration more efficiently as well as adhesion to endothelium and secretion of MMP-9 and VEGF. All of these processes are crucial for proplatelet formation and platelet shedding and are enhanced by C3 cleavage fragments during reactive thrombocytosis.

On the basis of this new information we foresee that inhibitors of the CXCR4–SDF-1 axis and C3 activation have clinical potential as drugs to control certain forms of thrombocytosis (e.g. as described in postoperative thrombocytosis in patients after coronary bypass grafting) (Christenson JT 1999). Furthermore, however statins show several pleiotropic effects on platelets (e.g. inhibit their activation), given a fact that platelet formation is dependent upon membrane lipid raft formation as we described here, statins, which perturb lipid raft formation, may also ameliorate stress-related thrombocytosis and thus could be considered as potential drugs to prevent bleeding-related post-trauma or post-operative thrombocytosis.

13 Acknowledgements

This work was supported by an NIH grant R01 DK074720-01 to MZR.

References

Allendorf DJ, Ostroff G, Baran JT, Dyke CW, Ratajczak MZ, Ross GD. (2003) Oral WGP β glucan treatment accelerates myeloid recovery and survival after radiation exposure. In: Unified Science & Technology for Reducing Biological Threats & Countering Terrorism, p. 104–113.

Avecilla ST, Hattori K, Heissig B, Tejada R, Liao F, Shido K, Jin DK, Dias S, Zhang F, Hartman TE, Hackett NR, Crystal RG, Witte L, Hicklin DJ, Bohlen P, Eaton D, Lyden D, de Sauvage F, Rafii S. (2004) Chemokine-mediated interaction of hematopoietic progenitors with the bone marrow vascular niche is required for thrombopoiesis. Nat Med.;**10**:64-71.

Bunting S, Widmer R, Lipari T, Rangell L, Steinmetz H, Carver-Moore K, Moore MW, Keller GA, de Sauvage FJ. (1997) Normal platelets and megakaryocytes are produced in vivo in the absence of thrombopoietin. Blood;**90**:3423-9.

Burstein SA, Mei RL, Henthorn J, Friese P, Turner K. (1992) Leukemia inhibitory factor and interleukin-11 promote maturation of murine and human megakaryocytes in vitro. J Cell Physiol. **153**:305-12.

Christenson JT. (1999) Preoperative lipid-control with simvastatin reduces the risk of postoperative thrombocytosis and thrombotic complications following CABG. Eur J Cardiothorac Surg.;**15**:394-349.

Coombe D.R., Watt S.M., Parish C.R (1994) Mac-1 (CD11b/CD18) and CD45 mediate the adhesion of hematopoietic progenitor cells to stromal cell elements via recognition of stromal heparan sulfate, *Blood* **84**; 739–752.

Crass T, Raffetseder U, Martin U, Grove M, Klos A, Kohl J, Bautsch W. (1996) Expression cloning of the human C3a anaphylatoxin receptor (C3aR) from differentiated U-937 cells, Eur J Immunol 26; 1944–1950

Gainsford T, Nandurkar H, Metcalf D, Robb L, Begley CG, Alexander WS. (2000) The residual megakaryocyte and platelet production in c-mpl-deficient mice is not dependent on the actions of interleukin-6, interleukin-11, or leukemia inhibitory factor. Blood; **95**:528-34.

Gewirtz AM. (1995) Megakaryopoiesis: state of the art. Thromb Haemost.**74**:204-209.

Goldfarb RD, Parrillo JE. (2005) Complement. Crit Care Med.; **33**:482-484.

Gurney AL, Carver-Moore K, de Sauvage FJ, Moore MW. (1994) Thrombocytopenia in c-mpl-deficient mice. Science. **265**:1445-1447.

Gurney AL, de Sauvage FJ. (1996) Dissection of c-Mpl and thrombopoietin function: studies of knockout mice and receptor signal transduction. Stem Cells.**14 Suppl 1**:116-23.

Hamada T, Mohle R, Hesselgesser J, Hoxie J, Nachman RL, Moore MA, Rafii S. (1998) Transendothelial migration of megakaryocytes in response to stromal cell-derived factor 1 (SDF-1) enhances platelet formation. J Exp Med.;**188**:539-48.

Hodohara K, Fujii N, Yamamoto N, Kaushansky K. (2000) Stromal cell-derived factor-1 (SDF-1) acts together with thrombopoietin to enhance the development of megakaryocytic progenitor cells (CFU-MK). Blood;**95**:769-75.

Kowalska MA, Ratajczak J, Hoxie J, Brass LF, Gewirtz A, Poncz M, Ratajczak MZ. (1999) Megakaryocyte precursors, megakaryocytes and platelets express the HIV co-receptor CXCR4 on their surface: determination of response to stromal-derived factor-1 by megakaryocytes and platelets. Br J Haematol.;**104**:220-229.

Lane WJ, Dias S, Hattori K, Heissig B, Choy M, Rabbany SY, Wood J, Moore MA, Rafii S. (2000) Stromal-derived factor 1-induced megakaryocyte migration and platelet production is dependent on matrix metalloproteinases. Blood.;**96**:4152-9.

Long MW, Hoffman R. (1999) Hematology: basic principles and practice. Eds. Churchill Livingstone. 245

Majka M, Ratajczak J, Villaire G, Kubiczek K, Marquez LA, Janowska-Wieczorek A, Ratajczak MZ. (2002) Thrombopoietin, but not cytokines binding to gp130 protein-coupled receptors, activates MAPKp42/44, AKT, and STAT proteins in normal human CD34+ cells, megakaryocytes, and platelets. Exp Hematol.;**30**:751-60

Majka M, Janowska-Wieczorek A, Ratajczak J, Kowalska MA, Vilaire G, Pan ZK, Honczarenko M, Marquez LA, Poncz M, Ratajczak MZ. (2000) Stromal-derived factor 1 and thrombopoietin regulate distinct aspects of human megakaryopoiesis. Blood.;**96**:4142-51.

Neben TY, Loebelenz J, Hayes L, McCarthy K, Stoudemire J, Schaub R, Goldman SJ. (1993) Recombinant human interleukin-11 stimulates megakaryocytopoiesis and increases peripheral platelets in normal and splenectomized mice. Blood.;**81**:901-908.

Nguyen DH, Taub D. (2002) CXCR4 function requires membrane cholesterol: implications for HIV infection. J Immunol.;**168**:4121-6.

Ratajczak MZ, Reca R, Wysoczynski M, Kucia M, Baran JT, Allendorf DJ, Ratajczak J, Ross GD. (2004) Transplantation studies in C3-deficient animals reveal a novel role of the third complement component (C3) in engraftment of bone marrow cells. Leukemia.;**18**:1482-90.

Ratajczak J, Reca R, Kucia M, Majka M, Allendorf DJ, Baran JT, Janowska-Wieczorek A, Wetsel RA, Ross GD, Ratajczak MZ. (2004) Mobilization studies in mice deficient in either C3 or C3a receptor (C3aR) reveal a novel role for complement in retention of hematopoietic stem/progenitor cells in bone marrow. *Blood* **103**:2071-8.

Ratajczak M.Z., Zuba-Surma E, Kucia M, Reca R, Wojakowski W, Ratajczak J (2006) The pleiotropic effects of the SDF-1-CXCR4 axis in organogenesis, regeneration and tumorogenesis. *Leukemia* **20**:1915-24.

Reca R, Mastellos D, Majka M, Marquez L, Ratajczak J, Franchini S, Glodek A, Honczarenko M, Spruce LA, Janowska-Wieczorek A, Lambris JD, Ratajczak MZ. (2003) Functional receptor for C3a anaphylatoxin is expressed by normal hematopoietic stem/progenitor cells, and C3a enhances their homing-related responses to SDF-1. Blood.;**101**:3784-93.

Riviere C, Subra F, Cohen-Solal K, Cordette-Lagarde V, Letestu R, Auclair C, Vainchenker W, Louache F. (1999) Phenotypic and functional evidence for the expression of CXCR4 receptor during megakaryocytopoiesis. Blood.;**93**:1511-23.

Sasaki H, Hirabayashic Y, Ishibashid T, Inoueb T, Matsudaa M, Kaia S, Ikutaa S, Yokoyamad K, Yokotac T, Maruyamad Y and Matsuyamaa S. (1995) Effects of erythropoietin, IL-3, IL-6 and LIF on a murine megakaryoblastic cell line: growth enhancement and expression of receptor mRNAs. Leuk Res.;**19**:95-102

Schafer AI. (2001) Thrombocytosis and thrombocythemia. Blood Rev.;**15**:159-66.

Teixido J., Hemler M.E., Greenberger J.S, Anklesaria P. (1992) Role of β 1 and β 2 integrins in the adhesion of human $CD34^{hi}$ stem cells to bone marrow stroma, *J Clin Invest* **90**:358–367.

Tornetta M.A., Foley J.J., Sarau H.M., Ames R.S. (1997) The mouse anaphylatoxin C3a receptor: molecular cloning, genomic organization, and functional expression, *J Immunol* **158**: 5277–5282.

Volanakis J.E., Frank M.M. (1998) The human complement system in health and disease, NetLibrary Inc., New York

Walport M.J. (2001) Complement. First of two parts, *N Engl J Med* **344:**1058–1066.

Walport M.J. (2001) Complement. Second of two parts, *N Engl J Med* **344:**1140–1144.

Wang JF, Liu ZY, Groopman JE. (1998) The alpha-chemokine receptor CXCR4 is expressed on the megakaryocytic lineage from progenitor to platelets and modulates migration and adhesion. Blood.;**92**:756-64

Wysoczynski M, Reca R, Ratajczak J, Kucia M, Shirvaikar N, Honczarenko M, Mills M, Wanzeck J, Janowska-Wieczorek A, Ratajczak MZ. (2005) Incorporation of CXCR4 into membrane lipid rafts primes homing-related responses of hematopoietic stem/progenitor cells to an SDF-1 gradient. Blood **105**:40-8.

Wysoczynski M, Kucia M, Ratajczak J, Ratajczak M.Z. (2007) Cleavage fragments of third complement component (C3) enhance SDF-1 mediated platelet production during reactive thrombocytosis. *Leukemia 21*:(in press).

17

In Vivo Biological Responses in the Presence or Absence of C3

J. Vidya Sarma and Peter A. Ward

The University of Michigan Medical School, Department of Pathology,
1150 West Medical Center Drive Ann Arbor, MI 48109-0602

1 Introduction

As is generally known, C3 is a paramount protein in the complement system and is required for complement activation that occurs via the classical, alternative and lectin pathways (Guo and Ward 2005). Of the activation products, our attention has been especially focused on C5a which is a potent anaphylatoxin inducing a variety of different biological responses. The classical C5a receptor, C5aR, is known to exist on many cell types in addition to myelopoietic cells. For instance, C5aR exists on a variety of epithelial cells such as endothelial cells, hepatocytes, lung alveolar epithelial cells and other cell types (Huber-Lang, Sarma, McGuire, Lu, Padgaonkar, Younkin, Guo, Weber, Zuiderweg, Zetoune and Ward 2003). C5aR is a rhodopsin type receptor with signaling pathways that include activation of MAP kinases (ERK 1/2, p38 and JNK 1/2) (Riedemann, Guo, Hollmann, Gao, Neff, Rueben, Speyer, Sarma, Wetsel, Zetoune and Ward 2004). The recently described C5a non-signaling receptor, C5L2, has been termed the "C5a default receptor" because of the absence of intracellular calcium responses to C5a and lack of evidence for other signaling pathway responses (Okinaga, Slattery, Humbles, Zsengeller, Morteau, Kinrade, Brodbeck, Krause, Choe, Gerard and Gerard 2003). The precise function of C5L2 is unknown. Its putative ability to bind non-C5a peptides such as C3a, C3a-desArg, etc, remains controversial. A recent publication describes that C5L2 does not bind C3a or C3a-desArg (Johswich, Martin, Thalmann, Rheinheimer, Monk and Klos 2006). It has been suggested that C3a ligation of C5L2 results in triglyceride synthesis, but this too is a subject of debate (Maslowska, Legakis, Assadi and Cianflone 2006).

In view of the centrality of C3 in the complement pathways, it was widely assumed that genetic absence or the presence of a non-functional mutant version of C3 would be embryonically lethal or associated with inability to survive post partum.

However, it is now known that humans who have genetically based C3 deficiency may have impaired host defense responses but in general are capable of relatively long life (Reis, Nudelman and Isaac 2004). Mice with genetic deficiency of C3 have little recognizable phenotypic characteristics when housed in the traditional animal facility setting, but it is clear that C3-deficient mice do indeed have problems with respect to defenses against bacterial infections (Mueller-Ortiz, Drouin and Wetsel 2004). The purpose of this report is to emphasize conditions in which the absence of C3 may or may not clearly affect a biological response.

2 Survival Advantage of C3 in Experimental Sepsis

It is now quite apparent that the genetic absence of C3 severely disposes rodents to adverse outcomes in the setting of experimental sepsis, whether after cecal ligation and puncture (CLP) or after infusion of live bacteria. It is known that bacterial challenge of C3-deficient mice leads to a high level of mortality after infusion of group B Streptococci (Wessels, Butko, Ma, Warren, Lage and Carroll 1995). Our studies have also investigated resistance to bacterial induced sepsis in rats using the CLP model in which perforation of the ligated cecum results in fecal release into the peritoneal cavity, leading to progressive sepsis (Czermak, Sarma, Pierson, Warner, Huber-lang, Bless, Schmal, Friedl and Ward 1999). As shown in Table 1, the survival in C3-depleted rats is greatly affected in an adverse manner when animals were depleted of C3 by the intraperitoneal injection of purified cobra venom factor. With C3-intact rats, CLP produced 50% survival rate by day 2, 20% survival by day 5 and 10% survival by day 10. In rats that were infused with blocking antibody to rat C5a, mortality rates were greatly reduced, with 85% of animals surviving by day 2, 70% surviving by day 5 and the same survival at day 10. In dramatic contrast, in C3-depleted rats that had less than 5% C3 as measured in sera, only 20% were alive at day 2, with 100% mortality by day 5. These data clearly indicate that under such experimental conditions survival after CLP is greatly enhanced by blockade of C5a and greatly reduced in animals depleted of C3.

3 Changes in C5a Receptor content in Blood Neutrophils (PMNs) after CLP or after In Vitro Exposure of Human PMNs to C5a

As shown in Table 2, surface expression of C5aR was evaluated by flow cytometry in blood PMNs isolated after induction of sepsis by CLP in rats (Guo, Riedemann, Bernacki, Sarma, Laudes, Reuben, Younkin, Neff, Paulauskis, Zetoune and Ward 2003). Plasma content of C5a was evaluated by ELISA. As shown, 4 hours after CLP there were dramatic reductions in C5aR content on blood PMNs, dropping by 71% when compared to sham blood PMNs. C5aR levels fell by 79% at 12 hours and by 88% at 24 hours, with a gradual restoration at 36 and 48 hours. However, at 48 hours

Table 1. Survival of Intact C3 and C3-depleted Rats after Cecal Ligation and Puncture (CLP)*.

Groups*	Survival (%) Days after CLP		
	2	5	10
C3 intact	50	20	10
C3 intact + anti-C5a	80	70	70
C3 depleted	20	0	0

n > 10 per group

there was still a 47% decrease in C5aR content on blood PMNs. Not shown in this table is the fact that interception *in vivo* of C5a with blocking antibody to C5a greatly attenuated reductions in C5aR content on blood PMNs from CLP rats. Accordingly, the dramatic drop in C5aR on blood neutrophil surfaces after CLP can be linked to the production of C5a. Changes in C5a content in plasma (as determined by ELISA) after CLP are also shown in Table 2. Very little increase in C5a was found 4 and 12 hours after CLP, at which time presumably most of the C5a generated likely resulted in binding of C5a to blood PMNs. However, at 24, 36 and 48 hours, the levels of C5a rose nearly threefold during this period of time and then gradually dropped. The reasons for the loss of C5aR on surfaces of blood PMNs after CLP might be due to the well known fact that C5aR ligation with C5a results in internalization of the ligand-receptor complex with the ultimate recycling of C5aR to the cell surface. However, in data not shown, we have demonstrated that, using lysates of blood PMNs from CLP rats, the total measurable cellular C5aR protein is greatly diminished based on determination by Western blot. This would suggest that in some manner C5aR is being released from the neutrophil surface as a result of *in vivo* ligation of the receptor with C5a during the course of sepsis induced by CLP. Alternatively, the N terminal region of C5aR to which the antibody is reactive has been lost, perhaps through proteolytic degradation.

We were interested in changes in C5a receptors occurring as a result of *in vitro* incubation of human blood PMNs with C5a (Huber-Lang, Sarma, Rittirsch, Schreiber, Weiss, Flierl, Younkin, Schneider, Suger-Wiedeck, Gephard, McClintock, Neff, Zetoune, Bruckner, Guo, Monk and Ward 2005). The data in Table 3 relate to this issue. Flow cytometry was used to measure surface expression of C5L2 on human PMNs. In cells that had been incubated *in vitro* at 37°C with 100 nM recombinant human C5a for 1-4 hours, at one hour there was little evidence of any change in C5L2 content on neutrophil surfaces, but by 3 hours and 4 hours the ability to detect surface C5L2 fell by 24% and 46%, respectively. Since it is known that ligation of C5L2 does not result in internalization of the receptor-ligand complex (Okinaga, et al 2003), the conclusion would be that either something has interfered with the detection of C5L2 or C5L2 has been shed from the surface of the PMN. Similar findings have been

Table 2. Changes in Blood PMN C5aR Content and Plasma C5a Level.

	Time (hr) after CLP				
Parameter	4	12	24	36	48
Change (%) in C5aR content on blood PMNs*	−71%	−79%	−88%	−76%	−47%
Change (fold) in plasma content of C5a**	+1.1	+1.1	+2.9	+3.0	+2.6

*Compared to C5aR content of sham blood PMNs.
**Compared to normal plasma which had a baseline C5a value of 2.6 nM.

obtained with C5aR content on blood PMNs both in rats after CLP (Table 2) and in companion studies in which blood PMNs were incubated *in vitro* with C5a (data not shown). The data in Table 4 (recently published (Guo, et al 2003)) indicate the changes in C5L2 of blood PMNs in humans with septic shock. When compared to PMNs from healthy controls, C5L2 levels on blood PMNs from septic humans were markedly reduced. Collectively, patients with septic shock ($n > 40$) showed dramatic reductions in detectable C5L2 on blood PMN surfaces, with a mean reduction of 89%. When these data were broken into groups with or without multi-organ failure (MOF), the group with no MOF had a mean reduction of C5L2 by 61%, while the septic group with MOF had a mean 95% reduction in C5L2, which was highly statistically significant. These data therefore suggest that C5L2 on blood PMNs is dramatically reduced in the course of sepsis in humans with septic shock. Similar changes have been found in the case of C5aR (data not shown). These data further suggest that changes in C5a receptors occurring in rodents with CLP are similar to changes found in humans with septic shock. The precise mechanisms to explain these changes in C5a receptors are poorly understood.

4 Sepsis Induced Defects in Innate Immune Function

It has been known for some time that humans with sepsis as well as septic animals have evidence of dysfunction involving phagocytic cells, although in the case of humans adequate information is not available regarding the time course and other details that might correlate with the intensity of innate immune dysfunction. Blood PMNs were obtained from septic rats 36 hours after the onset of CLP and examined for *in vitro* innate immune responses, including PMN chemotactic movement in response to 20 nM C5a, phagocytosis of opsonized zymosan particles and H_2O_2 production by PMA stimulated PMNs (Huber-Lang, Younkin, Sarma, McGuire, Lu,

Table 3. Changes in C5L2 in Human PMNs Exposed to C5a In Vitro.

Exposure of human PMNs	Change (%) in C5L2 compared to control PMNs
HBSS, 4 hr	0
10 nM C5a	
1 hr	+ 8%
3 hr	−24%
4 hr	−46%

Table 4. Changes in C5L2 on Human PMNs during Septic Shock.

Sepsis condition	**Change relative to normal PMNs**		
All patients	−89%	P<0.01 (All patients vs. Septic patients, no MOF)	P<0.05 (All patients vs. Septic patients, with MOF)
Septic patients, no MOF	−61%	P<0.05 (Septic patients, no MOF vs. with MOF)	
Septic patients, with MOF	−95%		

Guo, Padgaonkar, Curneutte, Erickson and Ward 2002). As shown in Table 5, chemotactic responses were reduced by 70% in the setting of CLP. It should be emphasized that this reduction in chemotactic responsiveness appeared to be global because similar defects were found in the response of the same cells to fMLP (Huber-Lang, Sarma, Lu, McGuire, Padgaonkar, Guo, Younkin, Kunkel, Ding, Erickson, Curnutte and Ward 2001). In addition, if CLP animals received blocking antibody to C5a at the beginning of CLP, chemotactic responses remain largely intact (Huber-Lang, et al 2001). In terms of phagocytosis, uptake of opsonized zymosan particles by blood PMNs was reduced by about 60% in PMNs obtained from CLP rats; most of the functional activity was preserved in anti-C5a-treated CLP rats (data now shown). Finally, H_2O_2 production induced by PMA treatment of CLP PMNs was diminished by 55%. These innate immune responses were largely preserved in animals receiving blocking antibody to C5a at the start of CLP (Huber-Lang, et al 2001).

Table 6 describes some of the defects in PMNs exposed *in vitro* to 100nM C5a. The details of these studies are described in ref (Huber-Lang, et al 2002). What is very clear is that developments of defects are dependent on the dose of C5a employed and duration of exposure (up to 90 minutes). There was a progressive loss in

Table 5. Sepsis-induced Defects in Innate Immune Functions of Blood PMNs from CLP Rats.

Parameter measured	Change in response*
Chemotaxis (to C5a)	−70%
Phagocytosis (opsonized zymosan particles)	−60%
H_2O_2 production (induced by PMA)	−55%

* All responses are relevant to responses of PMNs from CLP rats treated with anti-C5a at the time of CLP.

Table 6. *In Vitro* Induction by C5a of Defects in Human Blood PMNs.

1. Loss of phosphorylation of p42 MAPK.
2. Loss of phosphorylation of $p47^{phox}$.
3. Loss of translocation of $p47^{phox}$ to cell membrane.

responses of blood PMNs to PMA as defined by production of H_2O_2. Reasons for these defective innate immune responses can be linked to at least 3 defects as described in Table 6. Firstly, *in vitro* exposure of blood PMNs to C5a resulted in the loss of phosphorylation of p42 MAP kinase, which is an important signaling pathway in PMNs. Secondly, a defect occurring as a result of *in vitro* exposure of PMNs to C5a was the loss of phosphorylation of $p47^{phox}$ when cells were then exposed to PMA. $p47^{phox}$ is one of several cytosolic subunits that must be translocated to the cell membrane for full assembly of NADPH oxidase (Curnutte, Erickson, Ding and Badwey 1994), which is necessary for generation of H_2O_2 when PMNs are stimulated with PMA. The inability to phosphorylate the $p47^{phox}$ is associated with the loss of the translocation of $p47^{phox}$ to the cell membrane, thus preventing full assembly of NADPH oxidase on the PMN cell membrane. We have demonstrated that, in CLP rats, blood PMNs obtained 48 hours after the onset of CLP and stimulated *in vitro* with PMA had a markedly impaired ability to translocate $p47^{phox}$ to the cell membrane. In striking contrast, if animals received blocking antibody to C5a at the time of CLP, the H_2O_2 response to PMA was intact, and the translocation of $p47^{phox}$ to the cell membrane (as detected by Western blots of membrane fractions) was intact. These data indicate that innate immune functions of blood PMNs are adversely influenced by the generation of C5a in the course of sepsis. Such events perturb the signaling pathways and ultimately impair translocation of

Table 7. IgG Immune Complex Lung Responses in the Absence of C3.

1.	C5 but not C3 is required for full inflammatory response in $C3^{-/-}$ mice.
2.	FcγRs and C5aR are required for inflammatory injury.
3.	ATIII and hirudin block C5a production and inflammatory injury in $C3^{-/-}$ but not in $C3^{+/+}$ mice. Thrombin appears to be the C5 convertase in $C3^{-/-}$ mice[23].

cytosolic subunits to the cell membrane, interfering with full assembly of NADPH oxidase. These observations bear on the fact that when CLP rats were studied, colony forming units (CFUs) of bacteria in the blood, spleen and liver were greatly elevated 48 hours after the onset of CLP but CFUs were markedly suppressed in CLP rats initially protected with infusion of anti-C5a (Guo, et al 2003). Thus the ability in CLP to contain mixed anaerobic and aerobic bacteria is adversely affected by the presence of C5a.

5 Immune Complex Lung Inflamattory Responses in the Absence of C3

It was demonstrated a few years ago that acute lung inflammatory injury induced by intrapulmonary deposition of IgG immune complexes requires the participation of C5a, Fcγ receptors and C5aR (Skokowa, Ali, Felda, Kumar, Konrad, Shushakova, Schmidt, Piekorz, Nurnberg, Spicher, Birnbaumer, Zwirner, Claassens, Verbeek, van Rooijen, Kohl and Gessner 2005; Schmidt and Gessner 2005). Unexpectedly, the lung inflammatory response in $C3^{-/-}$ mice was equivalent to the response seen in $C3^{+/+}$ mice when albumin leak was used as a parameter of injury, PMN buildup as another, and hemorrhage as an additional indicator of inflammatory injury (Huber-Lang, Sarma, Zetoune, Rittirsch, Neff, McGuire, Lambris, Warner, Flierl, Hoesel, Gebhard, Younger, Drouin, Wetsel and Ward 2006) and Table 7. Not surprisingly, the roles of FcγR I and III as pro-inflammatory receptors has been documented (Ravetch 2002), while FcγR II appears to be an anti-inflammatory receptor (Shushakova, Skokowa, Shulman, Baumann, Zwirner, Schmidt and Gessner 2002). The data in the $C3^{-/-}$ mice were confusing and seemingly contradictory, since in the literature the depletion of C3 by cobra venom factor was known to markedly suppress the lung inflammatory response to immune complexes (Lukacs, Glovsky and Ward 2001). Accordingly, there was a great deal of confusion as to how the earlier observations and the more contemporary experiments could be interpreted. It is clear that FcγRs and C5aR are required for the full expression of lung inflammatory injury in response to deposition of IgG immune complexes. In the absence of FcγR I and III, the intensity of these inflammatory responses was greatly suppressed (Sylvestre, Clynes, Ma, Warren, Carroll, and Ravetch 1996). The use of $C5aR^{-/-}$ mice also indicated that the inflammatory responses were greatly reduced (Hopken, Lu, Gerard and Gerard 1997). Therefore, the earlier finding (Sylvestre, et al 1996) that complement was not required in immune complex triggered inflammatory responses and then the observations that the intensity of lung inflammatory responses in $C3^{-/-}$ lungs were equivalent to those of $C3^{+/+}$ mice (Sylvestre, et al 1996; Huber-Lang, et al 2006) are difficult to reconcile. The lung inflammatory response to IgG complexes deposition requires C5, similar to findings in a much earlier (Larsen, Mitchell and Henson 1981) as well as a more recent report (Peng, Hao, Madri, Su, Elias, Stahl, Squinto and Wang 2005). In the absence of C5, inflammatory injury was greatly attenuated and very little build up of PMNs developed in the lung. Similar to what Ravetch's group described (Sylvestre, et al 1996), we found that the parameters of lung and inflammatory injury after IgG immune complex deposition were equivalent in $C3^{-/-}$ and $C3^{+/+}$ lungs (Huber-Lang, et al 2006). However, we also demonstrated that the airway instillation of blocking antibody to mouse C5a greatly attenuated the parameters of lung inflammatory injury in both the $C3^{+/+}$and $C3^{-/-}$ mice, implicating a role for C5a in IgG immune complex mediated lung inflammatory responses. Our data suggested that, in the absence of C3, C5a could somehow be generated to facilitate the full inflammatory response of lung injury.

This led to an extended series of studies to determine why and how C5a can be generated in the absence of C3, since none of the traditional complement activation pathways should be capable of such an outcome. One possibility is the fact that we and others have demonstrated that serine proteases present in both PMNs (Ward and Hill 1970; Ward and Zvaifler 1973) and activated macrophages (Huber-Lang, Younkin, Sarma, Riedemann, McGuire, Lu, Kunkel, Younger, Zetoune and Ward 2002) can cleave C5 to produce C5a. Accordingly, phagocytic cells could be a source of proteases that could generate C5a from C5. However, our studies indicated that, when either $C3^{+/+}$ or $C3^{-/-}$ serum/plasma was activated by addition of glycogen particles, C5a as measured by functional activity (enzyme release from neutrophils and neutrophil chemotactic responses), could be generated both in $C3^{+/+}$ or $C3^{-/-}$ serum, with approximately 50% of the functional activity generated in $C3^{+/+}$ serum/plasma being generated in $C3^{-/-}$ serum (Huber-Lang, et al 2006). Furthermore, in both cases anti-C5a blocked these functional activities.

In the course of these studies, we tried to determine what the source of the C5 convertase might be in mouse $C3^{-/-}$ plasma and in mouse lungs undergoing IgG immune complex deposition. It became apparent that $C3^{-/-}$ plasma contains approximately three times more thrombin than $C3^{+/+}$ plasma. There had been earlier reports that thrombin might have C5 convertase activity but the earlier reports did not document that C5a had been generated (Wetsel and Kolb 1983). In the course of our studies, we also recognized that the liver, which is the known predominant source of prothrombin, showed much-intensified immunostaining for prothrombin/thrombin and that mRNA for prothrombin was greatly increased in extracts from $C3^{-/-}$ livers as compared to $C3^{+/+}$ livers. This suggested that, for whatever reason, in the absence of C3 the liver responds by generating accentuated levels of prothrombin.

The studies next turned to the model of IgG immune complex-induced acute lung inflammatory injury. Two different interventions were used, one being use of antithrombin III which is a serine protease inhibitor that blocks thrombin activity as well as several other activated clotting factors (as well as Kallikrein) and the other recombinant hirudin, which blocks the conversion of prothrombin to thrombin. These interventions were applied intratracheally together with the anti-BSA. Both lung leak of albumin (as an indicator of injury) as well as C5a levels in bronchoalveolar lavage (BAL) fluids were quantitated. Antithrombin III and hirudin had highly suppressive effects in the injury end points in $C3^{-/-}$ mice but statistically did not significantly affect lung injury in $C3^{+/+}$ mice. This would suggest that prothrombin/thrombin function as a C5 convertase predominantly in the $C3^{-/-}$ mice. In addition, when BAL fluids were examined from the same animals, it was clear that the presence of either antithrombin III or hirudin almost totally suppressed the increase of C5a in BAL fluids from $C3^{-/-}$ mice but had little effect on BAL fluids from $C3^{+/+}$ mice. These data strongly suggest that in the genetic absence of C3, thrombin can function as a C5 convertase to generate C5a in ways that will allow full development of lung injury in the $C3^{-/-}$ mice. The extent to which thrombin may also contribute to some of the injury in $C3^{+/+}$ mice remains to be determined. It should also be noted that when purified human thrombin was incubated with purified human C5, authentic C5a was generated with a molecular mass precisely equivalent to that of human C5a (Huber-Lang, et al 2006). It is unknown why prothrombin/thrombin levels increase in the genetic absence of C3, but it seems clear that such an accommodation has occurred and that under such conditions thrombin can now function effectively as a C5 convertase.

6 Conclusions

The data presented above indicate that a large array of changes develops during experimental sepsis, affecting C5aR, C5L2, and innate immune functions of circulating blood PMNs. It would appear that CLP-induced sepsis in rodents triggers unregulated production of C5a, which has detrimental effects on innate immune responsiveness of phagocytic cells, leading to a dysregulated inflammatory system. The data in septic rodents are similar to data obtained in humans with septic shock, suggesting that complement activation products and relevant receptors play a detrimental role in the setting of sepsis, especially when the ability to regulate activation of the complement system seems to be lost. The data also suggest that in the genetic deficiency of C3 an adaptation has occurred in which prothrombin/thrombin can now function as a C5 convertase to generate C5a and effectively override what would otherwise seem to be an absolute inability to generate C5a in the absence of C3. This provides additional evidence regarding the "plasticity" of the complement system and seems to underscore the fact that complement and its activation products play an important role in the protective functions linked to the innate immune system.

7 Acknowledgements

This work was supported by National Institutes of Health Grants GM 29507, HL 31963 and GM 01656 (PAW).

References

Curnutte, J.T., Erickson, R.W., Ding, J. and Badwey, J.A. (1994) Reciprocal interactions between protein kinase C and components of the NADPH oxidase complex may regulate superoxide production by neutrophils stimulated with a phorbol ester. J. Biol. Chem. 269, 10813-10819.

Czermak, B.J., Sarma, J.V., Pierson, C.L., Warner, R.L. Huber-Lang, M.S., Bless, N.M., Schmal, H., Friedl H.P. and Ward, P.A. (1999) Protective effects of C5a blockade in sepsis. Nat. Med. 5, 788-792.

Guo, R.F., Riedemann, N.C., Bernacki, K.D., Sarma, J.V., Laudes, I.J., Reuben, J.S., Younkin, E.M., Neff, T.A., Paulauskis, J.D., Zetoune, F.S. and Ward, P.A. (2003) Neutrophil C5a receptor and the outcome in a rat model of sepsis. FASEB J. 17, 1889-1891.

Guo, R.F. and Ward, P.A. (2005) Role of C5a in inflammatory responses. Annu. Rev. Immunol. 23, 821-852.

Hopken, U.E., Lu, B., Gerard, N.P. and Gerard, C. (1997) Impaired inflammatory responses in the reverse arthus reaction through genetic deletion of the C5a receptor. J. Exp. Med. 186, 749-756.

Huber-Lang, M.S., Sarma, J.V., Lu, K.T., McGuire, S.R., Padgaonkar, V.A., Guo, R.F., Younkin, E.M., Kunkel, R.G., Ding, J., Erickson, R.E., Curnutte, J.T. and Ward, P.A. (2001) Role of C5a in multiorgan failure during sepsis. J. Immunol. 166, 1193-1199.

Huber-Lang, M.S., Younkin, E.M., Sarma, J.V., Riedemann, N., McGuire, S.R., Lu, K.T., Kunkel, R.G., Younger, J.G., Zetoune, F.S. and Ward, P.A. (2002) Generation of C5a by phagocytic cells. Am. J. Pathol. 161, 1849-1859.

Huber-Lang, M.S., Younkin, E.M., Sarma, J.V., McGuire, S.R., Lu, K.T., Guo, R.F., Padgaonkar, V.A., Curnutte, J.T., Erickson, R. and Ward, P.A. (2002) Complement induced impairment of innate immunity during sepsis. J. Immunol. 169, 3223-3231.

Huber-Lang, M.S., Sarma, J.V. McGuire, S.R., Lu, K.T., Padgaonkar, V.A., Younkin, E.M., Guo, R.F., Weber, C.H., Zuiderweg, E.R., Zetoune, F.S. and Ward, P.A. (2003) Structure-function relationships of human C5a and C5aR. J. Immunol. 170, 6115-6124.

Huber-Lang, M.S., Sarma, J.V., Rittirsch, D., Schreiber, H, Weiss, M., Flierl, M., Younkin, E, Schneider, M., Suger-Wiedeck, H., Gephard, F., McClintock, S.D., Neff, T., Zetoune, F.S., Bruckner, Guo, R.F., Monk, P.N. and Ward, P.A. (2005) Changes in the novel orphan, C5a receptor (C5L2), during experimental sepsis and sepsis in humans. J. Immunol. 174, 1104-1110.

Huber-Lang, M.S., Sarma, J.V., Zetoune, F.S., Rittirsch, D., Neff, T.A., McGuire, S.R., Lambris, J.D., Warner, R.L., Flierl, M.A., Hoesel, L.M., Gebhard, F., Younger, J.G., Drouin, S.M., Wetsel, R.A. and Ward, P.A. (2006) Generation of C5a in the absence of C3: a new complement activation pathway. Nat. Med. 12, 682-687.

Johswich, K., Martin, M., Thalmann, J., Rheinheimer, C., Monk, P.N, and Klos, A. (2006) Ligand specificity of the anaphylatoxin C5L2 receptor and its regulation on myeloid and epithelial cell-lines. http://www.jbc.org/cgi/doi/10.1074/jbc.M609734200.

Larsen, G.L., Mitchell, B.C. and Henson, P.M. (1981) The pulmonary responses of C5 sufficient and deficient mice to immune complexes. Am. Rev. Respir. Dis. 123, 434-439.

Lukacs, N.W., Glovsky, M.M. and Ward, P.A. (2001) Complement-dependent immune complex-induced bronchial inflammation and hyperreactivity. Am. J. Physiol. Lung Cell Mol. Physiol .280, L512-L518.

Maslowska, M., Legakis, H., Assadi, F.,. and Cianflone, K. (2006) Targeting the signaling pathway of acylation stimulating protein. J. Lipid Res. 47, 643-652.

Mueller-Ortiz, S.L., Drouin, S.M.,. and Wetsel, R.A. (2004) The alternative activation pathway and complement component C3 are critical for a protective immune response against Pseudomonas aeruginosa in a murine model of pneumonia. Infect. Immun. 72, 2899-2906.

Okinaga, S., Slattery, D., Humbles, A., Zsengeller, Z., Morteau, O., Kinrade, M.B., Brodbeck, R.M., Krause, J.E., Choe, H.R., Gerard, N.P. and Gerard, C. (2003) C5L2 a non-signaling C5a binding protein. Biochemistry 42, 9406-9415.

Peng, T., Hao, L., Madri, J.A., Su, X., Elias, J.A., Stahl, G.L., Squinto, S. and Wang, Y. (2005) Role of C5 in the development of airway inflammation, airway hyperresponsiveness and ongoing airway response. J. Clin. Invest. 115, 1590-1600.

Ravetch, J.V. (2002) A full complement of receptors in immune complex diseases. J. Clin. Invest. 110, 1760-1761.

Reis, E.S., Nudelman, V. and Isaac, L. (2004) Nonsense-codon-mediated decay in human hereditary complement C3 deficiency. Immunogenetics 55, 667-6673.

Riedemann, N.C., Guo, R.F., Hollmann, T.J., Gao, H., Neff, T.A., Reuben, J.S., Speyer, C.L., Sarma, J.V., Wetsel, R.A., Zetoune, F.S. and Ward, P.A. (2004) Regulatory role of C5a in LPS-induced IL-6 production by neutrophils during sepsis. FASEB J. 18, 370-372.

Schmidt, R.E. and Gessner, J.E. (2005) Fc receptors and interaction with complement in autoimmunity. Immunol. Lett. 100, 56-67.

Shushakova, N., Skokowa, J., Schulman, J., Baumann, U., Zwirner, J., Schmidt, R.E. and Gessner, J.E. (2002) C5a anaphylatoxin is a major regulator of activating versus inhibitory FcgRs in immune complex-induced lung disease. J. Clin. Invest. 110, 1823-1830.

Skokowa, J., Ali, S.R., Felda, O., Kumar, V., Konrad, S., Shushakova, N., Schmidt, R.E., Piekorz, R.P., Nurnberg, B., Spicher, K., Birnbaumer, L., Zwirner, J., Claassens, J.W.C., Verbeek, J.S., van Rooijen, N., Kohl, J. and Gessner, J.E. (2005) Macrophages induce the inflammatory response in the pulmonary arthus reaction through $G\alpha_{i2}$ activation that controls C5aR and Fc receptor cooperation. J. Immunol. 174, 3041-3050.

Sylvestre, D., Clynes, R., Ma, M., Warren, H., Carroll, M.C. and Ravetch, J.V. (1996) Immunoglobulin G-mediated inflammatory responses develop normally in complement-deficient mice. J. Exp. Med. 184, 2385-2392.

Ward, P.A. and Hill. J.H. (1970) C5 chemotactic fragments produced by an enzyme in the lysosomal granules of neutrophils. J. Immunol. 104, 535-543.

Ward, P.A. and Zvaifler, N.J. (1973) Quantitative phagocytosis by neutrophils, II. Release of the C5 cleaving enzyme and inhibition of phagocytosis by rheumatoid factor. J. Immunol. 111, 1777-1782.

Wessels, M.R., Butko, P., Ma, M., Warren, H.B., Lage, A.L. and Carroll, M.C. (1995) Studies of group B streptococcal infection in mice deficient in complement C3 or C4 demonstrate an essential role for complement in both innate and acquired immunity. Proc. Natl. Acad. Sci. USA 92, 11490-11494.

Wetsel, R.A.,. and Kolb, W.P. (1983) Expression of C5a-like biological activities by the fifth component of human complement (C5) upon limited digestion with non complement enzymes without release of polypeptide fragments. J. Exp. Med. 157, 2029-2048.

18

Complement Activation of Drusen in Primate Model (*Macaca fascicularis*) for Age-Related Macular Degeneration

Takeshi Iwata

Takeshi Iwata, National Institute of Sensory Organs, National Hospital Organization Tokyo Medical Center, 2-5-1 Higashigaoka, Meguro-ku, Tokyo 152-8902 Japan

1 Introduction

Dysfunction of the visual system can alter normal human life style and lower quality of life. The most prevalent causes of visual impairment worldwide are cataracts, glaucoma, and age-related macular degeneration (AMD). These eye diseases are responsible for 69% of blindness globally. Although cataracts are the leading cause of blindness worldwide, recent advances in cataract surgery has significantly reduced the visual impairments caused by cataracts especially in developed countries. The most prevalent eye disease for elderly Europeans and Americans is AMD. This degenerative disease progresses from retinal deposits called drusen to neovascularization and retinal hemorrhages resulting in irreversible loss of central vision. In spite of the high incidence of AMD, a limited amount of information is available on the underlying pathological mechanisms causing these diseases. Obtaining tissues from the AMD donors is often difficult, and even when obtained, they are usually collected many hours or even days after death. Because of limitation for human tissue, the availability of animal models is becomes valuable because they can be used to investigate the molecular mechanisms of the disease and to test new therapeutic intervention.

The retina is composed of nine layers of neural and glial cells that are arranged concentrically at the posterior pole of the eye. Incoming light is focused on the central area of the retina called the fovea which is located in the center of the macula. In humans, the average size of the macula is only 6 mm in diameter. The outer surface of the retina is covered by a monolayer of retinal pigment epithelial (RPE) cells which forms a diffusion barrier between the neural retina and the choroidal blood supply. The RPE regulates the transport of proteins to the retina, and controls the hydration and ionic composition of the subretinal space. The physiological condition of the RPE is closely associated with the pathogenesis of AMD.

2 Introduction of AMD

AMD is a blinding disorder characterized by a marked decrease in central vision associated with RPE atrophy with or without choroidal neovascularization (CNV). Many factors including genetic, behavioral, and environmental, are involved in this disease. AMD is characterized by the degeneration of cone photoreceptors in the foveal region of the retina resulting in a decrease of central visual acuity. The progressive impairment of the retinal pigment epithelial (RPE) cells, and damage to Bruch's membrane and choriocapillaris results in retinal atrophy and photoreceptor dysfunction. In some cases, CNV develops, and the new vessels penetrate Bruch's membrane and pass into the subretinal space.

Two types of AMD are recognized; the non-neovascular type is called the dry-type AMD and includes more than 80% of the cases, and the neovascular type is called the wet-type AMD which is progressive with a higher probability of blindness. The prevalence of AMD differs considerably among the different ethnic groups, but the incidence increases with age in all groups. A lower prevalence of AMD has been reported in individuals of African ancestry than of Anglo-Saxon ancestry. Other risk factors for AMD are cigarette smoking, obesity, hypertension, and atherosclerosis.

3 Genetics of AMD

Epidemiological studies have shown that genetic factor play critical role for AMD. Twin studies have previously shown a higher concordance for AMD in monozygotic twins than in dizygotic twins (Heiba, Elston, Klein, and Klein 1994; Seddon, Ajani, and Mitchell 1997; Hammond, Webster, Snieder, Bird, Gilbert, and Spector 2002). In addition, first degree relatives of individuals with AMD have higher incidence of AMD over individuals without a family history of AMD. Genetic segregation studies have also shown a genetic effect that accounts for approximately 60% of AMD with a single major gene accounting for about 55% of the risk of developing AMD. Previous data have suggested that the etiology of AMD has a significant genetic component. Only a small proportion of the families with AMD show Mendelian inheritance, and the majority of the individuals inherit AMD in a complex multi-gene pattern. With the help of the haplotype marker project (HapMap Project), genome wide scanning has identified at least 13 loci linked to AMD on different chromosomes (Iyengar, Song, Klein, Klein, Schick, Humphrey, Millard, Liptak, Russo, Jun, Lee, Fijal, and Elston 2004; Schick, Iyengar, Klein, Klein, Reading, Liptak, Millard, Lee, Tomany, Moore, Fijal, and Elston 2003; Majewski, Schultz, Weleber, Schain, Edwards, Matise, Acott, Ott, and Klein 2003). Recently, a polymorphism of complement factor H (CFH) gene (*Y402H*) was shown to be associated with an increased risk for AMD (Klein, Zeiss, Chew, Tsai, Sackler, Haynes, Henning, SanGiovanni, Mane, Mayne, Bracken, Ferris, Ott, Barnstable, and Hoh 2005; Edwards, Ritter, Abel, Manning, Panhuysen, and

Farrer 2005; Haines, Hauser, Schmidt, Scott, Olson, Gallins, Spencer, Kwan, Noureddine, Gilbert, Schnetz-Boutaud, Agarwal, Postel, and Pericak-Vance 2005; Hageman, Anderson, Johnson, Hancox, Taiber, Hardisty, Hageman, Stockman, Borchardt, Gehrs, Smith, Silvestri, Russell, Klaver, Barbazetto, Chang, Yannuzzi, Barile, Merriam, Smith, Olsh, Bergeron, Zernant, Merriam, Gold, Dean, and Allikmets 2005).

These results were confirmed in many of the countries with large Caucasian populations but not in Japan (Okamoto, Umeda, Obazawa, Minami, Noda, Mizota, Honda, Tanaka, Koyama, Takagi, Sakamoto, Saito, Miyake, and Iwata 2006; Gotoh, Yamada, Hiratani, Renault, Kuroiwa, Monet, Toyoda, Chida, Mandai, Otani, Yoshimura, and Matsuda 2006). This gene is located on chromosome 1q25-31 where one of the candidate loci was identified by linkage studies. Another recent study reported that a haplotype association of tandemly located complement 2 and factor B was protective for AMD (Gold, Merriam, Zernant, Hancox, Taiber, Gehrs, Cramer, Neel, Bergeron, Barile, Smith, AMD Genetics Clinical Study Group, Hageman, Dean, Allikmets 2006). HTRA1, a serine protease 11 was recently discovered to be strongly associated with AMD. Unlike the CFH, our study shows strong association with this gene for Japanese AMD patients (Yoshita, Dewan, Zhang, Sakamoto, Okamoto, Minami, Obazama, Mizota, Tanaka, Saito, Takagi, Hoh, and Iwata, Yang, Camp, Camp, Sun, Tong, Gibbs, Cameron, Chen, Zhao, Pearson, Li, Chien, Dewan, Harmon, Bernstein, Shridhar, Zabriskie, Hoh, Howes, and Zhang 2006; Dewan, Liu, Hartman, Zhang, Liu, Zhao, Tam, Chan, Lam, Snyder, Barnstable, Pang, and Hoh 2006).

4 Biochemistry of AMD

The early stage of the dry type AMD is characterized by a thickening of Bruch's membrane, aggregation of pigment granules, and increasing numbers of drusen. The thickening of Bruch's membrane obstructs its function as a 'barrier' between the choroid and the RPE that protects the neural retina from the choriocapillary. Drusen are small yellowish-white deposits that are composed of lipids, proteins, glycoproteins, and glycosaminoglycans. They accumulate in the extracellular space and the inner aspects of Bruch's membrane. Drusen are not directly associated with visual loss but represent a risk factor for both the dry-type and wet-type AMD. The classification of hard and soft drusen is based on their size, shape, and color; hard drusen are yellowish with diameters <50 μm and are found in eyes that are less likely to progress to advanced stages of the disease, while soft drusen are darker yellow and larger in size, and are found in eyes more likely to progress to more advanced stages of AMD. A small percentage of dry-type AMD patients progress to the late stage of the wet-type AMD that is characterized by geographic atrophy or detachment of RPE and the development of CNV in the macular region. The presence of a CNV is the factor that most damages the neural retina because the newly developed vessels grow from the choriocapillaris through Bruch's membrane and extend laterally through the RPE cell layer (classic CNV) or extend between the inner Bruch's membrane and RPE (occult CNV). In advanced stages of AMD, the CNV and fluid leaked into the subretinal or intraretinal regions leads to cell death and retinal detachment.

Recent analyses of the progression of drusen have provided important clues that help understand the molecular pathology of AMD. Using both immunohistochemistry and proteomic techniques, the materials in drusen were found to be composed of molecules that mediate inflammatory and immune processes (Russell, Mullins, Schneider, and Hageman 2000; Mullins, Russell, Anderson, and Hageman 2000). These molecules include components of the complement pathway and modulators of complement activation, viz., vitronectin, clusterin, membrane cofactor protein, and complement receptor-1. In addition, molecules triggering inflammation, amyloid P component, α1-antitrypsin, and apolipoprotein E, were identified in drusen. Cellular debris from macrophages, RPE cells, and choroidal dendritic cells has also been identified in drusen. Additional proteins such as crystallins, EEFMP1, and amyloid-beta have been also found in drusen. The presence of immunoreactive proteins and the oxidative modifications of many proteins in drusen imply that both oxidation and immune functions are involved in the pathogenesis of AMD. These findings suggest that complement activation triggers innate immune responses in the subretinal space. The co-distribution of IgG and terminal complement complexes in drusen indicate that immune responses that directly target antigens in retinal cells might also be occurring. Anti-retinal autoantibodies have been reported in a number of ocular disorders, e.g., macular degeneration in an aged monkey model.

5 Primate Model for AMD

Over the past few years, genetic engineering techniques have generated a number of animal models of AMD in mice, rats, rabbits, pigs, and dogs (Chader 2002). However in mammals, a well-defined fovea is found only in primates (humans and monkeys), and a search for a monkey line affected with macular degeneration has been persistent for a long time. A monkey with macular degeneration was first described by Stafford et al in 1974. They reported that 6.6 % of the elderly monkeys they examined showed pigmentary disorders and drusen-like spots (Stafford, Anness, and Fine 1984). El-Mofty et al reported that the incidence of maculopathy was 50% in a colony of rhesus monkeys at the Caribbean Primate Research Center of the University of Puerto Rico (El-Mofty, Gouras, Eisner, and Balazs 1978). At the Tsukuba Primate Research Center (Tsukuba City, Japan), Suzuki et al found a single cynomolgus monkey (*Macaca fascicularis*) (Suzuki Monkeys) in 1986 with a large number of small drusen around the macular region (Nicolas, Fujiki, Murayama, Suzuki, Mineki, Hayakawa, Yoshikawa, Cho, Kanai 1996; Nicolas, Fujiki, Murayama, Suzuki, Shindo, Hotta, Iwata, Fujimura, Yoshikawa, Cho, Kanai 1996; Suzuki, Terao, and Yoshikawa 2003). This single affected monkey has multiplied to a large pedigree of more than 65 affect and 210 unaffected monkeys. Drusen were observed in the macular region as early as one year after birth, and the numbers increased and spread toward the peripheral retina throughout life. No histological abnormalities have been found in the retina, retinal vessels, or choroidal vasculatures of the eyes with drusen. However, abnormality in electroretinogram (ERG) were observed in sever case showing dysfunction of the macula.

Immunohistochemical and proteomic analyses of the drusen from these monkeys showed that the drusen were very similar to those in other monkeys with aged

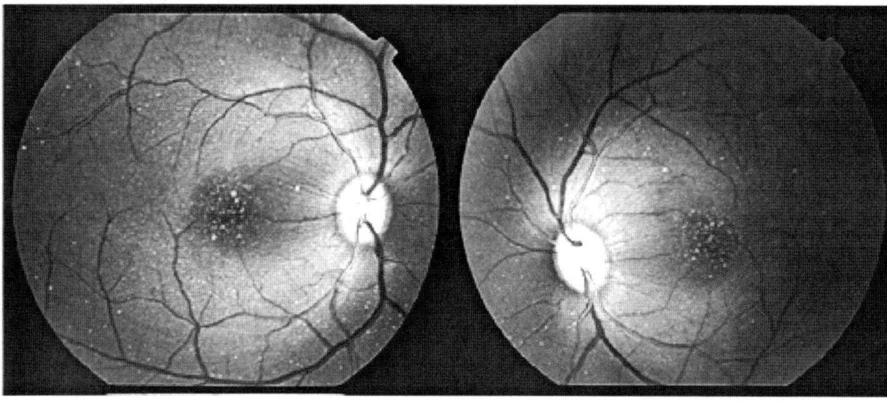

Fig. 1. Fundus photograph of both eyes of Suzuki Monkey showing accumulation of drusen (white spot) around the macular region.

macular degeneration sporadically found in older monkeys and also with human drusen (Umeda, Ayyagari, Allikmets, Suzuki, Karoukis, Ambasudhan, Zernant, Okamoto, Ono, Terao, Mizota, Yoshikawa, Tanaka, and Iwata 2005; Umeda, Suzuki, Okamoto, Ono, Mizota, Terao, Yoshikawa, Tanaka, and Iwata 2005; Ambati, Anand, Fernandez, Sakurai, Lynn, Kuziel, Rollins, and Ambati 2003). These observations have shown that the Suzuki Monkeys produce drusen that are biochemically similar to those in human AMD patients, but the development of the drusen occurs at an accelerated rate. More than 240 loci are being investigated to try to identify the disease causing gene and to understand the biological pathways leading to complement activation. Simultaneously, we have been studying a colony of aged monkeys which develop drusen after 15 years of birth.

Drusen components of these sporadically found affected monkeys were compared with human and Suzuki Monkeys by classical immunohistochemical techniques and by proteome analysis using mass spectrometer. Significant finding was that drusen contained protein molecules that mediate inflammatory and immune processes. These include immunoglobulins, components of complement pathway, and modulators for complement activation (e.g., vitronectin, clusterin, membrane cofactor protein, and complement receptor-1), molecules involved in the acute-phase response to inflammation (e.g., amyloid P component, α1-antitrypsin, and apolipoprotein E), major histocompatibility complex class II antigens, and HLA-DR antigens (Umeda et al. 2005). Cellular components have also been identified in drusen, including RPE debris, lipofuscin, and melanin, as well as processes of choroidal dendritic cells, which are felt to contribute to the inflammatory response. In addition to immune components, a number of other proteins were found in drusen. These appear to be vitronectin, clusterin, TIMP-3, serum amyloid P component, apolipoprotein E, IgG, Factor X, crystallins, EEFMP1, and amyloid-beta. The presence of immunoreactive proteins and oxidative modified proteins implicate both oxidation and immune functions in the pathogenesis of AMD.

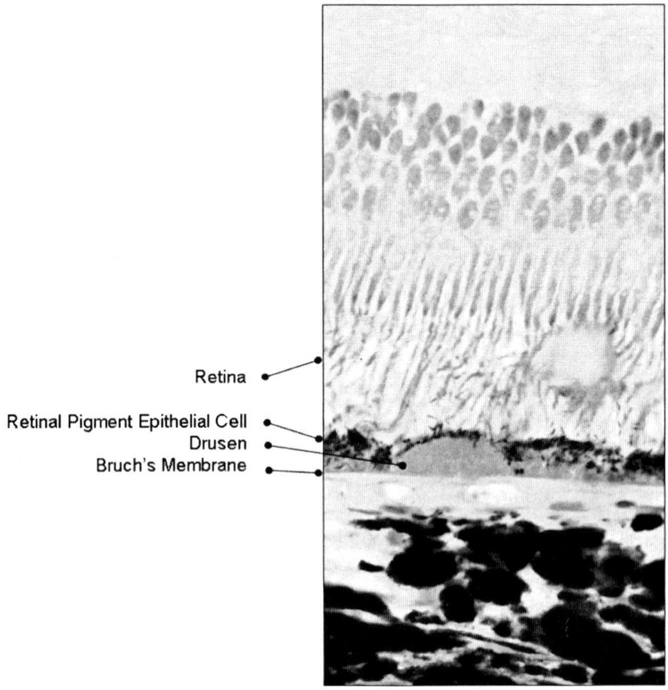

Fig. 2. Retinal histological section of affected Suzuki Monkey showing the accumulation of drusen between the retinal pigment epithelium and Bruch's membrane.

The eyes of monkey are structurally similar to human eyes which make them extremely valuable for AMD studies. However, there are limitations in using this species over other laboratory animals. Monkeys have a relatively longer life span, have a longer gestation period, have a lower birth numbers resulting in a slower expansion of the pedigree, more difficult to genetically manipulate, and the maintenance cost is high. In the other laboratory animals, the differences in the eye structure, lack of a fovea, and a low cone/rod ratio compared to humans have been considered to be a disadvantage for using them as AMD models. However, they are easier to manipulate genetically and easier and less expensive to maintain. This has made the development of a mouse model of AMD very attractive, and a number of mouse AMD models have been reported recently.

6 Mouse Model for AMD

The mouse model described by Ambati et al is deficient either in monocyte chemoattractant protein-1 or its cognate C-C chemokine receptor-2. These mice were found to develop the cardinal features of AMD including accumulation of lipofuscin in drusen beneath the RPE, photoreceptor atrophy, and CNV (Ambati et al. 2003). An impairment of macrophage recruitment allowed the accumulation of C5a and IgG, which leads to the production of vascular endothelial growth factor by the RPE cells and the development of CNVs. Another mouse model that has three known AMD risk factors: age, high fat cholesterol rich diet, and expression of human apolipoprotein E (apoE2, apoE3, apoE4) has been developed (Malek, Johnson, Mace, Saloupis, Schmechel, Rickman, Toth, Sullivan, and Bowes Rickman 2005). ApoE4-deficient mice are severely affected showing diffuse subretinal pigment epithelial deposits, drusen, thickened Bruch's membrane, and atrophy, hypopigmentation, and hyperpigmentation of the RPE.

Oxidative stress has long been linked to the pathogenesis of AMD. Imamura et al reported a Cu, Zn-superoxide dismutase (SOD1)-deficient mice that had features typical of AMD in human. Senescent Sod1 (-/-) mice had drusen, thickened Bruch's membrane, and choroidal neovascularization (Imamura, Noda, Hashizume, Shinoda, Yamaguchi, Uchiyama, Shimizu, Mizushima, Shirasawa, and Tsubota 2006). The number of drusen increased with age and also after exposure of young Sod1 (-/-) mice to excess light. The retinal pigment epithelial cells of Sod1 (-/-) mice showed oxidative damage, and their beta-catenin-mediated cellular integrity was disrupted. These findings suggested that oxidative stress may affect the junctional proteins necessary for the barrier integrity of the RPE. These observations strongly suggested that oxidative stress may play a major role in AMD.

The complement components, C3a and C5a, are present in drusen, and were observed in Bruch's membrane of a laser-induced CNV mice model. Neutralization of C3a or C5a by antibody or by blockade of their receptors by a complement inhibitor significantly reduced the CNV (Nozaki, Raisler, Sakurai, Sarma, Barnum, Lambris, Chen, Zhang, Ambati, Baffi, and Ambati 2006). These observations revealed a role for immunological mechanisms for the angiogenesis and provided evidence for future therapeutic strategies for AMD. Although the pathology of AMD is pronounced in the macula area, it is not confined to this region. Characteristics of human AMD such as thickening of Bruch's membrane, accumulation of drusen, and CNV have been observed in mouse models. Nevertheless, the primate model will still be the choice for AMD studies, especially at the stage when new therapeutic methods are tested and evaluated for the first time. However, it would be wise and more productive to study both primate and mouse models in AMD research. This will be necessary to learn the mechanisms underlying the disease and to identify clinical and molecular markers for the early stages of AMD. The findings from these studies will provide critical information needed to develop therapies for AMD.

7 Acknowledgements

This work was supported by the research grant from the Ministry of Health, Labour and Welfare of Japan.

References

Ambati, J., Anand, A., Fernandez, S., Sakurai, E., Lynn, B. C., Kuziel, W. A., Rollins, B. J., Ambati, B. K. (2003) An animal model of age-related macular degeneration in senescent Ccl-2- or Ccr-2-deficient mice. Nat. Med. 9, 1390-1397.

Chader, G. J. (2002) Animal models in research on retinal degenerations: Past progress and future hope. Vision Res 42, 393-399.

Dewan A., Liu, M., Hartman, S., Zhang, S. S., Liu, D. T., Zhao, C., Tam, P. O., Chan, W. M., Lam, D. S., Snyder, M., Barnstable, C., Pang, C. P., Hoh, J. (2006) HTRA1 promoter polymorphism in wet age-related macular degeneration. Science 314, 989-992.

Edwards, A. O., Ritter, R. 3rd., Abel, K. J., Manning, A., Panhuysen, C., Farrer, L. A. (2005) Complement factor H polymorphism and age-related macular degeneration. Science 308, 421-424.

El-Mofty A., Gouras, P., Eisner, G., Balazs, E. A. (1978) Macular degeneration in rhesus monkey (Macaca mulatta). Exp. Eye Res. 27, 499-502.

Gold, B., Merriam, J. E., Zernant, J., Hancox, L. S., Taiber, A. J., Gehrs, K., Cramer, K., Neel, J., Bergeron, J., Barile, G. R., Smith, R. T. (2006) AMD Genetics Clinical Study Group; G. S. Hageman, M. Dean, R. Allikmets. Variation in factor B (BF) and complement component 2 (C2) genes is associated with age-related macular degeneration. Nat. Genet. 38, 458-462.

Gotoh, N., Yamada, R., Hiratani, H., Renault, V., Kuroiwa, S., Monet, M., Toyoda, S., Chida, S., Mandai, M., Otani, A., Yoshimura, N., Matsuda, F. (2006) No association between complement factor H gene polymorphism and exudative age-related macular degeneration in Japanese. Hum. Genet. 120, 139-143.

Hageman, G. S., Anderson, D. H., Johnson, L. V., Hancox, L. S., Taiber, A. J., Hardisty, L. I., Hageman, J. L., Stockman, H. A., Borchardt, J. D., Gehrs, K. M., Smith, R. J., Silvestri, G., Russell, S. R., Klaver, C. C., Barbazetto, I., Chang, S., Yannuzzi, L. A., Barile, G. R., Merriam, J. C., Smith, R. T., Olsh, A. K., Bergeron, J., Zernant, J., Merriam, J. E., Gold, B., Dean, M., Allikmets, R. (2005) A common haplotype in the complement regulatory gene factor H (HF1/CFH) predisposes individuals to age-related macular degeneration. Proc. Natl. Acad. Sci. U S A 102, 7227-7232.

Haines, J. L., Hauser, M. A., Schmidt, S., Scott, W. K., Olson, L. M., Gallins, P., Spencer, K. L., Kwan, S. Y., Noureddine, M., Gilbert, J. R., Schnetz-Boutaud, N., Agarwal, A., Postel, E. A., Pericak-Vance, M. A. (2005) Complement factor H variant increases the risk of age-related macular degeneration. Science 308, 419-421.

Hammond, C. J., Webster, A. R., Snieder, H., Bird, A. C., Gilbert, C. E., Spector, T. D. (2002) Genetic influence on early age-related maculopathy: A twin study. Ophthalmology 109, 730-736.

Heiba, I. M., Elston, R. C., Klein, B. E., Klein, R. (1994) Sibling correlations and segregation analysis of age-related maculopathy: The Beaver Dam Eye Study. Genet. Epidemiol. 11, 51-67.

Imamura, Y., Noda, S., Hashizume, K., Shinoda, K., Yamaguchi, M., Uchiyama, S., Shimizu, T., Mizushima, Y., Shirasawa, T., Tsubota, K. (2006) Drusen, choroidal neovascularization, and retinal pigment epithelium dysfunction in SOD1-deficient mice: a model of age-related macular degeneration. Proc Natl Acad Sci U S A 103, 11282-11287.

Iyengar, S. K., Song, D., Klein, B. E., Klein, R., Schick, J. H., Humphrey, J., Millard, C., Liptak, R., Russo, K., Jun, G., Lee, K. E., Fijal, B., Elston, R. C. (2004) Dissection of genomewide-scan data in extended families reveals a major locus and oligogenic susceptibility for age-related macular degeneration. Am. J. Hum. Genet. 74, 20-39.

Klein, R. J., Zeiss, C., Chew, E. Y., Tsai, J. Y., Sackler, R. S., Haynes, C., Henning, A. K., SanGiovanni, J. P., Mane, S. M., Mayne, S. T., Bracken, M. B., Ferris, F. L., Ott, J., Barnstable, C., Hoh, J. (2005) Complement factor H polymorphism in age-related macular degeneration. Science 308, 385-389.

Schick, J. H., Iyengar, S. K., Klein, B. E., Klein, R., Reading, K., Liptak, R., Millard, C., Lee, K. E., Tomany, S. C., Moore, E. L., Fijal, B. A., Elston, R. C. (2003) A whole-genome screen of a quantitative trait of age-related maculopathy in sibships from the Beaver Dam Eye Study. Am. J. Hum. Genet. 72, 1412-1424.

Majewski, J., Schultz, D. W., Weleber, R. G., Schain, M. B., Edwards, A. O., Matise, T. C., Acott, T. S., Ott, J., Klein, M. L. (2003) Age-related macular degeneration--a genome scan in extended families. Am. J. Hum. Genet. 73, 540-550.

Malek, G., Johnson, L. V., Mace, B. E., Saloupis, P., Schmechel, D. E., Rickman, D. W., Toth, C. A., Sullivan, P. M., Bowes Rickman, C. (2005) Apolipoprotein E allele-dependent pathogenesis: a model for age-related retinal degeneration. Proc. Natl. Acad. Sci. U S A. 102, 11900-11905.

Mullins, R. F., Russell, S. R., Anderson, D. H., Hageman, G. S. (2000) Drusen associated with aging and age-related macular degeneration contain proteins common to extracellular deposits associated with atherosclerosis, elastosis, amyloidosis, and dense deposit disease. FASEB J 14, 835-846.

Nicolas, M. G., Fujiki, K., Murayama, K., Suzuki, M. T., Mineki, R., Hayakawa, M., Yoshikawa, Y., Cho, F., Kanai, A. (1996) Studies on the mechanism of early onset macular degeneration in cynomolgus (Macaca fascicularis) monkeys. I. Abnormal concentrations of two proteins in the retina. Exp. Eye Res. 62, 211-219.

Nicolas, M. G., Fujiki, K., Murayama, K., Suzuki, M. T., Shindo, N., Hotta, Y., Iwata, F., Fujimura, T., Yoshikawa, Y., Cho, F., Kanai, (1996) A. Studies on the mechanism of early onset macular degeneration in cynomolgus monkeys. II. Suppression of metallothionein synthesis in the retina in oxidative stress. Exp. Eye Res. 62, 399-408.

Nozaki, M., Raisler, B. J., Sakurai, E., Sarma, J. V., Barnum, S. R., Lambris, J. D., Chen, Y., Zhang, K., Ambati, B. K., Baffi, J. Z., Ambati J. (2006) Drusen complement components C3a and C5a promote choroidal neovascularization. Proc. Natl. Acad. Sci. U S A 103, 2328-2333.

Okamoto, H., Umeda, S., Obazawa, M., Minami, M., Noda, T., Mizota, A., Honda, M., Tanaka, M., Koyama, R., Takagi, I., Sakamoto, Y., Saito, Y., Miyake, Y., Iwata, T. (2006) Complement factor H polymorphisms in Japanese population with age-related macular degeneration. Mol. Vis. 12, 156-158.

Russell, S. R., Mullins, R. F., Schneider, B. L., Hageman, G. S. (2000) Location, substructure, and composition of basal laminar drusen compared with drusen associated with aging and age-related macular degeneration. Am J Ophthalmol 129, 205-214.

Seddon, J. M., Ajani, U. A., Mitchell, B. D. (1997) Familial aggregation of age-related maculopathy. Am. J. Ophthalmology 123, 199-206.

Stafford, T. J., Anness, S. H., Fine, B. S. (1984) Spontaneous degenerative maculopathy in the monkey. Ophthalmology 91, 513-521.

Yang, Z., Camp, N. J., Sun, H., Tong, Z., Gibbs, D., Cameron, D. J., Chen, H., Zhao, Y., Pearson, E., Li, X., Chien, J., Dewan, A., Harmon, J., Bernstein, P. S., Shridhar, V., Zabriskie, N. A., Hoh, J., Howes, K., Zhang, K.. (2006) A variant of the HTRA1 gene increases susceptibility to age-related macular degeneration. Science. 314, 992-993.

Yoshida, T., Dewan, A., Zhang, H., Sakamoto, R., Okamoto, H., Minami, M., Obazawa, M., Mizota, A., Tanaka, M., Saito, Y., Takagi, I., Hoh, J., Iwata, T. (2007) HTRA1 promoter polymorphism predisposes Japanese to age-related macular degeneration. Mol. Vis. 4, 13:545-548.

Umeda, S., Ayyagari, R., Allikmets, R., Suzuki, M. T., Karoukis, A. J., Ambasudhan, R., Zernant, J., Okamoto, H., Ono, F., Terao, K., Mizota, A., Yoshikawa, Y., Tanaka, Y., Iwata, T. (2005) Early-onset macular degeneration with drusen in a cynomolgus monkey (Macaca fascicularis) pedigree: exclusion of 13 candidate genes and loci. Invest. Ophthalmol. Vis. Sci. 46, 683-691.

Umeda, S., Suzuki, M. T., Okamoto, H., Ono, F., Mizota, A., Terao, K., Yoshikawa, Y., Tanaka, Y., Iwata, T. (2005) Molecular composition of drusen and possible involvement of anti-retinal autoimmunity in two different forms of macular degeneration in cynomolgus monkey (Macaca fascicularis). FASEB J. 19, 1683-1685.

Exploring the Complement Interaction Network Using Surface Plasmon Resonance

Daniel Ricklin[1] and John D. Lambris[2]

[1] University of Pennsylvania, Department of Pathology and Laboratory Medicine, ricklin@mail.med.upenn.edu
[2] University of Pennsylvania, Department of Pathology and Laboratory Medicine, lambris@mail.med.upenn.edu

1 Introduction

The human complement system relies on a well-balanced and tightly regulated network of protein interactions. A cascade of soluble and membrane-bound proteins, enzymes, cofactors, and regulators ensures the dynamic and specific answer to pathogenic conditions and makes the complement system to a central part of innate immunity. However, even small changes in this network may lead to severe disorders (e.g. auto-immune diseases), and some pathogens (mis-)use complement proteins for entering human cells. A detailed knowledge about the individual molecular interactions is essential for understanding of the complement cascade and for the development of therapeutic interventions against complement-related diseases. Surface plasmon resonance (SPR) biosensors are considered a key technology for characterizing such biomolecular interactions, since they not only offer a label-free determination of binding affinities, but also provide kinetic profiles. As a consequence, SPR is used for the analysis of complement interactions at a rapidly increasing pace. Based on various examples, this review discusses the large potential of SPR and illustrates its impact on complement research.

2 Surface Plasmon Resonance – The Key to Kinetic Constants

During the past ten years, surface plasmon resonance (SPR) has emerged as a key technology for the characterization of biomolecular interactions. Nowadays, SPR biosensors are widely used in a variety of areas (Rich and Myszka 2000, 2001, 2002, 2003, 2005a, 2005b, 2006) from life sciences and biology to proteomics

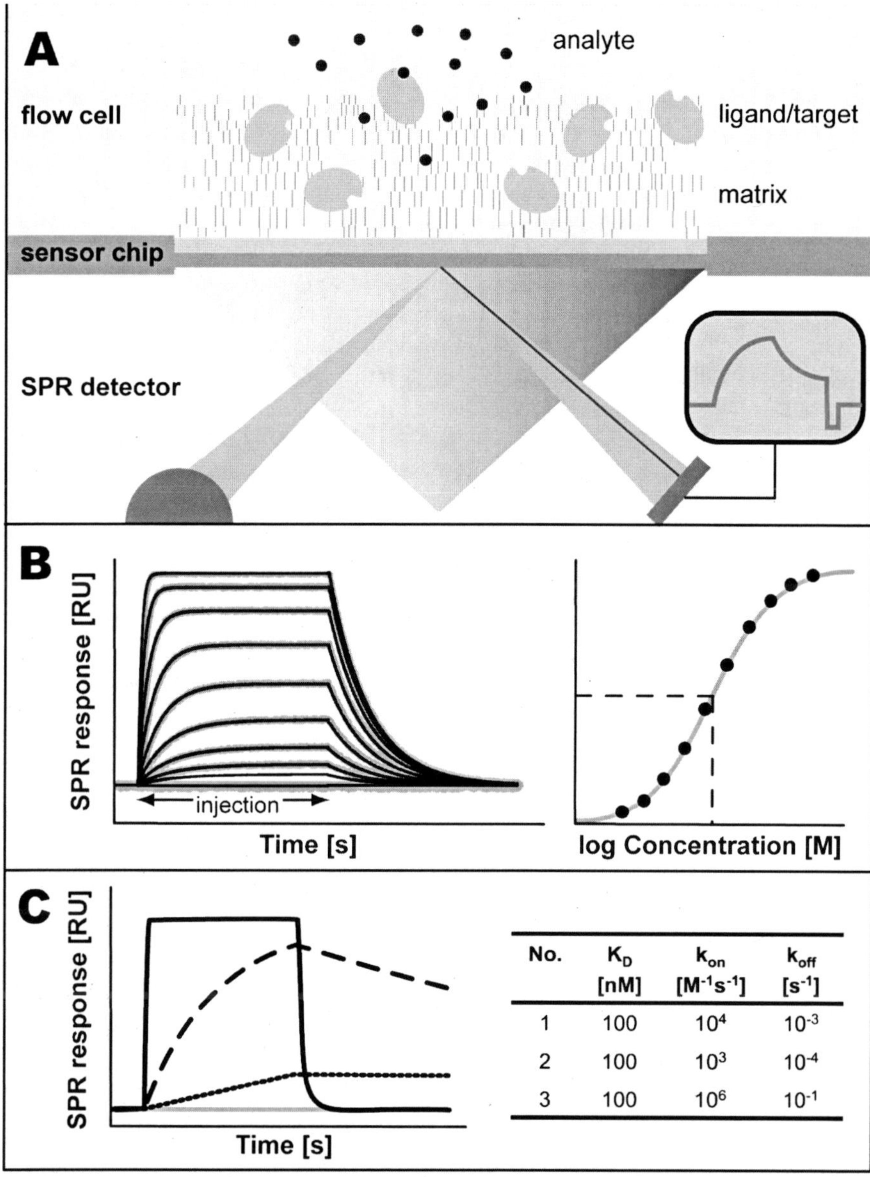

Fig. 1. Schematic representation of a typical SPR experiment (A). A target protein is immobilized on the hydrogel matrix of a gold sensor chips. Analyte molecules in solution interact with the target and generate a SPR response. The resulting sensorgrams (B) can be either kinetically fitted (left) or the affinity can be evaluated from the steady state responses (right). Different kinetic profiles may result in the same affinity constant (C).

(Nedelkov and Nelson 2003; Buijs and Franklin 2005; Yuk and Ha 2005), food analysis (Bergwerff and van Knapen 2006; Gambari and Feriotto 2006; Rasooly and Herold 2006), or drug discovery (Myszka and Rich 2000; Cooper 2002; Keusgen 2002; Myszka and Rich 2003). The success of SPR is founded on its unique ability to simultaneously acquire data about both affinity and kinetics of a molecular binding event. In addition, no labeling of the analyte molecule is required and typical experiments consume comparatively small sample amounts. While the principles of SPR detection have been already described in the late 1950s (Turbadar 1959), the first commercial SPR instruments were only introduced in 1990 (Jonsson, Fagerstam, Ivarsson, Johnsson, Karlsson, Lundh, Lofas, Persson, Roos, Ronnberg, Sjolander, Stenberg, Stahlberg, Urbaniczky, Ostlin and Malmqvist 1991; Malmqvist 1993). Even though a variety of biosensors based on SPR or similar technologies (e.g. plasmon waveguide resonance) had been developed since then, the vast majority of today's SPR analyses are performed with instruments produced by Biacore in Sweden. As a consequence, the following introduction and the examples are based on Biacore biosensors.

In a typical SPR experiment, one of the interaction partners (i.e. the ligand or target) is immobilized on the surface of a sensor chip, while the other partner (i.e. the analyte) is injected in solution (Fig. 1A). The sensor chip consists of a glass slide covered with a thin layer of gold. In order to avoid immobilization-derived denaturation of target proteins, a hydrogel matrix consisting of carboxymethyl dextran chains is attached to the surface. In addition to keeping the protein in a quasi-solvent environment, the matrix allows the introduction of functional groups for diverse immobilization chemistries and increases the coupling capacity by providing a three-dimensional network (Johnsson, Lofas and Lindquist 1991). Binding of the analyte in solution to the immobilized target leads to a change in refractive index (i.e. electron density) around the gold surface, which is detected as a change in the absorbance angle of totally reflected light by the SPR detector (Jonsson et al. 1991; Nagata and Handa 2000). It has been shown that this change, quantified as arbitrary 'resonance units' (RU), can be correlated with a mass increase on the chip surface, with 1 RU = 1 pg/mm^2 (Stenberg, Persson, Roos and Urbaniczky 1991). Upon injection of the analyte, the binding event can be observed in real-time, and the time course of the SPR signal is referred to as a sensorgram (Fig. 1B). As long as analyte molecules are injected, the sensorgram represents a changing ratio of associating and dissociating molecules until a steady state is reached. After injection end, the pure dissociation can be observed and extracted. The kinetic association and dissociation rate constants (k_{on} and k_{off}, respectively) are calculated by globally fitting the sensorgrams at various analyte concentrations to mathematical equations describing different binding models (Morton, Myszka and Chaiken 1995; Roden and Myszka 1996). In addition, the affinity (equilibrium dissociation constant K_D) of the interaction can be either calculated from the rate constants ($K_D = k_{off}/k_{on}$) or by fitting a concentration-dependent plot of the steady state signals (Fig. 1B). Additional yet less frequent applications of SPR assays include the determination of thermodynamic parameters (de Mol, Dekker, Broutin, Fischer and Liskamp 2005; Wear and Walkinshaw 2006), concentration analyses (Fagerstam, Frostell-Karlsson, Karlsson, Persson and Ronnberg 1992; Kikuchi, Uno, Nanami, Yoshimura, Iida, Fukushima and

Tsuchiya 2005), and competitive experiments (IC_{50}) (Karlsson, Kullman-Magnusson, Hamalainen, Remaeus, Andersson, Borg, Gyzander and Deinum 2000).

Although there are other technologies available for determining affinity constants, they often rely on analyte labeling (e.g. radioactive or fluorescence-based direct binding assays) or require large sample amounts as in the case of isothermal titration calorimetry (ITC). Alongside with stopped-flow spectrophotometry, SPR offers the unique advantage of acquiring kinetic rate constants. As illustrated in Fig. 1C, an interaction may show the same affinity constant but completely different kinetic profiles. While fast association and dissociation kinetics might be advantageous for high-throughput screening processes, an enzyme inhibitor should ideally show rapid association and very slow dissociation rates. In addition, changes in the kinetic profile may also indicate structural alterations and differential exposition of binding sites. As a consequence, kinetic profiling becomes increasingly important both in biological sciences and drug discovery (Markgren, Schaal, Hamalainen, Karlen, Hallberg, Samuelsson and Danielson 2002; de Mol et al. 2005).

While a label-free detection principle offers many advantages to the researcher, there are also some limitations and pitfalls to consider. Since SPR biosensors detect any change of electron density around the gold surface of the sensor chip, it may be hardly possible to discriminate between specific and non-specific signals. Along with secondary binding sites on the target protein itself, impurities of both target and analyte molecules may lead to further heterogeneities. Careful sample preparation is therefore a prerequisite for reliable SPR results. Even more, conformational changes of the target or contributions from the hydrogel matrix (e.g. by electrostatic effects) may be visible and overlaid with the binding signal (Winzor 2003). The choice of an appropriate reference surface and a critical evaluation of the raw data usually helps avoid such artifacts. Since the SPR response is directly correlated to the mass of the analyte, screening of small molecules often features unfavorable signal-to-noise ratios. However, improvements in both instrumental sensitivity and data processing expanded the range to lower molecular weight (< 100 Da) significantly (Myszka 2004). While no labeling of the analyte molecule is required, the experimental setup usually requires the target protein to be covalently attached to the surface. A variety of coupling chemistries is available for this purpose (Johnsson et al. 1991), most of which lead to a random and heterogeneous orientation on the surface. Dependent on the number and location of attachment sites, this may largely affect the target activity and the signal quality (Catimel, Nerrie, Lee, Scott, Ritter, Welt, Old, Burgess and Nice 1997; Peluso, Wilson, Do, Tran, Venkatasubbaiah, Quincy, Heidecker, Poindexter, Tolani, Phelan, Witte, Jung, Wagner and Nock 2003). Capturing by antibodies, streptavidin, or metal affinity surfaces (e.g. Ni^{2+}-NTA) is principally possible but often lacks surface stability or requires the expression of specific tags on the protein. Recently, a novel approach for the covalent yet oriented immobilization using a suicide enzyme reaction has been reported (Kindermann, George, Johnsson and Johnsson 2003), which might offer an alternative approach for coupling-sensitive proteins. The requirement of target attachment to a solid support raised some skepticism in the research community whether SPR-derived binding data are comparable to solution-based methods such as ITC. Meanwhile, a series of comparative studies could clearly demonstrate that SPR profiles are not only highly reproducible but indeed very comparable to data from ITC, stopped-flow spectrophotometry, or ultracentrifugation (Cannon, Papalia, Navratilova, Fisher, Roberts, Worthy, Stephen, Marchesini, Collins, Casper, Qiu, Satpaev, Liparoto, Rice, Gorshkova, Darling, Bennett, Sekar, Hommema, Liang, Day, Inman, Karlicek, Ullrich, Hodges, Chu, Sullivan, Simpson, Rafique, Luginbuhl, Westin, Bynum, Cachia, Li, Kao, Neurauter, Wong, Swanson and Myszka 2004; Katsamba, Navratilova, Calderon-Cacia, Fan, Thornton, Zhu, Bos, Forte, Friend, Laird-Offringa, Tavares, Whatley, Shi, Widom, Lindquist, Klakamp, Drake, Bohmann, Roell, Rose, Dorocke, Roth, Luginbuhl and Myszka 2006; Papalia, Leavitt, Bynum, Katsamba, Wilton, Qiu, Steukers, Wang, Bindu, Phogat, Giannetti, Ryan, Pudlak, Matusiewicz, Michelson, Nowakowski, Pham-Baginski, Brooks, Tieman, Bruce, Vaughn, Baksh, Cho, Wit, Smets, Vandersmissen, Michiels and Myszka 2006).

3 Applications in Complement Research

As a first line of human defense against pathogens and principal component of innate immunity, the complement system has to react dynamically and rapidly to changes in its environment. While the complement cascade is always kept on a low level of alertness (i.e. tick-over), activation of the system leads to a subsequent amplification of the response. In order to minimize the stress for host cells and tissues, the complement system has to be able to differentiate between healthy and pathogenic structures and to down-regulate the response after eliminating the threat. This complex and complicated task is fulfilled by a tailored and carefully balanced network of various soluble and membrane-bound proteins, which react with each other, with pathogenic proteins, or with surface structures (Walport 2001). For example, the complement component C3 alone is known to interact with more than 25 different molecules, making it one the most multifunctional and versatile proteins described so far (Lambris, Sahu and Wetsel 1998). During the regulation of the complement cascade, large conformational changes (as e.g. in the conversion from C3 to C3b), competition for binding sites, affinity enhancement through cofactors, and the formation of multi-molecule complexes play a crucial role. A detailed knowledge of the individual interactions is therefore essential for the understanding of complement function and regulation. In addition, these findings may build the base for the development of therapeutic interventions for many diseases that are connected with an erroneous regulation of the complement system (e.g. systemic lupus erythematosus, age-related macular degeneration). SPR technology is considered an ideal tool for investigating and describing interactions between complement components, since it allows not only a qualitative but also quantitative description of binding events without labeling requirements and with comparatively low sample amounts.

Comparable to many other research areas, SPR experienced increasing importance in the characterization of the complement system in recent years. As of writing this review, nearly 100 studies using SPR for investigating binding of

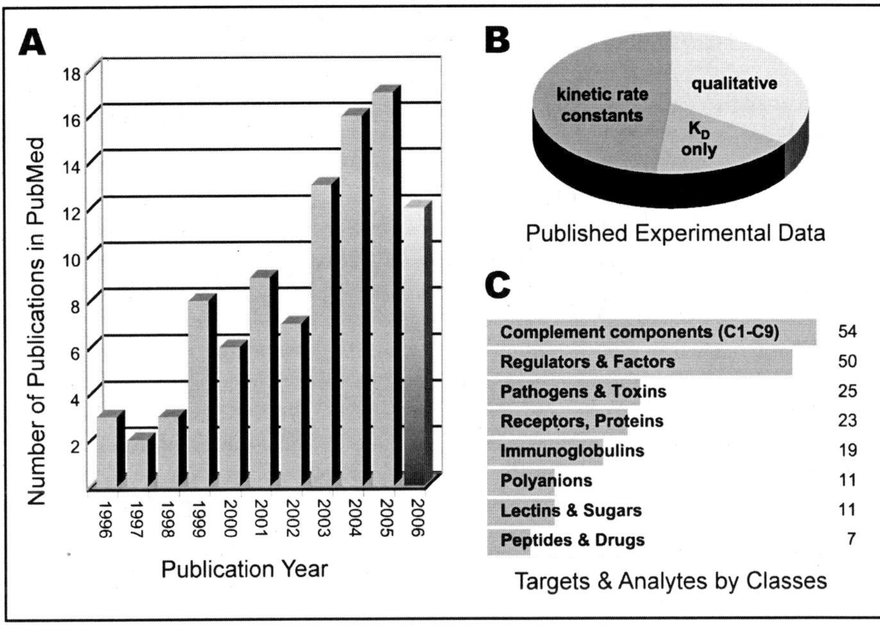

Fig. 2. Complement-related SPR studies in the PubMed database (as of November 2006) grouped by publication date (A), type of binding study (B), and by investigated classes of interacting molecules (C).

complement components are reported in the PubMed database. While first complement-related SPR studies appeared relatively late (more than 5 years after the introduction of the technique), a clear and steady increase in the number of publications could be observed in recent years (Fig. 2A). Almost half of the published studies took full advantage of the method and reported kinetic rate constants, while another 15% published affinity constants and a third of the publications only included qualitative descriptions of binding events (Fig. 2B). This ratio closely reflects the trend observed in the collectivity of SPR studies (Rich et al. 2005a). As expected, the majority of the investigated proteins belong to the complement components C1 to C9 and their fragments (Fig. 2C), with a clear emphasis on component C3 (more than 40%). Regulators of complement activation (RCAs) and complement factors contribute to a similar amount followed by various pathogenic as well as intrinsic proteins and receptors. Immunoglobulins and fragments thereof were often used for capturing, blocking, or identifying other components. However, some publication also screened antibodies for therapeutic purposes (see chapter 2.4). Besides the classical investigation of protein-protein interactions, SPR had also been used for characterizing the binding of complement components to carbohydrates (i.e. mannose-binding lectin [MBL] and MASPs of the lectin pathway), to polyanionic structures (e.g. heparin) and surfaces, or to drugs and therapeutic peptides (e.g. compstatin). In the following paragraphs, we illustrate the usefulness of SPR biosensors for answering crucial questions about the structure and function of proteins involved in the complement cascade.

3.1 Making connections – old and new ones

Obtaining biophysical profiles for individual interactions in the complement network has been the major aim for a large fraction of the published studies. While target immobilization is considered a major deviation from natural conditions for many biosensor experiments, the contrary is true in the case of C3 and its fragments. Upon activation of C3 to C3b, a previously buried thioester site is cleaved and exposed to the solvent. For a short period after activation, C3b can be covalently attached to polyhydroxyl surfaces, e.g. of bacterial cell walls, before the thioester site gets hydrolyzed. Coupled C3b is then able to form the C3 convertase and amplify the complement, finally leading to the formation of the membrane attack complex (MAC) and lysis of the cell. It has been shown that this mechanism can be mimicked on the sensor chip by in-situ formation of the convertase and that the carboxymethyl dextran matrix is accepted as a coupling surface by active C3b (Nilsson, Larsson, Hong, Elgue, Ekdahl, Sahu and Lambris 1998; see chapter 3.3, Fig. 3A).

Given its central role in the regulation of complement activation, it is not surprising that factor H (fH) has been investigated in large detail. This soluble plasma glycoprotein is mainly responsible for the down-regulation of complement activity on host surfaces (Pangburn 2000). As a member of the RCA family, fH consists of 20 modular structure elements called short consensus repeats (SCR) (Kirkitadze and Barlow 2001). Various binding sites for C3b, C3d, heparin, or pathogenic proteins are distributed over these SCRs leading to a complex binding behavior. In a series of mainly qualitative studies, SPR assays were used for dissecting and localizing individual binding sites on fH. For this purpose, truncated forms of fH containing only a limited number of SCRs were expressed and screened against different C3 fragments, heparin, and other polyanions (Jokiranta, Hellwage, Koistinen, Zipfel and Meri 2000; Jokiranta, Westin, Nilsson, Nilsson, Hellwage, Lofas, Gordon, Ekdahl and Meri 2001; Hellwage, Jokiranta, Friese, Wolk, Kampen, Zipfel and Meri 2002; Jokiranta, Cheng, Seeberger, Jozsi, Heinen, Noris, Remuzzi, Ormsby, Gordon, Meri, Hellwage and Zipfel 2005; Cheng, Hellwage, Seeberger, Zipfel, Meri and Jokiranta 2006). Combining these data, it was possible to detect or confirm three binding sites for each C3b (SCR 1-4, 8-15, 19-20) and heparin (SCR 7, 8-15, 20) on fH. Recently, Bernet *et al.* further characterized the fH-C3b interaction by kinetic means and detected a complex binding mode with a K_D of 50 nM for the primary binding site (Bernet, Mullick, Panse, Parab and Sahu 2004).

Heparin is not only known to interact with fH but also with a variety of other complement proteins. This makes this sulfated polysaccharide a potent regulator of both the classical and alternative complement pathway itself. In one of the most comprehensive kinetic profiling studies of a single target, Yu *et al.* analyzed the binding of a plethora of proteins (C1-C9, C1 inhibitor, factors B, H, I, and properdin) to immobilized, biotinylated porcine heparin (Yu, Munoz, Edens and Linhardt 2005). All of the screened components showed affinity constants in the nanomolar range, from 2 nM for factor B up to 320 nM in case of C2. Interestingly, the dissociation rate constant showed a nearly ten-times higher variability than the association rate. Since some of the investigated analytes showed a strong non-specific background signal on the streptavidin-carboxymethyl dextran chip, a novel sensor chip surface based on polyethylenglycol chains was developed (Munoz, Yu, Hallock, Edens and Linhardt 2005) and used for this study.

The close relationship between complement activation, inflammation, and wound healing makes a connection between the complement and the coagulation network not only possible, but very likely. However, the mechanisms how these cascades might be linked to each other are still poorly understood and just at the beginning of investigation (Markiewski, Nilsson, Nilsson-Ekdahl, Mollnes and Lambris 2007). Again, SPR studies have the potential to detect and describe important connections and bring more light into the pending questions. One of the most promising advances in this area was made by del Conde and coworkers (Del Conde, Cruz, Zhang, Lopez and Afshar-Kharghan 2005). By combining SPR with immunoblotting and flow cytometry, the authors could demonstrate an interaction between C3b and P-selectin, which is primarily expressed on platelets. In another study (Vaziri-Sani, Hellwage, Zipfel, Sjoholm, Iancu and Karpman 2005), factor H was found to interact with platelets via the GPIIb/IIIa receptor or thrombospondin, a glycoprotein known to induce platelet aggregation. SPR studies showed a direct interaction of fH with thrombospondin (K_D = 49 nM), which could be localized to the heparin binding area on SCR 20 by differential fH fragment binding and heparin-inhibition assays. These findings were connected by the authors to the pathogenesis of atypical hemolytic uremic syndrome (aHUS), where mutation in the C-terminal part of fH may lead to an insufficient down-regulation of the complement system. Finally, a series of studies used SPR to investigate the interaction between the C4-binding protein (C4BP) and the vitamin K-dependent anticoagulant protein S (He, Shen, Malmborg, Smith, Dahlback and Linse 1997; Evenas, Garcia De Frutos, Linse and Dahlback 1999; Giri, Linse, Garcia de Frutos, Yamazaki, Villoutreix and Dahlback 2002).

An involvement of the complement cascade in the pathogenesis of prion diseases, most likely through the classical pathway, has long been discussed. In a recent SPR-based study, Blanquet-Grossard *et al.* indeed reported a direct interaction between mouse prion protein (PrP) and human C1q (Blanquet-Grossard, Thielens, Vendrely, Jamin and Arlaud 2005). When PrP was covalently immobilized on the sensor chip, both full-length C1q and its isolated globular domain showed binding in the nanomolar range. However, the affinity of globular domain was reduced by more than a factor of 40, which could be almost exclusively attributed to a slower kinetic on rate. This effect was explained by an avidity effect, since intact C1q is a hexameric protein whereas the globular domain is monomeric. Interestingly, the binding affinity was essentially lost when soluble PrP was injected over immobilized C1q. While large differences upon reversal of the interaction partners usually indicate a problem in the assay format, they may indeed be attributed to changes in functionality in this case. By performing a series of ELISA the authors found evidence that soluble PrP undergoes significant conformational changes upon immobilization. Addition of Cu^{2+} ions (but not Mg^{2+}, Mn^{2+}, Zn^{2+}, or Ni^{2+}) led to a large increase in binding signal intensity, while the kinetic rate constants and the resulting affinity remained largely unchanged. These findings were consistent with a model that copper ions induce a pathogenic conversion of PrP and that more 'active' molecules are available on the chip surface upon addition of Cu^{2+}. Based on these SPR data, the authors therefore concluded that C1q might represent a natural sensor of converted prion protein.

3.2 The good and evil side of complement: interaction with pathogens

A central task of the complement system is the elimination of pathogenic intruders from the human body, which it usually fulfills dutifully and efficiently. However, some of the pathogens have found ways to avoid or even mis-use the complement system for their own fate. Cell-bound complement components and receptors may serve as target structures to some microorganisms for entering human cells (Lindahl, Sjobring and Johnsson 2000; Walport 2001). For example, the Eppstein-Barr virus (EBV) uses an interaction between its envelope glycoprotein 350 (gp350) and the complement receptor 2 (CR2) for this purpose (Nemerow, Mold, Schwend, Tollefson and Cooper 1987; Tanner, Weis, Fearon, Whang and Kieff 1987). As a consequence, B cells, which predominantly express CR2, are the major target of this virus. Using kinetic SPR profiles, the binding behavior of gp350 could be elucidated and compared to the natural CR2 ligands iC3b and C3d (Sarrias, Franchini, Canziani, Argyropoulos, Moore, Sahu and Lambris 2001). In contrast to the C3 fragment, which showed complex binding kinetics and affinities in the μM range, EBV gp350 featured 100 to 1000 times stronger affinities and the interaction followed a simple 1:1 binding model indicating a single gp350 binding site on CR2. Both the association and dissociation rate constants were responsible for the remarkable increase in affinity. In a recent SPR study, interferon α (IFN-α) and the immunoregulatory protein CD23 were also identified as ligands for CR2 (Asokan, Hua, Young, Gould, Hannan, Kraus, Szakonyi, Grundy, Chen, Crow and Holers 2006). Since both the infection with EBV (Ioannou and Isenberg 2000) and a therapeutic administration of IFN-α (James, Kaufman, Farris, Taylor-Albert, Lehman and Harley 1997) have been discussed as triggers for the autoimmune disease systemic lupus erythematosus (SLE), these findings may contribute to a better understanding of the pathological mechanisms behind SLE.

Due to its ability to differentiate between pathogenic and host surfaces, factor H is a major target for many bacterial intruders. Since an enrichment of fH on the bacterial surface leads to the same down-regulation of the complement cascade as on host surfaces, some bacteria (e.g. streptococci) produce fH-binding surface proteins as a part of their complement evasion strategy (Rautemaa and Meri 1999). SPR assays served as an important tool to describe or verify interactions between factor H and many pathogenic surface proteins (Table 1). In the case of the streptococcal protein Hic, kinetic profiles for the fH-Hic interaction were already included in the initial description of the novel protein (Janulczyk, Iannelli, Sjoholm, Pozzi and Bjorck 2000). By using a panel of fH fragments, the Hic binding site could be localized within SCR

Table 1. Pathogenic surface proteins that were analyzed by SPR for their interaction with fH.

Protein	Organism	Reference
Bac (β protein)	Streptococcus agalactiae	(Jarva et al. 2004)
Fba	Streptococcus pyrogenes	(Pandiripally et al. 2003)
Hic	Streptococcus pneumoniae	(Janulczyk et al. 2000) (Jarva et al. 2002) (Jarva et al. 2004)
LfhA	Leptospira interrogans	(Verma et al. 2006)
OspE	Borrelia burgdorferi	(Hellwage et al. 2001) (Alitalo et al. 2004)
OspE	Borrelia garinii	(Alitalo et al. 2005)
PspC	Streptococcus pneumoniae	(Dave et al. 2004)

8-11 (Jarva, Janulczyk, Hellwage, Zipfel, Bjorck and Meri 2002). Finally, it was demonstrated that Hic binds to more than one site in the middle part of fH (SCR 8-11, 12-14) and that its recognition is comparable to the structurally related streptococcal protein Bac (Jarva, Hellwage, Jokiranta, Lehtinen, Zipfel and Meri 2004). However, competition experiments with heparin suggested a slightly different binding mode since only Bac but not Hic binding was inhibited by heparin.

A viral strategy for evading an attack of the complement system is the production of RCA-like molecules (Mastellos, Morikis, Isaacs, Holland, Strey and Lambris 2003). Similar to the natural complement regulators, these proteins are also composed of SCR domains and are therefore able to act as cofactors or to interfere with complex formation. One of those proteins, the vaccinia virus complement control protein (VCP), has been extensively studied. While VCP was previously known to bind C3b, C4b, and heparin, SPR studies demonstrated that all four SCR domains are required for interaction with C3b/C4b, whereas heparin only bound to the C-terminal part of VCP (Smith, Sreenivasan, Krishnasamy, Judge, Murthy, Arjunwadkar, Pugh and Kotwal 2003). A more detailed SPR study by Bernet *et al.* allowed an even deeper insight in the molecular recognition pattern of VCP (Bernet et al. 2004). Despite some indications for multiple C3b/C4b binding sites in previous studies, the authors could demonstrate that VCP binding to these molecules follows a simple 1:1 binding mode with a nearly 5-fold stronger affinity for C4b. Competition experiments showed that both C3b and C4b interact within the same region of VCP. Furthermore, the VCP binding sites could be localized in the C4c and C3dg fragment, respectively. The study also illustrates the importance of the assay design: While an oriented immobilization of C3b/C4b on the sensor chip led to homogeneous signals following a simple 1:1 kinetic model, the reversed assay with randomly immobilized VCP resulted in significant heterogeneities and decreased affinity. Finally, the authors could demonstrate a large influence of the charge state by performing binding studies in buffers of increasing ionic strength. Latter was further investigated by Sfyroera *et al.*, who analyzed hybrid mutants between VCP and the smallpox inhibitor of complement enzymes (SPICE) in combination with electrostatic modeling (Sfyroera, Katragadda, Morikis, Isaacs and Lambris 2005). VCP and SPICE differ by only 11 amino acids but show a 1000-fold different regulatory activity. The study demonstrated that this activity could be largely influenced by modulating the electrostatic potential of the hybrid proteins.

The majority of the available SPR studies rely on isolated surface proteins of the pathogens for the binding experiments. However, in a series of publications about the interaction of echoviruses with the decay accelerating factor (DAF), whole virus particles were immobilized on the flow cell (Lea, Powell, McKee, Evans, Brown, Stuart and van der Merwe 1998; Goodfellow, Evans, Blom, Kerrigan, Miners, Morgan and Spiller 2005; Pettigrew, Williams, Kerrigan, Evans, Lea and Bhella 2006). Generally, all tested echoviruses featured rapid kinetic rate constants and K_D values in the low micromolar range. All steady state data sets could be fitted to a single-site binding model.

3.3 The complexity of complex formation

The ability to observe molecular interactions in real-time makes SPR technology an ideal tool for studying the formation of multi-molecular complexes. Typically, one central component is immobilized on the sensor chip, while the additional components are injected sequentially or as a mixture. Changes in signal intensity or dissociation half-times provide information about the formation and stability. By varying buffer conditions or adding cofactors, additional details about the essential requirements or driving forces of the complex formation can be obtained.

The C3 convertase is unquestionably the most important complex in the complement cascade. Upon hydrolysis of a buried thioester group in the native C3 molecule, the protein undergoes a large conformational change to C3(H_2O). During this process, binding sites for factor B are exposed, which leads to a C3(H_2O)B complex. The serine protease factor D, which is able to cleave complex-bound but not soluble factor B in its Ba and Bb fragments, leads to the formation of the initial convertase C3(H_2O)Bb and activation of a serine protease site on Bb. This complex cleaves native C3 (184 kDa) into the anaphylatoxin C3a (9 kDa) and C3b (175 kDa). The majority of C3b is covalently bound to polyanionic surfaces (e.g. bacterial cell walls) and forms the final C3 convertase (C3bBb) using factors B and D. The generation of additional C3b by the C3 convertase leads to the amplification of the response (amplification loop). The formation of the alternative pathway convertase on a SPR sensor chip was initially described by Nilson and coworkers for investigating the influence of the complement inhibitor compstatin (Nilsson et al. 1998). The authors immobilized purified C3b covalently on a carboxymethyl dextran sensor chip and sequentially injected factor B and D in the presence of Ni^{2+}. When native C3 was injected over the chip-bound complex, a signal increase caused by the deposition of newly formed C3b on the sensor chip via its thioester moiety could be observed (Fig. 3A). As proposed, no C3b deposition occurred when the C3 inhibitor compstatin was added. The in-situ convertase formation for the oriented immobilization of C3b was further developed and compared with standard amine coupling (Jokiranta et al. 2001). While probably closest to physiological conditions, this in-situ method is rather time

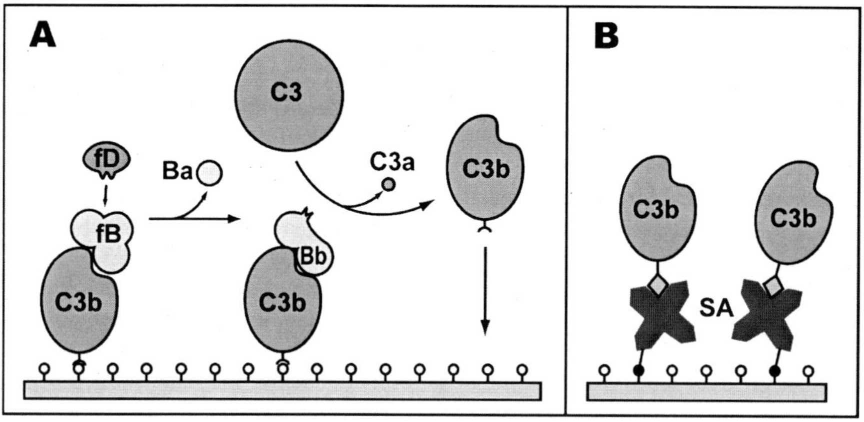

Fig. 3. Oriented immobilization of C3b via its thioester moiety by on-chip assembly of the C3 convertase (A) or by in vitro biotinylation and coupling to a streptavidin (SA)-coated sensor chip (B).

and protein consuming. Therefore, an alternative approach has been suggested by Sarrias *et al.*, in which the thioester domain of C3 is selectively biotinylated after activation and coupled to streptavidin-coated sensor chips (Sarrias et al. 2001). Biotinylated C3 samples can be further cleaved, e.g. to biotinylated C3b, stored, and freshly immobilized in an oriented manner (Fig. 3B).

The ability of SPR to monitor the assembly and stability of complexes in real-time was further used to study the influence of different regulators on the C3 convertase formation with an emphasis on either DAF (Harris, Abbott, Smith, Morgan and Lea 2005) or properdin (Hourcade 2006). Using immobilized C3b, it could be shown that the stability of the convertase complex (C3bBb) is significantly larger than the proenzyme complex (C3bB) in the presence of Mg^{2+} (Harris et al. 2005). Interestingly, only C3bB but not C3bBb was sensitive to removal of magnesium by EDTA indicating that Mg^{2+} is only important in the initial convertase formation step. On the other hand, only the fully assembled convertase was destabilized by injection of soluble DAF. The authors also demonstrated that DAF interacts with both C3b and factor B (via its vWF-A domain) in the micromolar affinity range. Kinetic analysis of the interaction between C3b and factor B indicated a complex binding mode, which might include a conformational change. In contrast to de-stabilizing proteins like DAF or factor H, properdin is the only known physiological promoter of the C3 convertase (Schwaeble and Reid 1999). In a recent publication (Hourcade 2006), Hourcade found evidence that properdin not only stabilizes pre-formed convertase but also promotes its formation. The author suggested that the multimeric properdin molecule is able to form a lattice, where multiple stages of the convertase (C3b, C3bB, C3bBb) are bound and stabilized. It was even found that iC3b, which was considered as deactivated C3b, may still contribute to convertase formation due to its affinity to properdin. A similar on-chip assembly approach has also been performed to investigate the formation of the membrane attack complex (Thai and Ogata 2005).

3.4 Developing therapeutic interventions

Despite the frequent use of SPR technology in drug discovery and development, only a very limited number of applications in the field of complement therapeutics is reported so far. Almost all of them involve macromolecules such as peptides and monoclonal antibodies. Only a single publication presents SPR data involving a small molecular therapeutic, the antibiotic drug chloramphenicol, as an inhibitor of the interaction between the E. coli adhesin DraE and decay accelerating factor (DAF) (Pettigrew, Anderson, Billington, Cota, Simpson, Urvil, Rabuzin, Roversi, Nowicki, du Merle, Le Bouguenec, Matthews and Lea 2004).

A whole series of independent publications deals with the neutralization of the anaphylatoxins C3a and C5a. Those small protein fragments are released from C3 and C5, respectively, upon complement activation and are known to induce various inflammatory and anaphylactic reactions. Specific blockage of these anaphylatoxins is therefore believed to be of substantial therapeutic significance. In one study, a panel of specific and non-specific IgG antibodies was screened on immobilized C3a and C5a (Basta, Van Goor, Luccioli, Billings, Vortmeyer, Baranyi, Szebeni, Alving, Carroll, Berkower, Stojilkovic and Metcalfe 2003). Surprisingly, all of the non-specific control antibodies also showed significant binding towards the anaphylatoxin. Since no interaction was detected with isolated Fc fragments and the specific antigens did not compete with C3a/C5a, the binding site was attributed to the constant region of the Fab fragment. Fung *et al.* described and characterized a C5a-neutralizing antibody, which detects both free C5a as well as the C5a region within C5 with subnanomolar affinity (Fung, Lu, Fure, Sun, Sun, Shi, Dou, Su, Swanson and Mollnes 2003). A different approach was taken by Fujita *et al.*, who designed C5a-complementary peptides by using molecular modeling (Fujita, Farkas, Campbell, Baranyi, Okada and Okada 2004). One of these peptides showed strong binding towards immobilized C5a.

One of the most potent and best characterized complement inhibitors so far is the cyclic tridecapeptide compstatin. While the initial discovery through screening of phage display libraries was performed by ELISA (Sahu, Kay and Lambris 1996), SPR data were involved in most stages of later development. For example, Nilsson could demonstrate that the on-chip formation of the C3 convertase (see also chapters 3.1 and 3.3, Fig. 3A) was successfully inhibited by compstatin (Nilsson et al. 1998). In a later study, detailed kinetic profiles for the interaction with C3, C3(H_2O) as well as their fragments C3b, C3c, and C3d have been evaluated (Sahu, Soulika, Morikis, Spruce, Moore and Lambris 2000). While binding to C3c could not be reliably detected by ELISA, SPR analysis showed a clear but much weaker interaction compared to native C3 (9 μM vs. 130 nM). Since compstatin did not bind to C3d at all, the authors concluded a limited exposition of the binding site upon cleavage of C3. Qualitative binding studies further revealed specificity towards component C3, with no binding to C4 and C5, as well as species specificity for primate C3 (human and baboon), while no binding was detected for mouse and rat C3 (Sahu, Morikis and Lambris 2003). Finally, a combination of SPR with various other assays was used for the characterization of different compstatin analogs (Soulika, Morikis, Sarrias, Roy, Spruce, Sahu and Lambris 2003).

4 Conclusions & Perspectives

Despite the comparatively small number of complement-related SPR publications, they cover an impressive range of applications; from qualitative screening up to detailed kinetic profiles, from small molecules and peptides up to multimolecular complexes and virus particles, and from picomolar up to millimolar affinity ranges. Besides their key application of measuring binding constants, SPR instruments are extensively used in complement research for the localization of binding sites. The reason for this may be due to the modular structure of many complement regulators (e.g. RCA) and the distinct fragmentation pattern of many complement components (e.g. C3). One of the major advantages of SPR technology for these applications is the ability to get a fairly detailed picture of changes in binding behavior upon cleavage or mutation, even with single injections and limited sample amounts. Another advantage is that the same assay can be easily repeated under different experimental conditions (e.g. buffer composition, pH, temperature). Since many of the interactions within the complement cascade are driven by electrostatic interactions, SPR assays were often used for detecting different binding patterns in buffers of increasing ionicity. However, the large involvement of charge effects may also lead to method-related artifacts when non-specific binding to the negatively charged carboxymethyl dextran matrix is involved. Careful assay design, control experiments and cross-validation with other methods are therefore highly important. The same is true for the coupling of the target molecule, which might significantly influence the homogeneity and affinity of the interaction. Fortunately, some of the complement components can be immobilized in a near-natural way with high activity.

With the recent publications of several crystal structures of complement proteins, such as C3 (Janssen, Huizinga, Raaijmakers, Roos, Daha, Nilsson-Ekdahl, Nilsson and Gros 2005), C3b (Abdul Ajees, Gunasekaran, Volanakis, Narayana, Kotwal and Krishna Murthy 2006; Janssen, Christodoulidou, McCarthy, Lambris and Gros 2006; Wiesmann, Katschke, Yin, Helmy, Steffek, Fairbrother, McCallum, Embuscado, Deforge, Hass and van Lookeren Campagne 2006), C3c (Janssen et al. 2005), C3d (Nagar, Jones, Diefenbach, Isenman and Rini 1998), or factor B and its fragment Bb (Ponnuraj, Xu, Macon, Moore, Volanakis and Narayana 2004), the mapping and characterization of interaction sites becomes even more exciting. SPR may play a major role in linking these static images to dynamic functions. While powerful as a technology by itself, combinations of SPR with other interaction technologies such as ITC may further potentiate its value. Increasing knowledge about the complement network and the availability of novel crystal structures are also believed to bring new fuel to the development of complement therapeutics, which may be another key area of SPR.

The almost unique ability to obtain kinetic profiles is a clear advantage of SPR. Especially when comparing similar analytes, differences in the kinetic on- and off-rates may e.g. indicate differential accessibility of a binding site or the formation of secondary contacts. However, even though nearly half of all complement-related SPR studies contained kinetic data, very few of them interpreted or connected the results in respect of structural or functional properties. Here again, the availability of crystal structures and the combination with complementary methods may help tapping the full potential of SPR technology.

The recent challenges in elucidating the structure and connections of the complement network seem to be a perfect match to the capabilities of SPR biosensors, and we are strongly looking forward to a continuous increase of related articles in the future.

References

Abdul Ajees, A., Gunasekaran, K., Volanakis, J.E., Narayana, S.V., Kotwal, G.J. and Krishna Murthy, H.M. (2006) The structure of complement C3b provides insights into complement activation and regulation. Nature 444, 221-225.

Alitalo, A., Meri, T., Chen, T., Lankinen, H., Cheng, Z.Z., Jokiranta, T.S., Seppala, I.J., Lahdenne, P., Hefty, P.S., Akins, D.R. and Meri, S. (2004) Lysine-dependent multipoint binding of the Borrelia burgdorferi virulence factor outer surface protein E to the C terminus of factor H. J Immunol 172, 6195-6201.

Alitalo, A., Meri, T., Comstedt, P., Jeffery, L., Tornberg, J., Strandin, T., Lankinen, H., Bergstrom, S., Cinco, M., Vuppala, S.R., Akins, D.R. and Meri, S. (2005) Expression of complement factor H binding immunoevasion proteins in Borrelia garinii isolated from patients with neuroborreliosis. Eur J Immunol 35, 3043-3053.

Asokan, R., Hua, J., Young, K.A., Gould, H.J., Hannan, J.P., Kraus, D.M., Szakonyi, G., Grundy, G.J., Chen, X.S., Crow, M.K. and Holers, V.M. (2006) Characterization of human complement receptor type 2 (CR2/CD21) as a receptor for IFN-alpha: a potential role in systemic lupus erythematosus. J Immunol 177, 383-394.

Basta, M., Van Goor, F., Luccioli, S., Billings, E.M., Vortmeyer, A.O., Baranyi, L., Szebeni, J., Alving, C.R., Carroll, M.C., Berkower, I., Stojilkovic, S.S. and Metcalfe, D.D. (2003) F(ab)'2-mediated neutralization of C3a and C5a anaphylatoxins: a novel effector function of immunoglobulins. Nat Med 9, 431-438.

Bergwerff, A.A. and van Knapen, F. (2006) Surface plasmon resonance biosensors for detection of pathogenic microorganisms: strategies to secure food and environmental safety. J AOAC Int 89, 826-831.

Bernet, J., Mullick, J., Panse, Y., Parab, P.B. and Sahu, A. (2004) Kinetic analysis of the interactions between vaccinia virus complement control protein and human complement proteins C3b and C4b. J Virol 78, 9446-9457.

Blanquet-Grossard, F., Thielens, N.M., Vendrely, C., Jamin, M. and Arlaud, G.J. (2005) Complement protein C1q recognizes a conformationally modified form of the prion protein. Biochemistry 44, 4349-4356.

Buijs, J. and Franklin, G.C. (2005) SPR-MS in functional proteomics. Brief Funct Genomic Proteomic 4, 39-47.

Cannon, M.J., Papalia, G.A., Navratilova, I., Fisher, R.J., Roberts, L.R., Worthy, K.M., Stephen, A.G., Marchesini, G.R., Collins, E.J., Casper, D., Qiu, H., Satpaev, D., Liparoto, S.F., Rice, D.A., Gorshkova, II, Darling, R.J., Bennett, D.B., Sekar, M., Hommema, E., Liang, A.M., Day, E.S., Inman, J., Karlicek, S.M., Ullrich, S.J., Hodges, D., Chu, T., Sullivan, E., Simpson, J., Rafique, A., Luginbuhl, B., Westin, S.N., Bynum, M., Cachia, P., Li, Y.J., Kao, D., Neurauter, A., Wong, M., Swanson, M. and Myszka, D.G. (2004) Comparative analyses of a small molecule/enzyme interaction by multiple users of Biacore technology. Anal Biochem 330, 98-113.

Catimel, B., Nerrie, M., Lee, F.T., Scott, A.M., Ritter, G., Welt, S., Old, L.J., Burgess, A.W. and Nice, E.C. (1997) Kinetic analysis of the interaction between the monoclonal antibody A33 and its colonic epithelial antigen by the use of an optical biosensor. A comparison of immobilisation strategies. J Chromatogr A 776, 15-30.

Cheng, Z.Z., Hellwage, J., Seeberger, H., Zipfel, P.F., Meri, S. and Jokiranta, T.S. (2006) Comparison of surface recognition and C3b binding properties of mouse and human complement factor H. Mol Immunol 43, 972-979.

Cooper, M.A. (2002) Optical biosensors in drug discovery. Nat Rev Drug Discov 1, 515-528.

Dave, S., Carmicle, S., Hammerschmidt, S., Pangburn, M.K. and McDaniel, L.S. (2004) Dual roles of PspC, a surface protein of Streptococcus pneumoniae, in binding human secretory IgA and factor H. J Immunol 173, 471-477.

de Mol, N.J., Dekker, F.J., Broutin, I., Fischer, M.J. and Liskamp, R.M. (2005) Surface plasmon resonance thermodynamic and kinetic analysis as a strategic tool in drug design. Distinct ways for phosphopeptides to plug into Src- and Grb2 SH2 domains. J Med Chem 48, 753-763.

Del Conde, I., Cruz, M.A., Zhang, H., Lopez, J.A. and Afshar-Kharghan, V. (2005) Platelet activation leads to activation and propagation of the complement system. J Exp Med 201, 871-879.

Evenas, P., Garcia De Frutos, P., Linse, S. and Dahlback, B. (1999) Both G-type domains of protein S are required for the high-affinity interaction with C4b-binding protein. Eur J Biochem 266, 935-942.

Fagerstam, L.G., Frostell-Karlsson, A., Karlsson, R., Persson, B. and Ronnberg, I. (1992) Biospecific interaction analysis using surface plasmon resonance detection applied to kinetic, binding site and concentration analysis. J Chromatogr 597, 397-410.

Fujita, E., Farkas, I., Campbell, W., Baranyi, L., Okada, H. and Okada, N. (2004) Inactivation of C5a anaphylatoxin by a peptide that is complementary to a region of C5a. J Immunol 172, 6382-6387.

Fung, M., Lu, M., Fure, H., Sun, W., Sun, C., Shi, N.Y., Dou, Y., Su, J., Swanson, X. and Mollnes, T.E. (2003) Pre-neutralization of C5a-mediated effects by the monoclonal antibody 137-26 reacting with the C5a moiety of native C5 without preventing C5 cleavage. Clin Exp Immunol 133, 160-169.

Gambari, R. and Feriotto, G. (2006) Surface plasmon resonance for detection of genetically modified organisms in the food supply. J AOAC Int 89, 893-897.

Giri, T.K., Linse, S., Garcia de Frutos, P., Yamazaki, T., Villoutreix, B.O. and Dahlback, B. (2002) Structural requirements of anticoagulant protein S for its binding to the complement regulator C4b-binding protein. J Biol Chem 277, 15099-15106.

Goodfellow, I.G., Evans, D.J., Blom, A.M., Kerrigan, D., Miners, J.S., Morgan, B.P. and Spiller, O.B. (2005) Inhibition of coxsackie B virus infection by soluble forms of its receptors: binding affinities, altered particle formation, and competition with cellular receptors. J Virol 79, 12016-12024.

Harris, C.L., Abbott, R.J., Smith, R.A., Morgan, B.P. and Lea, S.M. (2005) Molecular dissection of interactions between components of the alternative pathway of complement and decay accelerating factor (CD55). J Biol Chem 280, 2569-2578.

He, X., Shen, L., Malmborg, A.C., Smith, K.J., Dahlback, B. and Linse, S. (1997) Binding site for C4b-binding protein in vitamin K-dependent protein S fully contained in carboxy-terminal laminin-G-type repeats. A study using recombinant factor IX-protein S chimeras and surface plasmon resonance. Biochemistry 36, 3745-3754.

Hellwage, J., Jokiranta, T.S., Friese, M.A., Wolk, T.U., Kampen, E., Zipfel, P.F. and Meri, S. (2002) Complement C3b/C3d and cell surface polyanions are recognized by overlapping binding sites on the most carboxyl-terminal domain of complement factor H. J Immunol 169, 6935-6944.

Hellwage, J., Meri, T., Heikkila, T., Alitalo, A., Panelius, J., Lahdenne, P., Seppala, I.J. and Meri, S. (2001) The complement regulator factor H binds to the surface protein OspE of Borrelia burgdorferi. J Biol Chem 276, 8427-8435.

Hourcade, D.E. (2006) The role of properdin in the assembly of the alternative pathway C3 convertases of complement. J Biol Chem 281, 2128-2132.

Ioannou, Y. and Isenberg, D.A. (2000) Current evidence for the induction of autoimmune rheumatic manifestations by cytokine therapy. Arthritis Rheum 43, 1431-1442.

James, J.A., Kaufman, K.M., Farris, A.D., Taylor-Albert, E., Lehman, T.J. and Harley, J.B. (1997) An increased prevalence of Epstein-Barr virus infection in young patients suggests a possible etiology for systemic lupus erythematosus. J Clin Invest 100, 3019-3026.

Janssen, B.J., Christodoulidou, A., McCarthy, A., Lambris, J.D. and Gros, P. (2006) Structure of C3b reveals conformational changes that underlie complement activity. Nature 444, 213-216.

Janssen, B.J., Huizinga, E.G., Raaijmakers, H.C., Roos, A., Daha, M.R., Nilsson-Ekdahl, K., Nilsson, B. and Gros, P. (2005) Structures of complement component C3 provide insights into the function and evolution of immunity. Nature 437, 505-511.

Janulczyk, R., Iannelli, F., Sjoholm, A.G., Pozzi, G. and Bjorck, L. (2000) Hic, a novel surface protein of Streptococcus pneumoniae that interferes with complement function. J Biol Chem 275, 37257-37263.

Jarva, H., Hellwage, J., Jokiranta, T.S., Lehtinen, M.J., Zipfel, P.F. and Meri, S. (2004) The group B streptococcal beta and pneumococcal Hic proteins are structurally related immune evasion molecules that bind the complement inhibitor factor H in an analogous fashion. J Immunol 172, 3111-3118.

Jarva, H., Janulczyk, R., Hellwage, J., Zipfel, P.F., Bjorck, L. and Meri, S. (2002) Streptococcus pneumoniae evades complement attack and opsonophagocytosis by expressing the pspC locus-encoded Hic protein that binds to short consensus repeats 8-11 of factor H. J Immunol 168, 1886-1894.

Johnsson, B., Lofas, S. and Lindquist, G. (1991) Immobilization of proteins to a carboxymethyldextran-modified gold surface for biospecific interaction analysis in surface plasmon resonance sensors. Anal Biochem 198, 268-277.

Jokiranta, T.S., Cheng, Z.Z., Seeberger, H., Jozsi, M., Heinen, S., Noris, M., Remuzzi, G., Ormsby, R., Gordon, D.L., Meri, S., Hellwage, J. and Zipfel, P.F. (2005) Binding of complement factor H to endothelial cells is mediated by the carboxy-terminal glycosaminoglycan binding site. Am J Pathol 167, 1173-1181.

Jokiranta, T.S., Hellwage, J., Koistinen, V., Zipfel, P.F. and Meri, S. (2000) Each of the three binding sites on complement factor H interacts with a distinct site on C3b. J Biol Chem 275, 27657-27662.

Jokiranta, T.S., Westin, J., Nilsson, U.R., Nilsson, B., Hellwage, J., Lofas, S., Gordon, D.L., Ekdahl, K.N. and Meri, S. (2001) Complement C3b interactions studied with surface plasmon resonance technique. Int Immunopharmacol 1, 495-506.

Jonsson, U., Fagerstam, L., Ivarsson, B., Johnsson, B., Karlsson, R., Lundh, K., Lofas, S., Persson, B., Roos, H., Ronnberg, I., Sjolander, S., Stenberg, E., Stahlberg, R., Urbaniczky, C., Ostlin, H. and Malmqvist, M. (1991) Real-time biospecific interaction analysis using surface plasmon resonance and a sensor chip technology. Biotechniques 11, 620-627.

Karlsson, R., Kullman-Magnusson, M., Hamalainen, M.D., Remaeus, A., Andersson, K., Borg, P., Gyzander, E. and Deinum, J. (2000) Biosensor analysis of drug-target interactions: direct and competitive binding assays for investigation of interactions between thrombin and thrombin inhibitors. Anal Biochem 278, 1-13.

Katsamba, P.S., Navratilova, I., Calderon-Cacia, M., Fan, L., Thornton, K., Zhu, M., Bos, T.V., Forte, C., Friend, D., Laird-Offringa, I., Tavares, G., Whatley, J., Shi, E., Widom, A., Lindquist, K.C., Klakamp, S., Drake, A., Bohmann, D., Roell, M., Rose, L., Dorocke, J., Roth, B., Luginbuhl, B. and Myszka, D.G. (2006) Kinetic analysis of a high-affinity antibody/antigen interaction performed by multiple Biacore users. Anal Biochem 352, 208-221.

Keusgen, M. (2002) Biosensors: new approaches in drug discovery. Naturwissenschaften 89, 433-444.

Kikuchi, Y., Uno, S., Nanami, M., Yoshimura, Y., Iida, S., Fukushima, N. and Tsuchiya, M. (2005) Determination of concentration and binding affinity of antibody fragments by use of surface plasmon resonance. J Biosci Bioeng 100, 311-317.

Kindermann, M., George, N., Johnsson, N. and Johnsson, K. (2003) Covalent and selective immobilization of fusion proteins. J Am Chem Soc 125, 7810-7811.

Kirkitadze, M.D. and Barlow, P.N. (2001) Structure and flexibility of the multiple domain proteins that regulate complement activation. Immunol Rev 180, 146-161.

Lambris, J.D., Sahu, A. and Wetsel, R. (1998). The chemistry and biology of C3, C4, and C5. In: J.E. Volanikis and M. Frank (Eds.), *The human complement system in health and disease*. Marcel Dekker Inc., New York, pp. 83-118.

Lea, S.M., Powell, R.M., McKee, T., Evans, D.J., Brown, D., Stuart, D.I. and van der Merwe, P.A. (1998) Determination of the affinity and kinetic constants for the interaction between the human virus echovirus 11 and its cellular receptor, CD55. J Biol Chem 273, 30443-30447.

Lindahl, G., Sjobring, U. and Johnsson, E. (2000) Human complement regulators: a major target for pathogenic microorganisms. Curr Opin Immunol 12, 44-51.

Malmqvist, M. (1993) Biospecific interaction analysis using biosensor technology. Nature 361, 186-187.

Markgren, P.O., Schaal, W., Hamalainen, M., Karlen, A., Hallberg, A., Samuelsson, B. and Danielson, U.H. (2002) Relationships between structure and interaction kinetics for HIV-1 protease inhibitors. J Med Chem 45, 5430-5439.

Markiewski, M.M., Nilsson, B., Nilsson Ekdahl, K., Mollnes, T.E. and Lambris, J.D. (2007) Complement and Coagulation: Strangers or Partners in Crime? Immunology.

Mastellos, D., Morikis, D., Isaacs, S.N., Holland, M.C., Strey, C.W. and Lambris, J.D. (2003) Complement: structure, functions, evolution, and viral molecular mimicry. Immunol Res 27, 367-386.

Morton, T.A., Myszka, D.G. and Chaiken, I.M. (1995) Interpreting complex binding kinetics from optical biosensors: a comparison of analysis by linearization, the integrated rate equation, and numerical integration. Anal Biochem 227, 176-185.

Munoz, E.M., Yu, H., Hallock, J., Edens, R.E. and Linhardt, R.J. (2005) Poly(ethylene glycol)-based biosensor chip to study heparin-protein interactions. Anal Biochem 343, 176-178.

Myszka, D.G. (2004) Analysis of small-molecule interactions using Biacore S51 technology. Anal Biochem 329, 316-323.

Myszka, D.G. and Rich, R.L. (2000) Implementing surface plasmon resonance biosensors in drug discovery. Pharm. Sci. Technol. Today 3, 310-317.

Myszka, D.G. and Rich, R.L. (2003) SPR's Impact on Drug Discovery: Resolution, Throughput, and Versatility. Drug Discovery World 5, 1-5.

Nagar, B., Jones, R.G., Diefenbach, R.J., Isenman, D.E. and Rini, J.M. (1998) X-ray crystal structure of C3d: a C3 fragment and ligand for complement receptor 2. Science 280, 1277-1281.

Nagata, K. and Handa, H. (2000). *Real-Time Analysis of Biomolecular Interactions: Applications of BIACORE.* Springer, Tokio

Nedelkov, D. and Nelson, R.W. (2003) Surface plasmon resonance mass spectrometry: recent progress and outlooks. Trends Biotechnol 21, 301-305.

Nemerow, G.R., Mold, C., Schwend, V.K., Tollefson, V. and Cooper, N.R. (1987) Identification of gp350 as the viral glycoprotein mediating attachment of Epstein-Barr virus (EBV) to the EBV/C3d receptor of B cells: sequence homology of gp350 and C3 complement fragment C3d. J Virol 61, 1416-1420.

Nilsson, B., Larsson, R., Hong, J., Elgue, G., Ekdahl, K.N., Sahu, A. and Lambris, J.D. (1998) Compstatin inhibits complement and cellular activation in whole blood in two models of extracorporeal circulation. Blood 92, 1661-1667.

Pandiripally, V., Wei, L., Skerka, C., Zipfel, P.F. and Cue, D. (2003) Recruitment of complement factor H-like protein 1 promotes intracellular invasion by group A streptococci. Infect Immun 71, 7119-7128.

Pangburn, M.K. (2000) Host recognition and target differentiation by factor H, a regulator of the alternative pathway of complement. Immunopharmacology 49, 149-157.

Papalia, G.A., Leavitt, S., Bynum, M.A., Katsamba, P.S., Wilton, R., Qiu, H., Steukers, M., Wang, S., Bindu, L., Phogat, S., Giannetti, A.M., Ryan, T.E., Pudlak, V.A., Matusiewicz, K., Michelson, K.M., Nowakowski, A., Pham-Baginski, A., Brooks, J., Tieman, B.C., Bruce, B.D., Vaughn, M., Baksh, M., Cho, Y.H., Wit, M.D., Smets, A., Vandersmissen, J., Michiels, L. and Myszka, D.G. (2006) Comparative analysis of 10 small molecules binding to carbonic anhydrase II by different investigators using Biacore technology. Anal Biochem 359, 94-105.

Peluso, P., Wilson, D.S., Do, D., Tran, H., Venkatasubbaiah, M., Quincy, D., Heidecker, B., Poindexter, K., Tolani, N., Phelan, M., Witte, K., Jung, L.S., Wagner, P. and Nock, S. (2003) Optimizing antibody immobilization strategies for the construction of protein microarrays. Anal Biochem 312, 113-124.

Pettigrew, D., Anderson, K.L., Billington, J., Cota, E., Simpson, P., Urvil, P., Rabuzin, F., Roversi, P., Nowicki, B., du Merle, L., Le Bouguenec, C., Matthews, S. and Lea, S.M. (2004) High resolution studies of the Afa/Dr adhesin DraE and its interaction with chloramphenicol. J Biol Chem 279, 46851-46857.

Pettigrew, D.M., Williams, D.T., Kerrigan, D., Evans, D.J., Lea, S.M. and Bhella, D. (2006) Structural and functional insights into the interaction of echoviruses and decay-accelerating factor. J Biol Chem 281, 5169-5177.

Ponnuraj, K., Xu, Y., Macon, K., Moore, D., Volanakis, J.E. and Narayana, S.V. (2004) Structural analysis of engineered Bb fragment of complement factor B: insights into the activation mechanism of the alternative pathway C3-convertase. Mol Cell 14, 17-28.

Rasooly, A. and Herold, K.E. (2006) Biosensors for the analysis of food- and waterborne pathogens and their toxins. J AOAC Int 89, 873-883.

Rautemaa, R. and Meri, S. (1999) Complement-resistance mechanisms of bacteria. Microbes Infect 1, 785-794.

Rich, R.L. and Myszka, D.G. (2000) Survey of the 1999 surface plasmon resonance biosensor literature. J Mol Recognit 13, 388-407.

Rich, R.L. and Myszka, D.G. (2001) Survey of the year 2000 commercial optical biosensor literature. J Mol Recognit 14, 273-294.

Rich, R.L. and Myszka, D.G. (2002) Survey of the year 2001 commercial optical biosensor literature. J Mol Recognit 15, 352-376.

Rich, R.L. and Myszka, D.G. (2003) A survey of the year 2002 commercial optical biosensor literature. J Mol Recognit 16, 351-382.

Rich, R.L. and Myszka, D.G. (2005a) Survey of the year 2003 commercial optical biosensor literature. J Mol Recognit 18, 1-39.

Rich, R.L. and Myszka, D.G. (2005b) Survey of the year 2004 commercial optical biosensor literature. J Mol Recognit 18, 431-478.

Rich, R.L. and Myszka, D.G. (2006) Survey of the year 2005 commercial optical biosensor literature. J Mol Recognit 19, 478-534.
Roden, L.D. and Myszka, D.G. (1996) Global analysis of a macromolecular interaction measured on BIAcore. Biochem Biophys Res Commun 225, 1073-1077.
Sahu, A., Kay, B.K. and Lambris, J.D. (1996) Inhibition of human complement by a C3-binding peptide isolated from a phage-displayed random peptide library. J Immunol 157, 884-891.
Sahu, A., Morikis, D. and Lambris, J.D. (2003) Compstatin, a peptide inhibitor of complement, exhibits species-specific binding to complement component C3. Mol Immunol 39, 557-566.
Sahu, A., Soulika, A.M., Morikis, D., Spruce, L., Moore, W.T. and Lambris, J.D. (2000) Binding kinetics, structure-activity relationship, and biotransformation of the complement inhibitor compstatin. J Immunol 165, 2491-2499.
Sarrias, M.R., Franchini, S., Canziani, G., Argyropoulos, E., Moore, W.T., Sahu, A. and Lambris, J.D. (2001) Kinetic analysis of the interactions of complement receptor 2 (CR2, CD21) with its ligands C3d, iC3b, and the EBV glycoprotein gp350/220. J Immunol 167, 1490-1499.
Schwaeble, W.J. and Reid, K.B. (1999) Does properdin crosslink the cellular and the humoral immune response? Immunol Today 20, 17-21.
Sfyroera, G., Katragadda, M., Morikis, D., Isaacs, S.N. and Lambris, J.D. (2005) Electrostatic modeling predicts the activities of orthopoxvirus complement control proteins. J Immunol 174, 2143-2151.
Smith, S.A., Sreenivasan, R., Krishnasamy, G., Judge, K.W., Murthy, K.H., Arjunwadkar, S.J., Pugh, D.R. and Kotwal, G.J. (2003) Mapping of regions within the vaccinia virus complement control protein involved in dose-dependent binding to key complement components and heparin using surface plasmon resonance. Biochim Biophys Acta 1650, 30-39.
Soulika, A.M., Morikis, D., Sarrias, M.R., Roy, M., Spruce, L.A., Sahu, A. and Lambris, J.D. (2003) Studies of structure-activity relations of complement inhibitor compstatin. J Immunol 171, 1881-1890.
Stenberg, E., Persson, B., Roos, H. and Urbaniczky, C. (1991) Quantitative determination of surface concentration of protein with surface plasmon resonance using radiolabeled proteins. J Colloid Interface Sci 143, 513-526.
Tanner, J., Weis, J., Fearon, D., Whang, Y. and Kieff, E. (1987) Epstein-Barr virus gp350/220 binding to the B lymphocyte C3d receptor mediates adsorption, capping, and endocytosis. Cell 50, 203-213.
Thai, C.T. and Ogata, R.T. (2005) Recombinant C345C and factor I modules of complement components C5 and C7 inhibit C7 incorporation into the complement membrane attack complex. J Immunol 174, 6227-6232.
Turbadar, T. (1959) Complete Absorption of Light by Thin Metal Films. Proceedings of the Physical Society of London 73, 40-44.
Vaziri-Sani, F., Hellwage, J., Zipfel, P.F., Sjoholm, A.G., Iancu, R. and Karpman, D. (2005) Factor H binds to washed human platelets. J Thromb Haemost 3, 154-162.
Verma, A., Hellwage, J., Artiushin, S., Zipfel, P.F., Kraiczy, P., Timoney, J.F. and Stevenson, B. (2006) LfhA, a novel factor H-binding protein of Leptospira interrogans. Infect Immun 74, 2659-2666.
Walport, M.J. (2001) Complement. First of two parts. N Engl J Med 344, 1058-1066.
Wear, M.A. and Walkinshaw, M.D. (2006) Thermodynamics of the cyclophilin-A/cyclosporin-A interaction: A direct comparison of parameters determined by surface plasmon resonance using Biacore T100 and isothermal titration calorimetry. Anal Biochem 359, 285-287.
Wiesmann, C., Katschke, K.J., Yin, J., Helmy, K.Y., Steffek, M., Fairbrother, W.J., McCallum, S.A., Embuscado, L., Deforge, L., Hass, P.E. and van Lookeren Campagne, M. (2006) Structure of C3b in complex with CRIg gives insights into regulation of complement activation. Nature 444, 217-220.
Winzor, D.J. (2003) Surface plasmon resonance as a probe of protein isomerization. Anal Biochem 318, 1-12.
Yu, H., Munoz, E.M., Edens, R.E. and Linhardt, R.J. (2005) Kinetic studies on the interactions of heparin and complement proteins using surface plasmon resonance. Biochim Biophys Acta 1726, 168-176.
Yuk, J.S. and Ha, K.S. (2005) Proteomic applications of surface plasmon resonance biosensors: analysis of protein arrays. Exp Mol Med 37, 1-10.

Glycosylation as a Target for Recognition of Influenza Viruses by the Innate Immune System

Patrick C. Reading, Michelle D. Tate, Danielle L. Pickett, and Andrew G. Brooks

The University of Melbourne, Department of Microbiology and Immunology, Victoria, Australia, 3010

1 Introduction

Influenza is an important annual respiratory infection that remains a leading cause of illness and death throughout the world. Influenza A viruses bear two surface glycoproteins, the hemagglutinin (HA) and the neuraminidase (NA), both of which play a critical role in the ability of the virus to replicate in susceptible target cells. Oligosaccharides attached to the viral glycoproteins play a number of important biological roles, modulating stability and transport of the HA/NA to the cell surface during biosynthesis as well as contributing to the functional activity of each glycoprotein. Whilst it has long been recognized that the addition of oligosaccharide moieties can be an effective means of immune evasion via the modification or masking of antigenic epitopes on the virus, mannose-containing glycans on the viral HA and NA have also been shown to serve as targets for recognition by components of the innate immune system, including C-type lectins of the collectin family and the macrophage mannose receptor. This review will contrast the important biological role of glycosylation in the life cycle of influenza virus with the role it plays in recognition and destruction of virus by components of the innate immune system.

2 Influenza Viruses

Influenza viruses are enveloped viruses of the family *Orthomyxoviridae* and are classified into three distinct types, A, B and C, based on major genetic and antigenic differences. Influenza A viruses are responsible for the majority of influenza epidemics in humans and have the potential to cause severe pandemics. The disease caused by influenza during the seasonal epidemics is significant, however most morbidity and mortality seen is due to cardiopulmonary and lower-respiratory tract complications of the infection (Cox & Subbarao, 1999). Studies into the ways influenza viruses cause infection are important to aid in the control of the current circulating viruses and for the prevention and minimization of potential pandemics.

Influenza A is an enveloped virus with a genome comprising 8 segments of single-stranded negative sense RNA. These segments encode ten viral proteins, namely hemagglutinin (HA), neuraminidase (NA), and nuclear protein (NP), 3 proteins which form a RNA-dependent RNA polymerase complex (PB1, PB2 and PA), a matrix protein (M1) which forms an inner layer within the virion, a second matrix protein (M2) which forms a pH-dependent ion channel, and two non-structural proteins (NS1 and NS2). The two surface glycoproteins, HA and NA, protrude from the viral envelope as spikes.

Influenza A viruses are sub-typed according to the serological reactivity of the surface HA and NA molecules. Although 16 avian and mammalian serotypes of HA and 9 serotypes of NA have been identified only 3 HA subtypes (H1, H2 and H3) and 2 NA subtypes (N1 and N2) are known to have circulated in the human population. Although avian derived viruses (H5N1, H9N2 and H7N7) have infected humans they have not acquired the ability to spread from person to person (Suzuki, 2005).

3 Envelope Glycoproteins of Influenza Viruses

The HA is an envelope glycoprotein that mediates the attachment of virions to the cell surface and the fusion of the viral and endosomal membranes, thereby delivering the viral nucleocapsid into the cytoplasm to initiate viral replication (reviewed by Skehel & Wiley, 2000). HA recognizes terminal N-acetyl neuraminic acid (NeuAc) residues that are frequently expressed on oligosaccharide chains of cell surface glycoproteins and glycolipids. The HA monomer is synthesized as a polypeptide precursor protein (HA0) which undergoes post-translational cleavage by host cell proteases to yield polypeptide chains HA1 and HA2, which remain linked by a disulfide bond. Cleavage is usually mediated by serine protease/s produced by Clara cells of the bronchiolar epithelium, thereby confining infection to the respiratory tract in humans. Each HA monomer adopts a conformation consisting of a globular head and a stalk region that is anchored in the viral envelope via a short hydrophobic sequence. Each HA 'spike' on the virion consists of a trimer formed by the non-covalent association of 3 identical monomers.

The NA is a tetrameric glycoprotein with an ectodomain composed of a globular head and a slender stalk region. The globular head contains four catalytic sites and, like HA, NA also has a specificity for terminal NeuAc (reviewed by Saito et al., 1995). The viral NA acts to cleave terminal NeuAc moieties from cell surface oligosaccharides and aids in the ability newly synthesized virions to detach from infected cells and spread to neighbouring target cells. The activity of the NA molecule also cleaves the terminal sialic residues on glycans attached to the viral surface glycoproteins, thereby exposing the underlying sugars. As the HA and NA have opposing roles the balance between their activities can be vital in determining the infectivity of particular virus strains (Wagner et al., 2002).

4 Glycosylation of Influenza Virus HA and NA

Both the HA and NA carry potential sites for N-linked glycosylation, with oligosaccharide chains attached through N-glycosidic linkages to the Asn residues of the glycosylation site motif Asn-X-Ser/Thr, where X may represent any amino acid except proline (Kornfeld & Kornfeld, 1985). Glycosylation may be inhibited by certain combinations of Asn-X-Ser or when the glycosylation motif is followed by specific amino acid sequences (Kasturi et al., 1995, Mellquist et al., 1998).

4.1 Functions of glycosylation of influenza HA and NA glycoproteins.

The loss or gain of glycosylation sites from the HA and/or NA can have dramatic effects on the biosynthesis, stability and function of viral glycoproteins. Site directed mutagenesis has shown particular glycosylation sites to be important for the expression of functionally mature NA, while deletion of alternative sites from the head of NA altered the substrate specificity of the NA enzyme (Saito & Kawano, 1997). Glycosylation sites located in the stalk region of the HA are highly conserved between different HA subtypes and contribute to the stability of the HA molecule. Removal of stem-associated glycosylation was associated with impairments in trimerization, folding and transport of HA to the cell surface (Roberts et al., 1993) and modulation of pH stability of the HA (Ohuchi et al., 1997b). Oligosaccharide side chains influence a number of other biological properties of the HA molecule. Mutant influenza viruses of the H3 subtype lacking oligosaccharide chains from the distal tip of the HA spike differed from parent viruses in their ability to recognize sialylated cell surface receptors (Anders et al., 1990, Hartley et al., 1992). These and other studies (Ohuchi et al., 1997a, Gambaryan et al., 1998) have demonstrated the ability of oligosaccharides near the receptor-binding pocket to attenuate the affinity of HA for host cell receptors. Loss of a glycosylation site near the HA1/HA2 cleavage site of an avian influenza virus led to increased susceptibility of the HA to proteolytic cleavage *in vitro* and enhanced virulence of the virus for chickens *in vivo* (Kawaoka et al., 1984, Deshpande et al., 1987). Lack of glycosylation at site Asn 130 of the NA also plays a key role in the neurovirulence of strain A/WSN/33 in mice (Li et al., 1993).

It is well established that glycosylation can modulate immune recognition of influenza viruses by masking or modifying antigenic sites (Wilson et al., 1981, Caton et al., 1982, Skehel et al., 1984,). The HA glycoprotein serves as the major target for neutralizing antibody during influenza epidemics and pandemics. Variation in the number and location of glycosylation sites arising by antigenic drift may prevent recognition of major antigenic sites targeted by neutralizing antibodies elicited to previously circulating strains. Consistent with this notion, analysis of HA sequences of H3N2 viruses isolated from 1968 to 2002 has shown a progressive increase in numbers of glycosylation sites from 2 to 6/7 on the globular head of the molecule, although the 5 glycosylation sites in the stalk region were strictly conserved between strains (Skehel & Wiley, 2000, Abe et al., 2004). It should be noted, however, that the progressive increase in N-linked glycosylation sites observed over time in H3 HA1 does not appear to be a general feature of influenza virus evolution, and was not observed in human H1 HA1, avian H3 or influenza B HA1 (Zhang et al., 2004). Furthermore, there was no trend for increasing glycosylation sites in the avian or human N2, despite this glycoprotein showing variable numbers of glycoylation sites from year to year.

Considerable variation in number and location of potential glycosylation sites on the globular head of HA has been noted between different subtypes (Nobusawa et al., 1991, Inkster et al., 1993) and between strains (Ward & Dopheide, 1981, Abe et al., 2004), while those linked to the stalk region are highly conserved. Although recently circulating H3N2 viruses may have 6 to 7 glycoylation sites on the globular head of HA, it is not known whether all of these sites are glycosylated or not. A further factor contributing to glycan heterogeneity relates to the type of oligosaccharide chain attached at each particular site. The HA and NA glycoproteins of influenza viruses can bear a mixture of high-mannose (type I), complex (type II) or hybrid type oligosaccharides (Basak et al., 1981, Ward & Dopheide, 1981). The nature of the oligosaccharide moieties conferred to viral glycoproteins is determined by the extent to which these chains are processed in a particular host cell, and differences in both size and composition of viral glycans have been noted following growth of influenza viruses in different cell types (Nakamura & Compans, 1979, Deom & Schulze, 1985).

Terminal mannosylation of oligosaccharides is a feature common to many pathogenic organisms, including viruses, however terminal mannose residues are rarely expressed on mammalian glycoproteins and cell membranes. In response to patterns such as these, the immune response has evolved pattern recognition receptors (PRRs), a group of genetically encoded proteins expressed by cells of the innate immune system or secreted as soluble molecules in serum and body fluids. While glycosylation of HA and NA glycoproteins is of considerable importance in the life cycle of influenza viruses, there is growing evidence to indicate that mannose-containing oligosaccharides on the virus serve as a target for recognition and destruction by lectins of the innate immune system.

5 Mammalian C-Type Lectins

Mammalian C-type lectins are non-enzymatic proteins that selectively bind to specific carbohydrate structures that may be expressed by endogenous 'self' glycoproteins or by invading microbes. C-type lectins play a pivotal role in recognizing pathogens and regulating immunity to infection. All C-type lectins bear at least one highly conserved carbohydrate recognition domain (CRD) that contains Ca^{2+}-dependent binding pockets essential for carbohydrate recognition. Broadly speaking, CRDs exhibit specificity for either mannose and N-acetylglucosamine-type structures, or galactose and N-acetylgalactosamine-type structures (reviewed by Drickamer, 1999); such glycans are commonly expressed on the surface of bacteria, parasites, fungi and viruses. C-type lectins may be further subdivided into groups based on structural similarities; examples include secreted soluble proteins such as the collectins and transmembrane proteins such as the mannose receptor family.

5.1 Collectins and Influenza Viruses.

Collectins are a group of collagenous lectins that are present as large, multimeric proteins in serum and body fluids (Crouch & Wright, 2001, Lu et al., 2002, McCormack & Whitsett, 2002). Members of the family include the serum mannose-binding lectins (MBL), bovine serum proteins conglutinin and collectin-43 and lung surfactant proteins A (SP-A) and D (SP-D). Collectins mediate a range of antiviral activites against influenza A viruses *in vitro*, including direct inhibition of viral infectivity (Anders et al., 1994, Benne et al., 1995, Hartshorn et al., 1997, Hartshorn et al., 2002), formation of viral aggregates (Hartshorn et al., 1993, Hartshorn et al., 1997, Hartshorn et al., 2002) and lysis of virus-infected cells in the presence of complement (Reading et al., 1995). SP-A and SP-D are constitutively expressed in fluids lining the respiratory tract and as such are considered to be of most relevance in the context of influenzal disease.

Both SP-A and SP-D mediate antiviral activity against influenza A viruses through formation of viral aggregates and direct inhibition of viral infectivity. Furthermore, preincubation of influenza viruses with SP-A or SP-D protected neutrophils against the depressive effects of the virus on their respiratory burst responses (Hartshorn et al., 1994, Hartshorn et al., 1996) and this mechanism may be an important factor in host defence against bacterial superinfection during influenzal disease. SP-D was found to be particularly potent in this regard. The mechanisms by which SP-A and SP-D act against influenza viruses are different. SP-D acts as a classic β-inhibitor with Ca^{2+}-dependent binding via its CRD to oligosaccharides present on the HA and the NA glycoproteins (Anders et al., 1990, Hartley et al., 1992, Hartshorn et al., 1997). In contrast, SP-A inhibits influenza viruses in a Ca^{2+}-independent manner. The viral HA binds to terminal sialic acid expressed on the carbohydrate moiety of the SP-A molecule, thereby blocking the ability of HA to interact with cellular receptors (Benne et al., 1995, Hartshorn et al., 1997). In general, SP-D is a more potent inhibitor of influenza viruses than SP-A (Hartshorn et al., 1997).

More recent studies have formally addressed the *in vivo* roles for SP-A and SP-D in innate defence against influenza virus through the use of knock-out mice. SP-D knock-outmice exhibited enhanced viral replication inflammation and illness following infection with influenza virus (LeVine et al., 2001). SP-A knock-out mice also displayed enhanced susceptibility to influenza infection, however the effect was less pronounced than that observed in SP-D-deficient mice (LeVine et al., 2002). Furthermore, SP-D levels in lavage fluids increased several fold during the course of influenza infection of C57BL/6 mice, consistent with a role for this molecule in early host defence (Reading et al., 1997).

5.2 The Macrophage Mannose Receptor (MMR)

The MMR is a type I transmembrane glycoprotein expressed by macrophages, dendritic cells and some endothelial cells. The receptor contains a tandem array of eight CRDs that exhibit carbohydrate specificity for mannose, fucose and N-acetyl-D-glucosamine (East & Isacke, 2002). High affinity binding is achieved via the expression of multiple CRDs on a single polypeptide backbone. In contrast, oligomerization of subunits to promote CRD clustering allows members of the collectin family to mediate for high affinity binding to ligands.

The MMR has been implicated in binding and phagocytosis of a wide range of microorganisms, including bacteria, fungi and protozoa (Pontow et al., 1992, Martinez-Pomares et al., 2001, East & Isacke, 2002). Limited studies have addressed the role of the MMR in recognition and internalization of viruses by macrophages, although studies have implicated a role for MMR in binding and transmission of human immunodeficiency virus (HIV) (Nguyen & Hildreth, 2003). Furthermore, we have defined a role for the MMR in mediating infectious entry of influenza viruses into murine macrophages (Reading et al., 2000).

6 C-Type Lectins and Influenza Virus – Studies in a Mouse Model of Infection

The murine model of influenza infection allows for the correlation of the *in vitro* response of different influenza viruses to C-type lectins with the pathogenesis of influenzal disease following intranasal infection. While this model may provide important insights into the pathogenesis and the nature of influenzal disease, important distinctions must be made between human influenza and the mouse model. Mice are not naturally infected with influenza and therefore viruses must be adapted via sequential passaging in mouse lung. Intranasal infection of mice with mouse-adapted strains of influenza virus leads to a greater replication efficiency of the virus in the lung and the development of pneumonia similar to that seen in severe human influenza infections (Sweet & Smith, 1980, Brown, 1990). During mouse adaptation viruses acquire mutations to enhance replication efficiency in lung cells as well as other mutations that may allow the virus to evade particular innate immune defences. A comparison of mouse-adapted influenza viruses with less virulent strains can therefore provide important information to define which aspects of innate immunity are important in early host defence.

6.1 Detection of mannose-containing glycans on different influenza viruses

We hypothesized that glycosylation patterns of different influenza viruses would determine their sensitivity to C-type lectins *in vitro*, and this in turn would determine the course of influenzal disease *in vivo*. In the studies described we have used three influenza virus strains: A/Puerto Rico/8/34 (PR8; H1N1), and X-31 (H3N2) and

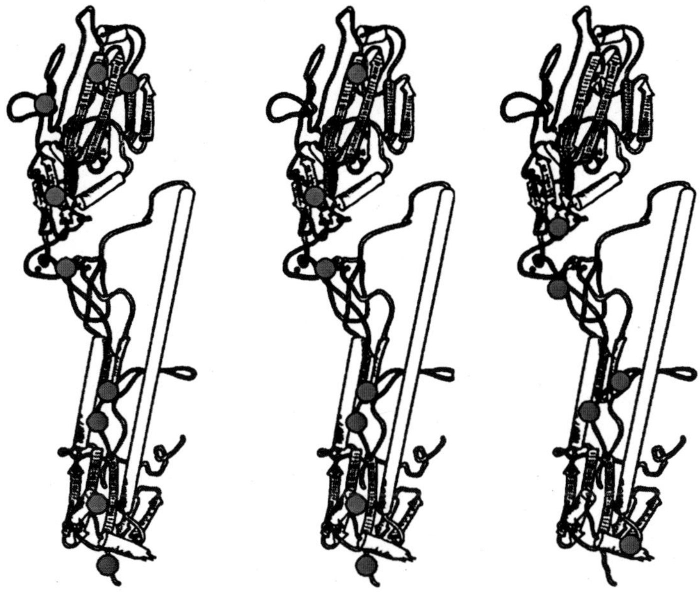

Fig. 1. Schematic diagram of the influenza virus HA monomer showing the location of potential sites of N-linked glycosylation on BJx109, X-31 and PR8 viruses. Diagram modified from (Wilson et al. 1981).

BJx109 (H3N2). The latter are high-yielding reassortants of PR8 with A/Aichi/2/68 and A/Beijing/353/89, respectively, and share a number of internal components derived from the PR8 virus in conjunction with the surface glycoproteins of their respective parent viruses. PR8 is a highly mouse-adapted virus and a particular feature of this strain is the absence of glycosylation sites on the globular head of the HA (Nakamura & Compans, 1979, Caton et al., 1982). In contrast, X-31 and BJx109 carry 2 and 4 potential glycosylation sites on the head of the HA molecule, respectively (Figure 1).

As mannose-containing glycans serve as a target for recognition by C-type lectins of the immune system, BJx109, X-31 and PR8 were assessed for reactivity with the plant lectin Concanavalin (Con-A). The number of potential N-linked glycosylation sites on the head of HA was found to correlate with detection of high mannose or hybrid-type glycans as indicated by the binding of ^{125}I-labelled Con-A to purified viruses of the three strains (Figure 2). BJx109 bound Con-A to high levels, X-31 to intermediate levels and PR8 bound the plant lectin very poorly. All binding was inhibited when ^{125}I-labelled Con-A was added in the presence of 100 mM α-methyl-D-mannoside.

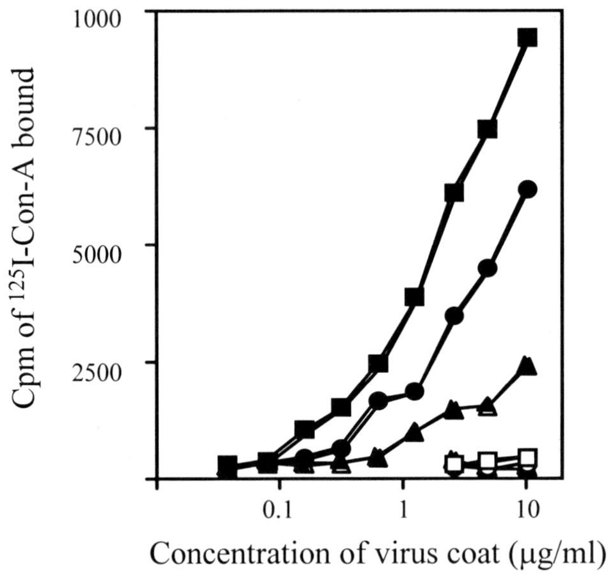

Fig. 2. Levels of mannose-containing oligosaccharide on purified influenza viruses. ELISA wells were coated with increasing concentrations of purified BJx109 (squares), X-31 (circles) or PR8 (triangles) viruses and binding of radiolabelled Concanavalin A was determined in binding buffer alone (black symbols) or in binding buffer supplemented with 100 mM α-methyl-D-mannoside (open symbols). Equivalent coating levels of the three viruses were confirmed using mAb 165 which binds to the host-derived carbohydrate antigen that is characteristic of egg-grown influenza viruses (data not shown).

6.2 Influenza viruses differ in their sensitivity to the antiviral activities of rat SP-D

BJx109, X-31 and PR8 viruses differ in their sensitivity to the antiviral activities of rat SP-D in a manner that correlated with their degree of mannose-containing glycans. The antiviral activity of SP-D was quantitated by the ability of different concentrations of rat SP-D to (i) inhibit agglutination of chicken erythrocytes (Figure 3A), (ii) inhibit the enzymatic activity of viral NA (Figure 3B), and (iii) neutralize virus infectivity (Figure 3C). In each assay BJx109 was found to be markedly more sensitive to SP-D than X-31, while PR8 was found to be resistant. In each assay, the antiviral activity of SP-D could be reversed by the inclusion of 5 mg/ml of mannan, a complex polymer of mannose residues (data not shown).

6.3 Influenza viruses differ in their ability to infect murine macrophages - a role for the mannose receptor

Early studies in our laboratory indicated that influenza viruses differed in their ability to infect murine macrophages, including alveolar macrophages. BJx109 infected

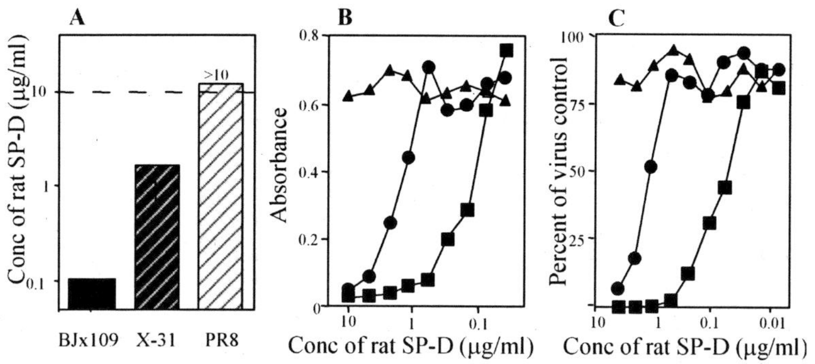

Fig. 3. Antiviral activity of rat SP-D against influenza viruses. (A) Inhibition of hemagglutination by recombinant rat SP-D. A standard dose (4 HAU) of BJx109, X-31 or PR8 was incubated with increasing concentrations of SP-D prior to the addition of 1% vol/vol chicken erythrocytes. Titres were determined as the lowest concentration of SP-D to completely inhibit hemagglutination. The dashed line indicates the highest concentration of SP-D tested. (B) Inhibition of influenza virus NA activity by SP-D. BJx109 (squares), X-31 (circles) and PR8 (triangles) were mixed with increasing concentrations of rat SP-D and assayed for viral NA activity as described (Reading et al., 1997). The dose of virus chosen was such that the 3 virus preparations displayed equivalent NA activity in the absence of SP-D. (C) Neutralization of BJx109 (squares), X-31 (circles) and PR8 (triangles) by rat SP-D. BJx109, X-31 or PR8 were added to increasing concentrations of rat SP-D and incubated for 30 mins at 37°C and the amount of infectious virus remaining was determined by immunofluorescence microscopy as described (Reading et al., 1997).

macrophages to high levels, X-31 to intermediate levels and strain PR8 infected macrophages very poorly. All three strains infected epithelial cell lines to similar levels. These findings were of particular interest, given epithelial cells and airway macrophages represent the primary targets of influenza infection following intranasal infecton of mice (Hennet et al., 1992). Whereas infection of epithelial cells leads to the production of viral progeny, infection of macrophages leads to virus destruction (Wells et al., 1978, Rodgers & Mims, 1981) and may represent an important antiviral mechanism. Given that (i) the ability of BJx109, X-31 and PR8 to infect murine macrophages correlated with their pattern of sensitivity to rat SP-D, and (ii) macrophages express the MMR, a C-type lectin with sugar specificity similar to that of the collectins, we investigated the role of the MMR in mediating infectious entry of influenza viruses into macrophages.

Yeast mannan, a high affinity ligand for the MMR, was found to act as a potent inhibitor of macrophage infection (Figure 4A) but had only negligible effects on the ability of influenza viruses to infect epithelial cell lines which lack the MMR (data not shown). Furthermore, a macrophage variant selected for high MMR expression, J774E (Diment et al., 1987), displayed an increased susceptibility to influenza virus infection compared to the parent cell line (Reading et al., 2000). To examine interactions between viral glycoproteins and the MMR, radiolabelled mannosylated-BSA (man-BSA) was used to detect the MMR on murine macrophages. Man-BSA bound to macrophages (but not epithelial cells) in a Ca^{2+}-dependent manner, and binding was inhibited by the chelation of Ca^{2+} or by the inclusion of mannan in the

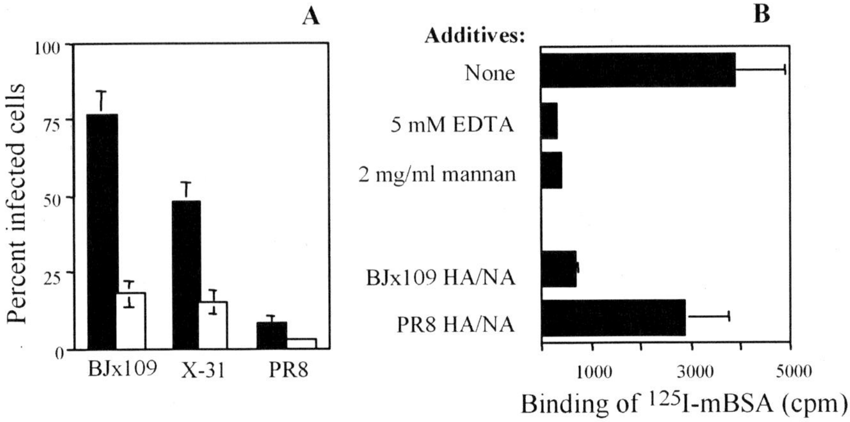

Fig. 4. A role for the MMR in infection of murine macrophages by influenza viruses. (A) Inhibition of macrophage infection by mannan. Monolayers of peritoneal macrophages were incubated for 20 min in serum-free media (black columns) or in media supplemented with 10 mg/ml mannan (white columns), prior to the addition of 10^6 plaque forming units (PFU) of influenza viruses. Mannan was maintained in the media for 6-7 hrs at which time cells were fixed and stained by immunofluorescence for expression of newly synthesized viral proteins, as described (Reading et al., 2000). (B) Inhibition of ^{125}I-mBSA binding to peritoneal macrophages by viral HA/NA glycoproteins. Aliquots of cells were pre-incubated on ice with buffer alone or buffer supplemented with 2 mg/ml mannan, 10 μg/ml HANA glycoproteins from BJx109 or PR8 or 5 mM EDTA. Results are expressed as the mean binding ($\pm$ 1 SD) of ^{125}I-mBSA from triplicate samples, as described (Reading et al., 2000).

binding buffer (Figure 4B). Purified preparations of influenza virus HA/NA glycoproteins were found to competitively inhibit the binding of mannosylated BSA to murine macrophages, however the HA/NA glycoproteins of strain PR8 were particularly poor in this capacity (Figure 4B). Given the known endocytic activity of the MMR, these findings suggested that direct interaction between the MMR and oligosaccharide chains expressed by viral HA/NA glycoproteins represents a major endocytic route for influenza viruses into macrophages.

6.4 Pathogenesis of different influenza viruses in a murine model of infection

The mouse model of infection provides a convenient model for examining both the pathogenesis and the nature of the immune response induced following influenza infections, allowing *in vitro* susceptibility of different influenza viruses to innate effector molecules to be correlated with the pathogenesis of influenzal disease following intranasal infection. A comparison of BJx109, X-31 and PR8 was of particular interest, given the marked differences we have observed between the viruses in their (i) sensitivity to the antiviral activities of rat SP-D, and (ii) ability to murine macrophages *in vitro*.

Following intranasal inoculation of C57BL/6 mice we observed major differences in disease progression (as assessed by weight loss) with BJx109, X-31 or

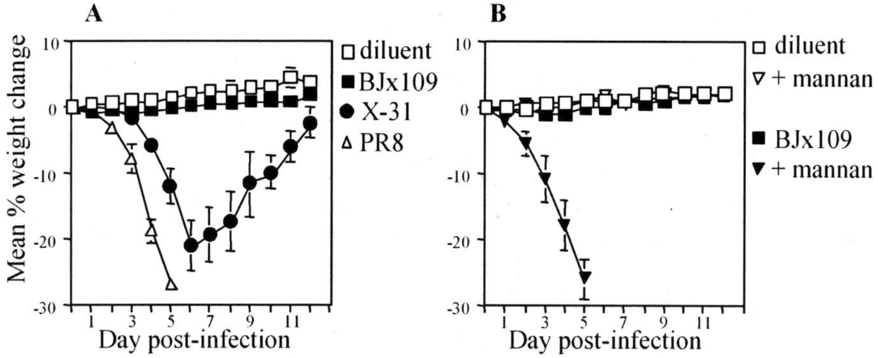

Fig. 5. Influenza A viruses differ in their virulence for C57BL/6 mice. Mice were inoculated via the intranasal route with 10^5 PFU of influenza viruses in a total volume of 50 μl. Weight loss was monitored daily over a 12 day period and mice that had lost >25% of their original body weight were euthanized. Data are presented as the mean percent weight change (± 1 SD). (A) Mice were inoculated with BJx109, X-31 or PR8. Control groups received an equivalent volume of virus diluent. (B) Mice were inoculated with virus diluent or with virus diluent supplemented with 10 mg/ml of mannan 24 hrs prior to infection with 10^5 PFU of BJx109. Every second day thereafter mice received additional inoculations with virus diluent alone, or virus diluent plus mannan, as appropriate..

PR8 (Figure 5A) such that the viruses may be classified as highly virulent (PR8), of intermediate virulence (X-31) and avirulent (BJx109) in this model. Furthermore, co-administration of mannan (a potent inhibitor of both collectins and MMR-mediated infection of macrophages *in vitro*) to BJx109-infected mice resulted in a profound exacerbation of influenzal disease (Figure 5B), generating a weight loss profile similar to that observed during PR8 infection. While mannan treatment of mock-infected mice had no effect on weight loss, mannan treatment of BJx109-infected mice converted a mild infection to severe clinical disease. These findings are consistent with a major role for mannose-specific lectins in controlling influenza virus infection *in vivo*.

7 Summary

Glycosylation clearly plays an important role in the life cycle of influenza viruses and certain glycosylation sites are required for the structural integrity and stability of the HA and NA glycoproteins during biosynthesis and formation of intact virions. Furthermore, glycosylation has been shown to modulate the functions of influenza glycoproteins, in particular the recognition of host cell receptors and in shielding antigenic epitopes on the viral HA. The addition of oligosaccharide moieties to the globular head of the HA does, however, correlate with an increased sensitivity to the antiviral activities of SP-D and to recognition and destruction of virus via the MMR on murine macrophages. Consequently, the degree of glycosylation appears to be an important factor in determining sensitivity to lectin-mediated defences, and therefore in determining the ability of a particular virus strain to replicate in the respiratory tract of mice following intranasal infection. The mouse-adapted PR8 strain which lacks mannose-containing glycans from the head of its HA molecule was largely resistant to the antiviral activities of SP-D and the MMR *in vitro* and induced severed clinical disease following intranasal infection of mice. The finding that mannan treatment of BJx109-infected mice facilitated an early and dramatic enhancement of disease severity is also consistent with a major role for mannose-specific lectins in limiting influenza virus growth and spread in the respiratory tract.

8 Acknowledgements

The authors would like to thank Dr. E. M. Anders for advice and helpful discussions regarding the manuscript. Patrick Reading is supported by a R. D. Wright Fellowship from the National Health and Medical Research Council (NHMRC) of Australia. This work was supported by the NHMRC, Australia.

References

Abe, Y., Takashita, E., Sugawara, K., Matsuzaki, Y., Muraki, Y. & Hongo, S. (2004). Effect of the addition of oligosaccharides on the biological activities and antigenicity of influenza A/H3N2 virus hemagglutinin. *J Virol* **78**, 9605-11.

Anders, E. M., Hartley, C. A. & Jackson, D. C. (1990). Bovine and mouse serum beta inhibitors of influenza A viruses are mannose-binding lectins. *Proc Natl Acad Sci U S A* **87**, 4485-9.

Anders, E. M., Hartley, C. A., Reading, P. C. & Ezekowitz, R. A. (1994). Complement-dependent neutralization of influenza virus by a serum mannose-binding lectin. *J Gen Virol* **75 (Pt 3)**, 615-22.

Basak, S., Pritchard, D. G., Bhown, A. S. & Compans, R. W. (1981). Glycosylation sites of influenza viral glycoproteins: characterization of tryptic glycopeptides from the A/USSR(H1N1) hemagglutinin glycoprotein. *J Virol* **37**, 549-58.

Benne, C. A., Kraaijeveld, C. A., van Strijp, J. A., Brouwer, E., Harmsen, M., Verhoef, J., van Golde, L. M. & van Iwaarden, J. F. (1995). Interactions of surfactant protein A with influenza A viruses: binding and neutralization. *J Infect Dis* **171**, 335-41.

Brown, E. G. (1990). Increased virulence of a mouse-adapted variant of influenza A/FM/1/47 virus is controlled by mutations in genome segments 4, 5, 7, and 8. *J Virol* **64**, 4523-33.

Caton, A. J., Brownlee, G. G., Yewdell, J. W. & Gerhard, W. (1982). The antigenic structure of the influenza virus A/PR/8/34 hemagglutinin (H1 subtype). *Cell* **31**, 417-27.

Cox, N. J. & Subbarao, K. (1999). Influenza. *Lancet* **354**, 1277-82.

Crouch, E. & Wright, J. R. (2001). Surfactant proteins a and d and pulmonary host defense. *Annu Rev Physiol* **63**, 521-54.

Deom, C. M. & Schulze, I. T. (1985). Oligosaccharide composition of an influenza virus hemagglutinin with host-determined binding properties. *J Biol Chem* **260**, 14771-4.

Deshpande, K. L., Fried, V. A., Ando, M. & Webster, R. G. (1987). Glycosylation affects cleavage of an H5N2 influenza virus hemagglutinin and regulates virulence. *Proc Natl Acad Sci U S A* **84**, 36-40.

Diment, S., Leech, M. S. & Stahl, P. D. (1987). Generation of macrophage variants with 5-azacytidine: selection for mannose receptor expression. *J Leukoc Biol* **42**, 485-90.

Drickamer, K. (1999). C-type lectin-like domains. *Curr Opin Struct Biol* **9**, 585-90.

East, L. & Isacke, C. M. (2002). The mannose receptor family. *Biochim Biophys Acta* **1572**, 364-86.

Gambaryan, A. S., Marinina, V. P., Tuzikov, A. B., Bovin, N. V., Rudneva, I. A., Sinitsyn, B. V., Shilov, A. A. & Matrosovich, M. N. (1998). Effects of host-dependent glycosylation of hemagglutinin on receptor-binding properties on H1N1 human influenza A virus grown in MDCK cells and in embryonated eggs. *Virology* **247**, 170-7.

Hartley, C. A., Jackson, D. C. & Anders, E. M. (1992). Two distinct serum mannose-binding lectins function as beta inhibitors of influenza virus: identification of bovine serum beta inhibitor as conglutinin. *J Virol* **66**, 4358-63.

Hartshorn, K. L., Crouch, E. C., White, M. R., Eggleton, P., Tauber, A. I., Chang, D. & Sastry, K. (1994). Evidence for a protective role of pulmonary surfactant protein D (SP-D) against influenza A viruses. *J Clin Invest* **94**, 311-9.

Hartshorn, K. L., Holmskov, U., Hansen, S., Zhang, P., Meschi, J., Mogues, T., White, M. R. & Crouch, E. C. (2002). Distinctive anti-influenza properties of recombinant collectin 43. *Biochem J* **366**, 87-96.

Hartshorn, K. L., Reid, K. B., White, M. R., Jensenius, J. C., Morris, S. M., Tauber, A. I. & Crouch, E. (1996). Neutrophil deactivation by influenza A viruses: mechanisms of protection after viral opsonization with collectins and hemagglutination-inhibiting antibodies. *Blood* **87**, 3450-61.

Hartshorn, K. L., Sastry, K., Brown, D., White, M. R., Okarma, T. B., Lee, Y. M. & Tauber, A. I. (1993). Conglutinin acts as an opsonin for influenza A viruses. *J Immunol* **151**, 6265-73.

Hartshorn, K. L., White, M. R., Shepherd, V., Reid, K., Jensenius, J. C. & Crouch, E. C. (1997). Mechanisms of anti-influenza activity of surfactant proteins A and D: comparison with serum collectins. *Am J Physiol* **273**, L1156-66.

Hennet, T., Ziltener, H. J., Frei, K. & Peterhans, E. (1992). A kinetic study of immune mediators in the lungs of mice infected with influenza A virus. *J Immunol* **149**, 932-9.

Inkster, M. D., Hinshaw, V. S. & Schulze, I. T. (1993). The hemagglutinins of duck and human H1 influenza viruses differ in sequence conservation and in glycosylation. *J Virol* **67**, 7436-43.

Kasturi, L., Eshleman, J. R., Wunner, W. H. & Shakin-Eshleman, S. H. (1995). The hydroxy amino acid in an Asn-X-Ser/Thr sequon can influence N-linked core glycosylation efficiency and the level of expression of a cell surface glycoprotein. *J Biol Chem* **270**, 14756-61.

Kawaoka, Y., Naeve, C. W. & Webster, R. G. (1984). Is virulence of H5N2 influenza viruses in chickens associated with loss of carbohydrate from the hemagglutinin? *Virology* **139**, 303-16.

Kornfeld, R. & Kornfeld, S. (1985). Assembly of asparagine-linked oligosaccharides. *Annu Rev Biochem* **54**, 631-64.

LeVine, A. M., Hartshorn, K., Elliott, J., Whitsett, J. & Korfhagen, T. (2002). Absence of SP-A modulates innate and adaptive defense responses to pulmonary influenza infection. *Am J Physiol Lung Cell Mol Physiol* **282**, L563-72.

LeVine, A. M., Whitsett, J. A., Hartshorn, K. L., Crouch, E. C. & Korfhagen, T. R. (2001). Surfactant protein D enhances clearance of influenza A virus from the lung in vivo. *J Immunol* **167**, 5868-73.

Li, S., Schulman, J., Itamura, S. & Palese, P. (1993). Glycosylation of neuraminidase determines the neurovirulence of influenza A/WSN/33 virus. *J Virol* **67**, 6667-73.

Lu, J., Teh, C., Kishore, U. & Reid, K. B. (2002). Collectins and ficolins: sugar pattern recognition molecules of the mammalian innate immune system. *Biochim Biophys Acta* **1572**, 387-400.

Martinez-Pomares, L., Linehan, S. A., Taylor, P. R. & Gordon, S. (2001). Binding properties of the mannose receptor. *Immunobiology* **204**, 527-35.

McCormack, F. X. & Whitsett, J. A. (2002). The pulmonary collectins, SP-A and SP-D, orchestrate innate immunity in the lung. *J Clin Invest* **109**, 707-12.

Mellquist, J. L., Kasturi, L., Spitalnik, S. L. & Shakin-Eshleman, S. H. (1998). The amino acid following an asn-X-Ser/Thr sequon is an important determinant of N-linked core glycosylation efficiency. *Biochemistry* **37**, 6833-7.

Nakamura, K. & Compans, R. W. (1979). Host cell- and virus strain-dependent differences in oligosaccharides of hemagglutinin glycoproteins of influenza A viruses. *Virology* **95**, 8-23.

Nguyen, D. G. & Hildreth, J. E. (2003). Involvement of macrophage mannose receptor in the binding and transmission of HIV by macrophages. *Eur J Immunol* **33**, 483-93.

Nobusawa, E., Aoyama, T., Kato, H., Suzuki, Y., Tateno, Y. & Nakajima, K. (1991). Comparison of complete amino acid sequences and receptor-binding properties among 13 serotypes of hemagglutinins of influenza A viruses. *Virology* **182**, 475-85.

Ohuchi, M., Ohuchi, R., Feldmann, A. & Klenk, H. D. (1997a). Regulation of receptor binding affinity of influenza virus hemagglutinin by its carbohydrate moiety. *J Virol* **71**, 8377-84.

Ohuchi, R., Ohuchi, M., Garten, W. & Klenk, H. D. (1997b). Oligosaccharides in the stem region maintain the influenza virus hemagglutinin in the metastable form required for fusion activity. *J Virol* **71**, 3719-25.

Pontow, S. E., Kery, V. & Stahl, P. D. (1992). Mannose receptor. *Int Rev Cytol* **137B**, 221-44.

Reading, P. C., Hartley, C. A., Ezekowitz, R. A. & Anders, E. M. (1995). A serum mannose-binding lectin mediates complement-dependent lysis of influenza virus-infected cells. *Biochem Biophys Res Commun* **217**, 1128-36.

Reading, P. C., Miller, J. L. & Anders, E. M. (2000). Involvement of the mannose receptor in infection of macrophages by influenza virus. *J Virol* **74**, 5190-7.

Reading, P. C., Morey, L. S., Crouch, E. C. & Anders, E. M. (1997). Collectin-mediated antiviral host defense of the lung: evidence from influenza virus infection of mice. *J Virol* **71**, 8204-12.

Roberts, P. C., Garten, W. & Klenk, H. D. (1993). Role of conserved glycosylation sites in maturation and transport of influenza A virus hemagglutinin. *J Virol* **67**, 3048-60.

Rodgers, B. & Mims, C. A. (1981). Interaction of influenza virus with mouse macrophages. *Infect Immun* **31**, 751-7.

Saito, T. & Kawano, K. (1997). Loss of glycosylation at Asn144 alters the substrate preference of the N8 influenza A virus neuraminidase. *J Vet Med Sci* **59**, 923-6.

Saito, T., Taylor, G. & Webster, R. G. (1995). Steps in maturation of influenza A virus neuraminidase. *J Virol* **69**, 5011-7.

Skehel, J. J., Stevens, D. J., Daniels, R. S., Douglas, A. R., Knossow, M., Wilson, I. A. & Wiley, D. C. (1984). A carbohydrate side chain on hemagglutinins of Hong Kong influenza viruses inhibits recognition by a monoclonal antibody. *Proc Natl Acad Sci U S A* **81**, 1779-83.

Skehel, J. J. & Wiley, D. C. (2000). Receptor binding and membrane fusion in virus entry: the influenza hemagglutinin. *Annu Rev Biochem* **69**, 531-69.

Suzuki, Y. (2005). Sialobiology of influenza: molecular mechanism of host range variation of influenza viruses. *Biol Pharm Bull* **28**, 399-408.

Sweet, C. & Smith, H. (1980). Pathogenicity of influenza virus. *Microbiol Rev* **44**, 303-30.

Wagner, R., Matrosovich, M. & Klenk, H. D. (2002). Functional balance between haemagglutinin and neuraminidase in influenza virus infections. *Rev Med Virol* **12**, 159-66.

Ward, C. W. & Dopheide, T. A. (1981). Amino acid sequence and oligosaccharide distribution of the haemagglutinin from an early Hong Kong influenza virus variant A/Aichi/2/68 (X-31). *Biochem J* **193**, 953-62.

Wells, M. A., Albrecht, P., Daniel, S. & Ennis, F. A. (1978). Host defense mechanisms against influenza virus: interaction of influenza virus with murine macrophages in vitro. *Infect Immun* **22**, 758-62.

Wilson, I. A., Skehel, J. J. & Wiley, D. C. (1981). Structure of the haemagglutinin membrane glycoprotein of influenza virus at 3 A resolution. *Nature* **289**, 366-73.

Zhang, M., Gaschen, B., Blay, W., Foley, B., Haigwood, N., Kuiken, C. & Korber, B. (2004). Tracking global patterns of N-linked glycosylation site variation in highly variable viral glycoproteins: HIV, SIV, and HCV envelopes and influenza hemagglutinin. *Glycobiology* **14**, 1229-46.

21

Immune Effects of Autoantigen-Associated RNA

Eric L. Greidinger

Miami Department of Veterans Affairs Medical Center and University of Miami Miller School of Medicine, Division of Rheumatology, egreidinger@med.miami.edu

1 Introduction

Ribonucleoproteins represent a major class of autoantigens targeted in human clinical autoimmune syndromes. While the protein components of these antigens have long been a focus of attention, the RNA components of ribonucleoproteins also have immune effects. These effects include the abilities to be recognized by elements of the adaptive immune system and to alter recognition of proteins by the adaptive immune system. More recently, ribonucleoprotein RNAs have also been shown to activate innate immunity pathways such as Toll-Like Receptors that appear to influence clinical autoimmune syndromes. These RNAs may play critical roles in the development of anti-ribonucleoprotein autoimmunity, and their actions may be relevant targets for therapeutic intervention. This review explores the mechanisms of immune recognition of self RNA, the factors that prevent recognition of self RNA, and pathways through with self RNA can participate in autoimmune disease pathogenesis.

RNA is a major component of human pathogens, as well as a major constituent of human cells. It is thus both a plausible activator of protective immune function and a potential target of autoimmunity. As early as the 1960's, RNA molecules had been observed to be targets of recognition by immunoglobulin (Schur and Monroe 1969). Even then, antibodies specific for RNA were remarkable for frequently being present in patients with Systemic Lupus Erythematosus (SLE) and other systemic rheumatic diseases, but seldom being present in normal individuals. In contrast, while immunoglobulin recognition of protein-RNA complexes is often found to be associated with specific syndromes of clinical autoimmunity such as SLE or Mixed Connective Tissue Disease (MCTD) (Sharp, Irvin, LaRoque, Velez, Daly, Kaiser, and Holman 1971) this kind of response can also be seen in a modest number of normal individuals (Marshall, McPherson, Greidinger, Hoffman, and Adler 2002). Thus, anti-RNA responses appear to be different from other autoantibody responses, since the others have also been described in substantial numbers of individuals without known autoimmune disease (Tan, Feltkamp, Smolen, Butcher, Dawkins, Fritzler, Gordon, Hardin, Kalden, Lahita, Maini, McDougal, Rothfield, Smeenk, Takasaki, Wiik, Wilson, and Koziol 1997). Moreover, in contrast to proteins and lipids, which are presented to T cells (by classical MHC molecules and CD1 molecules respectively), no system of antigen presentation of RNA to T cells has yet been described, and no RNA-specific T cells are known to exist. These differences raise questions: 1. Given the apparent absence of anti-RNA adaptive immune responses in individuals without autoimmune disease, how is host defense against RNA-based pathogens mediated? 2. Why are anti-self RNA responses apparently so infrequent among humans without autoimmune disease? And, 3. What is the relationship between the emergence of anti- RNA immunity and the development of autoimmunity?

2 Recognition and Immune Responses to RNA

Direct recognition of RNA by B cells or T cells is rare, and typically associated with the development of an autoimmune syndrome. To counterbalance the absence of non-autoimmune anti-RNA immune responses at the level of adaptive immunity, the innate immune system provides multiple pathways for the direct recognition of potentially pathogenic RNA. Innate immune recognition of RNA can be mediated by intracellular factors, secreted factors, or by membrane-bound receptors.

2.1 Intracellular RNA Recognition

Intracellular RNA recognition mechanisms are mediated by RNA helicases coupled to caspase recognition motifs, with the RIG-I and MDA5 RNA recognition pathways as prototypes of this mechanism (Yoneyama, Kikuchi, Matsumoto, Imaizumi, Miyagishi, Taira, Foy, Loo, Gale, Akira, Yonehara, Kato, and Fujita 2005). A growing family of homologs of these proteins and their co-regulatory factors involved in recognition and response to intracellular double-stranded RNA exists. At least some of the downstream effects of these pathways are mediated through the activation of interferon regulatory factor 3 (IRF3) (Yoneyama, Kikuchi, Natsukawa, Shinobu, Imaizumi, Miyagishi, Taira, Akira, and Fujita 2004). Intracellular RNA recognition by these systems has been shown to play important roles in host defense against several kinds of viral infections (Kato, Takeuchi, Sato, Yoneyama, Yamamoto, Matsui, Uematsu, Jung, Kawai, Ishii, Yamaguchi, Otsu, Tsujimura, Koh, Reis e Sousa, Matsuura, Fujita, and Akira 2006). RNA recognition along this type of pathway may induce apoptosis of the affected cell, or induce pro-inflammatory cytokine release. Given their ability to induce inflammation in response to RNA recognition, RIG-I-like helicases and other components of this pathway are potential candidates for mediating risk of anti-RNA or anti-RNP autoimmunity. However, the role of this relatively recently described gene family in autoimmune syndromes has not yet been explored. At least one gene product in this family, LGP2, with primarily down-regulatory activity has been reported (Rothenfusser, Goutagny, DiPerna, Gong, Monks, Schoenemeyer, Yamamoto, Akira, and Fitzgerald 2005).

2.2 RNA Recognition by Secreted Factors

The pentraxin C-reactive protein (CRP) is an acute phase response-secreted protein that recognizes self RNA (Du Clos 1989). CRP has a mixed set of effects on inflammation (Sjoberg, Trouw, McGrath, Hack, and Blom 2006). It can instigate inflammation by induction of the classical complement cascade, but it can also deliver inhibitors of the complement cascade to sites of RNA recognition. CRP also has mixed effects on cellular immunity, on the anti-inflammatory side ligating Fc gamma IIB receptors that have been found to have down-regulatory effects on inflammatory responses (Mineo, Gormley, Yuhanna, Osborne-Lawrence, Gibson, Hahner, Shohet, Black, Salmon, Samols, Karp, Thomas, and Shaul 2005), including regulatory effects on the development and expression of lupus (McGaha, Sorrentino, and Ravetch 2005), while on the pro-inflammatory side also activating immunity via Fc gamma IIA receptors (Mold and Du Clos 2006). Also, by efficiently clearing extracellular stores of RNAs (Szalai, Weaver, McCrory, van Ginkel, Reiman, Kearney, Marion, and Volanakis 2003), CRP and related "garbage disposal" molecules such as the collectins may reduce the ability of other pro-inflammatory RNA sensors to encounter RNA. Thus, the overall effect of exogenously administered CRP in models of spontaneous lupus has been reduction in disease severity (Rodriguez, Mold, Marnell, Hutt, Silverman, Tran, Du Clos 2006). Elevations in CRP have also been found to be a risk factor for atherosclerosis, a condition with increased prevalence in the setting of autoimmune diseases. It is has been hypothesized that increases in CRP clearance of self RNA could re-direct inflammatory signals toward atherosclerosis-associated endothelium or macrophages.

Deficiencies in the clearance of apoptotic material in general (such as with proximal complement component deficiencies) are also presumed to lead to increased availability of pro-inflammatory RNA species for propagation of anti-RNA-associated autoimmunity in lupus (Greidinger 2001). To date, though, other secreted factors that induce inflammatory responses after binding RNA have not been reported.

2.3 Receptor-Mediated RNA Recognition

Toll-Like Receptors (TLRs) are a series of TNF-receptor family molecules identified over the past 10 years to play major roles in innate immune responses (Pasare and Medzhitov 2005). TLR3, TLR7, and TLR8 are innate immune sensors that are specific for RNA motifs—double-stranded RNA in the case of TLR3 (Alexopoulou, Holt, Medzhitov, and Flavell 2001), and single-stranded RNAs for TLR7 and TLR8 (Heil, Hemmi, Hochrein, Ampenberger, Kirschning, Akira, Lipford, Wagner, and Bauer 2004). Activation of these receptors can be potent stimuli for pro-inflammatory responses. These TLRs induce pro-inflammatory cytokines including IL-6, IL-12, and TNF-alpha (Hubert, Voisine, Louvet, Heslan, Ouabed, Heslan, and Josien 2006). Single-stranded RNA-induced TLR activation has also been correlated with high production of Type I Interferons and thus implicated in the pathogenesis of lupus

(Vollmer, Tluk, Schmitz, Hamm, Jurk, Forsbach, Akira, Kelly, Reeves, Bauer, and Krieg 2005). RNA-binding TLR activation also promotes maturation of antigen presenting cells, with increased surface expression of MHC motifs and costimulatory molecules including CD86 and CD40 (Ito, Amakawa, Kaisho, Hemmi, Tajima, Uehira, Ozaki, Tomizawa, Akira, and Fukuhara 2002). In contrast to most other TLRs that are expressed on the plasma membrane of cells (except for the dsDNA-recognizing TLR9, which shares many characteristics with the RNA-recognizing TLRs), the RNA-recognizing TLRs are typically expressed on professional antigen presenting cell subpopulations only within endosomes (Wang, Shao, Bennett, Shankar, Wightman, Reddy 2006), or within similar intracellular compartments (de Bouteiller, Merck, Hasan, Hubac, Benguigui, Trinchieri, Bates, and Caux 2005). These TLRs are generally also limited in their expression to professional antigen presenting cells, with a few exceptions such as endometrial cells (Young, Lyddon, Jorgenson, and Misfeldt 2004), some fibroblasts (Matsumoto, Kikkawa, Kohase, Miyake, and Seya 2002), and NK cells where surface expression of some RNA-sensing TLRs has been described (Hart, Athie-Morales, O'Connor, and Gardiner 2005).

Toll-Like Receptors -3, -7, and -8 have activity on largely distinct sets of antigen presenting cells. TLR3 is expressed and functional primarily on myeloid dendritic cells, which generally do not express TLR7 (Edwards, Diebold, Slack, Tomizawa, Hemmi, Kaisho, Akira, and Reis e Sousa 2003). TLRs 7, 8, and 9 are typically co-expressed on the same cell types, including prominently on plasmacytoid dendritic cells. However, TLR7 agonists appear to be more potent at activating plasmacytoid dendritic cells while TLR8 agonists appear to be more monocyte-specific in inducing cell activation (Gorden, Gorski, Gibson, Kedl, Kieper, Qiu, Tomai, Alkan, and Vasilakos 2005). Perhaps not surprisingly, then, it is TLR7 that has been most closely linked with Systemic Lupus Erythematosus (Subramanian, Tus, Li, Wang, Tian, Zhou, Liang, Bartov, McDaniel, Zhou, Schultz, and Wakeland 2006), an anti-RNP-associated autoimmune disease in which type I interferons and plasmacytoid dendritic cells have been proposed to play important roles (Lovgren, Eloranta, Kastner, Wahren-Herlenius, Alm, and Ronnblom 2006).

Presumably, differences in cell-type expression and immune activation between the individual RNA-sensing TLRs have been driven by evolutionary pressures to shape the characteristics of immune responses to organisms that specifically recognize TLRs -3, -7, or -8 to induce optimal protective responses. The components of a protective response would differ based on the characteristics of the pathogen encountered. Some pathogens are dangerous and require the induction of a robust inflammatory response in some tissues but not others. Conversely, excess inflammation in particular tissues in response to a pathogen could lead to worse rather than better outcomes. Thus, regarding tissue specialization, TLR3 responses have been particularly associated with protection against pathogens such as pulmonary rhinovirus (Hewson, Jardine, Edwards, Laza-Stanca, and Johnston 2005). On the other hand, TLR3 has an adverse effect on influenza A infection due to excessive induction of pulmonary inflammation (Le Goffic, Balloy, Lagranderie, Alexopoulou, Escriou, Flavell, Chignard, and Si-Tahar 2006). TLR8 responses have been linked with solid organ inflammation, as in Coxsackie B virus-associated myocarditis (Triantafilou, Orthopoulos, Vakakis, Ahmed, Golenbock, Lepper, and Triantafilou 2005).

TLR7 responses appear to be efficacious against hematogenously circulating pathogens such as Hepatitis C (Horsmans, Berg, Desager, Mueller, Schott, Fletcher, Steffy, Bauman, Kerr, and Averett 2005). The effects of TLR7 in hematogenous HIV infection, though, appear to be less favorable due to induction of increased viral replication that counterbalances actions leading to acute viral inactivation (Schlaepfer, Audige, Joller, and Speck 2006).

Since myeloid dendritic cells (that are activated by TLR3 agonists) differ from plasmacytoid dendritic cells (that are activated by TLR7 agonists) and from monocytic cells (that are activated by TLR8 agonists) in their tissue distributions, responses to inflammatory signals, and interactions with other cell types, it follows that specific recognition by a particular TLR could lead to a distinct immune phenotype compared to recognition by a different TLR based on the cell type and tissue that it occupies, even if similar pro-inflammatory intracellular signals ensue from stimulation of each receptor. That said, the intracellular signals induced by the TLRs can also be quite different, particularly between TLR3 and the more conserved TLR7 and TLR8 (that share more sequence homology, tissue expression, and signaling features with each other and with the dsDNA-recognizing TLR9). Like most TLRs, TLR7 signals through the MyD88 adapter protein (Heil et al. 2004). TLR7 activation ultimately induces Type I Interferons through the effects of activated interferon regulatory factors 5 and 7 (IRF5 and IRF7) on gene transcription (Schoenemeyer, Barnes, Mancl, Latz, Goutagny, Pitha, Fitzgerald, and Golenbock 2005). In contrast, TLR3, acting through the MyD88-independent TRIF adaptor protein, activates IRF3 and IRF7 (Fitzgerald, Rowe, Barnes, Caffrey, Visintin, Latz, Monks, Pitha, and Golenbock 2003).

Among the cells that express the RNA-recognizing Toll-Like Receptors most strongly are dendritic cell subsets. This is remarkable since a major function/specialization of dendritic cells is the presentation of antigen to T cells, yet RNA itself is not known to be presented to T cells. The association of RNA recognition with protein antigen presentation, then, suggests an alternative to direct recognition of RNA in the host defense against RNA-containing pathogens: the recognition of RNA-protein complexes. In the responses to a protein-RNA complex, the RNA component can act as a physiological adjuvant due to its innate immune stimulatory ability to induce pro-inflammatory cytokines and cell surface molecules that increase the efficiency of induction of anti T cell (and anti-B cell) responses toward the linked protein(s). The direct association of TLR agonists with a protein antigen has also been shown to increase the efficiency of presentation by the dendritic cell of peptides from that protein to T cells (Blander and Medzhitov 2006). The ability of protein-RNA complexes to be recognized by autoantibodies, in turn, allows such an immune response to grow explosively in intensity, since autoantibodies binding to protein-RNA complexes can then opsonize the complexes, leading to more efficient uptake by antigen-presenting cells, and more pro-inflammatory activation of RNA-sensing TLRs (Fu, Xie, Chen, Zhu, Zhou, Thomas, Zhou, and Mohan 2006). Moreover, B cells expressing antibodies that recognize RNP structures on their B cell receptors also are susceptible to augmented inflammatory responses due to the co-ligation of their immunoglobulin target and the RNA-recognizing TLR7

(Lau, Broughton, Tabor, Akira, Flavell, Mamula, Christensen, Shlomchik, Viglianti, Rifkin, Marshak-Rothstein 2005).

3 Avoidance of Anti-Self RNA Responses

With the exception of relatively infrequent autoimmune diseases, the immune system appears to distinguish pathogen-associated RNA from self RNA successfully. Multiple pathways leading to avoidance of anti-self RNA responses have been proposed. At the adaptive immunity level, general mechanisms that contribute to tolerance, including autoreactive cell deletion and autoreactive cell anergy are likely to limit anti-RNA responses. Indeed, animals with defined defects in some of these tolerance pathways develop higher prevalences of anti-RNP responses and a higher rate of autoimmune disease (Monneaux, Dumortier, Steiner, Briand, and Muller 2001). T regulatory cells (Treg) may also have a role in limiting anti-RNP autoimmunity. Total Treg number has been reported to be reduced in individuals with established anti-RNP autoimmunity (Barath, Sipka, Aleksza, Szegedi, Szodoray, Vegh, Szegedi, and Bodolay 2006), and treatment of lupus-prone mice with ex vivo expanded Tregs ameliorates disease (Scalapino, Tang, Bluestone, Bonyhadi, and Daikh 2006). The lack of T cells specific for RNA antigens implies that Treg cells may also lack specificity for RNA, leading to a diminished role for such cells in anti-RNA autoimmunity compared with other autoimmune syndromes.

Anti-RNP responses at both the adaptive and innate immune levels may also be limited by control of the compartments in which self RNAs are typically present. RNA is generally maintained within cells. The extracellular environment is promptly cleared of potential sources of RNA by rapid uptake of apoptotic and necrotic cell debris (Potter, Cortes-Hernandez, Quartier, Botto, and Walport 2003), and by the effects of potent and widely expressed extracellular RNA-degrading RNase enzymes (El-Hefnawy, Raja, Kelly, Bigbee, Kirkwood, Luketich, Godfrey 2004). Experimental disruption of "garbage disposal" of cell debris (inclusive of RNA-containing structures) has been associated with the development of anti-RNP antibodies and lupus (Cohen, Caricchio, Abraham, Camenisch, Jennette, Roubey, Earp, Matsushima, Reap 2002). The absence of similar reports to date of RNase knockouts leading to increased induction of anti-RNP antibodies may have to do with the remarkable potency and redundancy of RNase activities observed in human tissues (Weickmann and Glitz 1982). Alternatively, it is possible that the relatively few self RNA molecules that are associated with relatively frequent autoimmune syndromes may be ones that exhibit resistance to endogenous RNase activities. In this regard, it is notable that exogenous administration of autoantigen-associated RNAs such as the snRNP-associated U1-RNA (Greidinger, Zang, Jaimes, Hogenmiller, Nassiri, Bejarano, Barber, and Hoffman 2006), as well as the Ro-associated Y RNAs (Kelly, Zhuang, Nacionales, Scumpia, Lyons, Akaogi, Lee, Williams, Yamamoto, Akira, Satoh, and Reeves 2006), have been shown to induce autoinflammatory effects in vitro and in vivo when administered in the absence of RNase inhibitors.

To some extent, the structures of self RNAs may also be less immunogenic than pathogen-derived RNAs. For example, post-transcriptional modifications of self RNAs have been shown to reduce the TLR stimulatory capacity of some self RNAs (Kariko, Buckstein, Ni, and Weissman D 2005). However, ribonucleoproteins derived from human cells have been shown to be agonists of TLRs (Hoffman RW, Gazitt T, Foecking MF, Ortmann RA, Misfeldt M, Jorgenson R, Young SL, and Greidinger EL 2004). This raises the additional question whether a paucity of modified residues that prevent induction of TLRs may predispose certain RNA molecules, such as U1-RNA (Branlant, Krol, Ebel, Lazar, Gallinaro, Jacob, Sri-Widada, and Jeanteur 1980), to participate in autoimmune responses. It is also potentially of interest that post-translationally modified portions of U1-RNA are cleaved off in apoptosis (Degen, Aarssen, Pruijn, Utz, and van Venrooij 2000).

Similarly, the intracellular systems of surveillance for pathogenic RNAs, such as the RIG-I system, also appear to be regulated by sequestration of self RNAs, elimination of potentially inappropriately immunogenic self RNAs, and possibly by additional structural cues that downregulate activation signals. To induce an innate immune response from the RIG-1 system, an intracellular RNA must have double-stranded structure, and must co-localize with this cytoplasmic sensing system (eliminating, for example, nuclear dsRNA structures such as the spliceosome as potential stimulators of this system). The mechanisms used by some cytoplasmic RNAs with double-stranded structure such as ribosomal RNAs and the Y RNAs to avoid activation by RNA-helicase cytoplasmic dsRNA sensing systems remain to be determined.

The cellular context in which RNA is recognized by the innate immune system may also help modulate disease. Circumstances may exist under which dendritic cells induce tolerogenic rather than immunogenic responses in response to TLR stimuation. Evidence for this includes both the ability of some subsets of dendritic cells that receive TLR-like maturation signals to mediate immunomodulatory responses (Dhodapkar, Steinman, Krasovsky, Munz, and Bhardwaj 2001), and the observation that mice with defects in TLR9 signaling develop more severe rather than more mild autoimmune manifestations (Christensen, Shupe, Nickerson, Kashgarian, Flavell, and Shlomchik 2006). We have also observed that TLR3-deficient mice develop more severe renal disease than TLR3-intact animals (Greidinger, et al. 2006). Given the tissue and organ specificity of particular dendritic cell subtypes, it is reasonable to speculate that the inhibitory effects of individual TLRs may also be tissue and organ-specific, and that TLRs that are pro-inflammatory in one tissue may mediate anti-inflammatory effects in another.

4 AntiRNA Responses and Autoimmunity

Several RNA-containing structures have been observed to be targets of autoimmune responses. This includes structures whose RNAs have already been mentioned such as snRNPs that contain U-RNAs (Sharp, et al. 1971), and the Ro complex that includes Y RNAs (Stein, Fuchs, Fu, Wolin, and Reinisch 2005). Additional nuclear

RNA-containing structures known to be targeted in autoimmunity include hnRNPs (Skriner, Sommergruber, Tremmel, Fischer, Barta, Smolen, and Steiner 1997), the Xist RNA-associated Barr Body (Brooks, Satoh, Hong, Reeves, and Yang 2002), the 7SL RNA-asociated Signal Recognition Particle (Walter and Blobel 1983), the Pm/Scl exosome complex (Raijmakers, Renz, Wiemann, Egberts, Seelig, van Venrooij, and Pruijn 2004), the Th/To endoribonuclease complex (Van Eenennaam, Vogelzangs, Lugtenberg, Van Den Hoogen, Van Venrooij, and Pruijn 2002), and other nucleolus-associated structures (Yamasaki, Honkanen-Scott, Hernandez, Ikeda, Barker, Bubb, Narain, Richards, Chan, Reeves, and Satoh 2006). Additional cytoplasmic RNA-containing structures known to be targeted in autoimmunity include ribosomes (Koffler, Faiferman, and Gerber 1977), GW bodies (Jakymiw, Ikeda, Fritzler, Reeves, Satoh, and Chan 2006), and tRNA synthetase complexes (Mathews and Bernstein 1983). Studies of the potential innate immune reactivity of the RNAs associated with these nuclear and cytoplasmic structures have for the most part not yet been reported. On the other hand, some RNA structures that have been found to have innate immune stimulatory activity in vitro (such as mRNA) are poorly documented to be autoantigen-associated structures in clinical autoimmune disease (Kariko, Ni, Capodici, Lamphier, and Weissman 2004). Other RNPs such as the Vault RNP have not been reported to date to be targets of autoimmunity or innate immunity agonists.

Multiple clinical autoimmune syndromes have been documented to occur along with immune responses to RNA-associated structures. With some structures, autoimmunity has been linked relatively tightly to a single clinical syndrome, such as with Signal Recognition Particle and Myositis (Kao, Lacomis, Lucas, Fertig, and Oddis 2004), the U3 snoRNP and scleroderma (Yang, Hildebrandt, Luderschmidt, and Pollard 2003), and ribosomal responses and lupus (Bonfa and Elkon 1986). In the cases of several other RNA-associated structures, the clinical syndromes induced either explicitly are typified by "overlap syndromes", as with snRNP (Sharp, et al. 1971), hnRNP (Skriner, et al. 1997), Pm/Scl (Warner and Greidinger 2004), and tRNA Synthetases (Marguerie, Bunn, Beynon, Bernstein, Hughes, So, and Walport 1990), or are at least known to be associated with multiple distinct clinical entities as with the Y RNA-associated Ro complex (Setty, Pittman, Mahale, Greidinger, and Hoffman 2002).

As an additional level of complexity, the kinetics of development of immune responses to RNA-associated autoantigens relative to clinical symptoms can also differ. Among patients that develop lupus, some RNA-associated autoantibodies (such as anti-Ro) are found to develop years before the onset of clinical disease, while others are closely temporally tied to the onset of clinical symptoms (such as anti-U1-RNP) (Arbuckle, McClain, Rubertone, Scofield, Dennis, James, Harley 2003). Anti-RNP antibodies have both been reported to be markers of a clinically mild subset of lupus-like disease (Reichlin and Van Venrooij 1991), yet also to be linked with more severe clinical disease (Kirou, Lee, George, Louca, Peterson, and Crow 2005). Thus, while anti-RNA responses at the level of adaptive immunity appear to be closely tied in general to the presence of autoimmune syndromes, the wide range of autoimmune manifestations of anti-RNA-associated autoimmunity suggests that considerable complexity exists in the specifics of anti-RNA responses.

With regard to the specificity of immune responses in any autoimmune syndrome, the diversity of autoantibodies that manifest even in syngeneic cohabitating animals with genetically determined spontaneous autoimmune syndrome demonstrates that stochastic factors may play a considerable role. The rate of development of anti-snRNP responses in New Zealand Mixed female mice, for example, has been reported to be approximately 30%, despite nearly 100% penetrance of lupus-like autoimmunity (Laderach, Koutouzov, Bach, and Yamamoto 2003). Moreover, a variety of other extrinsic variables may raise or lower the likelihood of immunity to any particular structure. These include the influence of major histocompatibility antigens in antigen presentation (O'Hanlon, Carrick, Targoff, Arnett, Reveille, Carrington, Gao, Oddis, Morel, Malley, Malley, Shamim, Rider, Chanock, Foster, Bunch, Blackshear, Plotz, Love, and Miller 2006), the impact of antecedent infectious processes on immune repertoire and response (McClain, Heinlen, Dennis, Roebuck, Harley, and James 2005), and the level of function of the variety of mechanisms that prevent anti-RNA immune responses as described above. Still, it is noteworthy that immune responses to protein-RNA complexes have demonstrated a tendency, once established, to induce antigenic spreading to other protein-RNA complexes, even if the complexes involved have different intracellular localization and lack obvious sequence and structural homology (Deshmukh, Kannapell, and Fu 2002). This suggests that once a context is established that favors induction of anti-RNA-associated autoimmunity, tonic ongoing autoimmune stimulation may be favored over the typical mechanisms that lead to downmodulation of an active immune response.

The development of anti-RNP antibodies that opsonize RNP autoantigens, rescuing the autoantigens and RNA from clearance or degradation and delivering them efficiently in a pro-inflammatory manner to antigen presenting cells to perpetuate and augment autoimmune responses, may be a critical factor in this process. The development of anti-RNP responses in animal models has been found to require B cells in addition to T cells and professional antigen-presenting cells (Monneaux, Briand, and Muller 2000). The ability of RNP-containing immune complexes to ligate pro-inflammatory receptors that facilitate endosomal Toll-Like Receptor signaling may also play an important role in this process (Boule, Broughton, Mackay, Akira, Marshak-Rothstein, Rifkin 2004). Additionally, in contrast to more typical inflammatory responses, where a productive response leads to elimination of the inciting antigen (for example, when all copies on an invading organism are destroyed), the opposite can happen in this case. Once autoimmune injury develops, more RNPs from damaged and dying cells are elaborated, adding additional antigen to fuel the autoimmune response. Moreover, in contrast to conventional protein antigens, an RNP antigen with the capability of inducing innate immune responses continues to provide tonic adjuvant-like "danger signals" that can downregulate otherwise immunomodulatory responses (Pasare and, Medzhitov 2003).

Since antibody opsonization of some RNA-containing complexes has been clearly shown to induce TLR7 signals and further autoimmunity (Lovgren, et al. 2006), the success of anti-B cell therapy in lupus may in part reflect the ability to this therapy to reduce innate immunity signaling. It is unresolved whether anti-B cell therapy would have similar benefits in TLR3 or TLR8-mediated processes.

While there are indications that non-TLR innate immune mechanisms also participate in anti-RNP immune responses (Barrat, Meeker, Gregorio, Chan, Uematsu, Akira, Chang, Duramad, Coffman 2005), even studies confined to the analysis of TLR effects on autoimmune responses provide indications for sources of complexity in the anti-RNA responses that can account for some of the clinical and immunological diversity of the responses that are seen.

The two common RNP targets of autoimmune responses studied to date, U1-RNA and Y-RNA, have been found to have TLR stimulatory properties that are compatible with the development of systemic autoimmunity. In an induced model of autoimmunity, U1-RNA has been shown to be as efficient as Complete Freund's Adjuvant in facilitating anti-peptide autoantibody responses and in the induction of clinical autoimmunity (Greidinger et al. 2006). This self RNA was found to stimulate both TLR3 and TLR7. TLR7 signaling has been found to be most important for the development of systemic lupus (Kelley, et al. 2006). Experiments in TLR3-knockout mice suggest that TLR3 signaling may in fact counter-regulate the induction of lupus manifestations (Greidinger, et al. 2006). On the other hand, pulmonary manifestations of autoimmunity that are typical of MCTD but not of lupus are exacerbated by TLR3 stimulation in this model. Thus, an identical antigenic stimulus can lead to different clinical autoimmune syndromes based on the innate immune context that is present. In the case of U1-RNA-associated autoimmunity, two major clinical syndromes (SLE and MCTD) may be distinguished in a mouse model by the balance of TLR7 versus TLR3-induced signaling. Studies to test for a similar relationship between human SLE and MCTD are warranted.

If we posit that TLR7 activation is associated with a lupus phenotype, and that TLR3 stimulation can be associated with an MCTD phenotype, it is reasonable to ask whether specific activation by RNA of other RNA recognizing sensors could also be linked to specific clinical syndromes. It is thus evocative to observe that TLR8-associated monocytic cells may be implicated in human Sjogren's Syndrome (Gottenberg, Cagnard, Lucchesi, Letourneur, Mistou, Lazure, Jacques, Ba, Ittah, Lepajolec, Labetoulle, Ardizzone, Sibilia, Fournier, Chiocchia, and Mariette 2006). A potential explanation for the clinical variation between presentations of anti-Ro autoimmunity could be the association of individual autoimmune syndromes with different patterns of innate immune activation by Ro-associated Y RNAs. Human cells express four different Y RNAs that associate with the Ro autoantigen. While human Y RNA complexes derived from intact cells have been shown to induce TLR7-like innate immune stimulation (Kelley, et al. 2006), some Ro-associated Y RNAs may also be capable of inducing TLR3 and/or TLR8-like signals, potentially accounting for Ro-associated subacute cutaneous lupus and Ro-associated Sjogren's Syndrome in addition to Ro-associated SLE. Previous difficulties in generating rodent models of anti-Ro autoimmunity that also express sialadenitis may be due to the reduced inflammatory activity of TLR8 in rodents compared to humans. Human TLR8 can be independently pro-inflammatory, but rodent TLR8 has only generated inflammatory responses when co-stimulated (Gorden, Qiu, Binsfeld, Vasilakos, and Alkan 2006).

Recent information has also implicated certain tRNAs as TLR agonists (Wang, Xiang, Shao, and Yuan 2006). These molecules associate with the tRNA synthetase proteins that are well-documented autoantigens in myositis and interstitial lung disease (Marguerie, et al. 1990). To date, the TLR inducing activity documented by tRNAs has been for alanyl tRNA with TLR3. To the extent that MCTD and the anti-tRNA Synthetase Syndrome (anti PL-12 in the case of alanyl tRNA synthetase) share clinical characteristics including infrequent renal disease, frequent lung disease, and associated myositis, this finding leaves open the question whether TLR3 activation is similarly influencing the tissue expression of disease in both MCTD and the tRNA Synthetase Syndrome.

If additional studies confirm the hypothesis that changes in the balance of innate immune responses to RNA determine the clinical phenotype of anti-RNP-associated autoimmune syndromes, then treatment strategies designed to alter the innate immune context present in a given patient would be potentially valuable. If, for example, one could upregulate TLR3 signals and downregulate TLR7 signals in anti-U1-RNA associated autoimmunity, this might shift the clinical pattern of disease from one of severe lupus to one of potentially more mild MCTD. On the other hand, specific inhibitors of individual innate immune pathways could lead to unbalanced activation of other innate immune effectors that could lead to more severe disease. Thus, even if TLR8 proved to be important in the development of Sjogren's Syndrome, a specific TLR8 inhibitor would have to be viewed with caution, lest the inhibition of TLR8 lead to unbalanced TLR7 activation and induction of lupus (Wang, et al. 2006).

To the extent that differences in innate immune activation determine the clinical syndromes induced by autoimmunity to specific RNA-associated autoantigens, additional predictions can be made about the role of RNA as an innate immune agonist in these cases. When a given antigen is associated with an overlap syndrome, one would predict that the RNA (and/or other innate immune effectors) linked to that antigen are capable of activating multiple innate immune pathways (as with U1-RNA activation of both TLR7 and TLR3). When a given antigen is associated with a single clinical syndrome, one would predict that the RNA (and other innate immune effectors) linked to that antigen are capable of activating only a single innate immune pathway. Furthermore, when an antigen is associated with multiple distinct clinical syndromes, one would predict that multiple RNAs or other innate immune effectors may exist in association with the antigen under some circumstances, and that these innate immune effectors are each specific for the activation of individual innate immune pathways. This would be the case in the Ro complex system, for example, if the innate immune stimulation pathways induced differed between the Y1, Y2, Y4, and Y5 RNAs.

5 Conclusion

The immune sensor and effector mechanisms that respond to RNA are qualitatively different from those that respond to protein and lipid antigens, based on the absence of direct T cell involvement and the predominance of multiple innate immune effector pathways. The ability for protein-RNA complexes encountering even low levels of anti-ribonucleoprotein complex-reactive B and T cell autoreactivity to autoamplify this autoreactivity by recruitment of RNA-induced innate immune pathways may help explain the apparently frequent representation of RNA-associated molecules as autoantigens in clinical autoimmune syndromes. Different innate immune RNA sensing mechanisms are associated with different tissue and cell type distributions, different downstream signaling pathways, and different expressions of pro-inflammatory versus anti-inflammatory effects. Preferential activation of specific innate immune receptors may play a key role in determining the tissue targeting and hence the clinical expression of anti-RNP-associated clinical syndromes.

References

Alexopoulou L., Holt A.C., Medzhitov R., Flavell R.A. (2001) Recognition of double-stranded RNA and activation of NF-kappaB by Toll-like receptor 3. Nature 413, 732-738.

Barath S., Sipka S., Aleksza M., Szegedi A., Szodoray .P, Vegh J., Szegedi G., Bodolay E. (2006) Regulatory T cells in peripheral blood of patients with mixed connective tissue disease. Scand. J. Rheumatol. 35, 300-304.

Barrat F.J., Meeker T., Gregorio J., Chan J.H., Uematsu S., Akira S., Chang B., Duramad O., Coffman R.L. (2005) Nucleic acids of mammalian origin can act as endogenous ligands for Toll-like receptors and may promote systemic lupus erythematosus. J. Exp. Med. 202, 1131-1139.

Blander JM, Medzhitov R. (2006) Toll-dependent selection of microbial antigens for presentation by dendritic cells. Nature 440, 808-812.

Bonfa E., Elkon K.B. (1986) Clinical and serologic associations of the antiribosomal P protein antibody. Arthritis Rheum. 29, 981-985.

Boule M.W., Broughton C., Mackay F., Akira S., Marshak-Rothstein A., Rifkin I.R. (2004) Toll-like receptor 9-dependent and -independent dendritic cell activation by chromatin-immunoglobulin G complexes. J. Exp. Med. 199, 1631-1640.

Branlant C., Krol A., Ebel J.P., Lazar E., Gallinaro H., Jacob M., Sri-Widada J., Jeanteur P. (1980) Nucleotide sequences of nuclear U1A RNAs from chicken, rat and man. Nucleic Acids Res. 8, 4143-4154.

Christensen S.R., Shupe J., Nickerson K., Kashgarian M., Flavell R.A., Shlomchik M.J. (2006) Toll-like receptor 7 and TLR9 dictate autoantibody specificity and have opposing inflammatory and regulatory roles in a murine model of lupus. Immunity 25, 417-428.

Cohen P.L., Caricchio R., Abraham V., Camenisch T.D., Jennette J.C., Roubey R.A., Earp H.S., Matsushima G., Reap E.A. (2002) Delayed apoptotic cell clearance and lupus-like autoimmunity in mice lacking the c-mer membrane tyrosine kinase. J. Exp. Med. 196, 135-140.

de Bouteiller O., Merck E., Hasan U.A., Hubac S., Benguigui B., Trinchieri G., Bates E.E., Caux C. (2005) Recognition of double-stranded RNA by human toll-like receptor 3 and downstream receptor signaling requires multimerization and an acidic pH. J. Biol. Chem. 280, 38133-38145.

Degen W.G., Aarssen Y., Pruijn G.J., Utz P.J., van Venrooij W.J. (2000) The fate of U1 snRNP during anti-Fas induced apoptosis: specific cleavage of the U1 snRNA molecule. Cell Death Differ. 7, 70-79.

Deshmukh U.S., Kannapell C.C., Fu S.M. (2002) Immune responses to small nuclear ribonucleoproteins: antigen-dependent distinct B cell epitope spreading patterns in mice immunized with recombinant polypeptides of small nuclear ribonucleoproteins. J. Immunol. 168, 5326-5332.

Dhodapkar M.V., Steinman R.M., Krasovsky J., Munz C., Bhardwaj N. (2001) Antigen-specific inhibition of effector T cell function in humans after injection of immature dendritic cells. J Exp Med. 193, 233-238.

Du Clos T.W. (1989) C-reactive protein reacts with the U1 small nuclear ribonucleoprotein. J. Immunol. 143, 2553-2559.

Edwards A.D., Diebold S.S., Slack E.M., Tomizawa H., Hemmi H., Kaisho T., Akira S., Reis e Sousa C. (2003) Toll-like receptor expression in murine DC subsets: lack of TLR7 expression by CD8 alpha+ DC correlates with unresponsiveness to imidazoquinolines. Eur J Immunol. 33, 827-833.

El-Hefnawy T., Raja S., Kelly .L, Bigbee W.L., Kirkwood J.M., Luketich J.D., Godfrey T.E. (2004) Characterization of amplifiable, circulating RNA in plasma and its potential as a tool for cancer diagnostics. Clin. Chem. 50, 564-573.

Fitzgerald K.A., Rowe D.C., Barnes B.J., Caffrey D.R., Visintin A., Latz E., Monks B., Pitha P.M., Golenbock D.T. (2003) LPS-TLR4 signaling to IRF-3/7 and NF-kappaB involves the toll adapters TRAM and TRIF. J. Exp. Med. 198, 1043-1055.

Fu Y., Xie C., Chen J., Zhu J., Zhou H., Thomas J., Zhou X.J., Mohan C. (2006) Innate stimuli accentuate end-organ damage by nephrotoxic antibodies via Fc receptor and TLR stimulation and IL-1/TNF-alpha production. J. Immunol. 176, 632-639.

Gorden K.B., Gorski K.S., Gibson S.J., Kedl R.M., Kieper W.C., Qiu X., Tomai M.A., Alkan S.S., Vasilakos J.P. (2005) Synthetic TLR agonists reveal functional differences between human TLR7 and TLR8. J. Immunol. 174, 1259-1268.

Gorden K.K., Qiu X.X., Binsfeld C.C., Vasilakos J.P., Alkan S.S.. (2006) Cutting Edge: Activation of Murine TLR8 by a Combination of Imidazoquinoline Immune Response Modifiers and PolyT Oligodeoxynucleotides. J. Immunol. 177, 6584-6587.

Gottenberg J.E., Cagnard N., Lucchesi C., Letourneur .F, Mistou S., Lazure T., Jacques S., Ba N., Ittah M., Lepajolec C., Labetoulle M., Ardizzone M., Sibilia J., Fournier C., Chiocchia G., Mariette X. (2006) Activation of IFN pathways and plasmacytoid dendritic cell recruitment in target organs of primary Sjogren's syndrome. Proc. Natl. Acad. Sci. U S A. 103, 2770-5.

Greidinger E.L. (2001) Apoptosis in lupus pathogenesis. Front. Biosci. 6, D1392-D1402.

Greidinger E.L., Zang Y., Jaimes K., Hogenmiller S., Nassiri M., Bejarano P., Barber G.N., Hoffman R.W. (2006) A murine model of mixed connective tissue disease induced with U1 small nuclear RNP autoantigen. Arthritis Rheum. 54, 661-669.

Hart O.M., Athie-Morales V., O'Connor G.M., Gardiner C.M. (2005) TLR7/8-mediated activation of human NK cells results in accessory cell-dependent IFN-gamma production. J. Immunol. 175, 1636-1642.

Heil F., Hemmi H., Hochrein H., Ampenberger F., Kirschning C., Akira S., Lipford G., Wagner H., Bauer S. (2004) Species-specific recognition of single-stranded RNA via toll-like receptor 7 and 8. Science 303, 1526-1529.

Hewson C.A., Jardine A., Edwards M..R, Laza-Stanca V., Johnston S.L. (2005) Toll-like receptor 3 is induced by and mediates antiviral activity against rhinovirus infection of human bronchial epithelial cells. J Virol. 79, 12273-12279.

Hoffman R.W., Gazitt T., Foecking M.F., Ortmann R.A., Misfeldt M., Jorgenson R., Young S.L., Greidinger E.L. (2004) U1 RNA induces innate immunity signaling. Arthritis Rheum. 50, 2891-2896.

Horsmans Y., Berg T., Desager J.P., Mueller T., Schott E., Fletcher S.P., Steffy K.R., Bauman L.A., Kerr B.M., Averett D.R. (2005) Isatoribine, an agonist of TLR7, reduces plasma virus concentration in chronic hepatitis C infection. Hepatology 42, 724-7231.

Hubert F.X., Voisine C., Louvet C., Heslan J.M., Ouabed A., Heslan M., Josien R. (2006) Differential pattern recognition receptor expression but stereotyped responsiveness in rat spleen dendritic cell subsets. J. Immunol. 177, 1007-1016.

Ito T., Amakawa R., Kaisho T., Hemmi H., Tajima K., Uehira K., Ozaki Y., Tomizawa H., Akira S., Fukuhara S. (2002) Interferon-alpha and interleukin-12 are induced differentially by Toll-like receptor 7 ligands in human blood dendritic cell subsets. J. Exp. Med. 195, 1507-1512.

Jakymiw A., Ikeda K., Fritzler M.J., Reeves W.H., Satoh M., Chan E.K. (2006) Autoimmune targeting of key components of RNA interference. Arthritis Res. Ther. 8, R87.

Kao A.H., Lacomis D., Lucas M., Fertig N., Oddis C.V. (2004) Anti-signal recognition particle autoantibody in patients with and patients without idiopathic inflammatory myopathy. Arthritis Rheum. 50, 209-215.

Kariko K., Buckstein M, Ni H., Weissman D. (2005) Suppression of RNA recognition by Toll-like receptors: the impact of nucleoside modification and the evolutionary origin of RNA. Immunity 23, 165-175.

Kariko K., Ni H., Capodici J., Lamphier M., Weissman D. (2004) mRNA is an endogenous ligand for Toll-like receptor 3. J. Biol. Chem. 279, 12542-12550.

Kato H., Takeuchi O., Sato S., Yoneyama M., Yamamoto M., Matsui K., Uematsu S., Jung A., Kawai T., Ishii K.J., Yamaguchi O., Otsu K., Tsujimura T., Koh C.S., Reis e Sousa C., Matsuura Y., Fujita T., Akira S. (2006) Differential roles of MDA5 and RIG-I helicases in the recognition of RNA viruses. Nature 441, 101-105.

Kelly KM, Zhuang H, Nacionales DC, Scumpia PO, Lyons R, Akaogi J, Lee P, Williams B, Yamamoto M, Akira S, Satoh M, Reeves WH. (2006) "Endogenous adjuvant" activity of the RNA components of lupus autoantigens Sm/RNP and Ro 60. Arthritis Rheum. 54, 1557-1567.

Kirou K.A., Lee C., George S., Louca K., Peterson M.G., Crow M.K. (2005) Activation of the interferon-alpha pathway identifies a subgroup of systemic lupus erythematosus patients with distinct serologic features and active disease. Arthritis Rheum. 52, 1491-1503.

Koffler D., Faiferman I., Gerber M.A. (1977) Radioimmunoassay for antibodies to cytoplasmic ribosomes in human serum. Science 198, 741-3.

Laderach D., Koutouzov S., Bach J.F., Yamamoto A.M. (2003) Concomitant early appearance of anti-ribonucleoprotein and anti-nucleosome antibodies in lupus prone mice. J. Autoimmun. 20, 161-170.

Lau C.M., Broughton C., Tabor A.S., Akira S., Flavell R.A., Mamula M.J., Christensen S.R., Shlomchik M.J., Viglianti G.A., Rifkin I.R., Marshak-Rothstein A. (2005) RNA-associated autoantigens activate B cells by combined B cell antigen receptor/Toll-like receptor 7 engagement. J. Exp. Med. 202, 1171-1177.

Le Goffic R, Balloy V, Lagranderie M, Alexopoulou L, Escriou N, Flavell R, Chignard M, Si-Tahar M. Detrimental contribution of the Toll-like receptor (TLR)3 to influenza A virus-induced acute pneumonia. PLoS Pathog. 2, e53.

Lovgren T., Eloranta M.L., Kastner B., Wahren-Herlenius M., Alm G.V., Ronnblom L. (2006) Induction of interferon-alpha by immune complexes or liposomes containing systemic lupus erythematosus autoantigen- and Sjogren's syndrome autoantigen-associated RNA. Arthritis Rheum. 54, 1917-1927.

Marguerie C., Bunn C.C., Beynon H.L., Bernstein R.M., Hughes J.M., So A.K., Walport M.J. (1990) Polymyositis, pulmonary fibrosis and autoantibodies to aminoacyl-tRNA synthetase enzymes. Q. J. Med. 77, 1019-1038.

Marshall B.C., McPherson R.A., Greidinger E., Hoffman R., Adler S.P. (2002) Lack of autoantibody production associated with cytomegalovirus infection. Arthritis Res. 4, R6.

Mathews M.B., Bernstein R.M. (1983) Myositis autoantibody inhibits histidyl-tRNA synthetase: a model for autoimmunity. Nature 304, 177-179.

Matsumoto M, Kikkawa S, Kohase M, Miyake K, Seya T. Establishment of a monoclonal antibody against human Toll-like receptor 3 that blocks double-stranded RNA-mediated signaling. Biochem. Biophys. Res. Commun. 293, 1364-1369.

McClain M.T., Heinlen L.D., Dennis G.J., Roebuck J., Harley J.B., James J.A. (2005) Early events in lupus humoral autoimmunity suggest initiation through molecular mimicry. Nat. Med. 11, 85-89.

McGaha T.L., Sorrentino B., Ravetch J.V.. (2005) Restoration of tolerance in lupus by targeted inhibitory receptor expression. Science 307, 590-593.

Mineo C., Gormley A.K., Yuhanna I.S., Osborne-Lawrence S., Gibson L.L., Hahner L., Shohet R.V., Black S., Salmon J.E., Samols D., Karp D.R., Thomas G.D., Shaul P.W. (2005) FcgammaRIIB mediates C-reactive protein inhibition of endothelial NO synthase. Circ Res. 97, 1124-1131.

Mold C., Du Clos T.W. (2006) C-reactive protein increases cytokine responses to Streptococcus pneumoniae through interactions with Fc gamma receptors. J Immunol 176, 7598-604.

Monneaux F., Briand J.P., Muller S. (2000) B and T cell immune response to small nuclear ribonucleoprotein particles in lupus mice: autoreactive CD4(+) T cells recognize a T cell epitope located within the RNP80 motif of the 70K protein. Eur. J. Immunol. 30, 2191-2200.

Monneaux F., Dumortier H., Steiner G., Briand J.P., Muller S. (2001) Murine models of systemic lupus erythematosus: B and T cell responses to spliceosomal ribonucleoproteins in MRL/Fas(lpr) and (NZB x NZW)F(1) lupus mice. Int Immunol. 13, 1155-1163.

O'Hanlon T.P., Carrick D.M., Targoff I.N., Arnett F.C., Reveille J.D., Carrington M., Gao X., Oddis C.V., Morel P.A., Malley J.D., Malley K., Shamim E.A., Rider L.G., Chanock S.J., Foster C.B., Bunch T., Blackshear P.J., Plotz P.H., Love L.A., Miller F.W. (2006) Immunogenetic risk and protective factors for the idiopathic inflammatory myopathies: distinct HLA-A, -B, -Cw, -DRB1, and -DQA1 allelic profiles distinguish European American patients with different myositis autoantibodies. Medicine (Baltimore). 85, 111-127.

Pasare C., Medzhitov R. (2003) Toll pathway-dependent blockade of CD4+CD25+ T cell-mediated suppression by dendritic cells. Science. 2003 Feb 14;299(5609):1033-1036.

Pasare C., Medzhitov R. (2005) Toll-like receptors: linking innate and adaptive immunity. Adv. Exp. Med. Biol. 560:11-18.

Potter P.K., Cortes-Hernandez J., Quartier P., Botto M., Walport M.J. (2003) Lupus-prone mice have an abnormal response to thioglycolate and an impaired clearance of apoptotic cells. J. Immunol. 170, 3223-3232.

Raijmakers R., Renz M., Wiemann C., Egberts W.V., Seelig H.P., van Venrooij W.J., Pruijn G.J. (2004) PM-Scl-75 is the main autoantigen in patients with the polymyositis/scleroderma overlap syndrome. Arthritis Rheum. 50, 565-569.

Reichlin M, Van Venrooij WJ. (1991) Autoantibodies to the URNP particles: relationship to clinical diagnosis and nephritis. Clin. Exp. Immunol. 83, 286-290.

Rodriguez W., Mold C., Marnell L.L., Hutt J., Silverman G.J., Tran D., Du Clos T.W.. Prevention and reversal of nephritis in MRL/lpr mice with a single injection of C-reactive protein. Arthritis Rheum. 54, 325-335.

Rothenfusser S., Goutagny N., DiPerna G., Gong M., Monks B.G., Schoenemeyer A., Yamamoto M., Akira S., Fitzgerald K.A. (2005) The RNA helicase Lgp2 inhibits TLR-independent sensing of viral replication by retinoic acid-inducible gene-I. J. Immunol. 175, 5260-5268.

Scalapino K.J., Tang Q., Bluestone J.A., Bonyhadi M.L., Daikh D.I. (2006) Suppression of disease in New Zealand Black/New Zealand White lupus-prone mice by adoptive transfer of ex vivo expanded regulatory T cells. J. Immunol. 177, 1451-1459.

Schlaepfer E., Audige A., Joller H., Speck R.F. (2006) TLR7/8 triggering exerts opposing effects in acute versus latent HIV infection. J Immunol. 2006 Mar 1;176(5):2888-95.

Schoenemeyer A., Barnes B.J., Mancl M.E., Latz E., Goutagny N., Pitha P.M., Fitzgerald K.A., Golenbock D.T. (2005) The interferon regulatory factor, IRF5, is a central mediator of toll-like receptor 7 signaling. J. Biol. Chem. 280, 17005-17012.

Schur, P.H. and Monroe, M. (1969) Antibodies to ribonucleic acid in systemic lupus erythematosus. Proc. Natl. Acad. Sci. U S A. 63, 1108-1112.

Setty Y.N., Pittman C.B., Mahale A.S., Greidinger E.L., Hoffman R.W. (2002) Sicca symptoms and anti-SSA/Ro antibodies are common in mixed connective tissue disease. J. Rheumatol. 29, 487-9.

Sharp G.C., Irvin W.S., LaRoque R.L., Velez C., Daly V., Kaiser A.D., and Holman H.R. (1971) Association of autoantibodies to different nuclear antigens with clinical patterns of rheumatic disease and responsiveness to therapy. J. Clin. Invest. 50, 350-359.

Sjoberg A.P., Trouw L.A., McGrath F.D., Hack C.E., Blom A.M. (2006) Regulation of complement activation by C-reactive protein: targeting of the inhibitory activity of C4b-binding protein. J. Immunol. 176, 7612-7620.

Skriner K, Sommergruber WH, Tremmel V, Fischer I, Barta A, Smolen JS, Steiner G. (1997) Anti-A2/RA33 autoantibodies are directed to the RNA binding region of the A2 protein of the heterogeneous nuclear ribonucleoprotein complex. Differential epitope recognition in rheumatoid arthritis, systemic lupus erythematosus, and mixed connective tissue disease. J. Clin. Invest. 100, 127-135.

Stein A.J., Fuchs G., Fu C., Wolin S.L., Reinisch K.M. (2005) Structural insights into RNA quality control: the Ro autoantigen binds misfolded RNAs via its central cavity. Cell. 121, 529-539.

Szalai A.J., Weaver C.T., McCrory M.A., van Ginkel F.W., Reiman R.M., Kearney J.F., Marion T.N., Volanakis J.E. (2003) Delayed lupus onset in (NZB x NZW)F1 mice expressing a human C-reactive protein transgene. Arthritis Rheum. 48, 1602-1611.

Tan E.M., Feltkamp T.E., Smolen J.S., Butcher B., Dawkins R., Fritzler M.J., Gordon T., Hardin J.A., Kalden J.R., Lahita R.G., Maini R.N., McDougal J.S., Rothfield N.F., Smeenk R.J., Takasaki Y., Wiik A., Wilson M.R., Koziol J.A. (1997) Range of antinuclear antibodies in "healthy" individuals. Arthritis Rheum. 40, 1601-1611.

Triantafilou K., Orthopoulos G., Vakakis E., Ahmed M.A., Golenbock D.T., Lepper P.M., Triantafilou M. (2005) Cell Microbiol. 7, 1117-1126.

Van Eenennaam H., Vogelzangs J.H., Lugtenberg D., Van Den Hoogen F.H., Van Venrooij W.J., Pruijn G.J. (2002) Identity of the RNase MRP- and RNase P-associated Th/To autoantigen. Arthritis Rheum. 46, 3266-3272.

Vollmer J., Tluk S., Schmitz C., Hamm S., Jurk M., Forsbach A., Akira S., Kelly K.M., Reeves W.H., Bauer S., Krieg A.M. (2005) Immune stimulation mediated by autoantigen binding sites within small nuclear RNAs involves Toll-like receptors 7 and 8. J Exp Med. 202, 1575-1585.

Walter P., Blobel G. (1983) Disassembly and reconstitution of signal recognition particle. Cell. 34:525-533.

Wang J, Shao Y, Bennett TA, Shankar RA, Wightman PD, Reddy LG. (2006) The functional effects of physical interactions among toll-like receptors 7, 8 and 9. J. Biol. Chem. Oct 13; [Epub ahead of print].

Wang Z., Xiang L., Shao J., Yuan Z. (2006) The 3' CCACCA sequence of tRNAAla(UGC) is the motif that is important in inducing Th1-like immune response, and this motif can be recognized by Toll-like receptor 3. Clin. Vaccine Immunol. 13, 733-739.

Warner N.Z, Greidinger E.L. (2004) Patients with antibodies to both PmScl and dsDNA. J. Rheumatol. 31, 2169-2174.

Weickmann J.L., Glitz D.G. (1982) Human ribonucleases. Quantitation of pancreatic-like enzymes in serum, urine, and organ preparations. J. Biol. Chem. 257(15):8705-8710.

Yamasaki Y., Honkanen-Scott M., Hernandez L., Ikeda K., Barker T., Bubb M.R., Narain S., Richards H.B., Chan E.K., Reeves W.H., Satoh M. (2006) Nucleolar staining cannot be used as a screening test for the scleroderma marker anti-RNA polymerase I/III antibodies. Arthritis Rheum. 54:3051-6.

Yang J.M., Hildebrandt B., Luderschmidt C., Pollard K.M. (2003) Human scleroderma sera contain autoantibodies to protein components specific to the U3 small nucleolar RNP complex. Arthritis Rheum. 48, 210-217.

Yoneyama M., Kikuchi M., Matsumoto K., Imaizumi T., Miyagishi M., Taira K., Foy E., Loo Y.M., Gale M. Jr., Akira S., Yonehara S., Kato A., Fujita T. (2005) Shared and unique functions of the DExD/H-box helicases RIG-I, MDA5, and LGP2 in antiviral innate immunity. J. Immunol. 175, 2851-2858.

Yoneyama M., Kikuchi M., Natsukawa T., Shinobu N., Imaizumi T., Miyagishi M., Taira K., Akira S., Fujita T. (2004) The RNA helicase RIG-I has an essential function in double-stranded RNA-induced innate antiviral responses. Nat. Immunol. 5, 730-737.

Young S.L., Lyddon T.D., Jorgenson R.L., Misfeldt M.L. (2004) Expression of Toll-like receptors in human endometrial epithelial cells and cell lines. Am J Reprod Immunol. 52, 67-73.

22

Induction and Evasion of the Type I Interferon Response by Cytomegaloviruses

Victor R. DeFilippis

Oregon Health and Science University, Vaccine and Gene Therapy Institute, defilipp@ohsu.edu

Abstract. Cytomegaloviruses represent supreme pathogens in that they are capable of occupying healthy mammalian hosts for life in the face of constant antiviral immune reactions. The inability of the host to eliminate the virus likely results from numerous counteractive strategies employed to disrupt the immune response. The role of type I interferon in the antiviral response has been well documented although only recently have the pathways of induction of this powerful cytokine been described. Cytomegaloviruses have been shown to both induce and be sensitive to the effects of type I interferon. Yet these viruses also possess numerous and varied phenotypes capable of inhibiting not only interferon induction but also interferon signaling and interferon-induced antiviral processes. The balance between induction and evasion of type I interferon responses by cytomegaloviruses is discussed in this review.

1 Introduction

Cytomegaloviruses (CMVs) are ancient, successful, and widespread pathogens belonging to the β-herpesvirus family (reviewed in (Mocarski, 2001, Pass, 2001)), a group believed to have emerged prior to the mammalian radiation (McGeoch, Cook, Dolan, Jamieson and Telford, 1995). CMVs are highly species-specific and as such their evolutionary course has paralleled mammalian divergence patterns such that each species has its own uniquely adapted CMV. Molecular epidemiological studies indicate that 60% to 100% of the adult human population is infected with human CMV (HCMV) (Pass, 2001, Vogel, Weigler, Kerr, Hendrickx and Barry, 1994). Infection is generally asymptomatic in healthy individuals, but the virus is a major cause of morbidity and mortality when the immune system is suppressed or underdeveloped, during which acute infection is largely uncontrolled. HCMV is thus an opportunistic pathogen during AIDS and represents a common, life-threatening post-transplant complication in allograft recipients (Nichols and Boeckh, 2000).

Congenital infection is also a serious problem with HCMV being the largest infectious cause of birth defects in the United States (Alford, Stagno, Pass and Britt, 1990). The persistent, life-long nature of HCMV infection has also been implicated in chronic diseases, such as atherosclerosis, restenosis, and transplant vascular sclerosis (reviewed in (Streblow, Orloff and Nelson, 2001)).

As with all herpesviruses, CMV persists in the host indefinitely following initial infection and viral shedding can occur years after primary exposure. CMV can manifest latent infection of its target cells (Söderberg-Naucler and Nelson, 1999) but various observations suggest that foci of active CMV replication occur frequently, if not continuously, within the infected host (Huff, Eberle, Capitanio, Zhou and Barry, 2003, Reinke, Prosch, Kern and Volk, 1999, Sylwester, Mitchell, Edgar, Taormina, Pelte, Ruchti, Sleath, Grabstein, Hosken, Kern, Nelson and Picker, 2005). The establishment and maintenance of this life-long yet typically inconsequential infection in immunocompetent hosts results from the multifaceted ability of CMV to manipulate disparate arms of the innate and adaptive immune systems. Unfortunately, due to the characteristically high degree of CMV species specificity, HCMV cannot replicate in host animals, even nonhuman primates. Therefore, laboratory models involving naturally adapted animal CMVs [e.g. murine CMV (MCMV), rat CMV, guinea pig CMV, rhesus macaque CMV (RhCMV)] are relied on for *in vivo* research into viral pathogenesis and immunobiology. Thus many of the examples discussed below include data obtained from these *in vivo* virus-host models.

2 The Type I Interferon System and its Effects on CMV Replication

Synthesis and secretion of interferon (IFN) by cells exposed to virus represents one of the earliest host antiviral reactions. Type I IFNs include a single IFNβ gene as well as at least 14 IFNα isotypes. Secretion of IFNα is largely restricted to macrophages, leukocytes, and dendritic cells (Colonna, Krug and Cella, 2002, Diebold, Montoya, Unger, Alexopoulou, Roy, Haswell, Al-Shamkhani, Flavell, Borrow and Reis e Sousa, 2003, Pestka, Langer, Zoon and Samuel, 1987) with some species being secondarily stimulated in fibroblasts by IFNβ (Erlandsson, Blumenthal, Eloranta, Engel, Alm, Weiss and Leanderson, 1998). In contrast, IFNβ is produced by most cell types, especially epithelial cells and fibroblasts. Antiviral effects are actually conferred by the activity of numerous IFN-stimulated gene (ISG) products. ISG-derived proteins target many different stages of virus replication (for review see (Samuel, 2001)) including gene transcription, protein translation, cleavage of single-stranded RNA (ssRNA), modulation of viral polymerase activity and trafficking, and protein processing.

ISG transcription is induced following signal transduction triggered via autocrine and paracrine binding to the type I IFN receptor by either IFNα or β. Contact with IFNα/β causes dimerization of the receptor subunits resulting in phosphorylation of associated tyrosine kinase 2 (Tyk2) and Janus kinase 1 (JAK) that phosphorylate STAT (signal transducers and activators of transcription) 2. Phosphorylation of

STAT2 leads to its recruitment of and heterodimerization with STAT1 followed by nuclear accumulation and association with IFN regulatory factor 9 (IRF9). This complex (termed IFN-stimulated gene factor 3; ISGF3) binds to IFN stimulated response elements (ISREs) contained in the promoters of numerous ISGs resulting in their transcriptional upregulation (Shuai and Liu, 2003).

CMV replication has been shown to be susceptible to the effects of IFN-stimulated proteins and IFN-induced cellular states. Pretreatment of human fibroblasts *in vitro* with type I IFNs was demonstrated to limit HCMV plaque formation and virus production by inhibiting expression of immediate early (IE) viral genes (Sainz, LaMarca, Garry and Morris, 2005) as well as other viral proteins (Torigoe, Campbell, Torigoe, Michelson and Starr, 1993). Likewise, MCMV IE gene transcription was also found to be impaired by pretreatment of cells with IFNα resulting in diminished viral replication (Gribaudo, Ravaglia, Caliendo, Cavallo, Gariglio, Martinotti and Landolfo, 1993). Furthermore, the IFN- and HCMV-inducible protein viperin has been shown to sharply reduce HCMV replication by blocking synthesis of viral structural proteins glycoprotein B (gB), pp65 and pp28 (Chin and Cresswell, 2001). Interestingly, eliminating IRF3 (and thus IRF3-dependent ISG and IFNβ expression; discussed below) using RNA interference from fibroblasts was also found to result in increased HCMV peak titers (DeFilippis, Robinson, Keck, Hansen, Nelson and Früh, 2006).

In vivo studies using MCMV further demonstrate the inhibitory effects of type I IFNs. Mice lacking the IFNα/β receptor were more sensitive to MCMV infection than wildtype mice (Salazar-Mather, Lewis and Biron, 2002). Yeow and colleagues (Yeow, Lawson and Beilharz, 1998) further observed that intramuscular application of various murine IFNα subtype expression constructs significantly decreased local MCMV titers. It was subsequently shown that similar injection of expression constructs also impaired systemic virus replication for some IFNα subtypes (Cull, Bartlett and James, 2002). While these studies show that the inhibition of CMV growth and replication by type I IFNs not total, it is probable that viral mechanisms of obstruction are at work limiting IFN's effects (discussed below). Badly needed to address the effects of IFN on CMV replication are investigations using mutant viruses lacking IFN-evasion genes and phenotypes.

3 Pattern Recognition and the Synthesis of Type I IFN

Pathways of initial IFNβ induction have been elucidated only in recent years (reviewed in (Hiscott, Grandvaux, Sharma, Tenoever, Servant and Lin, 2003)) and involve signaling independent of the JAK/STAT apparatus. Pathogens induce IFNβ transcription (along with a subset of ISGs and pro-inflammatory cytokines and chemokines) by triggering pattern recognition receptors (PRRs) which are host cell proteins that react to specific molecular components of pathogens (e.g. bacterial LPS, viral double stranded (dsRNA), unmethylated CpG DNA). The best-studied PRRs, Toll-like receptors (TLRs), are transmembrane glycoproteins (Martin and Wesche, 2002) of which eleven distinct human versions have been identified along with most of their respective ligands (Kawai and Akira, 2006). TLRs are expressed predominantly in immune surveillance cells such as dendritic cells (DCs) and macrophages and are fundamental to secretion of inflammatory cytokines and co-stimulatory molecules that control localized virus replication, facilitate DC maturation, and help coordinate adaptive immune responses by B and T lymphocytes (Takeuchi and Akira, 2001).

Induction of IFNβ expression via stimulated TLRs involves activation of two critical transcription factors: IFN regulatory factor 3 (IRF3) and nuclear factor kappa B (NFκB). Both are ubiquitously expressed in the cytoplasm and accumulate in the nucleus upon activation (Hiscott *et al.*, 2003, Wathelet, Lin, Parekh, Ronco, Howley and Maniatis, 1998). Stimulation of TLR3 or TLR4 leads to recruitment of the adaptor molecule TRIF (TIR-domain containing adaptor inducing IFNβ) which directs activation of the IRF3-phosphorylating kinases IKKε and TANK-binding kinase 1 (TBK1) (Fitzgerald, McWhirter, Faia, Rowe, Latz, Golenbock, Coyle, Liao and Maniatis, 2003, Sharma, tenOever, Grandvaux, Zhou, Lin and Hiscott, 2003). IRF3 phosphorylation leads to its homodimerization and association with transcriptional co-activators CBP and p300 that cause the complex to remain in the nucleus where it is required for transcription of IFNβ as well as a subset of ISGs containing specific positive regulatory domains (PRDIII and PRDI) (Kumar, McBride, Weaver, Dingwall and Reich, 2000, Lin, Heylbroeck, Pitha and Hiscott, 1998, Suhara, Yoneyama, Kitabayashi and Fujita, 2002, Weaver, Kumar and Reich, 1998, Yoneyama, Suhara, Fukuhara, Fukuda, Nishida and Fujita, 1998). TLRs 7, 8, and 9 also induce IFNα in plasmacytoid DCs yet via activation of IRF7, a protein closely related to IRF3 that is constitutively present in these cells but which can also be upregulated by IFNβ in other cells (reviewed in (Kawai and Akira, 2006)).

Unlike IRF3 and IRF7, activation of NFκB and activating protein 1 (AP-1) occurs via all TLRs. NFκB subunits (p50 and p65) are normally sequestered in the cytoplasm bound to the inhibitory factor κB molecule (IκB). TLR stimulation induces the inhibitory κ kinases (IKKs) by way of the adaptor molecule MyD88 (myeloid differentiation response protein 88) to phosphorylate IκB (with the exception of TLR3 which signals to IKKs via TRIF and RIP1) resulting in its degradation and the subsequent release of NFκB to accumulate in the nucleus where it binds to κB sites, including within the IFNβ promoter. Activation of AP-1 also occurs via MyD88 (or TRIF and TRAF6 in the case of TLR3) and involves MAP kinases such as c-Jun N-terminal kinase and p38. Secretion of IFNβ can also bring about a second phase of gene expression by inducing the JAK/STAT-mediated synthesis of IRF7, which interacts with IRF3 to induce transcription of IFNα subtypes as well as a broader array of ISGs (Levy, Marie, Smith and Prakash, 2002, Marie, Durbin and Levy, 1998).

A newly identified class of PRRs that detects virus-specific molecular patterns such as dsRNA or dsDNA is capable of similarly inducing IFNβ transcription (Kato, Sato, Yoneyama, Yamamoto, Uematsu, Matsui, Tsujimura, Takeda, Fujita, Takeuchi and Akira, 2005). The prototypes of this new family are retinoic acid inducible gene I (RIG-I), and melanoma differentiation-associated gene 5 (MDA-5) (Andrejeva, Childs, Young, Carlos, Stock, Goodbourn and Randall, 2004, Yoneyama, Kikuchi, Natsukawa, Shinobu, Imaizumi, Miyagishi, Taira, Akira and Fujita, 2004) which react to dsRNA and its analogue poly(I:C) (Gitlin, Barchet, Gilfillan, Cella, Beutler, Flavell,

Diamond and Colonna, 2006, Kato, Takeuchi, Sato, Yoneyama, Yamamoto, Matsui, Uematsu, Jung, Kawai, Ishii, Yamaguchi, Otsu, Tsujimura, Koh, Reis e Sousa, Matsuura, Fujita and Akira, 2006). Recent data also suggest the existence of an as yet unidentified molecular sensor that recognizes B-form dsDNA (Ishii, Coban, Kato, Takahashi, Torii, Takeshita, Ludwig, Sutter, Suzuki, Hemmi, Sato, Yamamoto, Uematsu, Kawai, Takeuchi and Akira, 2006, Okabe, Kawane, Akira, Taniguchi and Nagata, 2005, Stetson and Medzhitov, 2006). These exclusively cytoplasmic receptors signal to TBK1/IKKε via the adaptor molecule IFNβ promoter stimulator 1 (IPS-1) (Kawai, Takahashi, Sato, Coban, Kumar, Kato, Ishii, Takeuchi and Akira, 2005, Lin, Lacoste, Nakhaei, Sun, Yang, Paz, Wilkinson, Julkunen, Vitour, Meurs and Hiscott, 2006, Meylan, Curran, Hofmann, Moradpour, Binder, Bartenschlager and Tschopp, 2005, Seth, Sun, Ea and Chen, 2005, Xu, Wang, Han, Li, Zhai and Shu, 2005) and are thus independent from the TLR system.

4 Induction of IFN and ISGs by CMV

Despite the antiviral effects of type I IFNs, HCMV cellular binding and entry strikingly induce ISG expression and IFN synthesis. Numerous transcriptional studies have observed rapid and strong HCMV-mediated transcription of ISGs, IFNβ, and pro-inflammatory cytokines and chemokines (Abate, Watanabe and Mocarski, 2004, Boehme, Singh, Perry and Compton, 2004, Boyle, Pietropaolo and Compton, 1999, Browne, Wing, Coleman and Shenk, 2001, DeFilippis and Früh, 2005, DeFilippis *et al.*, 2006, Gravel and Servant, 2005, Netterwald, Jones, Britt, Yang, McCrone and Zhu, 2004, Simmen, Singh, Luukkonen, Lopper, Bittner, Miller, Jackson, Compton and Fruh, 2001, Taylor and Bresnahan, 2005, Zhu, Cong, Mamtora, Gingeras and Shenk, 1998, Zhu, Cong and Shenk, 1997). Initial induction of these genes by HCMV infection relies on virus-mediated activation of IRF3 and NFκB, events that have also been repeatedly demonstrated (Boehme *et al.*, 2004, Browne *et al.*, 2001, Collins, Noyce and Mossman, 2004, DeFilippis and Früh, 2005, DeFilippis *et al.*, 2006, Gravel and Servant, 2005, Navarro, Mowen, Rodems, Weaver, Reich, Spector and David, 1998, Preston, Harman and Nicholl, 2001, Yurochko and Huang, 1999, Yurochko, Hwang, Rasmussen, Keay, Pereira and Huang, 1997a, Yurochko, Kowalik, Huong and Huang, 1995, Yurochko, Mayo, Poma, Baldwin and Huang, 1997b). It has further been shown that elimination of IRF3 from fibroblasts using RNA interference or a dominant negative form of IRF3 blocks HCMV-mediated induction of IFNβ and a subset of ISGs (DeFilippis et al., 2006). While HCMV binding and entry are necessary for ISG and IRF3 activation (Netterwald et al., 2004), neither viral gene expression (Browne *et al.*, 2001, Navarro *et al.*, 1998, Noyce, Collins and Mossman, 2006, Preston *et al.*, 2001, Taylor and Bresnahan, 2005, Zhu *et al.*, 1998, Zhu *et al.*, 1997) nor host cell protein translation (Browne *et al.*, 2001, DeFilippis and Früh, 2005, DeFilippis *et al.*, 2006) are required for these to occur. The activation of ISGs by HCMV in the presence of the translation inhibitor cycloheximide further demonstrates that the initial induction of innate immune response is an IFN-independent (and directly IRF3-dependent) phenomenon.

While CMV-mediated IRF3- and NFκB-dependent transcriptional induction is well demonstrated, many questions remain unanswered surrounding the activated PRR(s), virus-associated triggering molecules, and ensuing signal transduction pathways leading to IFNβ and ISG transcription. Since transcriptionally inactive HCMV strongly stimulates this innate immune response (Browne *et al.*, 2001, Noyce *et al.*, 2006, Taylor and Bresnahan, 2005), it is evident that molecular component(s) of the virion are involved and studies have identified HCMV envelope glycoprotein B (gB) as crucial to IFNβ and ISG induction. A soluble form of gB has been shown in isolation to induce activation of IRF3 (Boehme et al., 2004), NFκB (Yurochko *et al.*, 1997a), transcription of IFNβ (Boehme et al., 2004), and transcription of ISGs (Boehme *et al.*, 2004, Boyle *et al.*, 1999, Simmen *et al.*, 2001). In fact, global expression analysis of HCMV-infected versus gB-treated cells revealed conspicuous similarities with respect to ISG induction possibly suggesting that the anti-CMV innate response is ultimately mediated by gB (Simmen et al., 2001). Yet virion-associated gB is actually a much stronger stimulus of ISG and IFNβ transcription than isolated gB perhaps indicating that during an infection, additional virion components are required for this host cell reaction (Boehme et al., 2004). Considering these observations in light of others showing that cellular entry by the HCMV particle (rather than simple virion-cell binding) is required for ISG induction (Netterwald et al., 2004) suggests that the response is likely not as simple as triggering of a single PRR by gB binding.

While it has been shown that HCMV and gB induce pro-inflammatory gene transcription via TLR2-dependent activation of NFκB (Compton, Kurt-Jones, Boehme, Belko, Latz, Golenbock and Finberg, 2003), the HCMV and gB signaling pathway(s) leading to IRF3 activation are not known. It has been proposed that other constituent molecule(s) making up the HCMV virion may also trigger specific TLR(s) (Boehme and Compton, 2006). For example, HCMV is known to incorporate viral and cellular ssRNA into the virion (Bresnahan and Shenk, 2000a, Greijer, Dekkers and Middeldorp, 2000) that could potentially trigger TLR7 or TLR8 (Diebold, Kaisho, Hemmi, Akira and Reis e Sousa, 2004, Heil, Hemmi, Hochrein, Ampenberger, Kirschning, Akira, Lipford, Wagner and Bauer, 2004). It is also possible that HCMV's CpG-rich genomic DNA could trigger TLR9 (Ahmad-Nejad, Hacker, Rutz, Bauer, Vabulas and Wagner, 2002), a receptor that has actually been identified as necessary for both IFNα secretion following MCMV infection and an overall effective immune response to the virus (Krug, French, Barchet, Fischer, Dzionek, Pingel, Orihuela, Akira, Yokoyama and Colonna, 2004, Tabeta, Georgel, Janssen, Du, Hoebe, Crozat, Mudd, Shamel, Sovath, Goode, Alexopoulou, Flavell and Beutler, 2004). TLR3 and TLR2 have also been found to play *in vivo* roles in MCMV-induced cytokine (including IFNα/β) secretion and infection control although the molecular stimuli in this case have yet to be defined (Szomolanyi-Tsuda, Liang, Welsh, Kurt-Jones and Finberg, 2006, Tabeta *et al.*, 2004). While the PRR involved in HCMV-mediated IFNβ induction remains unknown, recent work by Ishii and colleagues (Ishii et al., 2006) demonstrates that transfection of HCMV genomic DNA activates an as yet unidentified dsDNA innate sensor that leads to IFNβ transcription. It is thus possible that one reason the PRR responsible for HCMV-mediated IFNβ induction is unknown is that the molecule itself has yet to be described.

5 CMV Interference with the Induction of IFN and ISGs

Investigation into changes in the host cell transcriptome during HCMV infection has revealed that while live virus strongly and rapidly induces ISG and pro-inflammatory gene expression, viral particles rendered transcriptionally inactive by UV irradiation induce an even stronger such reaction (especially with respect to the IFNβ gene) (Browne *et al.*, 2001, Taylor and Bresnahan, 2005). This is also true of live HCMV infections performed in the presence of cycloheximide (Browne *et al.*, 2001, DeFilippis *et al.*, 2006). Yet even UV-inactivated virus does not induce full activation of the IFN pathway, as does treatment with type I IFN (Browne *et al.*, 2001, Simmen *et al.*, 2001). These observations have lead researchers to examine whether HCMV inhibits IFNβ and initial ISG transcription and perhaps even IFN-dependent JAK/STAT signaling by way of 1) A viral gene product synthesized soon after infection or; 2) A (likely virally-encoded) component of the virus particle.

Important clues to HCMV's strategic balance between IFN induction and evasion are perhaps provided by the closely related CMV of Rhesus macaques (RhCMV). Infection of Rhesus fibroblasts with RhCMV does not induce transcription of ISGs, IFNβ, or pro-inflammatory genes (DeFilippis and Früh, 2005). In addition, infection not only fails to activate IRF3 and NFκB but also prevents such activation by UV-inactivated HCMV particles (DeFilippis and Früh, 2005). Interestingly, these same results are seen when RhCMV particles were themselves UV inactivated and when infections were performed in the presence of cycloheximide thus indicating that viral gene expression is not necessary for IRF3 interference. These observations strongly suggest that a component of the RhCMV virion actively blocks virus-mediated IRF3 activation and subsequent IFNβ and ISG transcription.

For the above reasons, much attention has focused on proteins of the CMV virion tegument, a voluminous proteinaceous region located between the nucleocapsid and the envelope largely composed of virus-encoded proteins (Varnum, Streblow, Monroe, Smith, Auberry, Pasa-Tolic, Wang, Camp, Rodland, Wiley, Britt, Shenk, Smith and Nelson, 2004). A major tegument constituent which has been shown to possess immunomodulatory capability is phosphoprotein pp65 (Arnon, Achdout, Levi, Markel, Saleh, Katz, Gazit, Gonen-Gross, Hanna, Nahari, Porgador, Honigman, Plachter, Mevorach, Wolf and Mandelboim, 2005, Odeberg, Plachter, Branden and Soderberg-Naucler, 2003) (interestingly, RhCMV encodes two pp65 homologues; (Hansen, Strelow, Franchi, Anders and Wong, 2003)). Browne and colleagues (Browne and Shenk, 2003) as well as Abate and colleagues (Abate et al., 2004) have shown that *in vitro* infection with a replication-competent HCMV mutant from which the pp65 open reading frame (UL83) was deleted (termed RVAd65) actually induces higher levels of IFNβ and ISG transcription than infection with wild type virus. Unfortunately, the molecular basis for this effect was controversial. Browne et al. showed increased nuclear translocation of and DNA binding by NFκB as well as increased DNA binding by IRF1 following infection with RVAd65 relative to wild type virus with similar IRF3 activation being observed by both viruses. Contrary to numerous other reports (see above) Abate et al. saw no IRF3 activation by wild type

HCMV but infection with RVAd65 was found to induce IRF3 nuclear accumulation. These authors also found NFκB to be similarly activated by both viruses.

A potential resolution to these contradictory data is provided in the recent work of Taylor and Bresnahan (Taylor and Bresnahan, 2005, Taylor and Bresnahan, 2006). This research has identified the HCMV immediate early 86kDa protein (IE2-86) as capable of blocking transcription of IFNβ as well as the HCMV-induced chemokine RANTES-both transcriptional targets of IRF3 and NFκB. In addition, the authors have shown that deletion of the pp65 ORF from the viral genome, as was done in the case of RVAd65, actually inhibits expression of the adjacent related tegument protein pp71 open reading frame (UL82), which occurs with UL83 on a bicistronic messenger RNA (Ruger, Klages, Walla, Albrecht, Fleckenstein, Tomlinson and Barrell, 1987, Taylor and Bresnahan, 2006). The pp71 protein is known to be a positive regulator of the HCMV immediate early gene promoter leading to expression of IE2-86 (Bresnahan and Shenk, 2000b, Cantrell and Bresnahan, 2005) thus it is likely that the ISG inhibitory phenotype observed for RVAd65 may in fact result from an indirect transcriptional effect. Furthermore, a HCMV deletion mutant created by inserting stop codons into the pp65 open reading frame was shown to both properly express pp71 and impair IFNβ induction with the same efficiency as wild type virus (Taylor and Bresnahan, 2006). Ectopic expression of IE2-86 but not pp65 was also shown to block virus-induced expression of IFNβ and RANTES (Taylor and Bresnahan, 2006). Interference with NFκB has been speculated by the authors to be the mechanism of inhibition.

6 CMV Interference with JAK/STAT Signal Transduction

The antiviral potency of IFN-stimulated responses likely created the ancient selective pressure that lead to evolutionary acquisition of counteractive CMV phenotypes as well as their redundancy in that they target more than one step in the IFN reaction. Therefore, as with other herpesviruses, CMVs are known to not only impair IFN synthesis but additionally exhibit phenotypes that block JAK/STAT signaling. This is perhaps due to the necessity of infecting cells and tissues already exposed to IFN. HCMV has been shown to block IFNα-stimulated gene expression in endothelial cells, fibroblasts, and MRC-5 cells (Miller, Zhang, Rahill, Waldman and Sedmak, 1999, Paulus, Krauss and Nevels, 2006). This effect is not immediate, however, and only arises following viral gene expression as well as an initial (likely IRF3-dependent) ISG activation phase (Paulus et al., 2006). The molecular basis of this blockade is again complicated by contrasting experimental observations. Early work by Daniel Sedmak's group showed that in HCMV-infected cells protein levels of JAK1 and IRF9 are drastically reduced, the result of HCMV-mediated degradation (Miller, Rahill, Boss, Lairmore, Durbin, Waldman and Sedmak, 1998, Miller *et al.*, 1999). In contrast to these results, a recent study by Paulus and colleagues observed clear HCMV-mediated impairment of IFNα-induced gene expression but failed to detect any decrease in IRF9 levels (Paulus et al., 2006). This work went on to uncover an association between HCMV immediate early 72kDa protein (IE1-72) and STAT2 (and perhaps STAT1) in the nucleus that lead to the prevention of DNA binding by ISGF3 and subsequent ISG induction (Paulus et al., 2006).

The importance of obstructing IFN-dependent signaling for CMV replication is further illustrated by MCMV. Zimmerman and colleagues recently screened a transposon-insertion viral mutant library to identify a MCMV protein that blocks both type I and type II IFN signaling (Zimmermann, Trilling, Wagner, Wilborn, Bubic, Jonjic, Koszinowski and Hengel, 2005). The MCMV M27 protein was shown using a viral deletion mutant as well as ectopically expressed protein to be both necessary and sufficient for blocking the inductive signals of IFNα/β and IFNγ. The basis of this pleiotropic phenotype is M27's capacity to bind and degrade STAT2, a protein that until this study was not believed to be involved in IFNγ-dependent responses. The importance of the virus' ability to block IFN signaling is illustrated by the essential nature of the M27 gene for growth *in vivo* but not *in vitro*. MCMV viral mutants from which M27 is removed (termed ΔM27) are able to replicate like wild type virus *in vitro* on NIH 3T3 cells (Abenes, Lee, Haghjoo, Tong, Zhan and Liu, 2001) as well as IFNβ-deficient mouse fibroblasts (Zimmermann et al., 2005). Yet while *in vivo* growth of ΔM27 is severely impaired in wild type animals, in IFNα-receptor or IFNγ-receptor deficient mice the virus initially grows similarly to wild type MCMV prior to the onset of adaptive immune responses (Abenes *et al.*, 2001, Zimmermann *et al.*, 2005). These replication trends emphasize the specificity and overall importance of anti-IFN traits for CMV growth and survival. Interestingly, both HCMV and MCMV have thus been shown to target the same IFN signaling molecule (STAT2) albeit via different mechanisms and by proteins that share neither homology nor expression kinetics.

7 CMV Interference with IFN-Dependent Responses

The antiviral effects of the IFN response are ultimately the result of IFN (or IRF3) induced host cell proteins and activities. Therefore, IFN evasion phenotypes that target the actions and consequences of antiviral effector molecules also contribute to virus survival. While such a strategy is well represented in other virus groups (see (Weber, Kochs and Haller, 2004) for review) the ability of CMV to disrupt IFN-stimulated protein function has been woefully unexplored as a research topic. This reality is especially unfortunate in the case of HCMV, which, as described above, is known to induce rapid and intense expression of ISGs during *in vitro* infection yet still replicates to high titers (Abate *et al.*, 2004, Boehme *et al.*, 2004, Boyle *et al.*, 1999, Browne *et al.*, 2001, DeFilippis and Früh, 2005, DeFilippis *et al.*, 2006, Gravel and Servant, 2005, Netterwald *et al.*, 2004, Simmen *et al.*, 2001, Taylor and Bresnahan, 2005, Zhu *et al.*, 1998, Zhu *et al.*, 1997). Virus gene expression, protein synthesis, assembly, and egress thus all effectively, yet somewhat mysteriously, occur in the presence of myriad antiviral proteins.

One prominent example of HCMV-mediated inhibition of IFN-effector molecules has been described by Adam Geballe's group. Virus infection and IFN-stimulation can shut off protein synthesis via the actions of protein kinase R (PKR) and 2'-5' oligoadenylate synthetase (OAS)/RNaseL (reviewed in (Schneider and Mohr, 2003)). While constitutively present in cells, OAS and PKR are transcriptionally induced by type I and type II IFNs and the proteins are activated following exposure to viral dsRNA. Synthesis of short oligoadenylates by OAS activates RNase L which then degrades viral and cellular RNAs. PKR inhibits protein translation initiation by phosphorylating and thus inactivating eukaryotic translation initiation factor 2α (eIF2α). These activities can effectively block translation of viral and host messenger RNA. Child et al. first showed that a vaccinia virus (VV) dsRNA-binding phenotype that prevents activation of PKR and OAS could be functionally complemented by infection with live, but not UV-inactivated HCMV (Child, Jarrahian, Harper and Geballe, 2002). Subsequent work proceeded to identify closely related HCMV proteins pTRS1 and pIRS1 as both being independently able to prevent translational shutoff by 1) blocking OAS-mediated eIF2α phosphorylation and 2) reducing RNA degradation by RNase L (Child, Hakki, De Niro and Geballe, 2004). Another possible example of anti-ISG activity by CMV involves the HCMV- and IFN-inducible protein viperin. HCMV infection has been shown to change the normal subcellular localization of viperin from the endoplasmic reticulum (the IFNα-induced location) to perinuclear structures and vacuoles containing viral proteins (Chin and Cresswell, 2001). Whether this HCMV-induced redistribution represents a means of evading or disrupting viperin's antiviral effects is not known yet has been hypothesized (Chin and Cresswell, 2001).

IFN also plays pivotal roles in numerous immunomodulatory functions that extend beyond the individual virus-infected or susceptible cell. Such adaptive and innate responses are integral to controlling CMV infection and have thus elicited the evolution of immunoevasive or inhibitory traits by the virus. While these phenotypes do not directly target IFNs or their actions they nevertheless permit viral tolerance of responses that are enhanced or even controlled by IFN-stimulated gene products. For example, type I IFNs promote $CD8^+$ T-lymphocyte (CTL) responses by increasing antigen presentation through upregulation of MHC class I as well as other proteins involved in antigen presentation (Biron and Sen, 2001, Samuel, 2001). Multiple CMVs are able to block this mechanism, often by targeting more than one step in the process (e.g. MHC protein processing and trafficking, antigen loading, MHC gene downregulation) (reviewed in (Loenen, Bruggeman and Wiertz, 2001, Mocarski, 2002, Pinto and Hill, 2005)). IFNs also enhance natural killer (NK) cell cytotoxicity by upregulating perforins (see (Biron, Nguyen, Pien, Cousens and Salazar-Mather, 1999)). Both MCMV and HCMV have evolved multiple mechanisms for escape of NK cell-mediated responses (see (Arnon *et al.*, 2005, Rajagopalan and Long, 2005)). Therefore, CMVs employ redundant, multi-tiered mechanisms of immune evasion that may be directly or indirectly IFN-dependent.

8 Conclusion and Future Directions

Definition of the interaction between CMV and the type I IFN system has greatly expanded in the past few years yet numerous important questions still remain. With regard to IFN induction, the PRR and signaling pathway(s) triggered by CMV as well as the viral stimulatory mechanism or molecule(s) have yet to be elucidated. The increasingly frequent discovery of innate immune sensors (see (Werts, Girardin and Philpott, 2006)) makes it quite possible that the CMV receptor is currently unknown, yet perhaps not for long. Questions also surround the ability of HCMV to replicate successfully following strong virus-mediated induction of innate responses. While evidence suggests that at least some ISGs (Zhu, Cong, Yu, Bresnahan and Shenk, 2002) and transcription factors (Sambucetti, Cherrington, Wilkinson and Mocarski, 1989, Stein, Volk, Liebenthal, Kruger and Prosch, 1993) are proviral, studies are accumulating that demonstrate anti-CMV effects of this response (Benedict, Angulo, Patterson, Ha, Huang, Messerle, Ware and Ghazal, 2004, Chin and Cresswell, 2001, DeFilippis *et al.*, 2006, Eickhoff and Cotten, 2005, Sainz *et al.*, 2005). One central unknown in the investigation of CMV biology is the role of IFN evasion in virus latency and persistence. While the importance of IFN evasion for MCMV replication *in vivo* has been examined, very little is known regarding the relationship between IFN induction, inhibition, and tolerance for HCMV persistence. Unfortunately homology between MCMV and HCMV nonessential proteins is small and mostly analogous immune evasion and persistence strategies are employed by the two viruses. However, the development of the Rhesus macaque model to investigate HCMV immunobiology has the potential to yield invaluable knowledge. Aside from the biological similarities between humans and macaques, RhCMV and HCMV share significant homology within known immunomodulatory genes (Hansen et al., 2003) thereby making this a potentially more realistic experimental model.

References

Abate D.A., Watanabe S. and Mocarski E.S. (2004) Major human cytomegalovirus structural protein pp65 (ppUL83) prevents interferon response factor 3 activation in the interferon response. *J Virol*, **78**, 10995-11006.

Abenes G., Lee M., Haghjoo E., Tong T., Zhan X. and Liu F. (2001) Murine cytomegalovirus open reading frame M27 plays an important role in growth and virulence in mice. *J Virol*, **75**, 1697-1707.

Ahmad-Nejad P., Hacker H., Rutz M., Bauer S., Vabulas R.M. and Wagner H. (2002) Bacterial CpG-DNA and lipopolysaccharides activate Toll-like receptors at distinct cellular compartments. *Eur J Immunol*, **32**, 1958-1968.

Alford C.A., Stagno S., Pass R.F. and Britt W.j. (1990) Congenital and perinatal cutomegalovirus infection. *Rev Infect Dis*, **12**, 745-753.

Andrejeva J., Childs K.S., Young D.F., Carlos T.S., Stock N., Goodbourn S. and Randall R.E. (2004) The V proteins of paramyxoviruses bind the IFN-inducible RNA helicase, mda-5, and inhibit its activation of the IFN-beta promoter. *Proc Natl Acad Sci U S A*, **101**, 17264-17269.

Arnon T.I., Achdout H., Levi O., Markel G., Saleh N., Katz G., Gazit R., Gonen-Gross T., Hanna J., Nahari E., Porgador A., Honigman A., Plachter B., Mevorach D., Wolf D.G. and Mandelboim O. (2005) Inhibition of the NKp30 activating receptor by pp65 of human cytomegalovirus. *Nat Immunol*, **6**, 515-523.

Benedict C.A., Angulo A., Patterson G., Ha S., Huang H., Messerle M., Ware C.F. and Ghazal P. (2004) Neutrality of the canonical NF-kappaB-dependent pathway for human and murine cytomegalovirus transcription and replication in vitro. *J Virol*, **78**, 741-750.

Biron C.A., Nguyen K.B., Pien G.C., Cousens L.P. and Salazar-Mather T.P. (1999) Natural killer cells in antiviral defense: function and regulation by innate cytokines. *Annu Rev Immunol*, **17**, 189-220.

Biron C.A. and Sen G.C. (2001) Interferon and other cytokines. In: *Fields Virology* (ed B.N. Fields), pp. 321-351. Lippincott-Raven, Philadelphia.

Boehme K.W. and Compton T. (2006) Virus entry and activation of innate immunity. In: *Cytomegaloviruses: Molecular Biology and Immunology* (ed M.J. Reddehase), pp. 111-130. Caister Academic Press, Norfolk, UK.

Boehme K.W., Singh J., Perry S.T. and Compton T. (2004) Human cytomegalovirus elicits a coordinated cellular antiviral response via envelope glycoprotein B. *J Virol*, **78**, 1202-1211.

Boyle K.A., Pietropaolo R.L. and Compton T. (1999) Engagement of the cellular receptor for glycoprotein B of human cytomegalovirus activates the interferon-responsive pathway. *Mol Cell Biol*, **19**, 3607-3613.

Bresnahan W.A. and Shenk T. (2000a) A subset of viral transcripts packaged within human cytomegalovirus particles. *Science*, **288**, 2373-2376.

Bresnahan W.A. and Shenk T.E. (2000b) UL82 virion protein activates expression of immediate early viral genes in human cytomegalovirus-infected cells. *Proc Natl Acad Sci U S A*, **97**, 14506-14511.

Browne E.P. and Shenk T. (2003) Human cytomegalovirus UL83-coded pp65 virion protein inhibits antiviral gene expression in infected cells. *Proc Natl Acad Sci U S A*, **100**, 11439-11444.

Browne E.P., Wing B., Coleman D. and Shenk T. (2001) Altered cellular mRNA levels in human cytomegalovirus-infected fibroblasts: viral block to the accumulation of antiviral mRNAs. *J Virol*, **75**, 12319-12330.

Cantrell S.R. and Bresnahan W.A. (2005) Interaction between the human cytomegalovirus UL82 gene product (pp71) and hDaxx regulates immediate-early gene expression and viral replication. *J Virol*, **79**, 7792-7802.

Child S.J., Hakki M., De Niro K.L. and Geballe A.P. (2004) Evasion of cellular antiviral responses by human cytomegalovirus TRS1 and IRS1. *J Virol*, **78**, 197-205.

Child S.J., Jarrahian S., Harper V.M. and Geballe A.P. (2002) Complementation of vaccinia virus lacking the double-stranded RNA-binding protein gene E3L by human cytomegalovirus. *J Virol*, **76**, 4912-4918.

Chin K.C. and Cresswell P. (2001) Viperin (cig5), an IFN-inducible antiviral protein directly induced by human cytomegalovirus. *Proc Natl Acad Sci U S A*, **98**, 15125-15130.

Collins S.E., Noyce R.S. and Mossman K.L. (2004) Innate cellular response to virus particle entry requires IRF3 but not virus replication. *J Virol*, **78**, 1706-1717.

Colonna M., Krug A. and Cella M. (2002) Interferon-producing cells: on the front line in immune responses against pathogens. *Curr Opin Immunol*, **14**, 373-379.

Compton T., Kurt-Jones E.A., Boehme K.W., Belko J., Latz E., Golenbock D.T. and Finberg R.W. (2003) Human cytomegalovirus activates inflammatory cytokine responses via CD14 and Toll-like receptor 2. *J Virol*, **77**, 4588-4596.

Cull V.S., Bartlett E.J. and James C.M. (2002) Type I interferon gene therapy protects against cytomegalovirus-induced myocarditis. *Immunology*, **106**, 428-437.

DeFilippis V.R. and Früh K.J. (2005) Rhesus cytomegalovirus particles prevent activation of interferon regulatory factor 3. *J Virol*, **79**, 6419-6431.

DeFilippis V.R., Robinson B., Keck T.M., Hansen S.G., Nelson J.A. and Früh K. (2006) Interferon regulatory factor 3 is necessary for induction of antiviral genes during human cytomegalovirus infection. *J Virol*, **80**, 1032-1037.

Diebold S.S., Kaisho T., Hemmi H., Akira S. and Reis e Sousa C. (2004) Innate antiviral responses by means of TLR7-mediated recognition of single-stranded RNA. *Science*, **303**, 1529-1531.

Diebold S.S., Montoya M., Unger H., Alexopoulou L., Roy P., Haswell L.E., Al-Shamkhani A., Flavell R., Borrow P. and Reis e Sousa C. (2003) Viral infection switches non-plasmacytoid dendritic cells into high interferon producers. *Nature*, **424**, 324-328.

Eickhoff J.E. and Cotten M. (2005) NF-kappaB activation can mediate inhibition of human cytomegalovirus replication. *J Gen Virol*, **86**, 285-295.

Erlandsson L., Blumenthal R., Eloranta M.L., Engel H., Alm G., Weiss S. and Leanderson T. (1998) Interferon-beta is required for interferon-alpha production in mouse fibroblasts. *Curr Biol*, **8**, 223-226.

Fitzgerald K.A., McWhirter S.M., Faia K.L., Rowe D.C., Latz E., Golenbock D.T., Coyle A.J., Liao S.M. and Maniatis T. (2003) IKKepsilon and TBK1 are essential components of the IRF3 signaling pathway. *Nat Immunol*, **4**, 491-496.

Gitlin L., Barchet W., Gilfillan S., Cella M., Beutler B., Flavell R.A., Diamond M.S. and Colonna M. (2006) Essential role of mda-5 in type I IFN responses to polyriboinosinic:polyribocytidylic acid and encephalomyocarditis picornavirus. *Proc Natl Acad Sci U S A*, **103**, 8459-8464.

Gravel S.P. and Servant M.J. (2005) Roles of an IkappaB kinase-related pathway in human cytomegalovirus-infected vascular smooth muscle cells: a molecular link in pathogen-induced proatherosclerotic conditions. *J Biol Chem*, **280**, 7477-7486.

Greijer A.E., Dekkers C.A. and Middeldorp J.M. (2000) Human cytomegalovirus virions differentially incorporate viral and host cell RNA during the assembly process. *J Virol*, **74**, 9078-9082.

Gribaudo G., Ravaglia S., Caliendo A., Cavallo R., Gariglio M., Martinotti M.G. and Landolfo S. (1993) Interferons inhibit onset of murine cytomegalovirus immediate-early gene transcription. *Virology*, **197**, 303-311.

Hansen S.G., Strelow L.I., Franchi D.C., Anders D.G. and Wong S.W. (2003) Complete sequence and genomic analysis of rhesus cytomegalovirus. *J Virol*, **77**, 6620-6636.

Heil F., Hemmi H., Hochrein H., Ampenberger F., Kirschning C., Akira S., Lipford G., Wagner H. and Bauer S. (2004) Species-specific recognition of single-stranded RNA via toll-like receptor 7 and 8. *Science*, **303**, 1526-1529.

Hiscott J., Grandvaux N., Sharma S., Tenoever B.R., Servant M.J. and Lin R. (2003) Convergence of the NF-kappaB and interferon signaling pathways in the regulation of antiviral defense and apoptosis. *Ann N Y Acad Sci*, **1010**, 237-248.

Huff J.L., Eberle R., Capitanio J., Zhou S.S. and Barry P.A. (2003) Differential detection of B virus and rhesus cytomegalovirus in rhesus macaques. *J Gen Virol*, **84**, 83-92.

Ishii K.J., Coban C., Kato H., Takahashi K., Torii Y., Takeshita F., Ludwig H., Sutter G., Suzuki K., Hemmi H., Sato S., Yamamoto M., Uematsu S., Kawai T., Takeuchi O. and Akira S. (2006) A Toll-like receptor-independent antiviral response induced by double-stranded B-form DNA. *Nat Immunol*, **7**, 40-48.

Kato H., Sato S., Yoneyama M., Yamamoto M., Uematsu S., Matsui K., Tsujimura T., Takeda K., Fujita T., Takeuchi O. and Akira S. (2005) Cell type-specific involvement of RIG-I in antiviral response. *Immunity*, **23**, 19-28.

Kato H., Takeuchi O., Sato S., Yoneyama M., Yamamoto M., Matsui K., Uematsu S., Jung A., Kawai T., Ishii K.J., Yamaguchi O., Otsu K., Tsujimura T., Koh C.S., Reis e Sousa C., Matsuura Y., Fujita T. and Akira S. (2006) Differential roles of MDA5 and RIG-I helicases in the recognition of RNA viruses. *Nature*, **441**, 101-105.

Kawai T. and Akira S. (2006) TLR signaling. *Cell Death Differ*, **13**, 816-825.

Kawai T., Takahashi K., Sato S., Coban C., Kumar H., Kato H., Ishii K.J., Takeuchi O. and Akira S. (2005) IPS-1, an adaptor triggering RIG-I- and Mda5-mediated type I interferon induction. *Nat Immunol*.

Krug A., French A.R., Barchet W., Fischer J.A., Dzionek A., Pingel J.T., Orihuela M.M., Akira S., Yokoyama W.M. and Colonna M. (2004) TLR9-dependent recognition of MCMV by IPC and DC generates coordinated cytokine responses that activate antiviral NK cell function. *Immunity*, **21**, 107-119.

Kumar K.P., McBride K.M., Weaver B.K., Dingwall C. and Reich N.C. (2000) Regulated nuclear-cytoplasmic localization of interferon regulatory factor 3, a subunit of double-stranded RNA-activated factor 1. *Mol Cell Biol*, **20**, 4159-4168.

Levy D.E., Marie I., Smith E. and Prakash A. (2002) Enhancement and diversification of IFN induction by IRF-7-mediated positive feedback. *J Interferon Cytokine Res*, **22**, 87-93.

Lin R., Heylbroeck C., Pitha P.M. and Hiscott J. (1998) Virus-dependent phosphorylation of the IRF-3 transcription factor regulates nuclear translocation, transactivation potential, and proteasome-mediated degradation. *Mol Cell Biol*, **18**, 2986-2996.

Lin R., Lacoste J., Nakhaei P., Sun Q., Yang L., Paz S., Wilkinson P., Julkunen I., Vitour D., Meurs E. and Hiscott J. (2006) Dissociation of a MAVS/IPS-1/VISA/Cardif-IKKepsilon molecular complex from the mitochondrial outer membrane by hepatitis C virus NS3-4A proteolytic cleavage. *J Virol*, **80**, 6072-6083.

Loenen W.A., Bruggeman C.A. and Wiertz E.J. (2001) Immune evasion by human cytomegalovirus: lessons in immunology and cell biology. *Semin Immunol*, **13**, 41-49.

Marie I., Durbin J.E. and Levy D.E. (1998) Differential viral induction of distinct interferon-alpha genes by positive feedback through interferon regulatory factor-7. *Embo J*, **17**, 6660-6669.

Martin M.U. and Wesche H. (2002) Summary and comparison of the signaling mechanisms of the Toll/interleukin-1 receptor family. *Biochim Biophys Acta*, **1592**, 265-280.

McGeoch D.J., Cook S., Dolan A., Jamieson F.E. and Telford E.A. (1995) Molecular phylogeny and evolutionary timescale for the family of mammalian herpesviruses. *J Mol Biol*, **247**, 443-458.
Meylan E., Curran J., Hofmann K., Moradpour D., Binder M., Bartenschlager R. and Tschopp J. (2005) Cardif is an adaptor protein in the RIG-I antiviral pathway and is targeted by hepatitis C virus. *Nature*.
Miller D.M., Rahill B.M., Boss J.M., Lairmore M.D., Durbin J.E., Waldman J.W. and Sedmak D.D. (1998) Human cytomegalovirus inhibits major histocompatibility complex class II expression by disruption of the Jak/Stat pathway. *J Exp Med*, **187**, 675-683.
Miller D.M., Zhang Y., Rahill B.M., Waldman W.J. and Sedmak D.D. (1999) Human cytomegalovirus inhibits IFN-alpha-stimulated antiviral and immunoregulatory responses by blocking multiple levels of IFN-alpha signal transduction. *Journal of Immunology*, **162**, 6107-6113.
Mocarski E.S. (2001) Cytomegaloviruses and their replication. In: *Virology* (ed B.N. Fields), pp. 2629-2673. Lippincott-Raven, Philadelphia.
Mocarski E.S., Jr. (2002) Immunomodulation by cytomegaloviruses: manipulative strategies beyond evasion. *Trends Microbiol*, **10**, 332-339.
Navarro L., Mowen K., Rodems S., Weaver B., Reich N., Spector D. and David M. (1998) Cytomegalovirus activates interferon immediate-early response gene expression and an interferon regulatory factor 3-containing interferon- stimulated response element-binding complex. *Mol Cell Biol*, **18**, 3796-3802.
Netterwald J.R., Jones T.R., Britt W.J., Yang S.J., McCrone I.P. and Zhu H. (2004) Postattachment events associated with viral entry are necessary for induction of interferon-stimulated genes by human cytomegalovirus. *J Virol*, **78**, 6688-6691.
Nichols W.G. and Boeckh M. (2000) Recent advances in the therapy and prevention of CMV infections. *J Clin Virol*, **16**, 25-40.
Noyce R.S., Collins S.E. and Mossman K.L. (2006) Identification of a novel pathway essential for the immediate-early, interferon-independent antiviral response to enveloped virions. *J Virol*, **80**, 226-235.
Odeberg J., Plachter B., Branden L. and Soderberg-Naucler C. (2003) Human cytomegalovirus protein pp65 mediates accumulation of HLA-DR in lysosomes and destruction of the HLA-DR alpha-chain. *Blood*, **101**, 4870-4877.
Okabe Y., Kawane K., Akira S., Taniguchi T. and Nagata S. (2005) Toll-like receptor-independent gene induction program activated by mammalian DNA escaped from apoptotic DNA degradation. *J Exp Med*, **202**, 1333-1339.
Pass R.F. (2001) Cytomegalovirus. In: *Fields Virology* (ed P.M.H. David M. Knipe, Diane E. Griffin, Robert A. Lamb Malcolm A. Martin, Bernard Roizman and Stephen E. Straus), pp. 2675-2705. Lippincott Williams and Wilkins, Philadelphia.
Paulus C., Krauss S. and Nevels M. (2006) A human cytomegalovirus antagonist of type I IFN-dependent signal transducer and activator of transcription signaling. *Proc Natl Acad Sci U S A*, **103**, 3840-3845.
Pestka S., Langer J.A., Zoon K.C. and Samuel C.E. (1987) Interferons and their actions. *Annu Rev Biochem*, **56**, 727-777.
Pinto A.K. and Hill A.B. (2005) Viral interference with antigen presentation to CD8+ T cells: lessons from cytomegalovirus. *Viral Immunol*, **18**, 434-444.
Preston C.M., Harman A.N. and Nicholl M.J. (2001) Activation of interferon response factor-3 in human cells infected with herpes simplex virus type 1 or human cytomegalovirus. *J Virol*, **75**, 8909-8916.
Rajagopalan S. and Long E.O. (2005) Viral evasion of NK-cell activation. *Trends Immunol*, **26**, 403-405.
Reinke P., Prosch S., Kern F. and Volk H.D. (1999) Mechanisms of human cytomegalovirus (HCMV) (re)activation and its impact on organ transplant patients. *Transpl Infect Dis*, **1**, 157-164.
Ruger B., Klages S., Walla B., Albrecht J., Fleckenstein B., Tomlinson P. and Barrell B. (1987) Primary structure and transcription of the genes coding for the two virion phosphoproteins pp65 and pp71 of human cytomegalovirus. *J Virol*, **61**, 446-453.
Sainz B., Jr., LaMarca H.L., Garry R.F. and Morris C.A. (2005) Synergistic inhibition of human cytomegalovirus replication by interferon-alpha/beta and interferon-gamma. *Virol J*, **2**, 14.
Salazar-Mather T.P., Lewis C.A. and Biron C.A. (2002) Type I interferons regulate inflammatory cell trafficking and macrophage inflammatory protein 1alpha delivery to the liver. *J Clin Invest*, **110**, 321-330.
Sambucetti L.C., Cherrington J.M., Wilkinson G.W. and Mocarski E.S. (1989) NF-kappa B activation of the cytomegalovirus enhancer is mediated by a viral transactivator and by T cell stimulation. *Embo J*, **8**, 4251-4258.
Samuel C.E. (2001) Antiviral actions of interferons. *Clin Microbiol Rev*, **14**, 778-809, table of contents.

Schneider R.J. and Mohr I. (2003) Translation initiation and viral tricks. *Trends Biochem Sci*, **28**, 130-136.
Seth R.B., Sun L., Ea C.K. and Chen Z.J. (2005) Identification and Characterization of MAVS, a Mitochondrial Antiviral Signaling Protein that Activates NF-kappaB and IRF3. *Cell*, **122**, 669-682.
Sharma S., tenOever B.R., Grandvaux N., Zhou G.P., Lin R. and Hiscott J. (2003) Triggering the interferon antiviral response through an IKK-related pathway. *Science*, **300**, 1148-1151.
Shuai K. and Liu B. (2003) Regulation of JAK-STAT signalling in the immune system. *Nat Rev Immunol*, **3**, 900-911.
Simmen K.A., Singh J., Luukkonen B.G., Lopper M., Bittner A., Miller N.E., Jackson M.R., Compton T. and Fruh K. (2001) Global modulation of cellular transcription by human cytomegalovirus is initiated by viral glycoprotein B. *Proc Natl Acad Sci U S A*, **98**, 7140-7145.
Söderberg-Naucler C. and Nelson J.Y. (1999) Human cytomegalovirus latency and reactivation - a delicate balance between the virus and its host's immune system. *Intervirology*, **42**, 314-321.
Stein J., Volk H.D., Liebenthal C., Kruger D.H. and Prosch S. (1993) Tumour necrosis factor alpha stimulates the activity of the human cytomegalovirus major immediate early enhancer/promoter in immature monocytic cells. *J Gen Virol*, **74 (Pt 11)**, 2333-2338.
Stetson D.B. and Medzhitov R. (2006) Recognition of cytosolic DNA activates an IRF3-dependent innate immune response. *Immunity*, **24**, 93-103.
Streblow D.N., Orloff S.L. and Nelson J.A. (2001) Do pathogens accelerate atherosclerosis? *J Nutr*, **131**, 2798S-2804S.
Suhara W., Yoneyama M., Kitabayashi I. and Fujita T. (2002) Direct involvement of CREB-binding protein/p300 in sequence-specific DNA binding of virus-activated interferon regulatory factor-3 holocomplex. *J Biol Chem*, **277**, 22304-22313.
Sylwester A.W., Mitchell B.L., Edgar J.B., Taormina C., Pelte C., Ruchti F., Sleath P.R., Grabstein K.H., Hosken N.A., Kern F., Nelson J.A. and Picker L.J. (2005) Broadly targeted human cytomegalovirus-specific CD4+ and CD8+ T cells dominate the memory compartments of exposed subjects. *J Exp Med*, **202**, 673-685.
Szomolanyi-Tsuda E., Liang X., Welsh R.M., Kurt-Jones E.A. and Finberg R.W. (2006) Role for TLR2 in NK cell-mediated control of murine cytomegalovirus in vivo. *J Virol*, **80**, 4286-4291.
Tabeta K., Georgel P., Janssen E., Du X., Hoebe K., Crozat K., Mudd S., Shamel L., Sovath S., Goode J., Alexopoulou L., Flavell R.A. and Beutler B. (2004) Toll-like receptors 9 and 3 as essential components of innate immune defense against mouse cytomegalovirus infection. *Proc Natl Acad Sci U S A*, **101**, 3516-3521.
Takeuchi O. and Akira S. (2001) Toll-like receptors; their physiological role and signal transduction system. *Int Immunopharmacol*, **1**, 625-635.
Taylor R.T. and Bresnahan W.A. (2005) Human cytomegalovirus immediate-early 2 gene expression blocks virus-induced beta interferon production. *J Virol*, **79**, 3873-3877.
Taylor R.T. and Bresnahan W.A. (2006) Human cytomegalovirus immediate-early 2 protein IE86 blocks virus-induced chemokine expression. *J Virol*, **80**, 920-928.
Torigoe S., Campbell D.E., Torigoe F., Michelson S. and Starr S.E. (1993) Cytofluorographic analysis of effects of interferons on expression of human cytomegalovirus proteins. *J Virol Methods*, **45**, 219-228.
Varnum S.M., Streblow D.N., Monroe M.E., Smith P., Auberry K.J., Pasa-Tolic L., Wang D., Camp D.G., 2nd, Rodland K., Wiley S., Britt W., Shenk T., Smith R.D. and Nelson J.A. (2004) Identification of proteins in human cytomegalovirus (HCMV) particles: the HCMV proteome. *J Virol*, **78**, 10960-10966.
Vogel P., Weigler B.J., Kerr H., Hendrickx A.G. and Barry P.A. (1994) Seroepidemiologic studies of cytomegalovirus infection in a breeding population of rhesus macaques. *Lab Anim Sci*, **44**, 25-30.
Wathelet M.G., Lin C.H., Parekh B.S., Ronco L.V., Howley P.M. and Maniatis T. (1998) Virus infection induces the assembly of coordinately activated transcription factors on the IFN-beta enhancer in vivo. *Mol Cell*, **1**, 507-518.
Weaver B.K., Kumar K.P. and Reich N.C. (1998) Interferon regulatory factor 3 and CREB-binding protein/p300 are subunits of double-stranded RNA-activated transcription factor DRAF1. *Mol Cell Biol*, **18**, 1359-1368.
Weber F., Kochs G. and Haller O. (2004) Inverse interference: how viruses fight the interferon system. *Viral Immunol*, **17**, 498-515.
Werts C., Girardin S.E. and Philpott D.J. (2006) TIR, CARD and PYRIN: three domains for an antimicrobial triad. *Cell Death Differ*, **13**, 798-815.

Xu L.G., Wang Y.Y., Han K.J., Li L.Y., Zhai Z. and Shu H.B. (2005) VISA Is an Adapter Protein Required for Virus-Triggered IFN-beta Signaling. *Mol Cell*, **19**, 727-740.

Yeow W.S., Lawson C.M. and Beilharz M.W. (1998) Antiviral activities of individual murine IFN-alpha subtypes in vivo: intramuscular injection of IFN expression constructs reduces cytomegalovirus replication. *J Immunol*, **160**, 2932-2939.

Yoneyama M., Kikuchi M., Natsukawa T., Shinobu N., Imaizumi T., Miyagishi M., Taira K., Akira S. and Fujita T. (2004) The RNA helicase RIG-I has an essential function in double-stranded RNA-induced innate antiviral responses. *Nat Immunol*, **5**, 730-737.

Yoneyama M., Suhara W., Fukuhara Y., Fukuda M., Nishida E. and Fujita T. (1998) Direct triggering of the type I interferon system by virus infection: activation of a transcription factor complex containing IRF-3 and CBP/p300. *Embo J*, **17**, 1087-1095.

Yurochko A.D. and Huang E.S. (1999) Human cytomegalovirus binding to human monocytes induces immunoregulatory gene expression. *J Immunol*, **162**, 4806-4816.

Yurochko A.D., Hwang E.S., Rasmussen L., Keay S., Pereira L. and Huang E.S. (1997a) The human cytomegalovirus UL55 (gB) and UL75 (gH) glycoprotein ligands initiate the rapid activation of Sp1 and NF-kappaB during infection. *J Virol*, **71**, 5051-5059.

Yurochko A.D., Kowalik T.F., Huong S.M. and Huang E.S. (1995) Human cytomegalovirus upregulates NF-kappa B activity by transactivating the NF-kappa B p105/p50 and p65 promoters. *J Virol*, **69**, 5391-5400.

Yurochko A.D., Mayo M.W., Poma E.E., Baldwin A.S., Jr. and Huang E.S. (1997b) Induction of the transcription factor Sp1 during human cytomegalovirus infection mediates upregulation of the p65 and p105/p50 NF-kappaB promoters. *J Virol*, **71**, 4638-4648.

Zhu H., Cong J.P., Mamtora G., Gingeras T. and Shenk T. (1998) Cellular gene expression altered by human cytomegalovirus: global monitoring with oligonucleotide arrays. *Proc Natl Acad Sci U S A*, **95**, 14470-14475.

Zhu H., Cong J.P. and Shenk T. (1997) Use of differential display analysis to assess the effect of human cytomegalovirus infection on the accumulation of cellular RNAs: induction of interferon-responsive RNAs. *Proc Natl Acad Sci U S A*, **94**, 13985-13990.

Zhu H., Cong J.P., Yu D., Bresnahan W.A. and Shenk T.E. (2002) Inhibition of cyclooxygenase 2 blocks human cytomegalovirus replication. *Proc Natl Acad Sci U S A*, **99**, 3932-3937.

Zimmermann A., Trilling M., Wagner M., Wilborn M., Bubic I., Jonjic S., Koszinowski U. and Hengel H. (2005) A cytomegaloviral protein reveals a dual role for STAT2 in IFN-{gamma} signaling and antiviral responses. *J Exp Med*, **201**, 1543-1553.

23

Pore Formers of the Immune System

Eckhard R. Podack, Vadim Deyev and Motoaki Shiratsuchi

University of Miami, Department of Microbiology and Immunology, Miller School of Medicine, Miami, Florida 33136

1 Introduction

Multicellular organisms are under constant attack by microbial organisms, i.e. viruses, bacteria, protozoa and other parasites. It is critical, therefore, that the immune defense system evolved mechanisms to destroy invading pathogens. One of the most powerful defense mechanisms of the immune systems relies on physical destruction of invading microbes by blasting holes into their cell wall. On a molecular level these holes are gigantic craters that are incompatible with life unless repaired swiftly. The membrane lesions are further exploited by the immune system by using them as a gate for toxic components to gain access to the cell interior to aid the destruction of the pathogen.

The pore formers of the immune system are the molecular agents that cause physical damage to invading pathogens. Pore formers are monomeric molecules that interact with one another while inserting into membranes where they continue to polymerize further to form transmembrane pores with internal diameters of 10 nm to 16 nm. To date, two molecular pore formers of the mammalian immune system have been well recognized and described. A third pore former recently discovered is under intensive research in our laboratory. This report will summarize the experience with the known pore formers to allow conclusions and prediction for the third type.

As mentioned above pore formers are the most formidable weapons of the immune defensive armory. It is therefore absolutely necessary that targeting be carefully controlled in order to avoid collateral damage. Indeed uncontrolled release of each of the pore formers is accompanied by immune pathology and autoimmune syndromes.

The targeting of pore formers has to take into cognizance the location of the pathogen. Broadly speaking, certain pathogens may exist and replicate in the

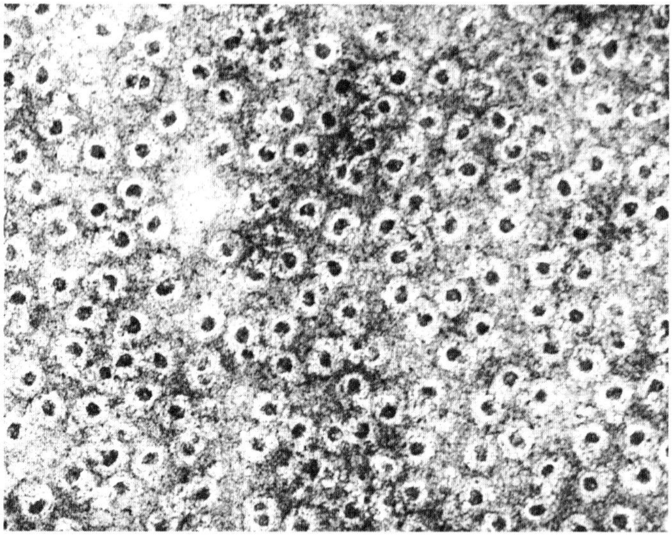

Fig. 1. Human complement lesions on rabbit erythrocyte membranes. After complement lysis, lysed membranes were collected and treated with trypsin and chymotrypsin. Samples were negatively stained with uranyl formate and visualized by electron microscopy. Final magnification: 300,000 fold.

extracellular space of multicellular organisms, while other pathogens invade cells and replicate intracellularly. Clearly the targeting mechanisms for these two types of pathogens must be very different. Intracellular pathogens are best removed by killing the host cell. On the other hand, extracellular pathogens need to be killed in a way that protects cells from accidental damage. The targeting systems that have evolved fulfill these requirements.

2 C9, the pore former of the complement system

Borsos, Dourmashkin and Humphrey in 1964 first described electron microscopically detectable images found on erythrocyte membranes lysed by complement that were designated as "complement lesions" (Borsos, T., Dourmashkin, R.R. and Humphrey, J.H. 1964). At the time there was a lively debate about the possible interpretation of these images and their relation to red cell lysis. Mayer (Mayer, M.M. 1972) was the first to suggest in his "doughnut hypothesis" of complement lysis that the electron-microscopic lesions actually presented pores, the central hole of the doughnut like structure seen in the microscope, in the targeted membrane. While this hypothesis was attractive it was unable to give a reasonable explanation for the molecular details of the structure of the symmetric complement lesion. It had been found based on work of Mueller-Eberhard's group (Kolb, W.P., Haxby, J.A., Arroyave, C.M. and Muller-Eberhard, H.J. 1972) that cell lysis was mediated by a complex (C5b-9, membrane attack complex – MAC) of the five terminal components C5b, C6, C7, C8 and several

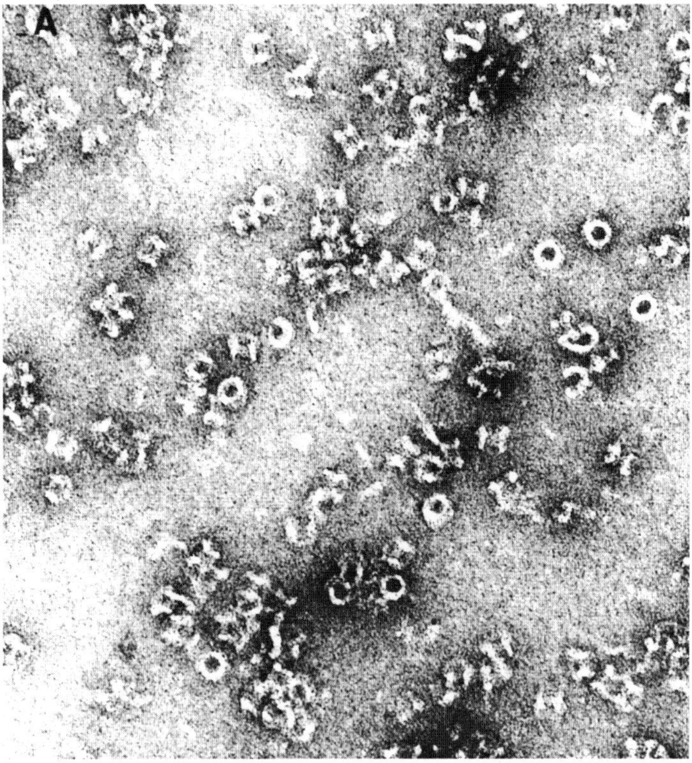

Fig. 2. Polymerized C9 (poly C9) in solution. Purified, monomeric C9 was incubated at 37C over night during which period it spontaneously polymerized. Poly C9 was stained with uranyl formate. Final magnification: 300,000.

molecules C9. Bhakdi and Tranum-Jensen confirmed in elegant electron-microscopic studies that the complement membrane lesions are indeed made by the C5b-9 complex supporting Meyer's hypothesis (Bhakdi, S. and Tranum-Jensen, J. 1978). However, they too were unable to provide a molecular model for the membrane lesion that would account for the five terminal complement components. It took until 1982 when Podack and Tschopp (Podack, E.R. and Tschopp, J. 1982) discovered that C9 alone was able to polymerize and form complexes (poly C9) that were able to create membrane pores and form membrane lesions resembling closely complement lesions.

When poly C9 assembles in solution, the polymerized complex is unable to insert into membranes. Membrane insertion and polymerization of C9 have to occur in a concerted fashion in order to create membrane pores. However, lipid bilayer membranes can be assembled around poly C9 by detergent dialysis. In this procedure both outside out (Fig. 4, left) and outside in (Fig. 4, right) poly C9 complexes can be visualized illustrating the transmembrane pore on either end of the poly C9 complex (Fig. 4). The full membrane attack complex (MAC) is a complex between C5b-8 and poly C9. C5b-8 serves as receptor for C9 and triggers C9's self polymerization and pore formation. C5b-8 fulfills the targeting function for C9, itself being deposited by

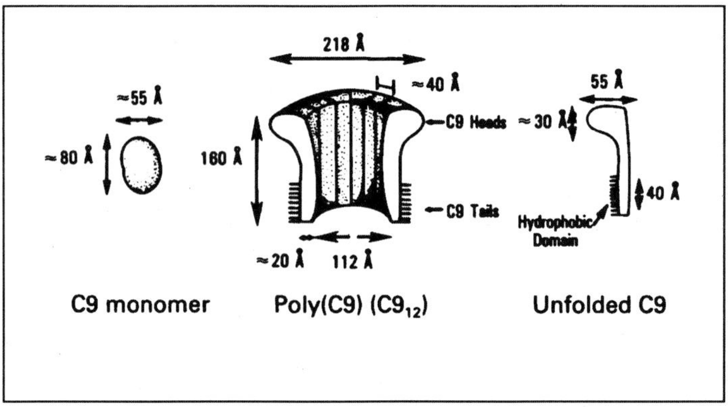

Fig. 3. Schematic presentation of C9 polymerization.

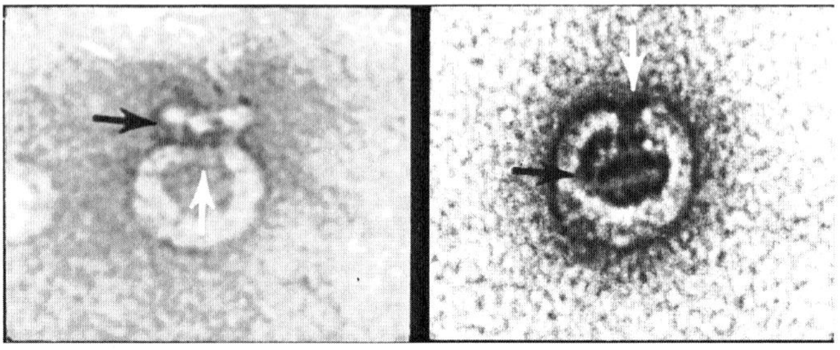

Fig. 4. Outside out (left) and outside in (right) poly C9 on lipid bilayer vesicles. Black arrows point to the torus of poly C9, white arrows to the bilayer penetrating end of poly C9. Note the continuous pore traversing the thickness of the bilayer.

the C5 convertase assembling on C3b molecules on target membranes through the recognition functions of complement. It is likely that C6, C7 and C8 arose by gene duplication from an ancestral C9 like poreformer. The C5b-7 complex acquires a membrane binding site after binding of C and inserts itself into the lipid bilayer without however forming a pore. In solution, C5b-7 complexes aggregate to irregular multimers through interaction of the hydrophobic membrane binding site (Fig. 5). Inserted into membranes, C5b-7 complexes also tend to aggregate unless C8 and C9 are simultaneously present. Fig. 6 shows the stepwise assembly of the MAC single bilayer phospholipid vesicles, beginning with C5b-7 and ending with the gross structural change upon binding and polymerization of C9. The biochemical principal for the generation of hydrophobic membrane inserting sites from monomeric hydrophilic proteins relies on protein/protein association -> conformational change and partial unfolding -> exposure of hydrophobic sites -> membrane insertion, if the

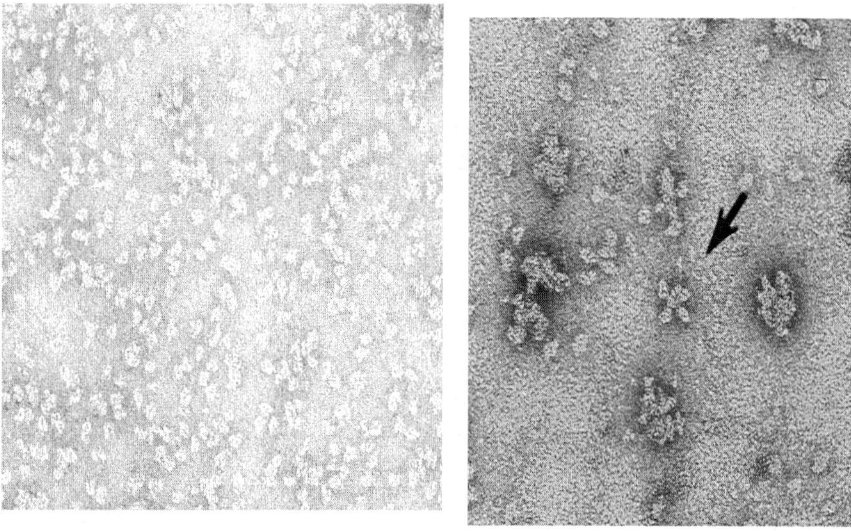

Fig. 5. Left electronmicroscopic image of the C5b-6 complex (left) and the C5b-7 complex (right) in solution. Note the aggregation of C5b-7 to a tetrameric complex (arrow) mediated by the hydrophobic membrane binding site. Uranyl formate. Magnification ~300,000 fold.

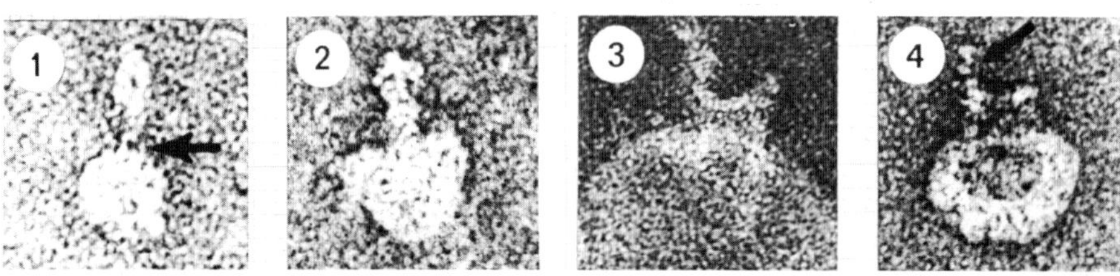

Fig. 6. Step wise assembly of the membrane attack complex (MAC) on single bilayer phopholipid vesicles. 1. C5b-7 on a small vesicle. Arrow points to the narrow membrane binding site of C5b-7. 2. C5b-8 complex on a single bilayer vesicle. Note that the vesicles in 1 and 2 exclude negative stain indicating that no pore has yet formed. 3. An incompletely assembled C5b-9 complex and 4. fully assembled MAC comprised of poly C9 and C5b-8 (arrow). Note also that the vesicle in 4 is filled with negative stain indicating pore formation. Magnification 600,000 fold.

reactions proceed adjacent to a membrane. If no membrane is present aggregation in solution takes place. C9 is specialized in that its self-association or association with C5b-8 leads to the exposure of new C9 binding sites that set into motion a self polymerization reaction that is, however, self limited through the formation of a hollow cylindrical molecule (poly C9). The inner surface of poly C9 is hydrophilic while the outer surface sitting in the lipid bilayer is hydrophobic.

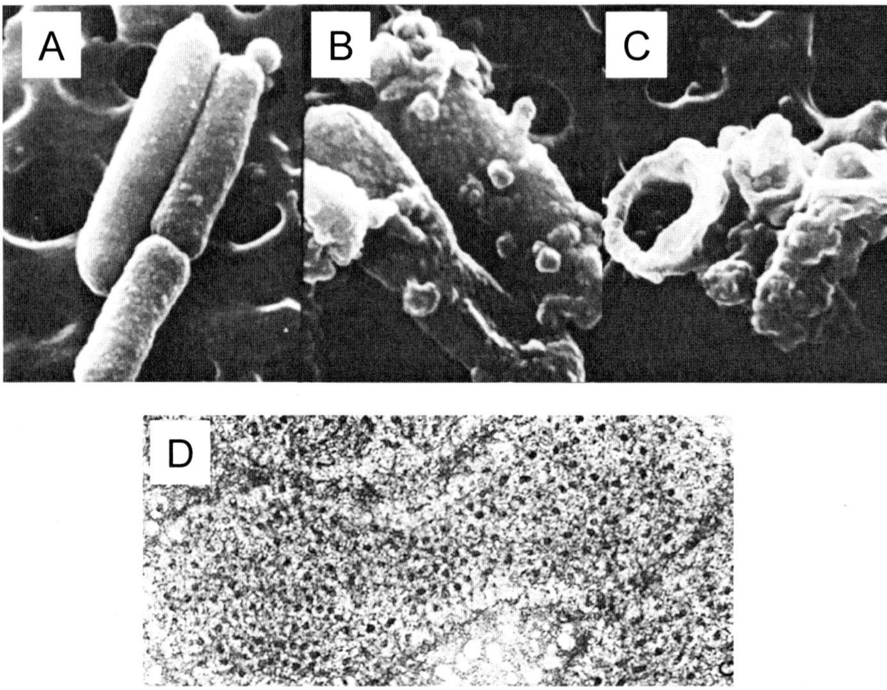

Fig. 7. Collapse of E. coli after complement and lysozyme treatment. A: E. coli B: E. coli + complement; C: E. coli + complement + lysozyme; scanning electron micrographs. D: Bacterial membrane fragments from bacteria treated with complement; negative staining transmission electron microscopy. Note high density of MACs.

While the MAC has cytolytic activity for red cells and other nucleated cells, its designated targets are bacterial membranes. Large numbers of MAC complexes probably can destroy the outer membrane of bacteria, but a small number of MACs is not expected to be very harmful. However, the MAC serves as conduit for lysozyme to pass through the MAC pore and digest the peptidoglycan, thereby leading to the collapse of the skeletal structure and the death of the bacteria. Thus, two effects of the MAC contribute to its cytocidal activity for bacteria. Physical disruption of the outer membrane through displacement of lipid and associated membrane proteins; and creation of pores allowing access of lysozyme to the peptidoglycan layer underlying the outer membrane. Lysozyme can easily diffuse.

3 Perforin 1, the pore former used by cytolytic lymphocytes

The finding that a single protein, C9, was able to assemble by polymerization into a complete transmembrane pore stimulated search for a similar protein in cytolytic

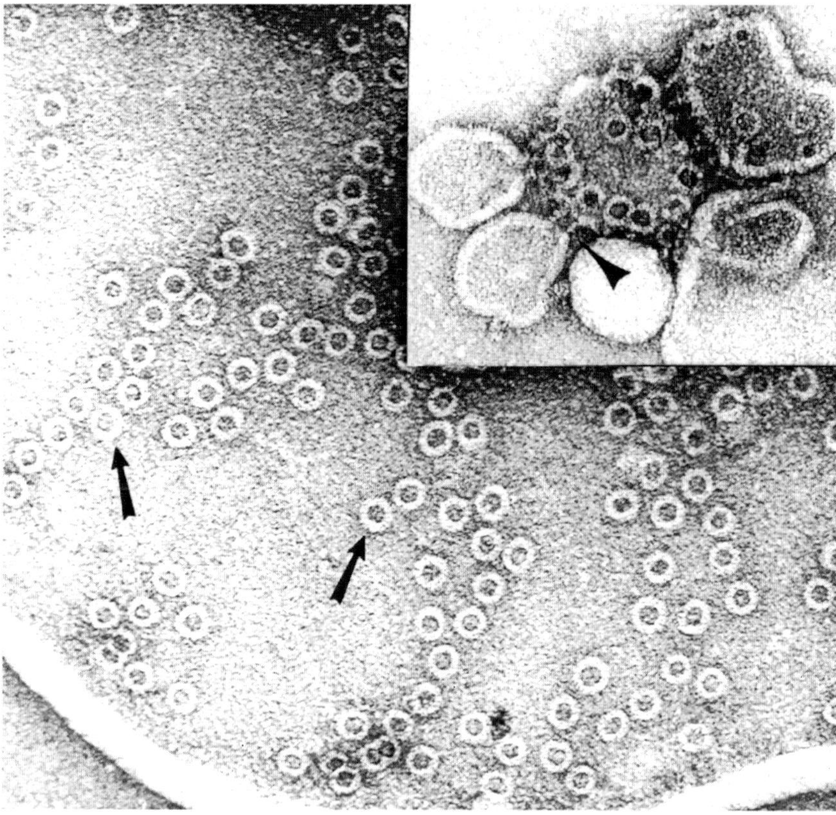

Fig. 8. Poly-Perforin complexes associated with membrane fragments in top view (larger panel, arrows) and side view (inset, arrow head). Uranyl formate; final magnification 250,000.

lymphocytes. The first evidence that there may be such a protein came again from electron microscopic studies of Dourmashkin (Dourmashkin, R.R., Deteix, P., Simone, C.B. and Henkart, P. 1980) who showed membrane lesions generated on red cells during antibody dependent cell mediated cytotoxicity. Using cloned CTL and NK cells Dennert and Podack (Dennert, G. and Podack, E.R. 1983; Podack, E.R. and Dennert, G. 1983) were able to define membrane associated polymeric complexes that had striking resemblance to poly C9. Under the hypothesis that the complex perforates the membrane and forms a transmembrane pore, the name Perforin was given for the monomeric protein and poly-Perforin for the complex. The pore size of poly-Perforin is 16nm diameter versus 10nm for polyC9. The receptors for Perforin are phospholipid head groups mediating binding via Ca. Ca is also the signal for Perforin polymerization and membrane insertion analogous to C9. The targeting of Perforin is accomplished by the cytotoxic cell recognizing the appropriate target and activating Perforin release. Perforin is stored in cytoplasmic storage granules of killer cells along with granzymes and lysosomal proteases (Fig. 10) (Masson, D. and Tschopp, J. Millard, P.J., Henkart, M.P., Reynolds, C.W. and Henkart, P.A. 1984; Podack, E.R.

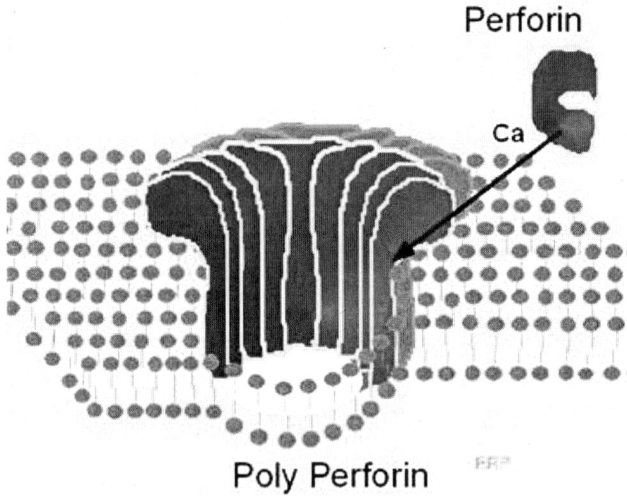

Fig. 9. Perforin undergoes Ca-dependent polymerization and membrane insertion. Phospholipid headgroups serve as Perforin receptor.

and Konigsberg, P.J. 1984). Upon contact with a cognate target cell the CTL secretes the granules vectorially into the intercellular space of the synapse. Perforin generates transmembrane pores on the target but not on the CTL membrane. Target cells need to repair these poly-Perforin pores or they would die by loss of membrane potential and by loss of intracellular components. To repair poly Perforin pores, the pore-bearing patch of membrane is internalized by endocytosis thereby sealing the plasma membrane. However, in the process of repair, endocytosis granzymes and lysosomal proteases are taken up within the endosome and released into the cytoplasm via the poly-Perforin pore. Multiple granzymes and proteases activate multiple apoptotic and non-apoptotic death pathways in the target resulting in its violent and rapid cell death (Heibein, J.A., Goping, I.S., Barry, M., Pinkoski, M.J., Shore, G.C., Green, D.R. and Bleackley, R.C. 2000; Lieberman, J. and Fan, Z. 2003; Pao, L.I., Sumaria, N., Kelly, J.M., van Dommelen, S., Cretney, E., Wallace, M.E., Anthony, D.A., Uldrich, A.P., Godfrey, D.I., Papadimitriou, J.M., Mullbacher, A., Degli-Esposti, M.A. and Smyth, M.J. 2005). Perforin/granzyme/protease mediated cell death is not an ordered apoptotic process running down along carefully planned genetic programs. Rather CTL mediated death is unprogrammed activation of multiple death pathways resulting disordered cell death with elements of apoptosis and necrosis.

Perforin-1 is the central mediator of cell death by CTL and is dedicated to eradication of parasitized cells that may not be compliant to apoptotic signaling pathways. Granzymes without Perforin are ineffective since they can not gain entry into the cell. Perforin deficiency is associated with the inability to clear certain viruses (such as LCMV) that are easily controlled in Perforin sufficiency. Perforin also has important functions in immuno-surveillance against tumors, since Perforin deficient

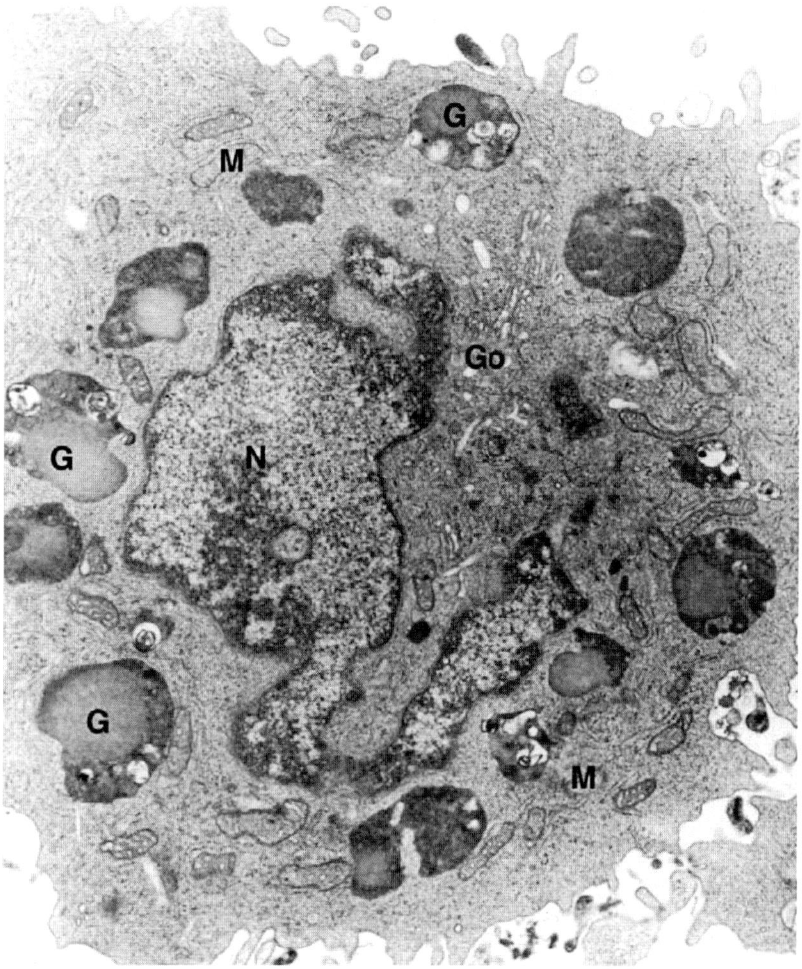

Fig. 10. Thin section electronmicrograph of cytotoxic T lymphocyte. Note the prominent cytoplasmic storage granules (G). M - mitochondria; Go – Golgi; N – nucleus.

mice exhibit upon aging a significantly increased tumor incidence (Smyth, M.J., Thia, K.Y., Street, S.E., MacGregor, D., Godfrey, D.I. and Trapani, J.A. 2000; Street, S.E., Cretney, E. and Smyth, M.J. 2001).

Perforin deficiency in humans is associated with the lethal syndrome of familial hemophagocytic lymphohistiocytosis (Stepp, S.E., Dufourcq-Lagelouse, R., Le Deist, F., Bhawan, S., Certain, S., Mathew, P.A., Henter, J.I., Bennett, M., Fischer, A., de Saint Basile, G. and Kumar, V. 1999). Without treatment, children die usually very young upon secondary viral infection due to uncontrollable CD8 CTL and macrophage expansion. A similar syndrome is seen in mice lacking Perforin and Fas-ligand (Spielman, J., Lee, R.K. and Podack, E.R. 1998). The likely cause of the disease-pathogenesis is the inability of CD8 CTL to control antigen presentation by

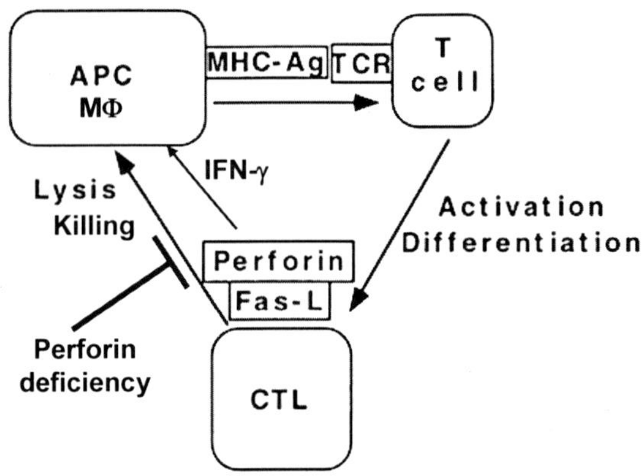

Fig. 11. Negative feed back regulation of antigen presentation by Perforin. APC activate pre CTL to express Perforin and granules. CTL kill APC by a Perforin dependent mechanism. In the absence of Perforin APC are not killed but continuously stimulated by IFN-γ.

killing APC (Fig. 11). APC continue to activated CD8 CTL resulting in uncontrolled expansion and cytokine production responsible for death.

4 Perforin-2, a novel pore former in macrophages and dendritic cells

The MAC proteins C6, C7, C8 and C9 and Perforin-1 share a conserved domain, the MAC-PF domain which is thought to be the part of the transmembrane tubule that inserts itself into the hydrophobic core of the target membrane.

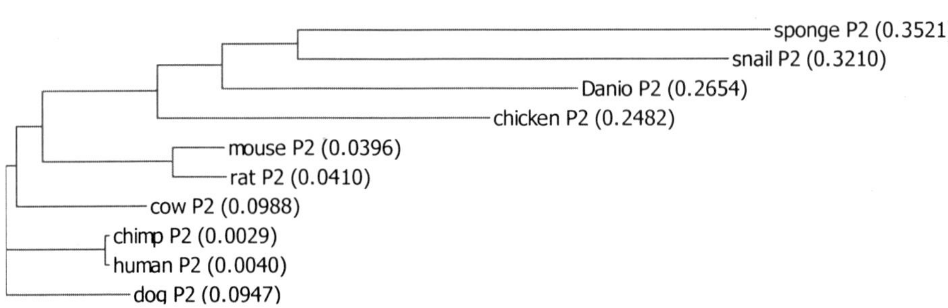

Fig. 12. Phylogenetic tree of Perforin-2. Danio – Zebrafish;

MAC-PF domain:

```
C9           MFLHVKGEIH LGRFVMRNR- --DVVLTTTF VDDIKALPTT YEKG------ -EYFAFLETY
Perforin-1   QYSFSTDTVE CRFYSFHVVH --TPPLHPDF KRALGDLPHH FNASTQP--- -AYLRLISNY
Perforin-2   QAVTTRVQVR NRIYTVKTTP --TSELSLGF TKALMDICDQ LEKNQTKMAT YLAELLILNY

C9           GTHYSSSGSL GGLYELIYVL DKASMKRKG- -VELKDIKRC LGYHLDVSL- ---AFSEISV
Perforin-1   GTHFIRAVEL GGRISALTAL RTCELALEG- -LTDNEVEDC LTVEAQV--- --NIGIHGSI
Perforin-2   GTHVITSVDA GAALVQEDHV -RSSFLLDN- -QNSQNTVTA SAGI------ AFLNIVNFKV

C9           GAEFNKDDCV KRGEGRAVNI TSE---NLID DVVSLIRGGT RKYAFELKEK LLRG-----T
Perforin 1  SAEAK--ACE EKKKHKMTA SFH---QTYR ERHSEVVGG- ---HHTSIND LLFG-----I
Perforin 2  ETD----YIS QTSLTKDYLS NRT------- NSRVQSFGG- ----VPFYPG ----------

C9           VIDVTDFVNW ASSINDA--P VLISQ-KLSP IYNLVPVKMK ------NAHL KKQNLERAIE
Perforin 1  QAGPEQYSAW VNSLPGS--P GLVDY-TLEP LHVLLDSQ-- --------DP RREALRRALS
Perforin 2  ----ITLETW QKGITNH--L VAIDR-AGLP LHFFIKPDKL PGLPGPLVKK LSKTVETAVR

C9           DYIN
Perforin 1  QYLT
Perforin 2  HYYT
```

Using this sequence as query, an mRNA expressed in macrophages was found predicting a protein with a MAC-PF domain (Kopacek, J., Sakaguchi, S., Shigematsu, K., Nishida, N., Atarashi, R., Nakaoke, R., Moriuchi, R., Niwa, M. and Katamine, S. 2000; Spilsbury, K., O'Mara, M.A., Wu, W.M., Rowe, P.B., Symonds, G. and Takayama, Y. 1995)

Conserved, predicted Perforin-2 sequence from man to sponge

```
 chimp P2    (1) -------------------------MNNFRATILFWAAAAWAK-------SGKPLGEMDEVGVQKCK-NALK
 human P2    (1) -------------------------MNNFRATILFWAAAAWAK-------SGKPSGEMDEVGVQKCK-NALK
   dog P2    (1) -------------------------MSSVRGAILFWVVVAWAK-------TDKPLEQTNETGFQKCK-NALK
   cow P2    (1) -------------------------MNSFRGAFLIWAVATWAE-------TDTSWGATDEPGFQNCK-NALK
 mouse P2    (1) ------------------MAKTICAMNSFMALVLIWMIIACAE-------ADKPLGETGTTGFQICK-NALK
   rat P2    (1) ------------------MAGNACTMNSFMAIALIWMMIACAE-------ADKPLRDPGMTGFQTCK-DTLK
 Danio P2    (1) --------------------------MKSRAFHLLMLCCFISV-------CNLHPLIRPNNGLRLCR-KNSS
 snail P2    (1) ----------------------MLCFVFGVSIVAGVIGGELL-------NTVQKPEFPKGDVRACYGDNKK
sponge P2    (1) MVCLTRQIGSLTDTHPTNKLVMVGGGGAGGLLSLVSGALSGGGGNIRAGAPRNMYPRGDPRNCLSGNPK
                 71                                                                  140
 chimp P2   (40) LPVLEVLPGGGWDNLRNVDMGRVMELTYSNCRTTEDGQYIIPDEIFTIPQKQSNLEMNSEILESWANYQS
 human P2   (40) LPVLEVLPGGGWDNLRNVDMGRVMELTYSNCRTTEDGQYIIPDEIFTIPQKQSNLEMNSEILESWANYQS
   dog P2   (40) LPVLEVLPGGGWDNLRNVDMGRVMDLTYRSCRTTEDGQYIIPDEITSIAQKQSNLEMNSEILESWVNYQS
   cow P2   (40) LPVLPVLPGGGWDNLRNVDTGRVMELAYSHCRTTEDGQYIVPDEIFTIPQKQSNLEMNSKILESWVNYQS
 mouse P2   (47) LPVLEVLPGGGWDNLRNVDMGRVMDLTYTNCKTTEDGQYIIPDEVYTIPQKESNLEMNSEVLESWMNYQS
   rat P2   (47) LPVLEVLPGGGWDNLRNIDMGRVIDLTYTNCKTTEDGQYIIPDEVYTIPQKESNLEMNSEILDSWVNYQS
 Danio P2   (39) LTALEVLPGGGWDNLRNIDMGRVMNLSYSQCQTTEDGVYLIPDEVFVIPQKVSGVETNSEIIMSWLEQKS
 snail P2   (43) LERFEVLPGQGWDNLRNVDAGLVVVYNYSRCRTTEDGRFLIPDTVNTIPLKASKLNVYAELISHWSNYTS
sponge P2   (71) LNILQVVPGIGWDNLRNSETGILTSFSYSQCKVTYDRRYLIPDETFAIPIKTSTIDYQAELFDHWDAYKS
                 141                                                                 210
 chimp P2  (110) STSYSINTELSLFS---KVNGKFSPEFQRMKTLQVKDQAITTRVQVRNLVYTVKINPTLELSSGFRKELL
 human P2  (110) STSYSINTELSLFS---KVNGKFSTEFQRMKTLQVKDQAITTRVQVRNLVYTVKINPTLELSSGFRKELL
   dog P2  (110) STSSSINLELSLYS---KVNGKFSSDFQQMKTLQVKDQAITTRVQIRNLIYTVKINSASKLSWGFKKDLM
   cow P2  (110) STSNSINMELSLFS---KVNGKFSLEFQRIKTLQVKDQAVTTQVQVRNLVYTVKINPDAELSLGFKKALM
```

```
 mouse P2  (117) TTSLSINTELALFS---RVNGKFSTEFQRMKTLQVKDQAVTTRVQVRNRIYTVKTTPTSELSLGFTKALM
   rat P2  (117) TTSFSINTELSLFS---KVNGKFSTEFQRMKTLQVKDQAVTTRVQVRNRIYTVKNSPTSELSFGFTNALM
 Danio P2  (109) STSSSVNADVSFFS---VLNAKFSTENQRMKTHQVKEGSVTARVQVRNHLYTVKAYPDFTLDSRFAKQAE
 snail P2  (113) TTAHGVNIDAGLKFGSVKVSGTFSSGYESVKSKQIGDKSYTTRVQLRYVRYSAKLQPDAALHPTFKSRLL
sponge P2  (141) VTSRSINAGFDFFG----------------------------------------------------KIGG
                 211                                                               280
 chimp P2  (177) DISDRLENNQTRMATYLAELLVLNYGTHVITSVDAGAALIQEDHLRASFLQDSQSSHSAVTASAGLAFQN
 human P2  (177) DISDRLENNQTRMATYLAELLVLNYGTHVTTSVDAGAALIQEDHLRASFLQDSQSSRSAVTASAGLAFQN
   dog P2  (177) DISDRLENNQTRMATYLAELLVLNYGTHVVTSVDAGAALLQEDHIRASFLQDSQSSHTAVTASAGVAFMN
   cow P2  (177) DISEQLENNQTRMATYLAELLVLNYGTHVITSVDAGAALIQEDHIRSSFLQDSQSSRSAVTASAGITFLN
 mouse P2  (184) DICDQLEKNQTKMATYLAELLILNYGTHVITSVDAGAALVQEDHVRSSFLLDNQNSQNTVTASAGIAFLN
   rat P2  (184) DICDQLEKNQTKMATYLAELLVLNYGTHVITSVDAGAALVQEDHIRSSFLLDNQNSENTVTASAGIAFLN
 Danio P2  (176) EIADAIENNQTRHANYLSEKLVLDYGTHVITSVDAGATLVQEDYLKMSYISNSQSDKSSVSASAGANFFD
 snail P2  (183) SIAGSLQLNKTDQARYDSELLVRDFGTHVVTSVDAGAALVQEDQVSSEFVNSRKFTKNQITAGASASFLG
sponge P2  (159) EILSLLKSNDTESADYAAQILIRDYGTHCITSIDAGAVLIKEDNLKSTIMSNYKGRADSLSTAAGVEFYD
                 281                                                               350
 chimp P2  (247) TVNFKFEENYTSQN--VLTKSYLSNRTNSRVQSIGGVPFYPG-ITLQAWQQGITNHLVAIDRSGLPLHFF
 human P2  (247) TVNFKFEENYTSQN--VLTKSYLSNRTNSRVQSIGGVPFYPG-ITLQAWQQGITNHLVAIDRSGLPLHFF
   dog P2  (247) VVNYKFEENYTSQN--ALTKSYLANRTHSRVRSIGGVPFYPG-ITLQAWQQSIANHLVAIDRAGLPLPFF
   cow P2  (247) IVNFKFEENYTSQN--TFTKSYLSNRTNSRVQSFGGLPFYPG-ITLQAWQQGVSNHLVAMDRAGLPLYFF
 mouse P2  (254) IVNFKVETDYISQT--SLTKDYLSNRTNSRVQSFGGVPFYPG-ITLETWQKGITNHLVAIDRAGLPLHFF
   rat P2  (254) IVNFKVETDHTSQT--TLTKSYLSNRTNSRVQSFGGIPFYPG-ITLETWQKGITNHLVAIDRAGLPLHFF
 Danio P2  (246) KVKFDIGGNTSQGS--SQSSSYQGNITYSLIQSHGGALFYPG-ITLQKWQQSTLNNLAAIDRSGLPLHYF
 snail P2  (253) IFSIDVSYHSSTSN--EVKTAYEKSRSSSQIDTLGGPMFKASNFTANDWTNEVDHELVAVDRSGDPLFFL
sponge P2  (229) MLKLRASAGFSSYSGDSDLKAYRQNRTSSRLYTYGGPPYKLG-MNLSRWENDLMNNLVATDRSGKPSHSL
                 351                                                               420
 chimp P2  (314) INPNMLP--DLPGPLVKKVSKTVETAVKRYYTFNTYPGCTDLNSPNFNFQAN-TDDGSCEGKMTNFSFGG
 human P2  (314) INPNMLP--DLPGPLVKKVSKTVETAVKRYYTFNTYPGCTDLNSPNFNFQAN-TDDGSCEGKMTNFSFGG
   dog P2  (314) ISPDTLP--ELPGPLVKKLSKTVEAAVRHYYAFNTYPGCTDANSPNFNFQAN-TDDGSCEGKMTNFSFGG
   cow P2  (314) INPERLP--DLPGPLVRKLSKTVEAAVRRYYAVNTYPGCTDLSSPNFNFQAN-TDDGSCEGKMTNFSFGG
 mouse P2  (321) IKPDKLP--GLPGPLVKKLSKTVETAVRHYYTFNTHPGCTNVDSPNFNFQAN-MDDDSCDAKVTNFTFGG
   rat P2  (321) IKPDKLP--GLPGPLVKKLSKTVETAVRHYYTFNTHPGCTNVDSPNFNFQAN-MEDDSCDAKVTNFTFGG
 Danio P2  (313) LNPSTFP--DLPTPTVNKLASTVRKAAERYYKVNTIPGCVNVDSPNFNFQAN-VDDASCEGPITNLSFGG
 snail P2  (321) INSASLP--ELPNSVLYQLQNLVEETILHYYEFNTYRGCTELDSPNFSPAAN-LDDGTCKSPYTNLTFGG
sponge P2  (298) STTQSLKPEVTTSQEVFLLRRLVKSAVSQYYHYNTHTGCKNPKAPNLDHQTNNGAPGVCKEPSANYTFGG
                 421                                                               490
 chimp P2  (381) VYQECTQLSGNSDVLLCQKLEQKNPLTGDFSCPSGYSPVHLLSQIHEEGYNHLECHRKCT------LLVF
 human P2  (381) VYQECTQLSGNRDVLLCQKLEQKNPLTGDFSCPSGYSPVHLLSQIHEEGYNHLECHRKCT------LLVF
   dog P2  (381) VFQECTQLSGKEAAQLCQTLEQRNPLTGAFSCPSGYSPIHLLSQVHEEGYNHLECRRKCT------LLVF
   cow P2  (381) VYQECTQFSGNEVVQLCQNLEQKNPLTGSVSCPSGYSPVQLLTQTHEEGYNHLECSRKCT------LYIF
 mouse P2  (388) VYQECTELSG---DVLCQNLEQKNLLTGDFSCPPGYTPVHLLSQTHEEGYSRLECKKKCT------LKIF
   rat P2  (388) LYQECTELSG---DALCQNLEQKNLLTGDFSCPPGYTPVHLLSQTHEEGYSRLECKKKCT------LKIF
 Danio P2  (380) IYQKCTPLTPDG-NIICDETAQKNPATGGYSCPQHYNTTLLHSEVVEKGFNHYECHTHCHSCGFLGLSTC
 snail P2  (388) VYQTCSMSSGSNNGDLCSGLDQVNPKTGGHTCPDGYESVELHTGRLSDSKSVHSCHSCWL----------
sponge P2  (368) VFQSCRSNGN----DICGKLLQKNPLTGGYSCPKNFKALLLQLGTERSHKMRRVCVWKRKCT-FFVFNCH
                 491                                                               560
 chimp P2  (445) CKTVCEDVFQVAKAEFRAFWCVASSQVPENSGLLFGGLFSSKSINPMTNAQSCPAGYFPLRLFEN-LKVC
 human P2  (445) CKTVCEDVFQVAKAEFRAFWCVASSQVPENSGLLFGGLFSSKSINPMTNAQSCPAGYFPLRLFEN-LKVC
   dog P2  (445) CKTVCEDVFRVAKAEFRAFWCVASGQIPENSGLLFGGLFSGKTINPLTNAQSCPAGYFPLRLFEN-LKVC
   cow P2  (445) CKTVCEDVFRVARAEFRAFWCAASGQVSENSGLLFGGLFSGKSINPLTNAQSCPAGYFQLKLFEN-LKVC
 mouse P2  (449) CKTVCEDVFRVAKAEFRAYWCVAAGQVPDNSGLLFGGVFTDKTINPMTNAQSCPAGYIPLNLFES-LKVC
   rat P2  (449) CKTVCEDVFRVAKAQFRAYWCVATGQVPDNSGLLFGGLFTDKSINPMTNAQSCPAGYIPLNLFES-LKVC
 Danio P2  (449) CDKTCGDSYHVRRAKLETLWCSSTHKTPENSGYLFGGLFGPGIQNPLTKSSSCPPSYFTQRFLSNGMMIC
 snail P2  (448) FFKCCHDNYYHSEATYIMHWCAATGPVSQDSGYLFGGLYTSQLNNPLTQGKTCPVNFYTRTLGKD-LHIC
sponge P2  (433) DVDDCTFVPSVEIASYQTYWCAPNKKNPPKFGYMFGGIYSNDIQNPITRSCSCPTHFLPLRMGER-ATVC
                 561                                                               630
 chimp P2  (514) VSQDYELGSRFAVPFGGFFSCTVGNPLVDPAISRDL----------------GAPSLKKCPGGFSQHPAL
 human P2  (514) VSQDYELGSRFAVPFGGFFSCTVGNPLVDPAISRDL----------------GAPSLKKCPGGFSQHPAL
   dog P2  (514) ASLDYELGFRFSVPFGGFFSCAVGNPLVNSAFT--E----------------GAPSLKKCPGGFSQHLAL
   cow P2  (514) ASLDYELGYRFSIPFGGFFSCAAGNPLVDSATSKDL----------------GAPSLRKCPGGFSQHLAL
 mouse P2  (518) VSLDYELGFKFSVPFGGFFSCIMGNPLVNSDTAKDV----------------RAPSLKKCPGGFSQHLAV
   rat P2  (518) VSLDYELGYKFSVPFGGFFSCIMGNPLVNSDTAKDI----------------GAPSLKKCPGGFSQHLAV
 Danio P2  (519) MSNDYEIGTRFSVPFAGFFSCQSGNPLSNGQSR--------------------------CPPQFSQHLAA
 snail P2  (517) ISDDYELGMKYSMPFGGFISCTTGNPLAMNPKPKSKGDMNSALPSLHSFFQGSKTWPKHCPKGYSQHLAY
sponge P2  (502) VSEDYELGHQFSLPFGGFFSCVSGNVLAGNGSSEFLNN--------------PKDWPMRCPGGFTQHLAL
                 631                                                               700
 chimp P2  (568) ISDGCQVSYCVKSGLFTGGSLPPARLPPFTRPPLMSQAATNTVIVTNSENARSWIKDSQTHQWRLGEPIE
```

```
  human P2   (568) ISDGCQVSYCVKSGLFTGGSLPPARLPPFTRPPLMSQAATNTVIVTNSENARSWIKDSQTHQWRLGEPIE
    dog P2   (566) ISDGCQVSYCVKSGLFTGGSLPPARLPPYTRPPLMSQAATNTVIVTNSETASSWIKDSQTRQWRLGEPLE
    cow P2   (568) ISDGCQVSYCVKAGLFTGGSLPPVRLPPYTRPPLMSQVATNTVLVTNHETASSWIKDPQTHQWRLGEPLE
  mouse P2   (572) ISDGCQVSYCVKAGIFTGGSLLPVRLPPYTKPPLMSQVATNTVIVTNSETARSWIKDPQTNQWKLGEPLE
    rat P2   (572) ISDGCQVSYCVKAGIFTGGSLLPVRLPPYTKPPLMSQVATNTVIVTSSETARSWIKDPQTNQWKLGEPLE
  Danio P2   (563) ISDGCQVLYCVQSGVFSGGHLKPVRLPPFTRPPVVGMIATNTVAVMTE-GERSWVRVGETKMWRLAKPGD
  snail P2   (587) VDQGCEINYCLLAGSLSEVGLPKIRRPPFQTAPLLLPSTENHVVFDPV--TLTWRKNQEAMQFISARGGD
 sponge P2   (558) TEKGCRVNFCVKAGSLLRASDLELVLPPFDPKPTLRKNSTSDLFSKPAVAPSGSLPIKGVPPPLDNNRVI
                   701                                                                   770
  chimp P2   (638) LRRAMNVIHGDGGGLSGGAAAGVTVGVTTILAVVITLAIYGTRKFKK-------KAYQAIEE--RQSLVP
  human P2   (638) LRRAMNVIHGDGGGLSGGAAAGVTVGVTTILAVVITLAIYGTRKFKK-------KAYQAIEE--RQSLVP
    dog P2   (636) LRRAMKVIRGDGGGLSGGAAAGVTMGVTTVLAAVIALAIYGTRKYKK-------RGYQALEDE-RQSLAA
    cow P2   (638) LRRAMRVVHGDGEGLSGGAAAGLTLGVTIALAGVVALAIYGARKSRK-------KGYQALQDE-KQSLAA
  mouse P2   (642) LRRAMTVIHGDSNGMSGGEAAGITLGVTIALGVVITLAIYGTRKYKK-------KEYQEIEE--QESLVG
    rat P2   (642) LHRAMTVIHGDGNGMSGGEAAGVTLGVIIALGIVITLAIYSTRKYKKE------KEYQEIEE--QESLVG
  Danio P2   (632) IKQMQSILDAS--EMSGGKKAGVAIGIIVLVALVVAGTVVIMKRRNRFSSLKLNRGYEEISEERNESSVE
  snail P2   (655) AT-----SSSTGSGMSAGAAAGIAVVATLGCVVISTIIIVLIKRRRKSS---AGYRHLAIDDPLLSSQSN
 sponge P2   (628) MYYPVLSSNTGQRDNGVGRITEPFFCHSWDWSVDLSVAHQAIDKFAFAK---------------------
                                                  Trans Membrane (yellow)

                   771                789
  chimp P2   (699) GTAATGDTTYQEQGQSPA-
  human P2   (699) GTAATGDTTYQEQGQSPA-
    dog P2   (698) GTAESGDAPGQEQEQSPA-
    cow P2   (700) GAAVNGDALDQEQAQNPA-
  mouse P2   (703) SLATDATVLNGEEDPSPA-
    rat P2   (704) SFATDASAPNGEQDPCPA-
  Danio P2   (700) IEQEQNEAANENPNQQLLS
  snail P2   (717) YGATGSDAVNVNVE-----
 sponge P2   (677) -------------------
```

Notes: The boxed sequences represent the MAC-PF domain. The yellow high-lighted sequences are predicted transmembrane domains. Conserved Y and S are outlined in purple in the predicted cytoplasmic domain.

Further database searches revealed that the predicted protein, which we named tentatively Perforin-2 or P2, was very highly conserved throughout the animal kingdom from mammals to sponges and even in sea anemones. A phylogentic tree showing the calculated ancestral relationships is shown in Fig. 12. The protein sequence predicts a transmembrane protein containing a typical leader sequence, a transmembrane domain and a short cytoplasmic tail with a conserved tyrosine and serine. The conserved domain structure is shown in Fig. 13. Two domains, designated P2 domain are highly conserved in addition to the MAC-PF domain shared with Perforin-1 and the MAC proteins.

We obtained the cDNA sequence and tagged it in frame with EGFP at thepredicted C terminus. Attempts at expression of the fusion cDNA resulted in cell

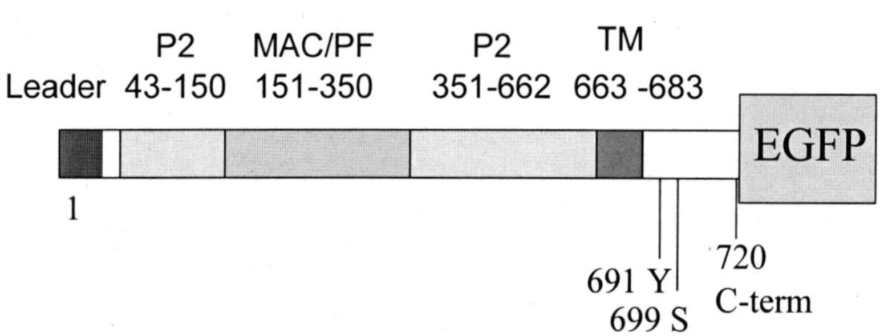

Fig. 13. Protein-domain structure of Perforin-2.

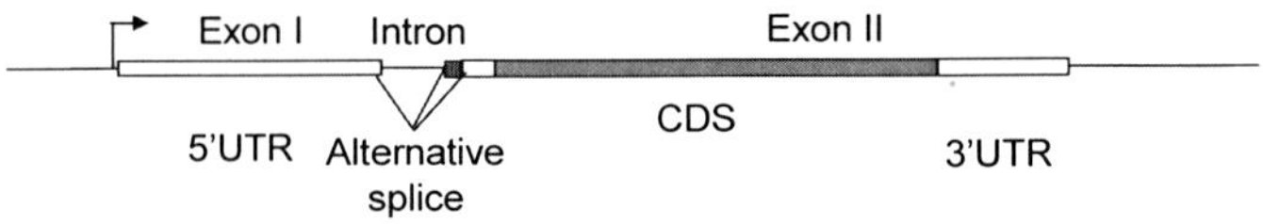

Fig. 14. Gene structure of murine Perforin-2.

Homology of 5'UTR and intron sequence of Perforin-2 in mammals

```
      dog   ACGAACCAGGGGTGATGG--AAAGGGAGGAGGGCGGGGGGTGGGGTGACTGGTTG---GC--GGGCACTGAGGGGGGCA
      mou   CAGGCCCCAACTTCAGTGTACCAATGAGACATCTAAGATTTGTCCTTGACACAGAG-TAGC---------CTGGAGGCGG
      hum   CAGCCTCCACCCTCAG-G--C-AAAGAGGAAACCCAGGGTTGGCCTTGACTAACAGCTTGCATAGGTATGGTGGAGCCAG
      rat   CAGCCTCCAACCTCGGTGTGCCAATGAGATATCTAAGATTTGTCCTCGACCCAGAG-AAGCACGCCTA--GTAGAGGCGG
      cow   AAACCTGAATGGTGAA-GACCCCTTTCAATTTTACACATCTGTCTAAGTTACATGC---ACTTAACCATGGAATGAAAAA
Consensus   CAGCCTCCA C TCAG G  CCAATGAGA ATC  AGATTTGTCCTTGAC CA AG   GC      A  GTGGAGGC G

      dog   CTTGACGGGATGAGCACT-GGTTGTTATTCTGTATGTTGGCAAATTGAACACCAATAAAAAATAAATTTATTATTTAAAA
      mou   GGTATGTAAATAAGCACAAGGTTGCTACT---AACTCCTGCGTTT---CCAGTAAAAAGAAACTAATACTGTGATTTCAC
      hum   GGTGTTTCAGTAAGGGTGGTGTGGTCATT---TGCCTCTGCATTT---ATAGTAAAAGAAAACTGATAATGGAGTCCCAA
      rat   GGCGTGTAAATAAGCACAAGGTTGCTACT---GACTCCTGCATTT---CCAGTAAAAAGAAACTAATATTGTGATTTCAC
      cow   ATTAAAGTTATAAGAGCAGCTCTCAAATT---ATTGCCAACTATACACAGAGGAAAAGGAGAGTGATATTGAAGTCCCAA
Consensus   GGTGT T AATAAGCACA GGTTG TATT    AC CCTGCATTT   ACAGTAAAAAGAAACTAATATTGTA TT CAA

      dog   AATAAGGGACATAGGGCTGCCTTTATATATCAA----GTACA-ACATGAGGTTTTCTTTTCTTTTTTTTTTTTATTCATG
      mou   AAAAGAAGAGAAAGGCAGGTTTTGAA-----------GTCAA-AGAGAGTGAAACAAAAGCC-----------AGACAGA
      hum   GAGACAGCAGTCAGGGAAAATATGAAACATCAA----GTCCA-AGAGAATGAGCAAAAAGCA-----------AAGCCAA
      rat   AAGAAAGGAGTAAGGCTGATTTTGAA-----------GTCAG-TTAAAGAGAGCCAAAAGCC-----------AGACAGA
      cow   CAGAGAGTCATTAGGGGAAAGCAGAACCATCAAGAGAGTGAATAAAAAATGAAACCAAAGCTTTCTTTCAA--AGCTGGA
Consensus   AAGA AGGAGT AGGG  A TTTGAA  ATCAA    GTCAA A A AATGA  CAAAAGC            AG CAGA

                           Splice
      dog   AGAGACACACAGAGAGAGGT---AGAGACACAGGCAGAGGGAGAAGCAGGCTCCTTG-CAGGGAGCCTGACGTGGGACTA
      mou   GCCTTCTGACAGAG------------------GTAAGTTGTTGAATCCACACTTTTA-CTGGGTGCCAAACCCCTGTGTA
      hum   ACTTTCTGGTGGAG------------GAAGCAGTAGGGTGTGGGGTCGGGATTTTTTTCTAAGTGCACACCCCTGCAGCA
      rat   GCCTTCAGACAGAGCCTTCTGACAGAGACAAGGTAAGGTGTTGAAACCGTGTTTTTA-CTGGGTGCCAAACGCTTCGGCA
      cow   ACGCTCTCATGGGG------------GTAACGGTAAGGTGTGTGGT---TGTTTTTT-CTGAGTGCCCACTGCTG-TGCA
Consensus   AC TTCTGACAGAG            GA AC GTAAGGTGT GAATC G  TTTTT  CTGGGTGCC AACGCTG  GCA
                           Splice
      dog   GATTCCGGATCCTGG-GATCACACCCTGAGCC---------GAAGGTGGACCCTCAACCG-CTGAGCCACCCAGGCGTCC
      mou   GAAGAA--AGGCAGG-G-TGAGGGACATAGGGTGTGTGTGGGGGGGGGGGATTGAAGAGTGTATCTTTGCATATCCAGA
      hum   GAGTAACCAGCCAGA-GCTGGGGGAAAAATTAGGATAGCTACCTGTTAGGCA-TGTAGGG-GTGTGTTTGCATGTTTAGT
      rat   GAAGAA--AGGCAGT-G-TGAGTGACACAGAA-----GAGGGAGGGATTGG--TGAAGAGTGTATATGTGCATATCCGGT
      cow   GAGTAACCAGCTGGGCGCTGGGGGGTCGGGGGC---AGCTGACTGTTAAGGA-TACAGGG-CTGGGTTT-CCTATCTAGT
Consensus   GA TAA  AGCCAGG G TGAGGGACA AG         G  GG  GGT GGGA TGAAG G GTGTGTTTGCATATC AGT
                                    Alternative Splice
      dog   CCAACATGAGGTTTTCAAAGTCCTGTTGGACCTTAGTTTTCCAATTGACTCTCATTTCTCTATTTTGCAATTAGCTCAGC
      mou   GTAGAATAATTTTTACAAGGTACTAATGGTCTCTAATATTCCAACCAGCTTTCACTTCTCTGTTTTGCAATTGGAGCAGC
      hum   ACGGCATAAATTCTTCAAAGACCTGATGGTCTTTAATATTCCAACCAACTCTCGTTTCCCCATTTTGTCATTAAATTAGC
      rat   GTAAAATAATTTTTACAAGGTACTAATGGTCTCTAATATTCCAGTCAACTTTCATTTCTCTATTTTGCAATTAGAGCAGC
      cow   GCCACACAATGTTTTCA--GTCCTGATGGACCTTAGAGTTCGAATTGACTCTCATTTCCCCATTTCGAGATTAGACTAGC
Consensus   GCAACATAATTTTTTCAA GTCCTGATGGTCTTTAATATTCCAATCAACTCTCATTTCTCTATTTTGCAATTAGA CAGC
                                    Alternative Splice
      dog   CTAAAGAGGAACTGGTGGCTTGGTGGCTTTGGAGAACTGTTGTAGGTGTTTGTTTCAGCAACTGCCTTCTTTCCCATCT
      mou   ATAAAGAGGAACTGATGGTTT--------TAGGAGAACTGTAG-AATTTCTGTTTCAACAACCGCTTTCTTTTCCATCC
      hum   TTAAAGAGGAACTTGTAGCTT--------TTAGAGAACTCATG-AGTTTTCCGCTTCATCATCTGCT-TCTGTTTTCTCC

      rat   ATAAAGAGGAACTGATGGGTT--------TAGGAGGACTGTAG-AAATTTCTGTTTCGGCAACTGCAATCTTTCCCATCC
      cow   CCAAAGAGGAACTGGTGGCTT--------TAGGAGAACTGCTA-AACTTTTTGTTTCAGCAACGGCTTTCTTTCCCGTCA
Consensus    TAAAGAGGAACTGGTGGCTT        TAGGAGAACTGTTG AA TTTCTGTTTCAGCAACTGCTTTCTTTCCCATCC
             Splice
      dog   GTTT-AGTTTGCACCCAGCTCTCTGGCTGCTGCTTAGGGCTGGCGTGC--AGCCAGGGCCGTCTGAGCC
      mou   TTCT--GTTTGCCCCAGTCTTGTTGGTGAATGCTTAGGGCTGGTATAT--GGCCAAAACCATCTGCGCC
      hum   ATCTTAGTTTGCCCAAAGCTTGCTGGCCGCTGTGTAGGGCTGGTGAGT--GGCTGGGGCTGTCTGAGCC
      rat   TTCT--GTTTGCCCTAGTCTTGCTGGTGAATGCCTGGGACTGGTATAT--GGCTGGAAACGCCTGCACC
      cow   GTTC-AGTCTGCCCCAAGCCCTCTGGCCACAGCTTAGGACCGGTGGGTCAGGCCGGGGCTGTCTGAGAC
Consensus    TCT AGTTTGCCCCAAGCTTGCTGGC ACTGCTTAGGGCTGGTGTGT  GGCCGGGGCCGTCTGAGCC
             Splice
```

Alignment of ~600bp 5'UTR of P2 upstream of ATG start translation signal.
Arrows indicate splice and alternative splice sites in mouse P2.

death of the transfected cells suggesting the Perforin-2 is cytotoxic for cells expressing it.

The gene structure of P2 is unusual. P2 is composed of two exons and one short intron containing also an alternative splice site (Fig. 14). The gene structure has been conserved from mouse to human. Perforin-2 mRNA obtained from the cytoplasmic space contains three species: Fully spliced, alternatively spliced and unspliced. This finding may be significant because the intron, especially the alternatively spliced region, is highly conserved among mammals.

Inclusion of the full length 5'UTR sequence of Perforin-2 in the gfp-tagged cDNA prevents expression of Perforin-2. The data suggest that the 5'UTR including the unspliced or alternatively spliced intron control Perforin-2 expression, most likely at the translational level.

5 Function of the three Pore Formers of the Immune System

The phylogenetic conservation of Perforin-2 down to sponges suggests a long ancestry of contemporary P2. Sponges use P2 as an antibacterial defense system (Wiens, M., Korzhev, M., Krasko, A., Thakur, N.L., Perovic-Ottstadt, S., Breter, H.J., Ushijima, H., Diehl-Seifert, B., Muller, I.M. and Muller, W.E. 2005). In macrophages and other APCs of vertebrates and mollusks (Mah, S.A., Moy, G.W., Swanson, W.J. and Vacquier, V.D. 2004) P2 is a transmembrane protein. We suggest that Perforin-2 is a membrane-tethered pore former that projects away from the membrane to which it is anchored, ready to polymerize and form pores on adjacent membranes. Expressed in phagocytes, Perforin-2 may attack phagocytosed bacteria or, when expressed on the plasma membrane, tumor cells or otherwise parasitized cells. The polymerization signal for Perforin-2 and its receptor is not known, however, it is possible that phosphorylation of the conserved tyrosine and serine are used for this purpose.

Complement component C9 may be a phylogenetyic descendent of Perforin-2. An antibacterial C9 like toxin circulating in blood or hemolymph may have been integrated into the opsonic complement system via several gene duplications generating C8, C7 and C6. The linkage of the C9-like pore former to complement expands the function of this anti-bacterial system beyond opsonization and chemoattraction of inflammatory cells to outright cytotoxic activity which appears to be useful and necessary for instance to eliminate the group of Neisseriae which cause severe disease in the absence of a functional MAC.

Perforin-1 has been recruited to NK cells and, with the development of the adaptive immune system, to CTL. Perforin-1 is used to kill host cells when they become parasitized by viruses, intracellular bacteria or by oncogenic transformation. Targeting of Perforin-1 is accomplished by the specificity of the killer cell and by the sophisticated vectorial secretion sytem of Perforin-1 and granzymes into the intercellular space of the CTL/target synapse. Perforin itself binds unspecifically to phospholipid head groups.

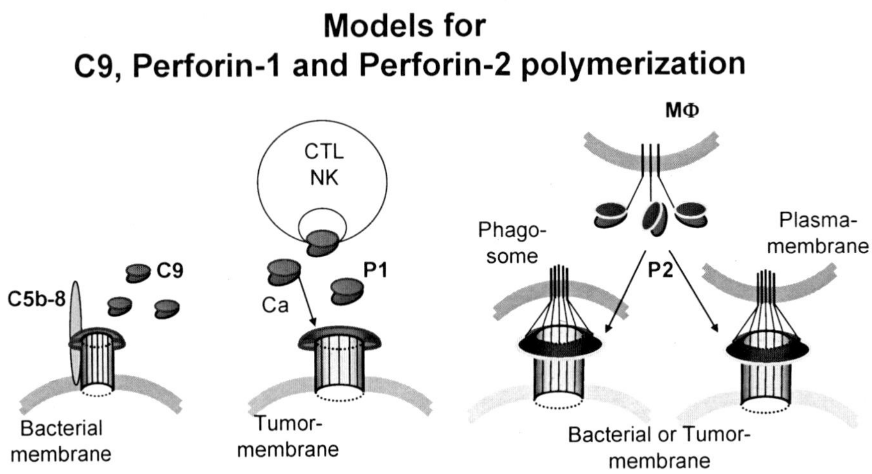

The three pore forming proteins of the immune system in this way have specialized functions and targeting systems to cope with the diversity and evolution of invading microbial and parasitic organisms.

6 Acknowledgements

This work was supported by grants from the NIH CA039201 and AI061807

References

Bhakdi, S. and Tranum-Jensen, J. (1978). Molecular nature of the complement lesion. Proc Natl Acad Sci U S A. 75(11), 5655-5659.

Borsos, T., Dourmashkin, R.R., et al. (1964). Lesions In Erythrocyte Membranes Caused By Immune Haemolysis. Nature. 202, 251-252.

Dennert, G. and Podack, E.R. (1983). Cytolysis by H-2-specific T killer cells. Assembly of tubular complexes on target membranes. J Exp Med. 157(5), 1483-1495.

Dourmashkin, R.R., Deteix, P., et al. (1980). Electron microscopic demonstration of lesions in target cell membranes associated with antibody-dependent cellular cytotoxicity.

Clin Exp Immunol. 42(3), 554-560.

Heibein, J.A., Goping, I.S., et al. (2000). Granzyme B-mediated cytochrome c release is regulated by the Bcl-2 family members bid and Bax. J Exp Med. 192(10), 1391-1402.

Kolb, W.P., Haxby, J.A., et al. (1972). Molecular analysis of the membrane attack mechanism of complement. J Exp Med. 135(3), 549-566.

Kopacek, J., Sakaguchi, S., et al. (2000). Upregulation of the genes encoding lysosomal hydrolases, a perforin-like protein, and peroxidases in the brains of mice affected with an experimental prion disease. J Virol. 74(1), 411-417.

Lieberman, J. and Fan, Z. (2003). Nuclear war: the granzyme A-bomb. Curr Opin Immunol. 15(5), 553-559.

Mah, S.A., Moy, G.W., et al. (2004). A perforin-like protein from a marine mollusk. Biochem Biophys Res Commun. 316(2), 468-475.

Masson, D. and Tschopp, J. (1987). A family of serine esterases in lytic granules of cytolytic T lymphocytes. Cell. 49(5), 679-685.

Mayer, M.M. (1972). Mechanism of cytolysis by complement. Proc Natl Acad Sci U S A. 69(10), 2954-2958.
Millard, P.J., Henkart, M.P., et al. (1984). Purification and properties of cytoplasmic granules from cytotoxic rat LGL tumors. J Immunol. 132(6), 3197-3204.
Pao, L.I., Sumaria, N., et al. (2005). Functional analysis of granzyme M and its role in immunity to infection. J Immunol. 175(5), 3235-3243.
Podack, E.R. and Dennert, G. (1983). Assembly of two types of tubules with putative cytolytic function by cloned natural killer cells. Nature. 302(5907), 442-445.
Podack, E.R. and Konigsberg, P.J. (1984). Cytolytic T cell granules. Isolation, structural, biochemical, and functional characterization. J Exp Med. 160(3), 695-710.
Podack, E.R. and Tschopp, J. (1982). Polymerization of the ninth component of complement (C9): formation of poly(C9) with a tubular ultrastructure resembling the membrane attack complex of complement. Proc Natl Acad Sci U S A. 79(2), 574-578.
Smyth, M.J., Thia, K.Y., et al. (2000). Perforin-mediated cytotoxicity is critical for surveillance of spontaneous lymphoma. J Exp Med. 192(5), 755-760.
Spielman, J., Lee, R.K., et al. (1998). Perforin/Fas-ligand double deficiency is associated with macrophage expansion and severe pancreatitis. J Immunol. 161(12), 7063-7070.
Spilsbury, K., O'Mara, M.A., et al. (1995). Isolation of a novel macrophage-specific gene by differential cDNA analysis. Blood. 85(6), 1620-1629.
Stepp, S.E., Dufourcq-Lagelouse, R., et al. (1999). Perforin gene defects in familial hemophagocytic lymphohistiocytosis. Science. 286(5446), 1957-1959.
Street, S.E., Cretney, E., et al. (2001). Perforin and interferon-gamma activities independently control tumor initiation, growth, and metastasis. Blood. 97(1), 192-197.
Wiens, M., Korzhev, M., et al. (2005). Innate immune defense of the sponge Suberites domuncula against bacteria involves a MyD88-dependent signaling pathway. Induction of a perforin-like molecule. J Biol Chem. 280(30), 27949-27959.

24

Pathogen-Specific Innate Immune Response

Ahmet Zeytun, Jennifer C. van Velkinburgh, Paige E. Pardington, Robert R. Cary, and Goutam Gupta

Los Alamos National Laboratory, Biosciences Division, Mail Stop M888, HRL-1, TA-43, Los Alamos, NM 87545. **E-mail:** gxg@lanl.gov

Abstract. This chapter summarizes our studies on the three toll-like receptor pathways, namely TLR4, TLR2, and TLR3, induced by lipopolysaccharides (LPS), peptidoglycan (PGN), and double-stranded RNA (dsRNA) in antigen presenting cells (APC). The particular emphasis is on the activation of human innate immune responses via cytokine and chemokine production. Three different measurements have been performed on monocytic and dendritic cells as model APCs: (i) the expression of various cytokine and chemokine genes by real-time PCR, (ii) the release of the cytokines and chemokines by ELISA, and (iii) gene expression analysis by cytokine and chemokine pathway-specific and whole genome microarrays. Real-time PCR and ELISA enable us to identify cytokines and chemokines that are produced specifically upon LPS, PGN, or dsRNA stimulation. Subsequently, microarray studies and appropriate validation experiments help us to identify genes involved in the upstream pathways that cause the induction of cytokines and chemokines. It is evident that TLR4-LPS, TLR2-PGN, and TLR3-dsRNA pathways are distinguished by the specific set of cytokines and chemokines they induce as well as by the upstream signaling events.

1 Introduction

Toll-like receptor (TLR) molecules on antigen presenting cells (APCs) play a critical role in host immune defense through specific recognition of pathogen-associated molecular patterns (PAMPs) (Takeda, Kaisho, and Akira 2003; Beutler, Hoebe, Du, and Ulevitch 2003). Different TLRs recognize different PAMPs. TLR4 recognizes lipopolysaccharides (LPS) derived from gram-negative bacteria whereas TLR2 recognizes peptidoglycan (PGN) from gram-positive bacteria. TLR3, on the other hand, recognizes double-stranded (ds) RNA from viruses. TLR4-LPS, TLR2-PGN, and TLR3-dsRNA recognitions initiate a cascade of intracellular signaling events

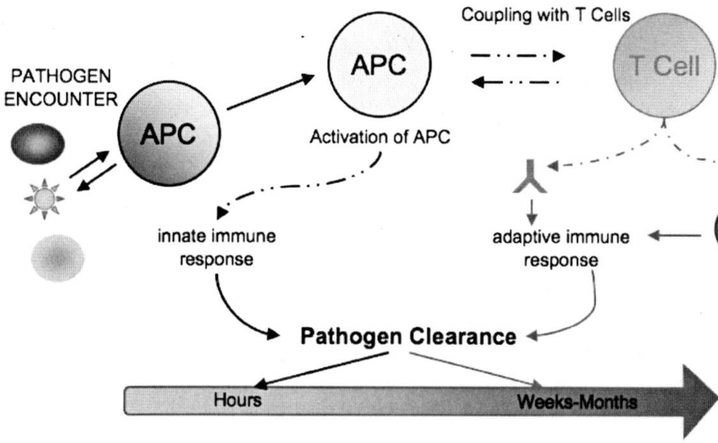

Fig. 1. Role of antigen presenting cells (APC) in anti-pathogen defense. Initial contact of APCs with the pathogen is the first step in host immune response. Upon pathogen encounter the APCs are activated and initiate the immediate and non-specific innate immunity which subsequently couples to pathogen- and antigen-specific adaptive immunity, probably through the expression of co-stimulatory molecules (Hoebe, Janssen, Kim, Alexopoulou, Flavell, Han, and Beutler 2003). The innate immune response takes only hours whereas the adaptive immunity develops after weeks and months. Activated APCs engulf and process foreign antigens from pathogens, then "present" them on the surface MHC Class II molecules, which, in turn, bind to the T cell receptors. The APC-T cell synapse leads to the production of antibodies by B cells and the generation of the cytotoxic T cells (CTC). Antibodies and CTC clear the invading pathogen.

resulting in the expressions of a vast array of immune response genes and proteins by APC (Hessle, Andersson, and Wold 2005; Ishii, Coban, and Akira 2005; Matsumoto, Funami, Oshiumi, and Seya 2004), including cytokines and chemokines that form an important component of the immune response. The induction of cytokine and chemokine genes and the subsequent release of the mature proteins can trigger innate immune response leading to pathogen clearance within hours. In addition, the release of cytokines and chemokines facilitates recruitment of T cells and, thereby, coupling of adaptive immunity and clearance of pathogens by antibodies or cytotoxic T cells (CTC) -- see Figure 1.

A thorough understanding of TLR-mediated induction of cytokine and chemokine pathways requires measurements of the following datasets: (i) mRNA levels of cytokine and chemokines specifically induced in APCs by PAMPs, (ii) level of mature cytokine and chemokine proteins released by APCs, and (iii) up-stream intracellular signaling events specifically triggered by PAMPs. The effectiveness of their pathways is determined by the level of released cytokines and chemokines and their binding to the cognate receptors on APCs or T cells. Since their release is primarily induced by transcription, the levels of released cytokine and chemokine proteins must be correlated with the corresponding mRNA levels. Finally, the level of mRNA must be mapped onto a signaling pathway(s) consisting of sequential events that are triggered upon TLR-PAMP ligation. A combined analysis of the three datasets should reveal the features of TLR pathways that are induced by more than one PAMP as well as the ones that are unique to a given PAMP.

It is fairly straightforward to determine the levels of cytokine and chemokine mRNAs and released proteins that are specifically induced by various PAMPs. However, comprehensive analyses of the upstream signaling events are yet to be achieved. This chapter focuses primarily on the signaling events related to TLR4, TLR2, and TLR3 pathways that map onto the following chronologically linked signaling components: formation of adapter complexes on the cytosolic side of the TLR, activation of signalosome complex, activation and nuclear transport of transcription factors. TLR2-4 signaling operates via the MyD88-dependent or MyD88-independent pathway (Hessle, *et al.* 2005; Ishii, *et al.* 2005; Matsumoto, *et al.* 2004). In the MyD88-dependent pathway, the loading of MyD88 on the cytosolic side of the TLR initiates the assembly of IRAK-1, IRAK-4, and TIRAP on the adapter complex, which leads to the phosphorylation of IRAKs. Upon their release, the phosphorylated IRAKs interact with TRAF-6 leading to the activation of TAK-1, which, in turn, phosphorylates the signalosome complexes including IKKα, β, γ and MAPK. Finally, transcription factors (such as, NFκB and AP-1) are activated and translocated to the nucleus. Many pro- and anti-inflammatory cytokines are induced via this pathway (Nagorsen, Marincola, and Panelli 2004; Kanehisa, Goto, Hattori, Aoki-Kinoshita, Itoh, Kawashima, Katayama, Araki, and Hirakawa 2006) --- see Figure 2. The MyD88-independent pathway utilizes an adapter complex involving TRIF and TRIP, which, in turn, activates the signalosome complex of IKKε/ι and TBK-1. Finally, the transcription factors IRF3/7 are activated and translocated to the nucleus. IFN-α and IFN–β are induced via MyD88-independent pathway.

2 Major Focus and Hypothesis

The main emphasis of this chapter is to summarize the signaling events that result in the induction of cytokine and chemokine genes in APCs upon exposure of LPS, PGN, and dsRNA. It is our hypothesis that a thorough understanding of the response by an APC requires both genome-scale and focused studies. The genome-scale studies are performed to identify novel and known genes and their associated pathways in the TLR-mediated induction of cytokine and chemokine genes during the early phase of innate immune response by APCs. Focused studies are done on a small subset of newly discovered and already known genes and their pathways to establish the mechanism of the induction of cytokine and chemokine genes during the early response.

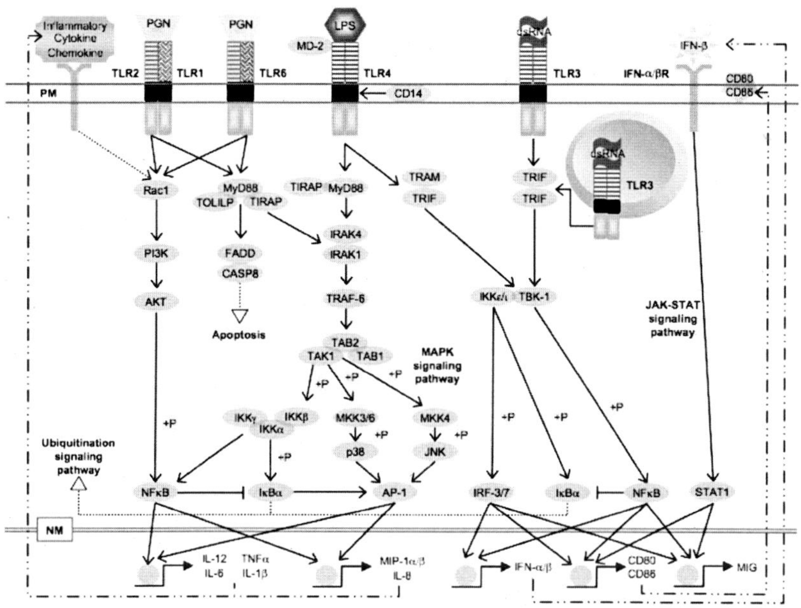

Fig. 2. TLR signaling pathways. The innate immune response is initiated by the recognition and physical interaction of APCs with PAMPs. Activation of the TLRs by different PAMPs triggers signaling pathways involving both common and distinct adapter molecules, protein kinases and transcription factors. Ultimately, activation of this signaling pathway causes expression of specific cytokine and chemokine genes. The release of mature cytokine and chemokine proteins can induce both autocrine and paracrine networks resulting in innate and adaptive immune responses. Note that, the expression of co-stimulatory molecules such as CD80 and CD86 may initiate the adaptive immune response (Hoebe, *et al.* 2003). TLR-specific pathways utilize adapter molecules in the cytosol to regulate gene expression necessary for immunity. A detailed knowledge of all factors, either specific to each TLR or shared among the three, will help identify specific markers of infection and/or therapeutic targets. TLR4 and TLR3 form homo-dimers whereas TLR2 may form TLR2/TLR1 or TLR2/TLR6 hetero-dimers. TLR dimers are color-coded. Association of different genes or proteins and their sequential connectivities are shown. TLR-activation triggers signaling through at least two distinct mechanisms: the MyD88-dependent or MyD88-independent pathways. Signaling through the MyD88-dependent pathway regulates the expression and release of inflammatory cytokines, whereas the MyD88-independent pathway leads to the expression and release of IFN-α/β, a key factor mediating cytotoxicity and apoptosis. Ligand-bound TLR2 and TLR4 signal through the MyD88-dependent pathway involving the activation of the signaling complex IKKα,β,γ either by TRAF or MAPK complexes leading to activation and nuclear transport of the transcription factors NFκB and AP-1. However, TLR4 can also operate via the MyD88-independent pathway. In the absence of MyD88, TLR4 is able to utilize the adapter complex TRIF/TRAM and induce activation of the signalosome IKKε/IKKι-TBK-1 complex and the transcription factor NFκB. The TLR3 pathway operates via the formation of the TRIF/TRIF adapter complex, activation of the signalosome IKKε/IKKι-TBK-1 complex and the transcription factors IRF3/7. Myd88-independent TLR3 and TLR4 pathways converge at the IKKε/IKKι-TBK-1 signalosome node, although these two receptors recognize two distinctly different PAMPs. Similarly, MyD88-dependent TLR2 and TLR4 pathways converge at the IRAK4/IRAK-1 signalosome node. Additional TLR-related pathways of interest include the alternative TLR2 network involving PI3K, AKT, and NFκB and the autocrine IFN-β network involving JAK-STAT.

3 Approach

First, the levels of cytokine and chemokine transcripts (or mRNA) are obtained by quantitative real-time PCR. *Second*, the amounts of released cytokines and chemokines are measured by quantitative enzyme-linked immunoabsorbent assay (ELISA). *Finally*, upstream signaling events that link the induction of specific cytokines and chemokines are identified using different methods (see Figure 3). One method involves the use of pathway-specific microarray, which contains probes for specific cytokine and chemokine genes and the genes that map onto their signaling pathways including those corresponding to adapter and signalosome complexes and transcription factors (see Figure 2). The pathway-specific microarray designed at LANL has three 70-mer oligonucleotide probes per gene, respectively for targeting 5′, middle, and 3′ regions of a gene. A 70-mer designed specifically to hybridize with the unique sequence of a gene has been proven to be more efficient in detecting the transcript. Each 70-mer probe is spotted in triplicate on the microarray, thereby allowing nine different measurements for recording the expression of a given gene. Thus, the LANL pathway-specific microarray offers sensitivity and dynamic range in the detection of transcripts, i.e., ability to (i) detect transcripts from 5-10 μg of total RNA and (ii) capture up to 30-fold change in up- or down-regulation of transcripts. Currently, the pathway-specific microarray contains probes for 136 experimental and 136 control genes. The experimental genes are expected to map on the pathways involving the induction of cytokines and chemokines due to TLR-PAMP interactions in APC. The control genes are chosen from the list of genes that are specifically induced in T cell and those that are unrelated to cytokine and chemokine pathways in APC. The other method involves the use of whole genome Agilent microarray with 60-mer oligonucleotide probes for 41,000 unique genes for the human genome. 70% of the probes are validated. The use of the Agilent microarray in conjunction with *SuperAmp* RNA amplification ensures minimal loss of signals and minimal introduction of bias in the gene expression analysis. The use of the Agilent microarray and *SuperAmp* allows reproducible detection of changes in the transcript levels down to 100 to 1,000 cells. While the pathway-specific microarray confirms the putative roles of genes in their associated pathways, the whole genome microarray lends to discovering the role of new genes and pathways, in addition to confirming the role of known genes and their pathways. The results from pathway-specific and whole genome microarrays are confirmed by real-time PCR and protein assays.

4 Results

Apart from their ligand specificity, TLR4, TLR2, and TLR3 possess several features that distinguish them from each other. For example, TLR4 and TLR3 show ligand

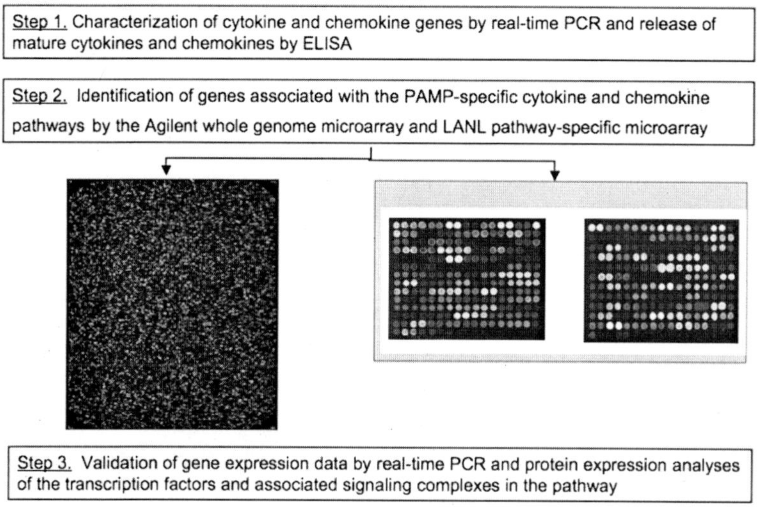

Fig. 3. Characterization of the cytokine and chemokine pathways induced by PAMPs: *two parallel approaches.* We have combined diverse RNA and protein expression assays to track the TLR2, TLR3, and TLR4 pathways. First, the TLR2, 4, and 3 pathways are stimulated by PGN, LPS, or dsRNA for 0 to 24 hours. The pattern of PAMP-specific induction of cytokines and chemokines are monitored at mRNA level by real-time PCR and at (secreted) protein level by ELISA. Second, the activation and nuclear transport of NFκB, AP-1, and IRF3/7 are examined by western blotting using antibodies specific for these transcription factors. Finally, the PAMP-specific induction of cytokine and chemokine genes is mapped to their upstream signaling pathways by identifying the appropriate genes by pathway-specific and whole genome microarray analyses.

induced dimerization (Martin and Wesche 2002; Yamamoto, Takeda, and Akira 2004) whereas TLR2 forms hetero-dimers with TLR6 or TLR1 (Martin, *et al.* 2002; Yamamoto, *et al.* 2004; Kopp and Medzhitov 2003). Additionally, studies from our laboratory (Zeytun, van Velkinburgh, and Gupta) and others (Matsumoto, Funami, Tanabe, Oshiumi, Shingai, Seto, Yamamoto, and Seya 2003) demonstrate that TLR3 is predominantly localized on the endosome whereas both TLR4 and TLR2 operate from the plasma membrane. Finally, surface expression of TLR4 shows cyclical up and down variation with time upon LPS and PGN stimulation whereas surface expression of TLR2 is unaffected under the same condition.

4.1 Induction of PAMP-Specific Cytokines and Chemokines

Distinctly different properties of TLR4, TLR2, and TLR3 suggest differential response upon stimulation by their respective PAMPs. Table 1 summarizes the results of induction of cytokine and chemokine genes by quantitative real-time PCR and their release by ELISA. TLR4 and TLR2 mediated production of cytokines and chemokines were monitored by stimulating monocytic THP-1 cells, respectively, by *E. coli* LPS and *Staphylococcal* PGN. TLR3 mediated production of cytokines and chemokines was monitored by stimulating THP-1 and primary human dendritic cells (DC) by dsRNA derived from human influenza virus. The objective was to examine whether the host innate immune responses by the three PAMPs can be distinguished by the pattern of cytokine and chemokine production. PGN is most potent in stimulating the induction of chemokines IL-8, MIP-1α, and MIP-1β in THP-1 at different time-points during 24 hours of post-exposure. Similarly, cytokine genes of TNF-α, IL-6, IL-10, and IL-12 are induced at a higher level by PGN stimulation than by LPS. Most of the cytokines and chemokines are released at a higher level from THP-1 by PGN stimulation than by LPS. The cytokine IL-6 is the only exception in that it is released at a higher level by LPS stimulation. Type I interferons, IFN-α/β, and chemokines MIP-1α/β are specifically induced in THP-1 or DC by dsRNA either at mRNA or at protein level.

4.2 Induction of PAMP-Specific Gene Networks

Subsequent to identifying the set of cytokines and chemokines, gene expression pattern of THP-1 cells stimulated for 0 to 24 hours with LPS and PGN was analyzed by the pathway-specific and sensitive microarray designed at LANL. Figure 4 shows (i) the genes that are significantly up- or down-regulated following 4- and 8-hour-LPS stimulations and (ii) their association with MyD88-dependent and MyD88-independent pathways. Only 4- and 8-hour time-points are chosen to capture the early response. The lightly shaded genes in the pathways correspond to those that show no detectable transcript or no change in the transcript level. Table 1 shows the up-regulation of the transcription factors, NFκB and IRF-7. Although no change was detected at the transcript level of AP-1, protein assays by western blotting show increase in the nuclear fractions of both NFκB and AP-1 at the 4- and 8-hour timepoint after LPS stimulation. The enhanced nuclear localization of NFκB and AP-1 is consistent with the induction of inflammatory cytokines and chemokines by LPS. Since LPS stimulation after 4-hours results in the up-regulation of IFN-α/β, we have tested by western blotting the nuclear fraction of phosphorylated (p)-STAT1, which is the expected outcome of the IFN-α/β autocrine network (see Figure 4). Although, there is little change in the level of STAT1 transcript, the nuclear fraction of (p)-STAT1 is enhanced 4-hours following LPS stimulation. We have also monitored the level of the signalosome complex, IKKα/IKKβ/IKKγ, by western blotting. We do not observe significant changes in the level of IKKα, IKKβ, or IKKγ. Although, we have not experimentally determined it, it has been shown that their kinase activity is enhanced upon LPS stimulation rather than their protein expression levels increasing

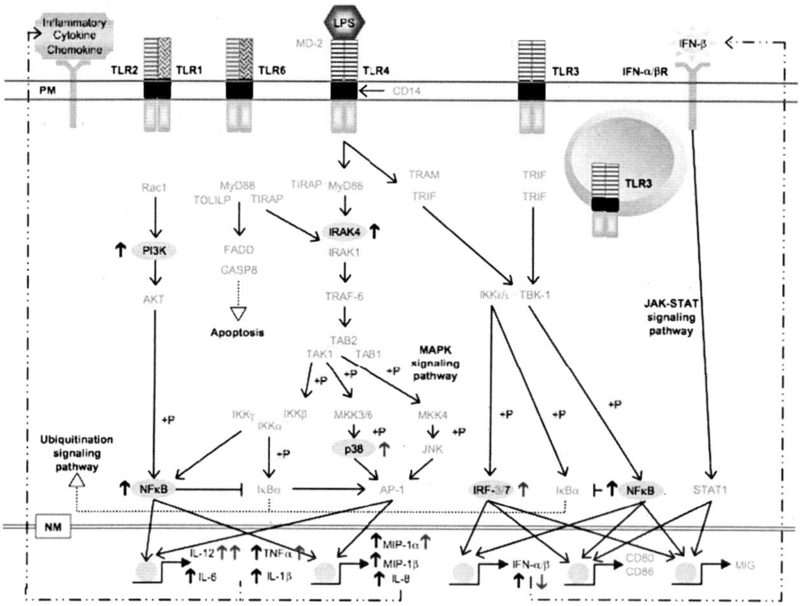

Fig. 4. The immune response due to the binding of TLR4 to its cognate ligand LPS. The endosome is represented as a pink sphere. Proteins are illustrated as solid teal spheres and labeled accordingly. Solid arrows indicate direct nodes in the pathway; pointillated arrows indicate alternative pathways; dashed arrows indicate coupling of the TLR4 pathway with cytokine/chemokine mediated signaling. Yellow circled "+P" indicates phosphorylation activity. Finally, resultant gene expression is illustrated beneath the nuclear membrane as unlabeled teal spheres (transcription factors) sitting atop a promoter region to regulate expression (black solid arrow) of particular genes. Proteins typically involved in the TLR pathways, but not affected by LPS treatment in our studies, are shown as transparent spheres. Bold solid black arrows indicate gene expression and protein changes (↑ induced; ↓ suppressed) at 4 hours post-stimulation with LPS, while red arrows indicate those changes observed at eight hours. Proteins that were found to be up-regulated between 2 and 10-fold are represented by a single arrow, whereas those regulated by >10-fold are represented by a double arrow. Down-regulation is represented, similarly, with arrows corresponding to the levels of negative regulation. Direct influences upon gene expression by LPS-bound TLR4 are indicated by the same color-coding and directionality of protein-associated arrows; however, the intensity of regulation is indicated by comparison with the levels observed as a result of PGN-bound TLR2. For example, IL-12 gene expression is substantially up-regulated by the TLR4 signaling pathway, when compared to the TLR2 signaling pathway.

(Baumgartner, Weber, Quirling, Fischer, Page, Adam, Von Schilling, Waterhouse, Schmid, Neumeier, and Brand 2002). We have also monitored the level of IRAK-1 associated with the MyD88-associated adapter complex in Figure 4. Western blotting shows a decrease in the level of IRAK-1 at 0.25, 0.5, 1, 4, and 8 hours following LPS stimulation. This is most likely a feedback response operating through IRAK-1 degradation against the release of inflammatory cytokines and chemokines (Carroll, Greene, Taggart, Bowie, O'Neill, and McElvaney 2005). The up-regulation of PI3

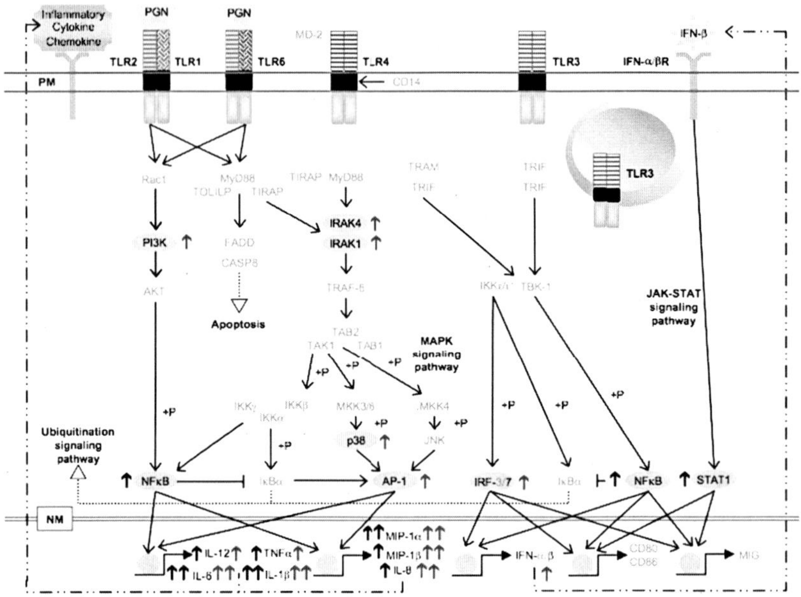

Fig. 5. The immune response due to the binding of TLR2 to its cognate ligand PGN. Proteins typically involved in the TLR pathways, but not affected by PGN treatment in our studies, are transparent spheres. Bold solid black arrows indicate gene expression and protein changes (↑ induced; ↓ suppressed) at four hours post-stimulation with LPS, while red arrows indicate those changes observed at eight hours. Proteins that were found to be up regulated between 2 and 10-fold are represented by a single arrow, whereas those regulated >10-fold are represented by a double arrow. Down-regulation is represented, similarly, with arrows corresponding to the levels of negative regulation. Direct influences upon gene expression by PGN-bound TLR2 are indicated by the same color-coding and directionality of protein-associated arrows; however, the intensity of regulation is indicated by comparison with the levels observed as a result of LPS-bound TLR4. For example, IL-6 gene expression is substantially up-regulated by the TLR2 signaling pathway, when compared to the TLR4 signaling pathway.

kinase (PI3K) in Figure 4 is also consistent with such a feedback response (Bian, Elner, Yoshida, and Elner 2004; Weiss-Haljiti, Pasquali, Ji, Gillieron, Chabert, Curchod, Hirsch, Ridley, van Huijsduijnen, Camps, and Rommel 2004). Although, all the nodes in the response network of Figure 4 have not been validated, the combination of pathway-specific microarray, real-time PCR, and ELISA provides a qualitative LPS-induced response network in THP-1 involving the primary MyD88-dependent pathway and secondary pathways induced by inflammatory cytokines and chemokines and type I interferons.

Figure 5 shows (i) the genes that are significantly up- or down-regulated following 4- and 8-hour-PGN stimulation in THP-1 and (ii) their association with MyD88-dependent and other pathways. Note that the pattern of cytokine and chemokine induction by PGN is distinct from that by LPS in the same THP-1. PGN

Table 1. Changes in mRNA (by real-time PCR) and/or protein (by ELISA) levels of the indicated cytokines and chemokines. Single asterisk indicates observations in THP-1 cells only; double asterisk indicates observations in DC alone. Shaded areas represent no measurement.

Cytocines	LPS		PGN		dsRNA	
	mRNA	Protein	mRNA	Protein	mRNA	Protein
TNF-α	Up (1-4 hr)	Up (↑)	Up	Up (↑↑)	No change	Not detected
IL-6	No change	Up	Up	No change	*Up (1-4 hr)	Not detected
IL-10	No change	No change	Up	Up	No change	Not detected
IL-12p40	Up	Up	Up (↑)	Up (↑↑)		
IL-1β	Up (1-4 hr)	Up	Up (↑)	Up (↑↑)	No change	Not detected
IFN-α					*Up (1-4 hr)	**Up (1-4 hr)
IFN-β					No change	Up
Chemokines	LPS		PGN		dsRNA	
	mRNA	Protein	mRNA	Protein	mRNA	Protein
IL-8	Up (↑) (1-8 hr)	No change	Up (↑↑)	No change	No change	*Up
MIP-1α	Up (1-4 hr)	Up	Up	Up	No change	Up
Mip-1β	Up (1-4 hr)	Up	Up	Up	Up	Up

stimulation shows significantly higher levels of transcripts as well as released mature proteins than LPS stimulation of the same THP-1. Accordingly, the transcripts of NFκB, AP-1, IRF-7, and STAT1 are expressed at higher levels in the THP-1 cells stimulated by PGN. The levels of NFκB, AP-1, IRF-7, and (p)-STAT1 in the nuclear fraction are increased significantly more by PGN than by LPS. Finally, a more pronounced decrease in the cytosolic protein level of IRAK-1 in the PGN stimulated cells is consistent with the fact that these cells release higher levels of cytokines and chemokines (see Table 1). Note that the IRAK-1 transcript is up-regulated (see Figure 5) but the inflammatory response of the cytokines and chemokines overrides the IRAK-1 up-regulation at the transcript level. Gene expression analyses by pathway-specific microarray and appropriate validation experiments help us confirm the involvement of selected genes in their pathways. However, since our goal is also to discover new genes and pathways, we have resorted to the whole human genome microarray to study the response of the human primary DC upon dsRNA stimulation.

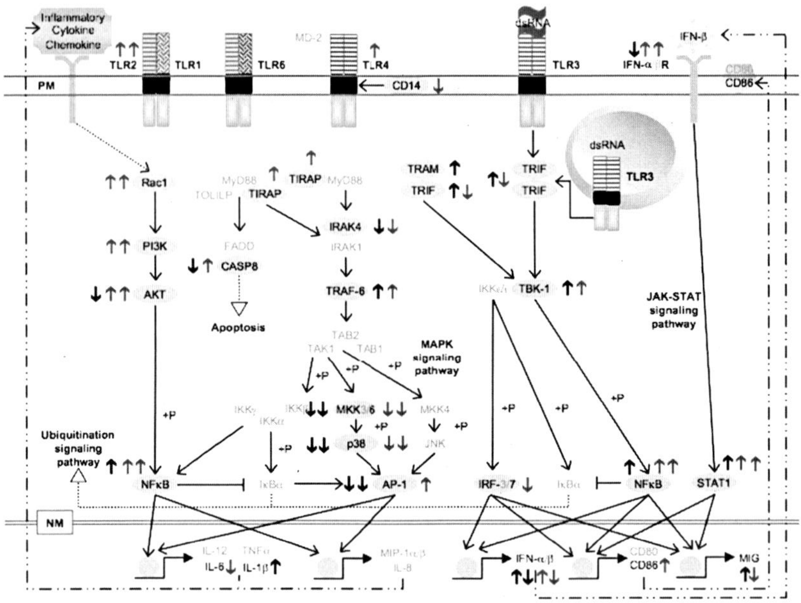

Fig. 6. The signaling pathway induced by the binding of TLR3 to its cognate ligand dsRNA. The endosome is represented as a pink sphere, an additional location of TLR3. Proteins typically involved in the TLR pathways, but not affected by dsRNA treatment in our studies, are transparent spheres. Bold solid black arrows indicate gene expression and protein changes (↑ induced; ↓ suppressed) at four hours post-stimulation with dsRNA, while red arrows indicate those changes observed at eight hours. Proteins and genes that were found to be up regulated between 2 and 10-fold are represented by a single arrow, whereas those regulated >10-fold are represented by a double arrow. Down regulation is represented similarly with arrows corresponding to the levels of negative regulation.

The DC is one of the most efficient APCs. Figure 6 shows the genes that are significantly up- or down-regulated following 4- and 8-hour-dsRNA stimulation in DC and their association with various TLR pathways. Although the host cells are different, dsRNA stimulation produces a significantly different response in the cytokine and chemokine profile. A smaller set of cytokines and chemokines are induced at the transcript and protein levels. MIP-1α/β and IFN-α/β are the most altered at gene or protein levels; note that the IFN-α/β transcripts show up and down fluctuations. The transcription factors NFκB and STAT1 transcripts are significantly up-regulated. IRF-7 transcript is slightly down-regulated. Protein assays by western blotting show increase in the nuclear fraction of NFκB and IRF-7, which are critical to the induction of MIP-1α/β and IFN-α/β (Julkunen, Sareneva, Pirhonen, Ronni, Melen, and Matikainen 2001). We have also analyzed the level of IKKε, a member of the signalosome IKKε, IKKι complex in the TLR3 pathway (Hacker and Karin 2006). An increase at the cytosolic level of IKKε protein is observed at the 4-hour timepoint. In contrast to IKKα and IKKβ, an increase in the cytosolic level of IKKε

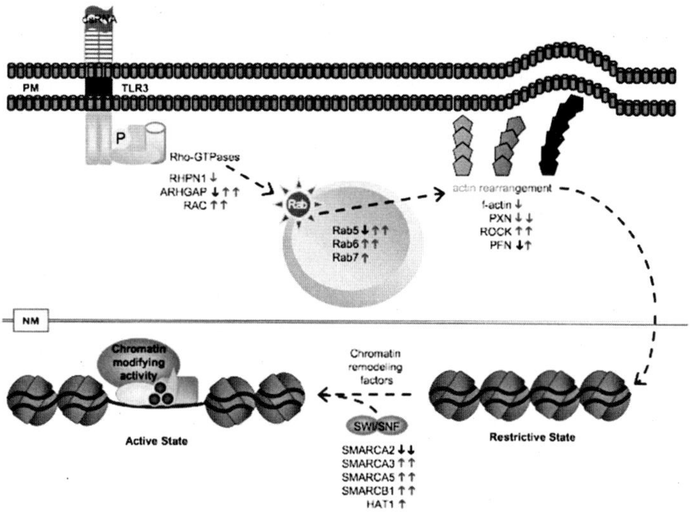

Fig. 7. TLR3 pathway may trigger novel intracellular effects. Upon dsRNA binding to TLR3, changes are observed in the transcript levels of Rho-GTPases which interact with the cyotplasmic portion of membrane receptors. In addition, known downstream effectors promoting actin rearrangement and chromatin remodeling are altered. Dashed arrows represent directionality of the signaling pathways involved. The endosome is represented as a pink sphere. Endosome-associated Rabs with levels altered upon TLR3 activation are listed. Bold solid black arrows indicate gene expression and protein changes (🡅 induced; 🡇 suppressed) at four hours post-stimulation with dsRNA, while red arrows indicate those changes observed at eight hours. Proteins and genes that are up-regulated between 2 and 10-fold are represented by a single arrow, whereas those up-regulated by >10-fold are represented by a double arrow. Down-regulation is represented, similarly, with arrows corresponding to the levels of negative regulation. Components of the actin cytoskeleton are illustrated as grey pentagons, with deepened shades corresponding to changes in actin filament conformation. A selection of cytoskeletal remodeling factors altered by TLR3 activation are listed. Chromatin is represented by nucleosomes (blue spheres), around which DNA (black solid lines) is wrapped to restrict access by transcriptional machinery. A selection of chromatin remodeling factors, required for transformation between the restrictive and active states, is listed with arrows indicating the directionality and intensity of dsRNA-TLR3 induced alterations.

correlates with the activation of NFκB and IRF-7 in the TLR3 pathway (Fitzgerald, McWhirter, Faia, Rowe, Latz, Golenbock, Coyle, Liao, and Maniatis 2003). At this time, we have no protein assay data to validate the role of TRIF or IRAK adapter complexes. The transcripts of Rac1, PI3K, and AKT, the three key members of the PI3K pathway, are all up-regulated by dsRNA stimulation. This pattern is not

Table 2. Acronyms

AKT	protein kinase B
AP-1	activating protein-1
APC	antigen presenting cell
ARHGAP	Rho GTPase activating protein
CASP	caspase, apoptosis-related cysteine protease
CD	cluster of differentiation
CTC	cytotoxic T cell
dsRNA	double-stranded ribonucleic acid
FADD	Fas-associated via death domain
HAT	histone acetyltransferase
IFN	interferon
IkB	inhibitor of kappa B
IKK	IkappaB kinase
IRAK	interleukin-1 receptor-associated kinase
IRF	interferon regulatory factor
JNK	c-Jun amino-terminal kinase
LPS	lipopolysaccharide
Mal	MyD88 adaptor-like protein
MAPK	mitogen activated protein kinase
MD	*myb*-regulated protein
MIG	monokine induced by interferon gamma
MKK	MAP kinase kinases
MyD88	myeloid differentiation primary response gene
NFkB	nuclear factor kappa B
NM	nuclear membrane
PAMP	pathogen-associated molecular pattern
PFN	profilin
PGN	peptidoglycan
PI3K	phosphoinositol-3-kinase
PM	plasma membrane
PXN	paxillin
Rab	ras-like proteins in brain
RAC	ribosome associated complex
RHPN	rhophilin, Rho GTPase binding protein
RIP	receptor interacting protein
ROCK	Rho-associated, coiled-coil containing protein kinase
SMARC	SWI/SNF related, matrix associated, actin dependent
	regulator of chromatin
STAT	signal transducer and activator of transcription
SWI/SNF	switching/sucrose non fermentation
TAB	TAK1 binding protein
TAK	transforming growth factor-ß (TGFß)-activated kinase
TANK	TRAF family member-associated NFκB activator
TBK	TANK-binding kinase 1
TICAM	toll-like receptor adaptor molecule
TIRAP	TIR domain-containing adapter protein
TLR	Toll-like receptor
TNF	tumor necrosis factor
TOLILP	Toll-interacting protein
TRAF	TNF receptor-associated factor
TRAM	Trif-related adaptor molecule
TRIF	Toll/IL-1R domain-containing adapter inducing IFN-beta

observed in THP-1 upon LPS or PGN stimulation (see Figures 4 and 5). It is possible that the excessive release of MIP-1α/β causes this effect, which needs to be proven by examining the effect of anti-MIP-1α/β upon dsRNA stimulation of DC. The gene expression analyses by whole human genome microarray also allow us to identify responding genes belonging to apoptosis, MAPK, ubiquitination, and several other relevant pathways that are expected to be altered by dsRNA. The most novel and potentially relevant pathway that we have discovered is the one that connects Ras-GTPase⇒endosomal Rab-GTPase⇒cytoskeleton⇒chromatin remodeling (Sieczkarski and Whittaker 2003; Johnson, Adkins, and Georgel 2005). Figure 7 shows the transcript levels we observed for various genes belonging within this novel pathway. This pathway is particularly relevant since the exposure to dsRNA results in intracellular signaling via TLR3-dsRNA complex translocated to the endosome of APCs. Additionally, this pathway may offer a mechanism of induction of a battery of genes upon viral infection (de Bouteiller, Merck, Hasan, Hubac, Benguigui, Trinchieri, Bates, and Caux 2005).

5 Conclusion

A diverse set of measurements allows identification of PAMP-specific induction of genes and proteins and their corresponding pathways. However, an appropriate number of experiments are needed to validate the responsive genes and proteins and their associations with relevant pathways. This chapter outlines an approach that permits delineation of early innate immune response mediated by TLR pathways. The early stage is defined by the coupling of primary and secondary responses during the first 8 hours of stimulation of APCs. The primary response is the induction of cytokine and chemokine genes and release of proteins upon TLR-PAMP ligation whereas the secondary response is the autocrine network established by the released cytokines and chemokines (see Figure 2). Note that the early innate immune response is induced within 8 hours and the coupling of the primary and secondary responses offers a mechanism of regulation of various host genes and proteins upon PAMP exposure. A detailed understanding of this regulation may provide valuable insight into the mechanism of innate immune response and how pathogen products may subvert host innate immunity during the early stage of infection.

6 Acknowledgements

We thank Byron Goldstein for providing technical and financial support for this project, Cheryl Kuske, Sue Barns, and Beth Cain for access to the real-time PCR instrument, Travis Wood and Steve Graves for access to the flow cytometry facility.

This work was funded by a DOE LDRD-DR program.

References

Baumgartner, B., Weber, M., Quirling, M., Fischer, C., Page, S., Adam, M., Von Schilling, C., Waterhouse, C., Schmid, C., Neumeier, D., and Brand, K. (2002) Increased IkappaB kinase activity is associated with activated NF-kappaB in acute myeloid blasts. Leukemia 16(10):2062-2071.

Beutler, B., Hoebe, K., Du, X., and Ulevitch, R.J. (2003) How we detect microbes and respond to them: the Toll-like receptors and their transducers. J Leukoc Biol 74(4):479-485.

Bian, Z.M., Elner, S.G., Yoshida, A., and Elner, V.M. (2004) Differential involvement of phosphoinositide 3-kinase/Akt in human RPE MCP-1 and IL-8 expression. Invest Ophthalmol Vis Sci 45(6): 1887-1896.

Carroll, T.P., Greene, C.M., Taggart, C.C., Bowie, A.G., O'Neill, S.J., and McElvaney, N.G. (2005) Viral inhibition of IL-1- and neutrophil elastase-induced inflammatory responses in bronchial epithelial cells. J Immunol 175(11):7594-7601.

de Bouteiller, O., Merck, E., Hasan, U.A., Hubac, S., Benguigui, B., Trinchieri, G., Bates, E.E., and Caux, C. (2005) Recognition of double-stranded RNA by human toll-like receptor 3 and downstream receptor signaling requires multimerization and an acidic pH. J Biol Chem 280(46):38133-38145.

Fitzgerald, K.A., McWhirter, S.M., Faia, K.L., Rowe, D.C., Latz, E., Golenbock, D.T., Coyle, A.J., Liao, S.M., and Maniatis, T. (2003) IKKepsilon and TBK1 are essential components of the IRF3 signaling pathway. Nat Immunol 4(5):491-496.

Hacker, H. and Karin, M. (2006) Regulation and function of IKK and IKK-related kinases. Sci STKE 2006(357):re13.

Hessle, C.C., Andersson, B., and Wold, A.E. (2005) Gram-positive and Gram-negative bacteria elicit different patterns of pro-inflammatory cytokines in human monocytes. Cytokine 30(6):311-318.

Hoebe, K., Janssen, E.M., Kim, S.O., Alexopoulou, L., Flavell, R.A., Han, J., and Beutler, B. (2003) Upregulation of costimulatory molecules induced by lipopolysaccharide and double-stranded RNA occurs by Trif-dependent and Trif-independent pathways. Nat Immunol 4(12):1223-1229.

Ishii, K.J., Coban, C., and Akira, S. (2005) Manifold mechanisms of toll-like receptor-ligand recognition. J Clin Immunol 25(6):511-521.

Johnson, C.N., Adkins, N.L., and Georgel, P. (2005) Chromatin remodeling complexes: ATP-dependent machines in action. Biochem Cell Biol 83(4):405-417.

Julkunen, I., Sareneva, T., Pirhonen, J., Ronni, T., Melen, K., and Matikainen, S. (2001) Molecular pathogenesis of influenza A virus infection and virus-induced regulation of cytokine gene expression. Cytokine Growth Factor Rev 12(2-3):171-180.

Kanehisa, M., Goto, S., Hattori, M., Aoki-Kinoshita, K.F., Itoh, M., Kawashima, S., Katayama, T., Araki, M., and Hirakawa, M. (2006) From genomics to chemical genomics: new developments in KEGG. Nucleic Acids Res 34(Database issue):D354-357.

Kopp, E. and Medzhitov, R. (2003) Recognition of microbial infection by Toll-like receptors. Curr Opin Immunol 15(4):396-401.

Martin, M.U. and Wesche, H. (2002) Summary and comparison of the signaling mechanisms of the Toll/interleukin-1 receptor family. Biochim Biophys Acta 1592(3):265-280.

Matsumoto, M., Funami, K., Oshiumi, H., and Seya, T. (2004) Toll-like receptor 3: a link between toll-like receptor, interferon and viruses. Microbiol Immunol 48(3):147-154.

Matsumoto, M., Funami, K., Tanabe, M., Oshiumi, H., Shingai, M., Seto, Y., Yamamoto, A., and Seya, T. (2003) Subcellular localization of Toll-like receptor 3 in human dendritic cells. J Immunol 171(6):3154-3162.

Nagorsen, D., Marincola, F.M., and Panelli, M.C. (2004) Cytokine and chemokine expression profiles of maturing dendritic cells using multiprotein platform arrays. Cytokine 25(1):31-35.

Sieczkarski, S.B. and Whittaker, G.R. (2003) Differential requirements of Rab5 and Rab7 for endocytosis of influenza and other enveloped viruses. Traffic 4(5):333-343.

Takeda, K., Kaisho, T., and Akira, S. (2003) Toll-like receptors. Annu Rev Immunol 21:335-376.

Weiss-Haljiti, C., Pasquali, C., Ji, H., Gillieron, C., Chabert, C., Curchod, M.L., Hirsch, E., Ridley, A.J., van Huijsduijnen, R.H., Camps, M., and Rommel, C. (2004) Involvement of phosphoinositide 3-kinase gamma, Rac, and PAK signaling in chemokine-induced macrophage migration. J Biol Chem 279(41):43273-43284.

Yamamoto, M., Takeda, K., and Akira, S. (2004) TIR domain-containing adaptors define the specificity of TLR signaling. Mol Immunol 40(12):861-868.

Zeytun, A., van Velkinburgh, J.C., and Gupta, G. Unpublished Data.

Flagellin Signalling in Plant Immunity

Delphine Chinchilla[1], Thomas Boller[1] and Silke Robatzek[1, 2]

[1] Basel-Zurich-Plant Science Center, Botanical Institute, University Basel, Hebelstrasse 1, 4056 Basel, Switzerland
[2] Max-Planck-Institute for Plant Breeding Research, Carl-von-Linne-Weg 10, 50829 Cologne, Germany

1 Introduction

Like all higher living organisms, plants are constantly exposed to microbes that either grow epiphytically on the organ surface, establish beneficial interactions in specific tissues, or infect host tissues as pathogens and cause disease. In order to infect, pathogens must attach to the plant surface and break physical barriers to enter the tissue interior. Thus, to penetrate leaf tissue successfully, phytopathogenic bacteria have developed a mechanism that actively promotes aperture of stomata, openings in the leaf tissue that enable gas exchange of the plant with the environment. In contrast to the situation in animals, phytopathogenic bacteria do not enter host cells but proliferate within the apoplast, the intracellular space within the leaf tissue. Once phytopathogens are in contact with host cells, they are exposed to cell surface resident immune receptors that recognize microbial-derived molecules and trigger post-invasive defense responses. In order to access nutrients from host cells phytopathogens have to further subvert plant immune responses.

Plant cells are connected across their cell walls and membranes by plasmodasmata that provide cell-to-cell communication. Long distance signaling involves the vasculature comprised of living cells of the phloem and dead cells of the xylem. Generally, the phloem transports assimilates produced in leaves and the xylem is the transportation vessel for water and minerals taken up by the roots. Some phytopathogenic bacteria specifically colonize the phloem e.g. *Ralstonia*, or the xylem e.g. *Xylella*. Upon a local immune response plants may acquire systemic immunity in distant tissues that protects it from secondary infections. This phenomenon is called Systemic Acquired Resistance (SAR) and demonstrates that plants not only rely on cell-autonomous immune responses to trigger resistance (Dong, 2004).

1.1 PAMP-Triggered Plant Immune Responses

Unlike animals, plants lack a circulatory system and an adaptive immune system involving antibody production. Thus, they rely entirely on their innate immune capability to combat infection by microbes. During evolution, this has resulted in a large repertoire of single transmembrane-spanning receptor-like kinases (RLKs) and receptor-like proteins (RLPs), potentially functioning as immune receptors. Additionally, a large number of intracellular membrane-associated and cytosolic immune receptors have been evolved that are classified as CC-NB-LRR or TIR-NB-LRR proteins and are related to animal NOD-like proteins (Shiu et al., 2001; Tor et al., 2004; Chisholm et al., 2006).

Arabidopsis thaliana is a widely used plant species to investigate immune responses upon infections of the bacterium *Pseudomonas syringae*. In *A. thaliana*, two RLKs have been shown to act as immune receptors detecting so-called pathogen-associated molecular patterns (PAMPs), molecules that are highly conserved in several classes of microbes and are essential for microbial survival (Zipfel and Felix, 2006). As in animals, PAMPs perceived by plants can be microbial cell wall components such as lipopolysaccharides (LPS) from gram-negative bacteria, peptidoglycans from gram-positive bacteria, chitin, ergosterol or heptaglucosides from oomycetes or fungi. Additionally, plants are able to recognize microbial patterns that are important for pathogen virulence e.g. bacterial flagellin, the main building unit of the bacterial motility organ. PAMP-triggered immunity has a major role in plant basal resistance.

1.2 Effector-Triggered Plant Immune Responses

In order to overcome PAMP-triggered immunity, pathogens have evolved strategies involving the secretion of virulence or effector molecules to manipulate plant immune responses. Phytopathogenic bacteria express the so-called Hrp-pilus to penetrate the plant plasma membrane and using the type-III-secretion system (TTSS) they inject a battery of effector molecules into the host cytoplasm (Chisholm et al., 2006). Several pathogen-derived effectors have been shown to down-regulate and suppress PAMP-triggered immune responses (Grant et al., 2006). For example, the TTSS-delivered effector protein HopPtoD2 of *Pseudomonas syringae* exhibits tyrosine-phosphatase activity and affects the signaling capacities of Mitogen-Activated Protein Kinase (MAPK) cascades (Espinosa et al., 2003). In such a situation phytopathogenic microbes have bypassed both layers of pre- and post-invasive plant immunity and are able to colonize plant tissues, proliferate and cause disease. However, plants have evolved an additional surveillance layer of immunity. In a cultivar-specific manner they express intracellular immune receptors of the CC-NB-LRR or TIR-NB-LRR type that recognize strain-specific pathogen-derived effector molecules (Chisholm et al., 2006). In formal terms, this corresponds to the classic *gene-for-gene* interaction, in which matching avirulence genes of the pathogen and resistance genes of the host plant lead to resistance (Flor, 1971). This effector-triggered immunity, as it is called today, is thought to override the suppressive function of the pathogen-secreted effectors and potentiate the PAMP-triggered immune response (Chisholm et al., 2006;

Tao et al., 2003). In contrast to PAMP-triggered immunity, effector-triggered immunity results in rapid and localized host cell death at the site of attempted pathogen infection that is referred to as the Hypersensitive Response (HR). Plants use this system of localized host cell suicide to prevent further pathogen ingress.

1.3 Other Mechanisms

Another strategy to bypass PAMP-triggered immunity consists in "camouflage". The elicitor-active epitope of a PAMP may be modified in a way to avoid recognition without loss-of-function. For example, flagellin derived from the phytopathogen *Ralstonia solanacearum* is not recognized by *Arabidopsis* (Pfund et al., 2004). Similarly, the elicitor-active region of flagellin is altered in the flagellin of *Agrobacterium tumefaciens* and does not trigger any immune responses (Bauer et al., 2001). Interestingly, strain-specific differences in the elicitor activity of flagellin proteins were detected among *Xanthomonas campestris* pv. *campestris* (Sun et al., 2006). Thus, *X. campestris* avoids flagellin recognition to promote its virulence.

2 PAMP Perception in Plants

2.1 Perception of Bacterial Flagellin and EF-Tu

Bacterial flagellin is recognized as a PAMP in animals as well as plants. Many different plant species including *A. thaliana* have been shown to respond to bacterial flagellin. The epitope recognized by plants spans the most conserved region of the N-terminus of the protein (Felix et al, 1999). The synthetic peptide flg22 corresponding to this region stimulates all the typical immune responses known in plants. Remarkably, the flg22-region is distinct from the flagellin epitope that is recognized in mammals. However, both are buried within the internal conserved part of the flagellum filament (Zipfel and Felix, 2006). This suggests that during evolution plants and animals have independently acquired a system for flagellin detection based on a similar principle. In plants, recognition of surface exposed variable parts of the flagellum filament might occur also, but in a different way, thereby behaving like an effector in a plant cultivar-bacterial strain-specific manner (Fujiwara et al., 2004).

Recently, it has been discovered that the elongation factor Tu (EF-Tu) from bacteria functions as a PAMP in *A. thaliana* (Kunze et al., 2004). Its elicitor-active epitope could be mapped to the conserved N-terminus of the protein and a synthetic peptide stretching this region, elf26, was shown to trigger plant immune responses. In this case, only *Arabidopsis* and related plant species (members of the *Brassicaceae*) appear to posses the capability to recognize EF-Tu, members of other plant families seem not to respond.

2.2 Flg22 and elf26 Signaling

Both flg22 and elf26 peptides label a specific high-affinity binding site in *A. thaliana*, which appear membrane resident but are different in size. In competition experiments using radiolabelled peptide variants the elicitors each bind with an IC_{50} (IC_{50} = agonist concentration, that is needed to produce 50% of the maximum possible response) of ~4 nM (Bauer et al., 2001; Zipfel et al., 2006). Flg22 labels a band of ~175 kDa and elf26 labels a distinct band of ~150 kDa (Chinchilla et al., 2006; Zipfel et al., 2006). Comparing defense responses mounted in *Arabidopsis* upon flg22 or elf26 challenge revealed a striking similarity. Both were found to elicit an alkalinization of the extracellular medium exhibiting similar EC_{50} (EC_{50} = antagonist concentration that is needed to reduce an effect by 50%) values of ~0,3 nM for flg22 and ~0,2 nM for elf26 (Felix et al, 1999; Kunze et al, 2004).

Immediate early responses upon flg22 or elf26 challenge comprise transient changes in ion fluxes, as measured by alkalinization of the extracellular medium, and the generation of reactive oxygen species (ROS), nitric oxide (NO) as well as accumulation of the plant stress hormone ethylene (Felix et al., 1999; Zeidler et al., 2004). Furthermore, flg22 treatment leads to rapid changes in the phosphorylation pattern of proteins (Peck et al., 2001; Nuhse et al., 2003) and transient activation of a MAPK cascade (Nuhse et al., 2000; Zipfel et al., 2006). MPK6 has been identified as the major MAPK activated upon flg22 elicitation (Nuhse et al., 2000). Asai et al. (2002) proposed the flg22-activated MAPK cascade comprised of the MAPK kinase kinase MEKK1, the MAPK kinases MKK4/MKK5 and the MAPKs MPK3/MPK6 triggering activity of the plant specific transcription factors WRKY22/WRKY29 downstream of FLS2. However, recently Ichimura et al. (2006) reported that MEKK1 is not required for flg22-induced activation of MPK3/MPK6 but instead MPK4 that negatively regulates SAR (Andreasson et al., 2005). Furthermore, Meszaros et al. (2006) demonstrated that in addition to MPK3/MPK6 flg22 stimulates activation of MPK4. They also showed that the MAPK kinase MKK1 is activated upon flg22 elicitation and is required for MPK3/MPK4/MPK6 activation. The authors postulate parallel MAPK cascades involved in the flg22/FLS2 pathway.

Flg22 signaling results in numerous changes in gene expression. In *A. thaliana*, expression of ~1000 genes appears to be up-regulated and of ~200 genes down-regulated upon flg22 treatment (Zipfel et al., 2004). Among the up-regulated genes, RLKs and receptors recognizing effector proteins are overrepresented, pointing to a priming of defense to potentially invading pathogens. Several genes were also found to code for proteins with potential antimicrobial activities. Furthermore, flg22 treatment induces deposition of callose, a cell wall glucan involved in plant immunity (Felix et al., 1999; Gomez-Gomez et al., 1999). An interesting phenotype caused by flg22 and elf26 treatment is the inhibition of seedling growth. *Arabidopsis* seedlings grown in the presence of either PAMP are much smaller than normal grown seedlings (Gomez-Gomez et al., 1999). A simple explanation could be that the seedlings in the presence of PAMP undergo continuous immune stimulation that requires most energy resources. Nevertheless, it has been found that among the flg22 down-regulated genes are a large number involved in the auxin pathway, a plant hormone responsible for cell division and expansion (Zipfel et al., 2004). Interestingly, Navarro et al. (2006)

identified a flg22 up-regulated miRNA that targets transcripts of the auxin receptor for degradation, thereby down-regulating auxin signaling.

Among the genes induced by the flg22/elf26 are WRKY transcription factors. Within their DNA-binding domain they contain the conserved WRKY motif, which recognizes the W-box, a *cis*-element containing the core sequence TGAC (Ulker and Somssich, 2004). They comprise a large gene family and have been shown to play a role in plant immune responses. Their involvement in flg22 signaling could was implied by a survey of promoter sequences of flg22 up-regulated genes showing an overrepresentation of W-boxes. Expression of about 80% of the flg22 up-regulated genes was also induced upon cycloheximide treatment (Navarro et al., 2004). This is indicative of immediate early response genes that are under control of negative regulators. Comparing sets of flg22 (PAMP) and Avr9 (effector) regulated genes revealed a substantial overlap (Navarro et al., 2004). Further comparison of flg22 regulated genes with expression data obtained from *A. thaliana* plants infected with either non-host, virulent or avirulent strains of *P. syringae* bacteria support the fact that flg22/elf26 mediated defense responses are highly related to infections by non-host bacteria, and that there is a large overlap between all the different defense responses (Navarro et al., 2004; Tao et al., 2003).

Both, flg22 and elf26 trigger extracellular medium alkalinization, the generation of ROS and ethylene in a similar manner (Gomez-Gomez et al., 1999; Kunze et al., 2004). Even on the level of signaling MAPK cascades there is a striking overlap in the identity and time course of MAP kinase activation. They trigger activation of MAPKs labeling a major band at ~48 kDa (likely MPK6) and showing the maximum of activation at about 10 min (Zipfel et al., 2006). Furthermore, both peptides induce common sets of genes and are capable of inducing disease resistance to a similar extend against virulent *P. syringae* bacteria. It is interesting to note that elicitation with both flg22 and elf26 peptides simultaneously did not lead to any striking additive or synergistic effect. Only at subnanomolar concentrations a small additive effect was detected (Zipfel et al., 2006). Although flg22 and elf26 are perceived by distinct receptors (Gomez-Gomez and Boller, 2000; Zipfel et al., 2006), they stimulate a common set of immune response probably involving a common MAPK cascade. This argues for convergence of pathways and that one pathway may trigger the other in order to achieve an enhanced recognition capacity for infections.

While flg22 and elf26 appear to trigger a common set of defense responses, there are minor differences in kinetics to be observed. Whereas the lag-phase of flg22 stimulated medium alkalinization was about 30 s, it was about 70 s for elf26 (Zipfel et al., 2006). In addition, the maximum response appeared to be higher for flg22 than for elf26. Furthermore, flg22 induced MAP kinase activation preceded that stimulated by elf26. This slight difference in the timing of flg22 and elf26 elicitation was reflected in expression studies using whole genome arrays. Whereas flg22 treatment resulted in an up-regulation of about 1000 genes, expression of about only 430 was induced upon elf26 challenge within 30 min (Zipfel et al., 2004; Zipfel et al., 2006). Most of the elf26 up-regulated genes could be further induced within 60 min of incubation. Comparison of the flg22 (30 min) and elf26 (60 min) up-regulated genes with an induction value more than 2-fold revealed a nearly congruent set of most the 700 genes (Zipfel et al., 2006). This set of genes comprises elements of the flg22/elf26

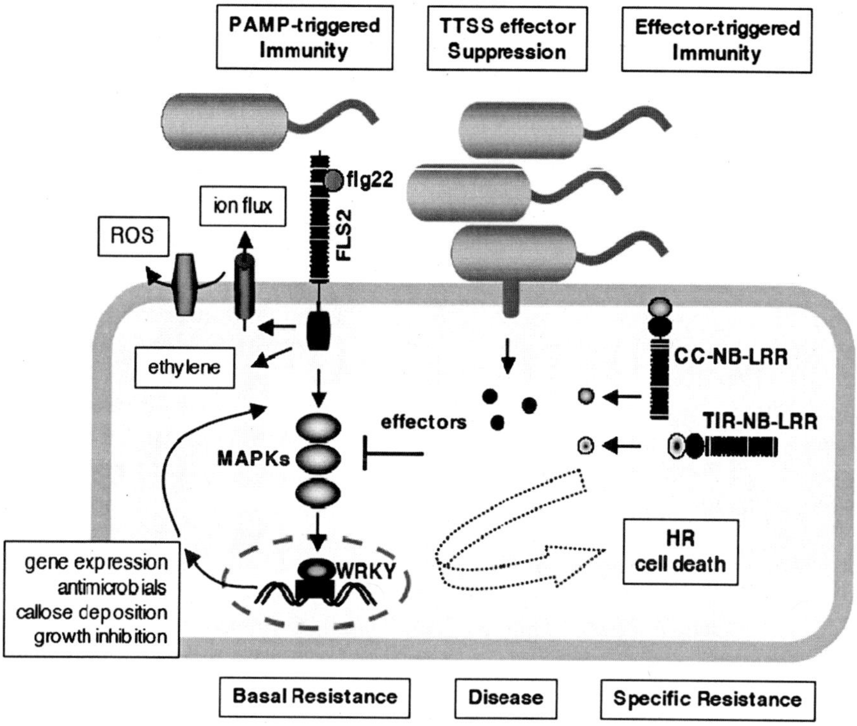

Fig. 1. Plant immune responses. Perception of flagellin (flg22) by FLS2 elicits ion fluxes, the generation of reactive oxygen species (ROS), production of the stress hormone ethylene and activates a MAP kinase cascade. WRKY transcription factors mediate numerous changes in gene expression, including those encoding components of the flg22/FLS2 pathway itself. Phytopathogenic bacteria inject effector proteins via their type-III-secretion system (TTSS) into the cell where they exert suppressive functions on PAMP-triggered immunity. Additionally, plants use intracellular immune receptors (CC-NB-LRR or TIR-NB-LRR type) that recognize bacterial effector proteins in a plant-cultivar/pathogen-strain specific manner. Effector-triggered immunity appears to be a potentiation of PAMP-triggered responses leading to a rapid localized cell death (HR).

signaling pathway including MAPK cascade components and members of the family of WRKY transcription factors (Zipfel et al., 2004; Navarro et al., 2004).

3 The Plant Flagellin Receptor

In mammals, most important PRRs mediating innate immunity are the Toll-like receptors (TLRs). TLRs are cell-surface or endosome-localized immune receptors comprised of an extracellular leucine-rich repeat (LRR) domain, a single transmembrane spanning region and a short cytoplasmic tail. PAMP-perception involves the adaptor molecule MyD88 linking TLRs to the signaling kinase IRAK (Akira and Takeda, 2004). To date, 11 TLRs have been identified, and TLR5 was shown to recognize bacterial flagellin by direct physical interaction (Smith et al., 2003).

Identification of the flagellin receptor in plants was achieved by a genetic approach. A mutagenized *A. thaliana* population was screened for seedlings resistant to flg22-triggered inhibition of growth. This led to the cloning of the *FLS2* gene (for *Flagellin Sensing 2*) encoding a RLK comprised of an extracellular domain consisting of a signal peptide and 28 LRRs, a single transmembrane spanning domain, and an intracellular serine/threonine kinase domain (Gomez-Gomez and Boller, 2000). Unlike TLR5, this demonstrates that the plant flagellin receptor is a single molecule composed of a receiver and a transmitter domain. Mutant plants devoid of the *FLS2* gene are impaired in binding and appear to be insensitive to flg22. Recently, a physical interaction between FLS2 and its ligand flg22 could be demonstrated (Chinchilla et al., 2006). The authors showed that FLS2 itself is responsible for specificity of flagellin recognition indicating that FLS2 is the *bona fide* receptor for bacterial flagellin. The flg22/FLS2 pathway including elicited plant defense responses is summarized in Fig. 1.

3.1 FLS2-Mediated Plant Immunity

In order to ascertain the contribution of FLS2 to plant immunity, *fls2* mutant and wild type plants were infected with bacteria. Only when the bacteria were sprayed onto the leaf surface, a difference between *fls2* mutants and wild type could be detected (Zipfel et al., 2004). Upon spray infection with pathogenic *P. syringae* bacteria *fls2* mutants exhibited enhanced disease susceptibility. Furthermore, the *A. thaliana* accession Ws-0, originally identified as flg22 insensitive and termed *fls1*, turned out to be a natural *fls2* mutant that when supplied with a functional FLS2 allele showed an enhanced disease resistance towards spray-inoculated bacteria (Gomez-Gomez et al., 1999; Zipfel et al., 2004). Spray-inoculation mimics that of natural infection where bacteria require flagellum-driven motility to actively penetrate into leaf tissues. It has been shown that phytopathogenic bacteria are able to sense plant-derived amino and organic acids that directs them towards plant tissues (Yao and Allen, 2006). Moreover, *P. syringae* was found to move towards leaf stomata when applied on the leaf surface (Melotto et al, 2006).

Melotto et al. (2006) discovered that PAMP perception induces stomata closure as shown for flg22 and LPS. Moreover, FLS2 is expressed in leaf stomata (Robatzek et al., 2006). Thus, flg22 perception plays an important role in pre-invasive immunity. In contrast to flg22-induced disease resistance, flg22-triggered stomata closure seemed to involve the salicylic acid (SA) and the abscisic acid (ABA) pathways (Zipfel et al., 2004; Melotto et al., 2006). Phytopathogenic bacteria have developed a strategy to overcome PAMP-triggered stomata closure. In contrast to *E. coli*, *P. syringae* produces the compound coronatine, an octadecanoid acid that is highly similar to the plant hormone jasmonic acid. Coronatine was shown to counteract the effect of PAMP perception and to mediate re-opening of stomata (Melotto et al., 2006).

FLS2 was recently isolated in a different manner. Several bacterial-derived effector proteins were individually introduced into a *P. syringae* strain that normally infects bean but not *A. thaliana* and tested for their ability to cause disease on various *Arabidopsis* accessions (de Torres et al., 2006). The effector protein AvrPtoB was found to promote disease of this non-host bacterial strain most efficiently on the accession Ws-0. In contrast, the *Arabidopsis* accession Nd-1 was resistant. Genetic mapping of the responsible locus controlling disease in Ws-0 revealed it to be FLS2. This indicates that the virulence function of AvrPtoB is most effective in the absence of flagellin signaling. It is important to note, that AvrPtoB is a very potent elicitor of HR in tomato and tobacco and that AvrPtoB exhibits E3 ligase activity (Janjusevic et al., 2006). However, the domain exhibiting E3 ligase activity and mediating HR appeared to be not responsible for the virulence effect in *A. thaliana* (de Torres et al., 2006). Additionally, AvrPtoB was shown to suppress flg22-triggered deposition of callose (de Torres et al., 2006). Thus, AvrPtoB counteracts flg22-triggered immunity as demonstrated for many others (Grant et al., 2006). In a different study, AvrPto and AvrPtoB were shown to suppress flg22-stimulated MAPK activation likely upstream of the MAPK kinase kinase (He et al., 2006).

3.2 FLS2 Receptor Endocytosis

In order to gain further insights to the molecular mechanism of flg22-mediated receptor activation, signaling and attenuation, FLS2 subcellular localization has been investigated, using FLS2 fused to the green fluorescence protein (GFP). This construct was functional and found to localize to plasma membranes (Robatzek et al., 2006). Upon flg22 incubation for 30 min, FLS2 disappeared from plasma membranes and accumulated in numerous intracellular mobile vesicles. Prolonged incubation to 60 min led to a complete loss of any FLS2-GFP derived fluorescence signal. In contrast, a pulse-treatment of flg22 for 20 min followed by washing triggered first the accumulation of vesicles and resulted at later time points in a plasma membrane resident fluorescence signal. This re-appearance of FLS2 in the plasma membrane was due to protein *de novo* synthesis and not a consequence of vesicle recycling. To address specificity of flg22-mediated FLS2 internalization, flg22 peptide variants and elf26 have been tested. The peptide variant $flg22^{A.tum}$, which corresponds to the sequence of *Agrobacterium tumefaciens* flagellin and does not elicit any defense response in *A. thaliana*, did not trigger FLS2 accumulation into vesicles (Felix et al., 1999; Robatzek et al., 2006). Furthermore, the shortened peptide variant flg22Δ2 devoid of the two most C-terminal amino acids acts as a potent antagonist of flg22 and also did not cause FLS2 internalization. Instead, the plasma membrane resident fluorescence signal slightly increased which might be due to a "locking" of FLS2 and provides evidence for a flg22 independent recycling process of non-activated FLS2 at the plasma membrane. In addition, elf26, which is recognized by the receptor kinase EFR and elicits a common set of immune responses than flg22, was not capable of inducing FLS2 internalization (Zipfel et al., 2006; Robatzek et al., 2006). These data demonstrate that FLS2 internalization requires binding of its ligand flg22 as well as receptor activation.

Flg22-triggered FLS2 accumulation into vesicles is reminiscent of receptor endocytosis. Further support was obtained by interference using chemical compounds. Brefeldin A, which affects Golgi-derived vesicles, did not constrict flg22-induced internalization of FLS2 (Robatzek et al., 2006). In contrast, Wortmannin, which impairs formation of plasma membrane-derived prevacuolar compartments, inhibits flg22-induced accumulation of FLS2 into vesicles. This provides good evidence that the endocytic process follows well-established pathways. In addition, inhibitors of cytoskeleton function (Oryzalin and Latrunculin) were shown to impair FLS2 endocytosis. Furthermore, chemical compounds targeting signaling components have been tested. K252a, a general kinase inhibitor, abolished FLS2 endocytosis, which indicates that the kinase function of the FLS2 receptor itself is required and that phosphorylation of the FLS2 cytoplasmic domain might play a role in endocytosis. Mutational analysis of potentially phosphorylated serine or threonine residues within the cytoplasmic part of FLS2 revealed three threonine residues that when substituted with alanine rendered FLS2 non-functional while preserving its flg22 binding capacity (Robatzek et al., 2006). The role of threonine-867, which appears to be a highly conserved residue of the juxta membrane region among a high number of RLKs, has been further investigated by transgenically expressing a respective FLS2 mutant variant fused to GFP. Confocal microscopy revealed that $FLS2^{T867V}$-GFP localized to plasma membranes. However, FLS2 endocytosis triggered by flg22 treatment was strongly impaired suggesting a role for FLS2 juxta membrane phosphorylation during endocytosis. Interestingly, *fls2* mutant plants transgenically expressing the $FLS2^{T867V}$-GFP variant were as susceptible as *fls2* mutants when infected with virulent *P. syringae*. This implies that flg22 signaling and FLS2 endocytosis are interconnected.

3.3 Endocytosis and flg22 Signaling

In order to address a possible link between FLS2 endocytosis and flg22 signaling Wortmannin was used in several bioassays for potential interference with flg22-elicited responses. *Arabidopsis* cell culture was shortly pre-treated with Wortmannin, subsequently elicited with flg22 for different time points and subjected for in gel MAP kinase assay. Whereas no MAPK signal could be detected without flg22 treatment, a strong MAPK activation was observed after 10 min of flg22 elicitation (Fig. 2A and B). In the presence of Wortmannin this was reduced, suggesting that MAPK activation requires an endocytic process involving FLS2. In contrast, Wortmannin appeared to have no effect on flg22-induced generation of ROS (Fig. 2C). This indicates that endocytic processes might not play a role in the activation of early flg22-triggered defense responses but might be required for downstream flg22 signaling. However, a regulatory function of endocytic processes for attenuating flg22-triggered early defense responses could not be excluded.

Another interesting observation was made using proteasome inhibitors. In the presence of MG132 or LNLL, flg22-triggered endocytosis of FLS2 was impaired, pointing to the possibility that E3 ligases are involved during FLS2 internalization.

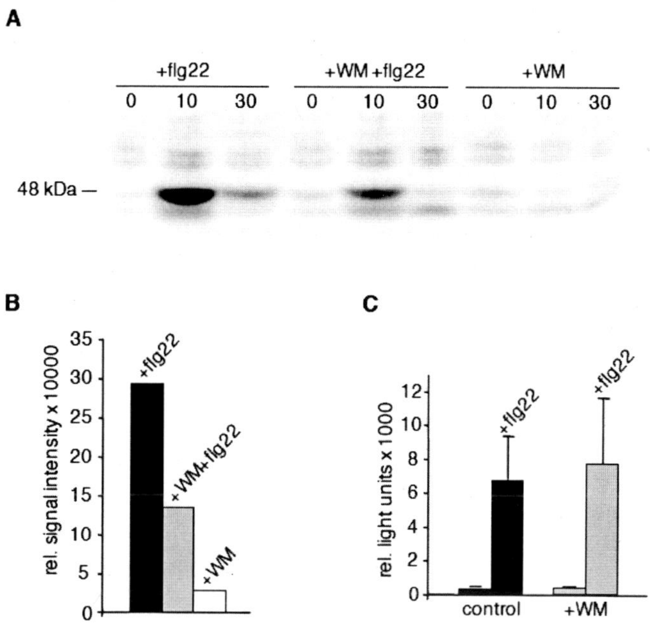

Fig. 2. Effect of Wortmannin on flg22-triggered responses. A, In gel MAP kinase assay of *Arabidopsis* cell culture untreated or treated for 10 and 30 min with 100 nM flg22 after a pre-incubation period of 15 min with 33 μM Wortmannin (WM). The predominant band appears at 48 kDa with a maximum of activation at 10 min. B, Quantification of the predominant band at the 10 min time point. C, Oxidative burst. *Arabidopsis* leaf pieces were either untreated or treated with 100 nM flg22 after a pre-incubation period of 15 min with 33 μM WM. Production of ROS measured by luminescence.

This is supported by the fact that a number of genes encoding potential E3 ligases appear to be up-regulated by flg22 (Navarro et al, 2004). Furthermore inspection of the FLS2 cytoplasmic domain could not identify the well-characterized tyrosine-based tetrapeptide motif YxxΦ mediating receptor endocytosis (Kurten, 2003), but revealed a conserved PEST-like motif (Robatzek et al., 2006). In addition to poly-ubiquitination and labeling proteins for proteasomal degradation, PEST-like motifs have been described to also mediate mono-ubiquitination that serve as tags for endocytosis of transmembrane proteins (Haglund et al., 2003). The role of the PEST-like motif in FLS2 has been addressed by mutational analysis. Transgenic expression of the mutant variant $FLS2^{P1076A}$-GFP in *fls2* mutant plants revealed that the flg22-induced generation of ROS remained unaffected, while flg22-triggered inhibition of seedling growth was impaired. Moreover, these plants exhibited an intermediate disease susceptibility phenotype. Whereas the expression pattern of $FLS2^{P1076A}$-GFP appeared to be normal, no flg22-triggered endocytosis could be observed (Robatzek et al., 2006). Comparison of results obtained from both mutant FLS2 variants revealed that early flg22 responses (ROS) seem not to involve FLS2 endocytosis.

3.4 Other Immune Receptors

Recently, the receptor recognizing bacterial EF-Tu was identified (Zipfel et al., 2006). Based on the overlapping expression profiles between the flg22 and elf26 peptides, the authors proposed that the EF-Tu receptor is among those with flg22/elf26 up-regulated gene expression and is likely encoded by a LRR-RLK. Respective T-DNA knock-out lines in *A. thaliana* were selected and one line displaying complete insensitivity to elf26 elicitation was termed EFR. Expression of EFR in normally elf26 non-responsive *Nicotiana benthamiana* resulted in elf26-responsiveness. Thus, *N. benthamiana* gained the elf26 perception system demonstrating that EFR encodes the EF-Tu receptor. Additionally, *efr* mutant plants did not show binding of elf26. Analysis of the *efr* mutant phenotype revealed increased susceptibility towards *A. tumefaciens* as measured by transient *Agrobacterium*-mediated T-DNA transfer (Zipfel et al., 2006). It is important to note that the elf26 peptide derived from *A. tumefaciens* EF-Tu acts as a potent elicitor compared to that of *P. syringae* (Kunze et al., 2004).

The receptor recognizing the fungal PAMP xylanase, LeEix has been identified from tomato (Ron and Avni, 2004). In contrast to FLS2 and EFR, LeEix is a member of transmembrane RLPs lacking the intracellular kinase domain. Classically, RLPs in plants have been described to function as immune receptors recognizing microbial effector proteins, e.g. the tomato Cf proteins (Rivas and Thomas, 2005). LeEix contains the tyrosine-based endocytic motif YxxΦ. Mutation of this motif resulted in non-functional LeEix, which suggests that endocytosis might be required for xylanase signaling. In addition to RLPs, RLKs were shown to play a role in effector perception. The RLK Xa21 from rice recognizes the bacterial effector protein AvrXa21 of *X. campestris* in a cultivar/strain-specific manner and triggers HR (Song et al., 1995).

4 Conclusion and Perspectives

Research on elicitor and PAMP signaling was a long-standing tradition in plant science. Results obtained from the flg22/FLS2 pathway provided novel insights into plant immunity. In addition, comparative studies of flagellin and EF-Tu signaling in *A. thaliana* revealed largely overlapping responses, indicating a point of convergence in the signaling cascades. Some differences were observed in the kinetics, which might suggest an order in timing of how plants perceive distinct PAMPs. Moreover, EF-Tu is only recognized by plants of the family *Brassicacea* which includes *A. thaliana*. Whether this is an evolutionary young event remains to be determined. It appears that plant immune receptors are members of polymorphic families (Bakker et al., 2006). Thus, even devoid of an adaptive immune system, plants are able to rapidly evolutionary adapt to changing environmental biotic stresses.

Ligand-induced endocytosis of a PAMP immune receptor appears to be a newly emerging field and was first discovered for the flagellin receptor FLS2 in plants. A similar observation has been also recently made for mammalian TLR4 recognizing LPS (Husebye et al., 2006). Upon addition of LPS, plasma membrane resident TLR4 is internalized and further targeted for lysosomal degradation. Moreover, LPS was found to be co-internalized. Similarly in plants, LPS has been detected in intracellular vesicles suggesting a mechanism of endocytosis (Gross et al., 2005). Binding of flg22 appears to be irreversible (Bauer et al., 2001); however, whether flg22 would be co-internalized remains to be determined. Furthermore, it will be interesting to know whether mammalian TLR5 recognizing flagellin would be subjected to endocytosis. Despite the divergence between plants and animals, both follow similar principles of non-self recognition detecting microbial-derived PAMPs, and the cognate transmembrane receptors seem to be involved in endocytic processes (Robatzek, 2006). For example, tomato LeEix seems to require the endocytic motif YxxΦ for xylanase signaling. Similarly, *Arabidopsis* EFR contains the YxxΦ motif and might be a target for endocytosis (Zipfel et al., 2006). Thus, PRR endocytosis might be a more general phenomenon and it is important to understand its role for the outcome of host immunity and how pathogens could interfere with this process.

5 Acknowledgements

We thank Dr. J. Parker for critically reading the manuscript. This work was supported by a grant of the Swiss National Science Foundation to T.B., and by a grant of the DFG to S.R. (SFB670).

References

Akira, S., and Takeda, K., 2004, Toll-like receptor signaling, *Nat. Rev. Immunol.* **4**: 499-511.

Andreasson, E., Jenkins, T., Brodersen, P., Thorgrimsen, S., Petersen, N.H., Zhu, S., Qiu, J.L., Micheelsen, P., Rocher, A., Petersen, M., Newman, M.A., Bjorn, Nielsen, H., Hirt, H., Somssich, I., Mattsson, O., and Mundy, J., 2005, The MAP kinase substrate MKS1 is a regulator of plant defense responses, *EMBO J.* **24**: 2579-2589.

Asai, T., Tena, G., Plotnikova, J., Willmann, M.R., Chiu, W.L., Gomez-Gomez, L., Boller, T., Ausubel, F.M., and Sheen, J., 2002, MAP kinase signalling cascade in Arabidopsis innate immunity, *Nature* **415**: 977-983.

Bakker, E.G., Toomajian, C., Kreitman, M., Bergelson, J., 2006, A genome-wide survey of R gene polymorphisms in Arabidopsis, *Plant Cell* **18**: 1803-1818.

Bauer, Z., Gomez-Gomez, L., Boller, T., and Felix, G., 2001, Sensitivity of different ecotypes and mutants of Arabidopsis thaliana toward the bacterial elicitor flagellin correlates with the presence of receptor-binding sites, *J. Biol. Chem.* **267**: 45669-45676.

Chinchilla, D., Bauer, Z., Regenass, M., Boller, T., and Felix, G., 2006, The Arabidopsis receptor kinase FLS2 binds flg22 and determines the specificity of flagellin perception, *Plant Cell* **18**: 465-476.

Chisholm, S.T., Coaker, G., Day, B., and Staskawicz, B.J., 2006, Host-microbe interactions: shaping the evolution of the plant immune response, *Cell* **124**: 803-814.

de Torres, M., Mansfield, J.W., Grabov, N., Brown, I.R., Ammouneh, H., Tsiamis, G., Forsyth, A., Robatzek, S., Grant, M., and Boch, J., 2006, Pseudomonas syringae effector AvrPtoB suppresses basal defence in Arabidopsis, *Plant J.* **47**: 368-382.

Dong,X., 2004, NPR1, all things considered, *Curr Opin Plant Biol.* **7**: 547-52.

Espinosa, A., Guo, M., Tam, V.C., Fu, Z.Q., and Alfano, J.R., 2003, The Pseudomonas syringae type III-secreted protein HopPtoD2 possesses protein tyrosine phosphatase activity and suppresses programmed cell death in plants, *Mol. Microbiol.* **49**: 377-387.

Felix, G., Duran, J.D., Volko, S., and Boller, T., 1999, Plants have a sensitive perception system for the most conserved domain of bacterial flagellin, *Plant J.* **18**: 265-276.

Fujiwara, S., Tanaka, N., Kaneda, T., Takayama, S., Isogai, A., and Che, F.S., 2004, Rice cDNA microarray-based gene expression profiling of the response to flagellin perception in cultured rice cells, *Mol Plant Microbe Interact.* **17**: 986-998.

Gomez-Gomez, L., and Boller, T., 2000, FLS2: an LRR receptor-like kinase involved in the perception of the bacterial elicitor flagellin in Arabidopsis, *Mol. Cell.* **5**: 1003-1011.

Gomez-Gomez, L., Felix, G. and Boller, T., 1999, A single locus determines sensitivity to bacterial flagellin in Arabidopsis thaliana. *Plant J.* **18**, 277-284

Grant, S.R., Fisher, E.J., Chang, J.H., Mole, B.M., and Dangl, J.L., 2006, Subterfuge and Manipulation: Type III Effector Proteins of Phytopathogenic Bacteria, *Annu. Rev. Microbiol.* **60**:425-449.

Gross, A., Kapp, D., Nielsen, T., and Niehaus, K., 2005, Endocytosis of Xanthomonas campestris pathovar campestris lipopolysaccharides in non-host plant cells of Nictotiana tabacum, *New Phytologist.* **165**: 215-226.

Haglund, K., Di Fiore, P.P., and Dikic, I., 2003, Distinct monoubiquitin signals in recptor endocytosis., *Trends Biochem. Sci.* **28**: 598-603.

Ichimura, K., Casais, C., Peck, S.C., Shinozaki, K., and Shirasu, K., 2006, MEKK1 is requried for MPK4 activation and regulates tissue specific and temperature dependent cell death in Arabidopsis, *J. Biol. Chem.*

Janjusevic, R., Abramovitch, R.B., Martin, G.B., and Stebbins, C.E., 2006, A bacterial inhibitor of host programmed cell death defenses is an E3 ubiquitin ligase, *Science* **311**: 222-226.

Kunze, G., Zipfel, C., Robatzek, S., Niehaus, K., Boller, T., and Felix, G., 2004, The N terminus of bacterial elongation factor Tu elicits innate immunity in Arabidopsis plants, *Plant Cell* **16**: 3496-3507.

He, P., Shan, L., Lin, N.C., Martin, G.B., Kemmerling, B., Nürnberger, T., and Sheen, J., 2006, Specific bacterial suppressors of MAMP signaling upstream of MAPKKK in Arabidopsis innate immunity, *Cell* **125**: 563-575.

Husebye, H., Halaas, O., Stenmark, H., Tunheim, G., Sandanger, O., Bogen, B., Brech, A., Latz, E., and Espevik, T., 2006, Endocytic pathways regulate Toll-like receptor 4 signaling and link innate and adaptive immunity, *EMBO J.* **25**: 683-692.

Kurten, R.C., 2003, Sorting mortifs in receptor trafficking, *Adv. Drug Deliv. Rev.* **55**: 1405-1419.

Melotto, M., Underwood, W., Koczan, J., Nomura, K., and He, S.Y., 2006, Plant stomata function in innate immunity against bacterial invasion, *Cell* **126**: 969-980.

Meszaros, T., Helfer, A., Hatzimasoura, E., Magyar, Z., Serazetdinova, L., Rios, G., Bardoczy, V., Teige, M., Koncz, C., Peck, S., and Bogre, L., 2006, The Arabidopsis MAP kinase kinase MKK1 participates in defence responses to the bacterial elicitor flagellin, *Plant J.*

Navarro, L., Dunoyer, P., Jay, F., Arnold, B., Dharmasiri, N., Estelle, M., Voinnet, O., and Jones, J.D., 2006, A plant miRNA contributes to antibacterial resistance by repressing auxin signaling, *Science* **312**: 436-439.

Navarro, L., Zipfel, C., Rowland, O., Keller, I., Robatzek, S., Boller, T., Jones, J.D., 2004, Interplay between fg22 innate immune response, Avr9-race-specific defense response and other bacterial interactions, *Plant Phys.* **135**: 1-16.

Nuhse, T.S., Peck, S.C., Hirt, H. and Boller, T., 2000, Microbial elicitors induce activation and dual phosphorylation of the Arabidopsis thaliana MAPK 6, *J. Biol. Chem.* **275**, 7521-7526.

Nuhse, T.S., Boller, T., and Peck, S.C., 2003, A plasma membrane syntaxin is phosphorylated in response to the bacterial elicitor flagellin, *J. Biol. Chem.* **278**: 45248-45254.

Peck, S.C., Nuhse, T.S., Hess, D., Iglesias, A., Meins, F., and Boller, T., 2001, Directed proteomics identifies a plant-specific protein rapidly phosphorylated in response to bacterial and fungal elicitors, *Plant Cell* **13**: 1467-1475.

Pfund, C., Tans-Kersten, J., Dunning, F.M., Alonso, J.M., Ecker, J.R., Allen, C., and Bent, A.F., 2004, Flagellin is not a major defense elicitor in Ralstonia solanacearum cells or extracts applied to Arabidopsis thaliana, Mol. *Plant Microbe Interact.* **17**: 696-706.

Rivas, S., and Thomas, C.M., 2005, Molecular interactions between tomato and the leaf mold pathogen Cladosporium fulvum, *Annu. Rev. Phytopathol.* **43**: 395-436.

Robatzek, S., 2006, Vesicle trafficking in plant immune responses, *Cell. Microbiol.* In press.

Robatzek, S., Chinchilla, D., and Boller, T., 2006, Ligand-induced endocytosis of the pattern recognition receptor FLS2 in Arabidopsis, *Genes Dev.* **20**: 537-542.

Ron, M., and Avni, A., 2004, The receptor for the fungal elicitor ethylene-inducing xylanase is a member of a resistance-like gene family in tomato, *Plant Cell* **16**: 1604-1615.

Shiu, S.H., and Bleecker, A.B. (2001) Receptor-like kinases from Arabidopsis form a monophyletic gene family related to animal receptor kinases, *Proc. Natl. Acad. Sci. USA* **98**: 10763-10768.

Song, W.Y., Wang, G.L., Chen, L.L., Kim, H.S., Pi, L.Y., Holsten, T., Gardner, J., Wang, B., Zhai, W.X., Zhu, L.H., Fauquet, C., and Ronald, P., 1995, A receptor kinase-like protein encoded by the rice disease resistance gene, Xa21, *Science* **270**: 1804-1806.

Sun, W., Dunning, F.M., Pfund, C., Weingarten, R., and Bent, A.F., 2006, Within-species flagellin polymorphism in Xanthomonas campestris pv campestris and its impact on elicitation of Arabidopsis FLAGELLIN SENSING2-dependent defenses, *Plant Cell* **18**: 764-779.

Tao, Y., Xie, Z., Chen, W., Glazebrook, J., Chang, H.S., Han, B., Zhu, T., Zou, G., and Katagiri, F., 2003, Quantitative nature of Arabidopsis responses during compatible and incompatible interactions with the bacterial pathogen Pseudomonas syringae, *Plant Cell* **15**: 317-330.

Tor, M., Brown, D., Cooper, A., Woods-Tor, A., Sjolander, K., Jones, J.D.G., Holub, E.B., 2004, Arabidopsis downy mildew resistance gene RPP27 encodes a receptor-like protein similar to CLAVATA2 and tomato CF-9, *Plant Phys.* **135**: 1100-1112.

Ulker, B. and Somssich, I.E., 2004, WRKY transcription factors: from DNA binding towards biological function, *Curr. Opin. Plant Biol.* **7**: 491-498.

Yao, J., and Allen, C. ,2006, Chemotaxis is required for virulence and competitive fitness of the bacterial wilt pathogen Ralstonia solanacearum, *J. Bacteriol.* **188**: 3697-3708.

Zeidler, D., Zahringer, U., Gerber, I., Dubery, I., Hartung, T., Bors, W., Hutzler, P., and Durner, J., 2004, Innate immunity in Arabidopsis thaliana: lipopolysaccharides activate nitric oxide synthase (NOS) and induce defense genes, *Proc. Natl. Acad. Sci. USA* **101**: 15811-15816.

Zipfel, C., and Felix, G., 2005, Plants and animals: a different taste for microbes?, *Curr Opin Plant Biol.* **8**: 353-360.

Zipfel, C., Kunze, G., Chinchilla, D, Caniard, A., Jones, J.D., Boller, T., and Felix, G., 2006, Perception of the bacterial PAMP EF-Tu by the receptor EFR restricts Agrobacterium-mediated transformation, *Cell* **125**: 749-760.

Zipfel, C., Robatzek, S., Navarro, L., Oakeley, E.J., Jones, J.D., Felix, G., and Boller, T., 2004, Bacterial disease resistance in Arabidopsis through flagellin perception, *Nature* **428**: 764-767.

26

Ancient Origin of the Complement System: Emerging Invertebrate Models

Maria Rosaria Pinto[1], Daniela Melillo[1], Stefano Giacomelli[1], Georgia Sfyroera[2] and John D. Lambris[2]

[1] Stazione Zoologica "Anton Dohrn", Laboratory of Cell Biology, pinto@szn.it
[2] University of Pennsylvania, Department of Pathology and Laboratory Medicine, lambris@mail.med.upenn.edu

1 Introduction

All metazoans are endowed with defense mechanisms against invading pathogens. The defense machinery has increased in complexity during evolution, with the addition of new and diverse components that have acquired the ability to cooperate with each other to provide an efficient and prompt immune response. The highest level of functionality has been reached in higher vertebrates, in the form of integrated action by the innate and adaptive immune systems.

In this context, the complement system, the major effector arm of vertebrate innate immunity, represents a link between innate and adaptive immunity (Song et al. 2000). In higher vertebrates, it involves more than 30 humoral and cell membrane proteins that are organized into different activation and effector pathways. The effector function of the complement system can be activated through three different activation pathways: the alternative pathway, triggered by the binding of a complement component to the pathogen surface; the mannose-binding lectin (MBL) pathway, initiated by the binding of serum lectins to a pathogen; and the classical pathway, triggered by the binding of an antibody to antigen. All these pathways lead to the activation of C3, the acknowledged molecular pillar of the complement system. Its proteolytic cleavage by C3 convertase triggers the effector function of the complement, leading to the recruitment of inflammatory cells and opsonization of the pathogen, or to its lysis through the formation of the membrane attack complex (MAC) (Lambris 1990).

In the past decade, in the context of the renewed interest in innate immunity, the complement system has been investigated in increasing depth. One successful approach to analyzing the complement system has involved the study of its evolutionary origin. The search for complement components has been carried out in

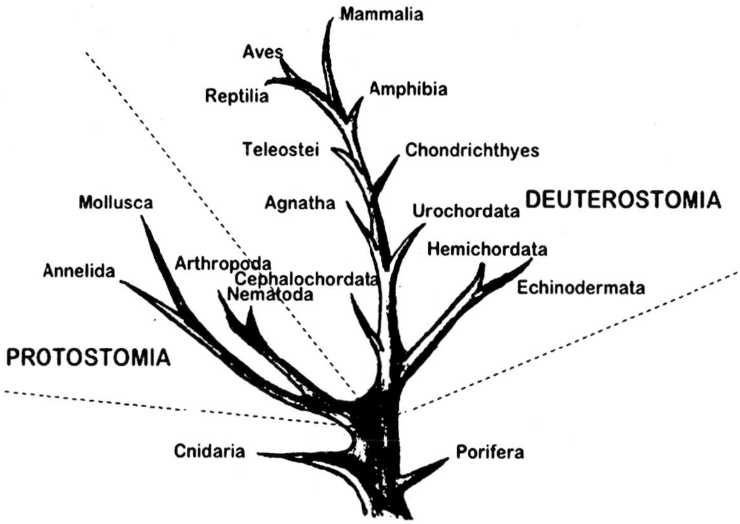

Fig. 1. Phylogenetic tree.

very divergent species, aided by the powerful tools provided by computational biology and the genome projects that are ongoing in many invertebrate species. A surprising result of these endeavors has been the identification of complement genes in very ancient species belonging to the phylum Cnidaria, which have been in existence since about 1,300 million years ago (mya) (Dishaw et al. 2005; Nonaka and Kimura 2006). Several authors have previously reviewed the results of these investigations, focusing their attention mainly on the molecular evolution of the invertebrate complement components. Indeed, data on the biological role of these components are extremely limited because of a number of factors, including the slow development, until recently, of interest in the invertebrate complement system, the paucity of the available biological material, and the relative newness of the available functional assays.

In this review we summarize the published data on invertebrate complement components, giving priority to the functional aspects of these components when known, and analyzing the molecular signatures and domain structures that suggest particular complement-associated activities. The phyla are discussed in the order of their phylogenetic position (Fig. 1); the accession numbers of the cited complement molecules are given in Table I.

Table I. Accession numbers of the amino acid sequences cited in the text

Swiftia exserta C3	AAN86548
Carcinoscorpius rotundicauda C3	AF517564
Strongylocentrotus purpuratus C3	AAC14396
Branchiostoma belcheri C3	BAB47146
Ciona intestinalis C3-1	CAC85959
Ciona intestinalis C3-2	CAC85958
Halocynthia roretzi C3	BAA75069
Carcinoscorpius rotundicauda Bf	AAV65032
Strongylocentrotus purpuratus Bf	AAC79682
Ciona intestinalis Bf-1	BAD89299
Ciona intestinalis Bf-2	BAD89300
Ciona intestinalis Bf-3	BAD89301
Halocynthia roretzi MBL	BAB69891
Halocynthia roretzi ficolin1	BAB60704
Halocynthia roretzi ficolin2	BAB60705
Halocynthia roretzi ficolin3	BAB60706
Halocynthia roretzi ficolin4	BAB60707
Branchiostoma belcheri MASP1	BAC75886
Branchiostoma belcheri MASP3	BAC75887
Halocynthia roretzi MASPa	BAA19762
Halocynthia roretzi MASPb	BAA19763
Branchiostoma belcheri C6	BAB47147
Ciona intestinalis C3aR	CAI84650
Halocynthia roretzi Integrin α Hr1	BAB21479
Halocynthia roretzi Integrin α Hr2	BAB21480
Halocynthia roretzi Integrin β Hr1	BAD15077
Halocynthia roretzi Integrin β Hr2	BAD15078
Strongylocentrotus purpuratus Sp5	AAR87482
Strongylocentrotus purpuratus Sp5013	AAR87483

2 Cnidaria

A search for the presence of complement genes has been carried out in the genome of the sea anemone *Nematostella vectensis.* This organism belongs to the phylum Cnidaria, which diverged from the Bilateria about 1,300 mya, before the protostome/deuterostome divergence, which occurred approximately 1,000 mya. This analysis was based on a search for predicted domain structures, taking into account the fact that at least five complement gene families exhibit a unique domain combination found only among human complement genes. This search produced evidence for the presence of two complement components, C3 and factor B (Bf) (Nonaka and Kimura 2006).

This finding is in agreement with the report of Dishaw and collaborators (2005), who cloned from the gorgonian coral *Swiftia exserta* a full-length cDNA for a C3-like gene, *SeC3*. The analysis of the deduced amino acid sequence of this cDNA revealed a 24 and 45% identity and similarity, respectively, with the human C3 molecule. Many molecular features of mammalian C3, such as the canonical thioester site, the associated catalytic histidine, the β-α cleavage site, and the C3a anaphylatoxin domain, were present in SeC3. The existence of a putative α-γ cleavage site suggested that SeC3, unlike mammalian C3, is a three-chain molecule similar to mammalian C4 (Karp et al. 1981), lamprey C3 (Nonaka 1994), horseshoe crab C3 (Zhu et al. 2005), and cobra venom factor (Vogel et al. 1996). All the canonical cysteine residues of mammalian C3 were conserved in SeC3, with the exception of those bridging the α and β chains. The authors suggested that the β-chain could be either released by the functional protein or associated with the α-chain through different interactions. They also suggested that the absence of the disulfide bond leaves the anaphylatoxin region more exposed to the convertase, facilitating its enzymatic cleavage. This site in SeC3 shows a context for the R-S target bond (RTR-S) that is different from that specific to mammalian C3 (LAR-S). In a phylogenetic analysis, carried out using the minimum evolution distance method on 52 thiol ester proteins (TEPs) from different vertebrate and invertebrate species, SeC3 clustered with the deuterostome invertebrate C3-like proteins, the sister group of the C3, C4 and C5 vertebrate proteins.

3 Arthropoda

The horseshoe crab *Carcinoscorpius rotundicauda* (Arthropoda) can be considered a "living fossil," as it first appeared about 550 mya. Among the protostome species analyzed thus far, *C. rotundicauda* is the only one that possesses complement system components (Zhu et al. 2005). In fact, a search for complement system genes in the available genomes of the protostomes *Anopheles gambiae*, *Drosophila melanogaster*, and *Caenorhabditis elegans* did not produce any results, and this absence has been interpreted as a secondary loss that has occurred many times at various stages during protostome evolution (Nonaka and Kimura 2006).

Using different bacterial matrices, Zhu et al. (2005) have identified several proteolytic fragments of a C3 homolog, CrC3, in the plasma of *C. rotundicauda*. A cDNA encoding the corresponding molecule has been cloned and sequenced. Analysis of the deduced amino acid sequence of CrC3 revealed the presence of all the canonical features of C3, including a domain showing high sequence similarity to that of vertebrate C3/C4/C5 anaphylatoxins. However, the vertebrate C3 convertase cleavage site LXR/S was replaced in CrC3 by EGR/F, which is more similar to the motif QGR/S of *Ciona intestinalis* C3-1 (Marino et al. 2002). A three-chain structure of the mature CrC3 was predicted on the basis of the presence of the C3/C4/C5 β-α cleavage site RKKR and two α-γ processing motifs. This structure, which is characteristic of higher vertebrate C4 molecule, is common to SeC3 and lamprey C3. C3-mediated opsonization was also observed in form of the attachment of CrC3 to the surfaces of different groups of microbes and by its subsequent proteolytic fragmentation.

Two other major proteins, homologs of the horseshoe crab tachylectins and vertebrate ficolins (Kairies et al. 2001), have been recovered from bacterial surfaces. These two proteins, the carcinolectins CL5a and CL5b, act synergistically in recognizing various pathogens and initiating the activation of the complement system in the lectin pathway (Matsushita and Fujita 2002).

A homolog of vertebrate C2 and Bf (CrC2/Bf) has also been identified in *C. rotundicauda* by subtractive hybridization of hepatopancreas cDNAs from naïve and bacteria-challenged horseshoe crabs. CrC2/Bf showed the highest identity to sea urchin C2/Bf and exhibited the main canonical sequence motifs of all C2/Bf molecules: the factor D cleavage site, Mg^{2+}-binding site, and serine protease catalytic triad (H, D, S), together with the expected overall domain architecture, consisting of multiple complement control protein (CCP) modules, a von Willebrand factor (VWF) domain and a trypsin serine protease domain. In contrast to vertebrate C2/Bf, which contains three CCP modules, CrC2/Bf, like lower deuterostome C2/Bf, contains five CCP modules. It has also been suggested that the Mg^{2+}-dependent serine protease activity found in LPS-treated horseshoe crab plasma can participate in the CrC3 activation that is mediated by CrC2/Bf. This activity is also correlated with the production of a peptide homologous to the vertebrate C3a fragment, whose presence in horseshoe crab plasma was inferred from the comparison of fragments from naïve and microbe-incubated plasma samples. These results, together with the EDTA- or protease inhibitor-mediated inhibition of bacteria phagocytosis by hemocytes, suggested the presence in this protostome species of a sophisticated C3-centered opsonic defense system homologous to the deuterostome complement system.

4 Echinodermata

Echinoderms, which belong to the deuterostome lineage, diverged from chordates about 900 mya. The first suggestion that an immune response mediated by complement molecules might be present in this phylum dates back to the pioneering work of Bertheussen and coworkers (Bertheussen 1982 and 1983; Bertheussen and Seljelid 1982). In a series of experiments, they demonstrated that phagocytosis of red blood cells by echinoderm coelomocytes was increased by human C3 and decreased by inhibitors of complement opsonization. These results represented the first evidence suggesting the presence of an alternative complement pathway in an invertebrate species. These functional observations were later confirmed and extended by the identification in a *Strongylocentrotus purpuratus* LPS-activated coelomocyte EST library of two genes (*Sp152* and *Sp064)* encoding SpBf and SpC3, which are homologs, respectively, of the vertebrate complement components Bf/C2 and C3 (Smith et al. 1996).

Sp152 is present in the genome as a single-copy gene that encodes a protein with a deduced molecular mass of 91 kDa and a conserved cleavage site for a putative factor D protease. Analysis of the deduced amino acid sequence revealed that SpBf is a mosaic protein, including five short consensus repeats (SCR), a VWF domain, and a serine protease domain (Smith et al. 1998). *Sp152* is constitutively expressed in coelomocytes, in which four types of splice variants have been identified by RT-PCR analysis: full-length messages with five SCRs, two kinds of spliced variants with four SCRs (with SCR1 or SCR4 spliced out), and messages with three SCRs (Smith et al. 1998; Smith et al. 2001). No experimental data are yet available with regard to the size and function of SpBf protein/s or their expression pattern in phagocytes. It has been speculated that, by analogy to the mammalian alternative pathway, SpBf interacts with SpC3 to form a C3 convertase.

This hypothesis was supported by the finding in a coelomocytes EST library of a second clone, the aforementioned *Sp064, which* encodes SpC3, a molecule exhibiting all the critical molecular and functional features of the vertebrate C3 molecules (Smith et al. 1996; Al-Sharif et al. 1998; Smith et al. 2001). In sea urchins, SpC3 is synthesized as a pro-protein in the coelomocytes. These cells are the only source of C3 in *S. purpuratus*, which lacks organs corresponding to the liver and hepatopancreas. The processed protein, partially purified from coelomic fluid and analyzed by SDS-PAGE, showed a single band of about 210 kDa under non-reducing conditions and two bands of about 130 and 80 kDa under reducing conditions. These bands correspond, respectively, to the typical α and β chains of the vertebrate C3 (Al-Sharif et al. 1998; Smith et al. 2001). The SpC3 α chain has a conserved thioester site (GCGEQ) in the context of a hydrophobic region, as well as a catalytic histidine located about 100 amino acids toward the C terminus. Experiments carried out with methylamine have suggested that SpC3 exhibits functional characteristics that are typical of thioester-mediated opsonic activity, such as methylamine binding and autolysis (Smith 2002).

More recently, the finding that SpC3 represents a major humoral opsonin in *S. purpuratus* coelomic fluid has extended these results. In fact, the phagocytic activity of coelomocytes increases after the incubation of yeast target cells with coelomic fluid containing SpC3, and this activity can be specifically inhibited with an anti-SpC3 antibody (Clow et al. 2004).

Among the four phagocyte types identified in the total coelomocyte population of *S. purpuratus* (Johnson 1969), only a large discoidal type and a smaller polygonal form express SpC3, with different subcellular localization (Gross et al. 2000). An analysis using confocal microscopy revealed that only a single coelomocyte type, the polygonal phagocyte, is able to ingest the opsonized yeast cells (Clow et al. 2004). The expression of SpC3 in the coelomic fluid is increased in response to LPS injection or to injury, with a slightly greater increase in response to LPS. Coelomocytes exhibit similar behavior: in response to LPS injection or to injury, their concentration increases in coelomic fluid, as does the number of SpC3+ coelomocytes (Clow et al. 2000).

Screening efforts to detect *Sp064* transcripts in *S. purpuratus* during embryonic development have revealed the presence of message in unfertilized eggs and throughout embryogenesis, with peak levels at the mesenchyme blastula and gastrula stages. Continuous exposure of the embryos to a heat-killed marine pathogen, *Vibrio diazotrophicus*, from the hatched blastula stage to the pluteus stage, produces a significant increase in *Sp064* transcripts when compared to the unexposed embryos. These results suggest that in the case of the sea urchin, the developing embryo has a defense system against pathogens that is operated by the complement system (Shah et al. 2003).

In addition to *Sp152* and *Sp064*, two other complement-related cDNAs, *SP5* and *SP5013*, have been identified in *S. purpuratus* coelomocytes. These genes encode two mosaic proteins, SpCRL (*Strongylocentrotus purpuratus* complement related protein, long form) and SpCRS (*Strongylocentrotus purpuratus* complement related protein, short form), which possess structural domains that are also found in the regulatory proteins factor H and factor I and the terminal pathway molecules C6 and C7. The functional roles of these two genes, which are constitutively expressed in all sea urchin tissues, are still unknown (Multerer and Smith 2004).

5 Chordata

The phylum Chordata includes three subphyla: the Cephalochordata, which, according to the most recent phylogenetic analyses (Delsuc et al. 2006; Vienne and Pontarotti 2006), diverged first from a common chordate ancestor around 890 mya; and the Urochordata and Vertebrata, which diverged from each other about 790 mya (Nonaka and Kimura 2006).

Despite the fact that cephalochordates and urochordates occupy a key phylogenetic position in the evolution of immune-related molecules and mechanisms, very little information was available until recently about the function, molecules, and pathways of the complement system in these chordates. While this lack of information still holds true for cephalochordates, the situation has changed significantly for urochordates, in part because of the impetus provided by the sequencing of the *Ciona intestinalis* genome (Dehal et al. 2002) and the ongoing genome projects in *Ciona savignyi* and *Oikopleura dioica*.

5.1 Cephalochordata

The information available regarding the cephalochordate complement system pertains mainly to the identification and sequencing of two mannose-binding lectin-associated serine proteases (MASPs) (MASP-1 and MASP-3; Endo et al. 2003) and of C3-like (amphiC3) and C6-like (amphiC6) cDNA clones from a notochord cDNA library of the amphioxus *Branchiostoma belcheri* (Suzuki et al. 2002).

The amphioxus MASP-1/3 gene structure is very similar to that of human MASP-1/3 (Fujita 2002). It consists of a region of eight exons encoding a heavy (H) chain, followed by a single exon encoding a MASP-3 light (L) chain and a five-exon region encoding a MASP-1 L chain. MASP-1 and MASP-3 are generated as a result of alternative splicing of the primary mRNA, producing two pro-enzymes with identical H chains and distinct L chains, and with the conserved six-domain structure of the human MASPs (Endo et al. 2003). According to the primary structure and exon organization of the genes and the codon encoding the active serine site, the MASP gene family has been divided into two lineages, the TCN and the AGY types (Endo et al. 1998). The amphioxus MASP-1 and MASP-3 serine protease domains are characterized by an active site encoded by an AGY codon, unusually combined with a disulfide bridge forming a histidine loop (Endo et al. 2003). The combination found in amphioxus is intermediate between the TCN-type, characteristic of the human MASP-1 and ascidian MASPs (Ji et al., 1997) (possessing a TCN codon at the active serine site and a histidine loop disulfide bridge in the serine protease domain) (Arlaud and Gagnon 1981), and the AGY-type, which includes MASP-2, MASP-3, C1r/C1s, and lower vertebrate MASPs (characterized by the absence of a histidine loop disulfide bridge and the presence of an AGY codon at the active serine site). Another peculiar feature of amphioxus MASPs is the presence of an aspartic acid residue at position –6 upstream of the active serine site, which suggests trypsin-type substrate specificities (Kraut 1977).

The *amphiC3* full-length clone in *B. belcheri* encodes a protein of 1792 amino acids, exhibiting 29% identity with human C3 and containing a possible β-α processing site, a canonical thioester site, and the downstream catalytic histidine. A canonical anaphylatoxin domain was also predicted, with six cysteine residues in conserved positions (Suzuki et al. 2002).

Conversely, the full-length *amphiC6* clone encodes a C6-like molecule with the highest identity to human C6. This C6-like molecule has a conserved modular structure in the central portion of the molecule, including two thrombospondin type1 (TSP), one low-density lipoprotein-receptor class A (LDLRA), one MAC/perforin, one epidermal growth factor (EGF), and one TSP module, as well as sequences in the N- and C-terminal regions with no significant similarity to any known sequence. The N- and C-terminal regions also present unique features not found in any molecule of the lytic pathway: arginine and proline stretches at the N-terminus, and nine consecutive repeats of the heptapeptide DA(D/E)TSPG at the C-terminus (Suzuki et al. 2002).

The only functional data available concern the hemolytic activity exhibited by *B. belcheri* humoral fluid toward rabbit erythrocytes; this activity is both Mg^{2+}-dependent and heat-sensitive. The fact that C3 is present in the humoral fluid and that the hemolytic activity can be inhibited by a rabbit anti-human C3 antiserum, zymosan, methylamine, hydrazine, and phenylmethylsulfonyl fluoride provides indirect evidence for the presence of a complement system-mediated immune response (Zhang et al. 2003).

5.2 Urochordata

The urochordate subphylum has been organized into three classes: Larvacea, Thaliacea and Ascidiacea. Species belonging to the Ascidiacea include well-established models that are used by a large community of researchers worldwide. In fact, because of their wide geographical distribution and phylogenetic position, which have allowed researchers to explore the evolutionary origins of the vertebrate lineage, ascidians are among the most extensively studied animal models, especially in the fields of developmental biology and, more recently, immunology.

For studies of the complement system, two main models have been used: the Japanese species *Halocynthia roretzi* and the temperate seawater cosmopolitan species *Ciona intestinalis*. More limited information has also come from two Australian species, *Pyura stolonifera* and *Styela plicata*.

Genomic analyses to detect immune-related genes in *C. intestinalis* (Dehal et al. 2002; Azumi et al. 2003) have provided the most comprehensive picture of complement genes in urochordates, confirming data already reported in the literature and drastically expanding our knowledge at the molecular level. By using a pattern-based search method (Azumi et al. 2002), a number of genes encoding components of the lectin and alternative activation pathways have been identified: nine mannose-binding lectins, nine ficolins, two C1q molecules, four members of the C3/C4/C5/alpha-2-macroglobulin (α2M) family, two factorB/C2 components, and four MASPs. This analysis, extended to the late complement components, has allowed the identification of eleven gene models containing the MAC/perforin domain. Nine of these, exhibiting domain structures similar to those of human late components, are potential complement components. A search of the *Ciona* genome for complement regulatory components, characterized in mammals by SCR domain repeats, has resulted in the identification of 132 gene models (Azumi et al. 2002).

The most relevant result of this analysis concerns the unpredictable molecular complexity of the ascidian complement system, which exhibits an expansion in gene number that is comparable to or, in some cases, even higher than that of its mammalian counterparts. This phylogenetic analysis indicates that gene expansion was generated by duplication events that occurred independently in the ascidian and vertebrate lineages.

Detailed information on individual ascidian complement components at both the molecular and functional levels has come from the studies summarized in the following paragraphs.

5.2.1 C3, C3a, C3aR

C3 genes have been sequenced and characterized in two different urochordate species, the ascidians *H. roretzi* (Nonaka et al. 1999) and *C. intestinalis* (Marino et al. 2002), which contain one (*AsC3*) and two C3-like genes (*CiC3-1* and *CiC3-2*), respectively. The deduced amino acid sequences of *H. roretzi* and *C. intestinalis* C3s exhibit a canonical processing site for α and β chains, as well as a thioester site, with a catalytic histidine located downstream. A convertase cleavage site is present in a conserved position in AsC3 and CiC3-1, while the presence of a long sequence insertion, rich in threonine residues in the same region, makes it difficult to locate the site in CiC3-2.

In mammals, the main site of C3 production is the liver; in ascidians, which lack a true liver, no screening has been done to identify C3-producing tissues. However, Northern blot analysis in *H. roretzi* has revealed C3 transcripts in the hepatopancreas and blood cells (Nonaka et al. 1999), and *CiC3-1* and *CiC3-2* have been cloned from *Ciona* hemocyte RNA (Marino et al. 2002).

The presence of the C3 gene products in the circulating hemolymph of *H. roretzi* and *C. intestinalis* has been established using specific antibodies, which in *Ciona* also identified compartment cells and granular amoebocytes as the C3-producing hemocyte cell types in sections of the tunic of LPS-injected animals (Pinto et al. 2003). Similar results have been obtained in *Styela plicata*, in which heterologous antibodies revealed in the hemolymph a C3-like protein (Raftos et al. 2002) that was synthesized by phagocytic hemocytes and rapidly exocytosed after stimulation with pathogen-associated antigens (Raftos et al. 2004).

The *H. roretzi* C3 acts as a humoral opsonin. In fact, incubation of yeast target cells with hemolymph containing *H. roretzi* C3 leads to a significant increase in phagocytic activity by hemocytes. This activity is completely abolished when the hemolymph is depleted of C3 by incubation with an anti-AsC3 antibody or with a chelating agent such as EDTA. The removal of either C3 or divalent cations from the hemolymph also abolishes the binding of C3 to the yeast target cells, thus demonstrating that phagocytosis proceeds through the yeast opsonization mediated by C3, whose activation requires divalent cations (Nonaka et al. 1999).

In mammals, production of the C3b fragment of the opsonic pathway is associated with the release of the anaphylatoxin C3a, a potent mediator of inflammatory reactions. Mammalian C3a includes six conserved cysteines that form three disulfide bonds which stabilize a tightly packed core consisting of four antiparallel helical regions (Ember et al. 1998; Wetsel et al. 2000). The ascidian C3s sequenced thus far have only four conserved cysteines and do not possess a canonical anaphylatoxin domain. Furthermore, the ascidian C3a C-terminal sequences, like those of all other known invertebrates and lower vertebrates, differ from the mammalian C-terminal consensus sequence GLAR, which participates in binding to the C3a receptor (Ember et al. 1998; Wetsel et al. 2000). Despite these differences in structure, C3a-mediated inflammatory activity has been demonstrated in *C. intestinalis* (Pinto et al. 2003). In fact, both the recombinant C3-1a fragment and synthetic peptides reproducing the C-terminus, with and without the terminal arginine, are able to promote *in vitro* hemocyte chemotaxis in a dose-dependent manner. This activity can be inhibited by an anti-*Ciona* C3-1a-specific antibody, as well as by the pretreatment of hemocytes with pertussis toxin. These results clearly indicate that the chemotactic activity is dependent on the interaction between the ligand C3a and a G protein-coupled receptor.

C. intestinalis C3a activity has been further characterized by immunohistochemical and *in situ* hybridization studies of tunic sections of LPS-injected animals. These analyses showed a total hemocyte number that was 5 times higher in the injured area in LPS-injected animals than in the controls and that the number of granular amoebocytes was increased 15-fold in the injured area. At the same time, granular amoebocytes and compartment cells were actively engaged in producing C3, which reached its highest level of expression at 48 h after LPS injection. Taken together, these findings indicate that C3-1a acts as a chemotaxin for *C. intestinalis* hemocytes and that C3-1a-mediated hemocyte recruitment to sites of injury may play an important role in inflammatory processes (Pinto et al. 2003).

Similar results have been obtained using *Pyura stolonifera*, another species of Ascidiacea. Activation of serum by LPS and zymosan in this species generates an 8.5-kDa proteolytic fragment that confers on serum a chemotactic activity toward hemocytes, as demonstrated by *in vitro* chemotaxis experiments (Raftos et al. 2003).

In mammals, the bioactive fragment C3a binds specifically to cell-surface G protein-coupled seven-transmembrane receptors (Ember et al. 1998; Wetsel et al. 2000). Recently, the receptor molecule involved in C3a-mediated chemotaxis (CiC3aR) has been cloned and characterized in *C. intestinalis* (Melillo et al. 2006). Its expression profile, as evaluated by Northern blot analysis, indicates that like the mammalian C3a receptors, it is broadly expressed in different tissues and organs. It encodes a 95-kDa seven-transmembrane protein that is characterized by a long hydrophilic region of 162 amino acids located between the fourth and fifth transmembrane domains, a common feature of all mammalian C3a receptors. The secondary structure prediction and alignment with other C3aRs evidenced the presence of a very long insertion in the third cytoplasmic loop and an elongation of the cytoplasmic tail. Immunostaining of circulating hemocytes performed with three polyclonal antibodies raised against synthetic peptides reproducing sequences of the first and second extracellular loops and the third intracellular loop have revealed that CiC3aR is constitutively expressed in only two kinds of phagocytic hemocytes, hyaline and granular amoebocytes. In chemotactic assays, the antibodies against the first and second extracellular loops can inhibit the directional migration of hemocytes toward the synthetic peptide reproducing the CiC3a C-terminal sequence, thus providing compelling evidence that *C. intestinalis* expresses a functional C3aR homologous to the mammalian receptor (Melillo et al. 2006).

5.2.2 Factor B

Thus far, factor B (Bf) genes have been recognized, sequenced and analyzed in detail at molecular level only in *C. intestinalis* (Dehal et al. 2002; Azumi et al. 2003; Yoshizaki et al. 2005).

Three genes identified in this species, *CiBf-1*, *CiBf-2*, and *CiBf-3*, encode proteins with identical domain structures that resemble the basic domain structure of the vertebrate Bf/C2 gene family. They are characterized by the presence of three SCR domains, a von Willebrand factor type A domain, and a serine protease domain. In addition to these, in *Ciona* two LDLR domains and one SCR domain are present at the N-terminus. The active site of the three CiBf serine protease domains is of the AGY type, a feature shared with vertebrate MASP-2, MASP-3, and C1r/C1s, together with the absence of a histidine loop disulfide bridge. The amino acid sequence identity between CiBf-1 and CiBf-2 is 88%, and between CiBf-3 and CiBf-1 or CiBf-2 is 49%.

Phylogenetic analysis, including a detailed characterization of the genomic organization and intron/exon composition, has indicated that CiBf genes are the result of duplication and gene conversion events that occurred within the urochordate lineage after the divergence from the vertebrate subphylum. CiBf genes, distributed within a 50-kb genomic region, mapped to a different chromosome than do the

CiC3-1 and CiC3-2 loci, suggesting that the linkage among the MHC class III complement genes was established in vertebrates.

5.2.3 MBL/Ficolins

Clear evidence of the presence of a lectin activation pathway in ascidians has been provided by the molecular and functional characterization of an *H. roretzi* 36-kDa lectin (GBL) that binds specifically to glucose. Sequence analysis revealed that GBL contains a C-type lectin domain but lacks the collagen-like domain of mammalian MBLs.

To determine whether GBL acts as a recognition molecule in the lectin pathway of the ascidian complement system, the *Halocynthia* plasma was depleted of GBL or C3 with specific antibodies and used in assays of yeast phagocytosis by hemocytes. Incubation of plasma with either of the antibodies resulted in a significant decrease in the phagocytic activity, suggesting that GBL recognizes carbohydrates on the yeast surface and in turn activates C3 through the associated MASPs (Sekine et al. 2001).

In mammals, the lectin pathway can be triggered by pathogens and involves recognition by lectins other than MBL. These lectins belong to the ficolin family and are characterized by the presence of an NH_2-terminal domain containing cysteine residues, a fibrinogen-like domain, and a collagen-like domain (Matsushita et al. 2000). In *H. roretzi,* four cDNA clones encoding orthologs of mammalian ficolins have been sequenced and their carbohydrate binding specificity has been assessed. It has been suggested that their association with MASPs contributes to the activation of C3, but thus far there is no experimental evidence to support this hypothesis (Kenjo et al. 2001).

5.2.4 MASPs

Two MASPs have been identified in an *H. roretzi* cDNA library, MASPa and MASPb (Endo et al. 2003). The deduced amino acid sequences of these two molecules exhibit the two main specific features of mammalian MASP1: the codon of the serine in the active site is of the TCN type, and the catalytic histidine is present in a disulfide bridge loop. The residue determining the substrate specificity is located at the -6 position with reference to the active site; this residue is aspartic acid in MASPa, as in all vertebrate MASPs, and threonine in MASPb. This finding suggests a trypsin-like activity for MASPa and a different substrate specificity for MASPb. The absence of experimental data clarifying the activity of the two ascidian MASPs means that it is not currently possible to interpret the sequence analysis results.

5.2.5 Complement Receptors Type 3 and Type 4

In mammals, complement receptors type 3 and type 4 (CR3 and CR4) are members of the β_2-integrin family. These molecules are membrane-bound heterodimers, each consisting of a different α subunit non-covalently associated with the same β subunit.

Among the apparently unrelated functions documented for CR3 and CR4 is the ability of both molecules, expressed mainly on myeloid cells, to stimulate phagocytosis by binding iC3b-opsonized bacteria.

In *H. roretzi*, four integrin subunits, two α (α_{Hr1} and α_{Hr2}) and two β (β_{Hr1} and β_{Hr2}), have been cloned, sequenced and partially characterized (Miyazawa et al. 2001; Miyazawa and Nonaka 2004). Analysis of the deduced amino acid sequences indicated that all of them have the typical domain structures of the mammalian α and β integrin subunits. Involvement of the integrin α_{Hr1} subunit in the C3-dependent phagocytic activity of *H. roretzi* hemocytes has been demonstrated in phagocytosis assays using a specific antibody against a recombinant protein reproducing the extracellular region of α_{Hr1} (Miyazawa et al. 2001).

To investigate the heterodimer composition in ascidians, insect cell lines were co-infected with two recombinant baculovirus species, the first containing the α_{Hr1} gene and the second containing either the β_{Hr1} or β_{Hr2} gene. Immunoprecipitation of the cell line extracts with anti-α_{Hr1} antibody demonstrated that both the β_{Hr1}, and β_{Hr2} subunits are associated with the α_{Hr1} subunit. The association of α_{Hr1} with β_{Hr1} was also confirmed in western blot analysis of ascidian hemocytes.

The type of pairing found in ascidians, namely the same integrin α subunit (α_{Hr1}) paired with different integrin β subunits (β_{Hr1} and β_{Hr2}), is different from the mammalian CR3 and CR4 pairing pattern and resembles that of α_v integrins. Because of this type of pairing, the authors referred to these ancestral forms of complement receptors as the "α_{Hr1} integrin family" or "hemocyte integrin family."

6 Concluding Remarks

The search for complement components in the invertebrate species analyzed thus far has clearly demonstrated the presence of C3-like and Bf-like molecules in very ancient animal phyla and has identified these molecules as the most basic complement component assembly. In particular, comparative analysis of C3 amino acid sequences has identified the most important molecular signatures in very primitive species. In fact, the thioester site with its associated catalytic histidine, the β-α chain processing site, the C3 convertase site, and the anaphylatoxin domain, are features shared by all C3 molecules. These findings strongly indicate the recruitment, very early in evolution, of ancestral C3 and Bf into a primordial alternative pathway of C3 activation, with the consequent formation of the peptide fragment C3a and the opsonin fragment C3b, which binds to the microbial surface through a thioester bond.

In this context, it is remarkable that the human C3a fragment has antimicrobial properties. It binds to bacterial membranes and directly kills the microorganisms by inducing breaks in the membrane. These results suggest an unforeseen ancestral role for a C3-related peptide and point to the C3a molecule as a link between two important arms of innate immunity, the complement system and antimicrobial peptides (Andersson Nordahl et al. 2004). Among the many different roles played by the C3a peptide in mammals (Mastellos and Lambris 2002), antimicrobial activity could represent the simplest and most primordial function. Studies of invertebrate complement components, which lack the complexity of the mammalian complement system, could assist us in exploring this activity and discovering new molecular mechanisms underlying the central event of the complement reaction cascade, namely C3 activation. The study of invertebrates can further improve our knowledge of the complement system, which is surprisingly flexible, as shown by the very recent finding in mouse of a novel C3 activation pathway mediated by SIGN-R1, a membrane-bound lectin, which initiates a classical but Ig-independent pathway (Kang et al. 2006). The analysis of the urochordate *C. intestinalis* genome has confirmed the absence of the pivotal genes for adaptive immunity in this species, which is phylogenetically at the base of the vertebrate lineage. At the same time, it has confirmed the presence of many complement genes, including two C1q-like genes. The ascidian C1q could either act as a lectin, like the lamprey C1q (Matsushita et al. 2004), or with other unknown membrane bound receptor/s, as in the case of murine SIGN-R1. Future investigation of ascidian C1q expression and biological function could help to shed light on the origin of the classical pathway of the complement system.

It is noteworthy that the studies carried out on the invertebrate complement system are still fragmentary, limited to a few species and available genomes, and mainly devoted to genomic and cDNA sequence analyses. Greater efforts in investigating the biological activities of invertebrate complement components and their mutual functional relationships should provide answers to the many basic questions in this field that remain to be answered.

7 Acknowledgements

We thank Dr. Deborah McClellan for editorial assistance.

References

Al-Sharif, W.Z., Sunyer, J.O., Lambris, J.D. and Smith, L.C. (1998) Sea urchin coelomocytes specifically express a homologue of the complement component C3. J. Immunol. 160, 2983-2997.

Andersson Nordahl, E., Rydengård, V., Nyberg, P., Nitsche, D.P., Mörgelin, M., Malmsten, M., Björck, L. and Schmidtchen, A. 2004 Activation of the complement system generates antibacterial peptides. Proc. Natl. Acad. Sci. USA 101, 16879-16884.

Azumi, K., De Santis, R., De Tomaso, A., Rigoutsos, I., Yoshizaki, F., Pinto, M.R., Marino, R., Shida, K., Ikeda, M., Ikeda, M., Arai, M., Inoue, Y., Shimizu, T., Satoh, N., Rokhsar, D.S., Du Pasquier, L., Kasahara, M., Satake, M., and Nonaka, M. (2003). Genomic analysis of immunity in a Urochordate and the emergence of the vertebrate immune system: "waiting for Godot". Immunogenetics 55, 570-581.

Bertheussen, K. (1982) Receptors for complement on echinoid phagocytes. II. Purified human complement mediates echinoid phagocytosis. Dev. Comp. Immunol. 6, 635-642.

Bertheussen, K. and Seljelid R. (1982) Receptors for complement on echinoid phagocytes. I. The opsonic effect of vertebrate sera on echinoid phagocytosis. Dev. Comp. Immunol. 6, 423-431.

Bertheussen, K. (1983) Complement-like activity in sea urchin coelomic fluid. Dev. Comp. Immunol. 7, 21-31.

Clow, L.A., Gross, P.S., Shih, C-S. and Smith, L.C. (2000) Expression of SpC3, the sea urchin complement component, in response to lipopolysaccharide. Immunogenetics 51, 1021-1033.

Clow, L.A., Raftos, D.A., Gross, P.S. and Smith, L.C. (2004) The sea urchin complement homologue, SpC3, functions as an opsonin. J. Exp. Biol. 207, 2147-2155.

Dehal P., Satou, Y., Campbell, R.K., Chapman, J., Degnan, B., De Tomaso, A. Davidson, B., Di Gregorio, A., Gelpke, M., Goodstein D.M. et al. 2002. The draft genome of *Ciona intestinalis*: insights into chordate and vertebrate origins. Science 298, 2157-2167.

Delsuc, F., Brinkmann, H., Chourrout, D. and Philippe, H. (2006) Tunicates and not cephalochordates are the closest living relatives of vertebrates. Nature 439, 965-968.Edds, K. (1993) Cell biology of echinoid coelomocytes. I Diversity and characterization of cell types. J. Invert. Pathol. 61, 173-178.

Dishaw, L.J., Smith, S.L. and Bigger, C.H. 2005 Characterization of a C3-like cDNA in a coral: phylogenetic implications. Immunogenetics 57, 535-548.

Ember, J.A., Jagels, M.A. and Hugli, T.E. (1998). Characterization of complement anaphylatoxins and their biological responses. In: J.E. Volanakis and M.M. Frank, (Eds.) *The Human Complement System in Health and Disease*. Marcel Dekker, New York, pp. 241-284.

Endo, Y., Nonaka, M., Saiga, H., Kakinuma, Y., Matsushita, A., Takahashi, M., Matsushita, M. and Fujita, T. (2003) Origin of mannose-binding lectin-associated serine protease (MASP)-1 and MASP-3 involved in the lectin complement pathway traced back to the invertebrate, amphioxus. J Immunol. 170, 4701-4707.

Fujita, T. (2002) Evolution of the lectin-complement pathway and its role in innate immunity. Nat. Rev. Immunol. 2, 346-353.

Gross, P.S., Clow, L.A. and Smith, L.C. (2000) SpC3, the complement homologue from the purple sea urchin, *Strongylocentrotus purpuratus*, is expressed in two subpopulations of the phagocytic coelomocytes. Immunogenetics 51, 1034-1044.

Ji, X, Azumi, K, Sasaki, M. and Nonaka M. (1997) Ancient origin of the complement lectin pathway revealed by molecular cloning of mannan binding protein-associated serine protease from a urochordate, the Japanese ascidian, *Halocynthia roretzi*. Proc. Natl. Acad. Sci. USA 94, 6340-6345.

Johnson, P.T. (1969) The coelomic elements of sea urchins (*Strongylocentrotus*). I. The normal coelomocytes; their morphology and dynamics in hanging drops. J. Invert. Pathol. 13, 25-41.

Kairies, N., Beisel, H.G., Fuentes-Prior, P., Tsuda, R., Muta, T., Iwanaga, S., Bode, W., Huber, R. and Kawabata, S. (2001) The 2.0-A crystal structure of tachylectin 5A provides evidence for the common origin of the innate immunity and the blood coagulation systems. Proc. Natl. Acad. Sci. U S A 98, 13519-13524.

Kang, Y-S., Do, Y., Lee, H-K., Park, S.H., Cheong, C., Lynch, R.M., Loeffler, J.M., Steinman, R.M. and Park, C.G. 2006 A dominant complement fixation pathway for pneumococcal polysaccharides initiated by SIGN-R1 interacting with C1q. Cell 125, 47-58.

Karp, D.R., Parker, K.L., Shreffler, D.C. and Capra, J.D. (1981) Characterization of the murine C4 precursor (pro-C4): evidence that the carboxyterminal subunit is the C4 gamma-chain. J. Immunol. 126, 2060-2061.

Kelly, K.L., Cooper, E.L. and Raftos, D.A. (1992) Purification and characterization of a humoral opsonin from the solitary urochordate *Styela clava*. Comp. Biochem. Physiol. 103B, 749-753.

Kenjo, a., Takahashi, M., Matsushita, M., Endo, Y., Nkata, M., Mizuochi, T. and Fujita, T. (2001) Cloning and characterization of novel ficolins from the solitary ascidian, *Halocynthia roretzi*. J. Biol. Chem.276, 19959-19965.

Kraut, J. (1977) Serine proteases: structure and mechanism of catalysis. Annu. Rev. Biochem. 46, 331-358.

Lambris, J.D. (1990) The Third Component of Complement - Chemistry and Biology. Springer-Verlag, Heidelberg.

Marino R., Kimura, Y., De Santis, R., Lambris, J.D. and Pinto M.R. (2002). Complement in urochordates: cloning and characterization of two C3-like genes in the ascidian *Ciona intestinalis*. Immunogenetics 53, 1055-1064.

Mastellos, D. and Lambris, J.D. (2002) Complement: more than a ‘guard’ against invading pathogens? Trends Immunol. 23, 485-91.

Matsushita, M., Endo, Y., Nonaka, M. and Fujita, T. (1998) Complement-related serine proteases in tunicates and vertebrates. Curr. Opin. Immunol. 10, 29-35.

Matsushita, M., Endo, Y. and Fujita, T. (2000) Cutting edge: complement-activating complex of ficolin and mannose-binding lectin-associated serine protease. J. Immunol. 164, 2281-2284.

Matsushita, M. and Fujita, T. (2002) The role of ficolins in innate immunity. Immunobiology 205, 490-497.

Matsushita, M., Matsushita, A., Endo, Y., Nakata, M., Kojima, N., Mizuochi, T. and Fujita, T. (2004) Origin of the classical complement pathway: lamprey orthologue of mammalian C1q acts as a lectin. Proc. Natl. Acad. Sci. USA 101, 10127-10131.

Melillo, D., Sfyroera, G., De Santis, R., Graziano, R., Marino, R., Lambris, J.D. and Pinto, M.R. (2006) First identification of a chemotactic receptor in an invertebrate species: structural and functional characterization of *Ciona intestinalis* C3a Receptor. J. Immunol. 177, 4132-4140.

Miyazawa, S., Azumi, K., and Nonaka, M. (2001) Cloning and characterization of integrin α subunits from the solitary ascidian, *Halocynthia roretzi*. J. Immunol. 166, 1710-1715.

Miyazawa, S. and Nonaka, M. (2004) Characterization of novel ascidian β integrins as primitive complement receptor subunits. Immunogenetics 55, 836-844.

Multerer, K.A. and Smith, L.C. (2004) Two cDNAs from the purple sea urchin, *Strongylocentrotus purpuratus*, encoding mosaic proteins with domains found in factor H, factorI, and complement components C6 and C7. Immunogenetics 56, 89-106.

Nonaka, M. (1994) Molecular analysis of the lamprey complement system. Fish Shellfish Immunol. 4, 437-446.

Nonaka, M., Azumi, K., Ji, X., Namikawa-Yamada, C., Sasaki, M., Saiga, H., Dodds, A.W., Sekine, H., Homma, M.K., Matsushita, M., Endo, Y. and Fujita, T. (1999) Opsonic complement component C3 in the solitary ascidian, *Halocynthia roretzi*. J. Immunol. 162, 387-391.

Nonaka, M. and Kimura, A. (2006) Genomic view of the evolution of the complement system. Immunogenetics 58, 701-713.

Pearce, S., Newton, R.A., Nair, S.V. and Raftos, D.A. (2001) Humoral opsonins of the tunicate, *Pyura stolonifera*. Dev. Comp. Immunol. 25, 377-385.

Pinto M.R., Chinnici, C.M., Kimura, Y., Melillo, D., Marino, R., Spruce, L.A., De Santis, R., Parrinello, N. and Lambris, J. D. (2003) CiC3-1a-mediated chemotaxis in the deuterostome invertebrate *Ciona intestinalis* (Urochordata). J Immunol. 171, 5521-5528.

Raftos, D.A., Nair, S. V., Robbins, J., Newton, R. A. and Peters. R. (2002) A complement component C3-like protein from the tunicate, *Styela plicata*. Dev. Comp. Immunol. 26, 307-312.

Raftos, D.A., Robbins, J. Newton, R.A. and Nair, S. V. (2003) A complement component C3a-like peptide stimulates chemotaxis by hemocytes from an invertebrate chordate-the tunicate, *Pyura stolonifera*. Comp. Biochem. Physiol. Part A 134, 377-386.

Raftos D.A., Stillman, D.L. and Cooper, E. L. (1998) Chemotactic responses of tunicate (Urochordata, Ascidiacea) hemocytes in vitro. J. Invertebr. Pathol. 72, 44-49.

Sekine, H., Kenjo, A., Azumi, K., Ohi, G., Takahashi, M., Kasukawa, R., Ichikawa, N., Nakata, M., Mizuochi, T., Matsushita M., Endo, Y. and Fujita, T. (2001) An ancient lectin-dependent complement system in an ascidian: novel lectin isolated from the plasma of the solitary ascidian, *Halocynthia roretzi*. J. Immunol. 167, 4504-4510.

Shah, M., Brown, K.M. and Smith, L.C. (2003) The gene encoding the sea urchin complement protein, SpC3, is expressed in embryos and can be upregulated by bacteria. Dev. Comp. Immunol. 27, 529-538.

Smith, L.C. (2002) Thioester function is conserved in SpC3, the sea urchin homologue of the complement component C3. Dev. Comp. Immunol. 26, 603-614.

Smith, L.C., Britten, R.J. and Davidson E.H. (1995) Lipopolysaccharide activates the sea urchin immune system. Dev. Comp. Immunol. 19, 217-224.

Smith, L.C., Chang, L., Britten, R.J. and Davidson, E.H. (1996) Sea urchin genes expressed in activated coelomocytes are identified by expressed sequence tags. J. Immunol. 156, 593-602.

Smith, L.C., Clow, L.A. and Terwilliger, D.P. (2001) The ancestral complement system in sea urchins. Immunol. Rev. 180, 16-34.

Smith, L.C., Shih, C-S. and Dachenhausen, S.G. (1998) Coelomocytes express SpBf, a homologue of factor B, the second component in the sea urchin complement system. J. Immunol. 161, 6784-6793.

Song, W-C., Sarrias, M.R. and Lambris, J.D. (2000) Complement and innate immunity. Immunopharmacology 49, 187-198.

Suzuki, M.M., Satoh, N. and Nonaka, M. (2002) C6-like and C3-like molecules from the cephalochordate amphioxus, suggest a cytolytic complement system in invertebrates. J. Mol. Evol. 54, 671-679.
Vienne, A. and Pontarotti, P. (2006) Metaphylogeny of 82 gene families sheds a new light on chordate evolution.Int. J. Biol. Sci. 2, 32-37.
Vogel, C.W., Bredehorst, R., Fritzinger, D.C., Grunwald, T., Ziegelmuller, P. and Kock, M.A. (1996) Structure and function of cobra venom factor, the complement-activating protein in cobra venom. Adv. Exp. Med. Biol. 391, 97-114.
Wetsel, R.A., Kildsgaard, J. and Haviland, D.L. (2000) Complement anaphylatoxins (C3a, C4a, C5a) and their receptors (C3aR, C5aR/CD88) as therapeutic targets in inflammation. In: J.D. Lambris and V.M. Holers, (Eds.), *Therapeutics Interventions in the Complement System.* The Humana Press, New York, pp. 113-153.
Yoshizaki, F.Y., Ikawa, S., Satake, M, Satoh, N., Nonaka, M. (2005) Structure and the evolutionary implication of the triplicated complement factor B genes of a urochordate ascidian, *Ciona intestinalis.* Immunogenetics 56, 930-942.
Zhu, Y., Thangamani, S., Ho, B. and Ding, J.L. (2005) The ancient origin of the complement system. EMBO J. 24, 382-394.

27

Biological Roles of Lectins in Innate Immunity: Molecular and Structural Basis for Diversity in Self/Non-Self Recognition

Gerardo R. Vasta, Hafiz Ahmed, Satoshi Tasumi, Eric W. Odom, and Keiko Saito,

University of Maryland Biotechnology Institute, Center of Marine Biotechnology, 701 East Pratt Street, Baltimore, MD 21202. vasta@umbi.umd.edu

1 Introduction

Lectins and other pattern recognition proteins are critical components of innate immune mechanisms in invertebrates and vertebrates. Unlike immunoglobulins, TCRs, and VLRs, which generate diversity in recognition by genetic recombination, lectins like most innate immune receptors are "hard-wired" in the germline. Therefore, one of the outstanding questions is how the innate immune system is able to cope with the great diversity of potential microbial infectious challenges. Although the concept of pattern recognition proposes that only a handful of microbial conserved surface molecules need to be recognized for successful innate immune defense, the highly diversified microbial communities to which all organisms are exposed to and the dynamic changes in surface expression components suggests that a substantial diversity in non-self recognition mechanisms may be required for immune protection. The detailed analysis of the structural basis of lectin ligand binding and the diversity and complexity of the lectin repertoires in taxa that lack adaptive immunity, such as invertebrates, strongly suggests that this is the case. Further, recent studies have extended these observations to ectothermic vertebrates. In this review we focus on C-type and F-type lectins to illustrate how in the absence of genetic rearrangement, a substantial degree of diversity in recognition and effector functions is still achieved. The presence of multigene families, tandemly arrayed polymorphic recognition domains, formation of chimeric structures by exon shuffling, and a considerable "plasticity" of their carbohydrate binding sites further contribute to expand the ligand recognition spectrum and functional diversification of lectins as innate immune receptors. Spatial and temporal changes in expression profiles of lectins further enhance the functional capacity of these receptors during infectious challenge. Further, in the course of evolution some members of both C- and F- lectin families have been co-opted to carry out functions unrelated to innate immunity. The invaluable contribution of ongoing genome, transcriptome and proteome projects on non-mammalian model organisms towards a more realistic assessment of the lectin repertoire diversity, coupled to the structural analysis of selected components, and the use of forward and reverse genetic approaches in those model organisms amenable to genetic manipulation, continues to uncover novel structural features and further insight into their functional aspects.

Interactions between cells, or between cells and the extracellular matrix (ECM), can be modulated by both soluble and cell surface-associated receptor proteins that decode information from cell surface and ECM glycans, such as glycoproteins, polysaccharides and glycolipids. The recognition of endogenous ligands by lectins have been proposed to mediate developmental processes, as well as indirect roles in innate (Khan and Kubes 2003; Ley and Kansas 2004) and adaptive immunity (Barrionuevo, Beigier-Bompadre, Ilarregui, Toscano, Bianco, Isturiz, and Rabinovich 2007; Rabinovich, Toscano, Ilarregui, and Rubinstein 2004). Lectins can also decode information present in exogenous glycans, as well as "modified" endogenous glycans, such as those displayed by tumor cells. Thus, the efficient, specific, discrimination between "self" and "non-self" achieved by decoding of information present in cell surface sugar moieties constitutes a critical component of innate immune mechanisms, particularly when associated to equally efficient effector mechanisms, such as opsonization or complement activation, that lead to the destruction of the infectious challenge.

Unlike immunoglobulins, T cell receptors (TCRs), and variable lymphocyte receptors (VLRs), which generate diversity in recognition by genetic recombination, lectins like most innate immune receptors, are "hard-wired" in the germline. Therefore, one of the critical outstanding questions has been how the innate immune system is able to cope ith the great diversity of potential microbial challenges. Although the concept of "pattern recognition" proposes that only a handful of microbial conserved surface molecules need to be recognized for successful innate immune defense, the inherent diversity of the microbial world (and thus, of potential pathogens), and their dynamic changes in surface expression suggests that a substantial diversity in non-self recognition is required. The detailed analysis of the structural basis of lectin ligand binding and the diversity and complexity of the lectin repertoires in taxa that lack adaptive immunity, such as invertebrates, strongly suggests that this is the case. Further, recent studies have extended these observations to ectothermic vertebrates. The invaluable information from ongoing genome, transcriptome and proteome projects on non-mammalian model organisms towards a more realistic assessment of the lectin repertoire diversity, coupled to the structural analysis of selected components, and the use of forward and reverse genetic approaches in those model organisms amenable to genetic manipulation, has uncovered novel structural features and provided further insight into their functional aspects. In addition to the identification of multigene families, and multiple isoforms for each family member, the formation of chimeric structures by CRDs with other recognition and effector domains, presence of tandemly arrayed polymorphic CRDs, and in some cases, their considerable "plasticity" of their carbohydrate binding sites further contribute to expand the ligand recognition spectrum and functional diversification of lectins as innate immune receptors. Spatial and temporal changes in expression profiles of lectins further enhance the functional capacity of these receptors during infectious challenge.

2 Current classification of animal lectins

Although monomeric lectins have been described, most soluble or membrane-associated lectins are covalently or non-covalently bound oligomeric assemblages of peptide subunits characterized by the presence of one or more CRDs (Taylor and Drickamer 2003). Current classification of animal lectins are based on the presence of conserved amino acid sequence motifs within the CRD, distinct properties such as requirement of divalent cations or a reducing environment for ligand binding, and most importantly the structural fold (Table 1). Several major families are now recognized: C-, F-, P-, X-, and I-types, galectins (formerly S-type) and heparin-binding. Pentraxins, such as C-reactive protein (CRP) and serum amyloid P, are now considered a lectin family. CRP is a prototypical pentraxin that recognizes and opsonizes bacteria and parasites, and activates complement, a property that shares with collectins, ficolins, and other lectins that mediate innate immune mechanisms (Bottazzi, Garlanda, Salvatori, Jeannin, Manfredi, and Mantovani 2006; Vasta, Quesenberry, Ahmed, and O'Leary 2001). Other pattern recognition molecules such as the peptidoglycan receptor proteins (PGRPs), which are specific receptors that recognize the bacterial cell wall peptidoglycan (Kim, Byun, and Oh 2003), are broadly distributed in vertebrates and invertebrates alike (Kang, Liu, Lundstrom, Gelius, and Steiner 1998). The recently identified F-type lectin family (Cammarata, Benenati, Odom, Salerno, Vizzini, Vasta, and Parrinello 2007; Odom and Vasta 2006) will be discussed in further detail later in this review.

3 Lectins as recognition and effector factors in innate immunity, and modulators of adaptive immune responses

Lectin-carbohydrate interactions mediate a variety of biological processes, such as glycoprotein folding, trafficking, and clearance, triggering of signalling pathways, fertilization, and microbial attachment, (Varki, Cummings, Esko, Freeze, Hart, and Marth 1999). Equally relevant is the lectin-mediated recognition of exogenous or endogenous that can, directly or indirectly, mediate immune defense mechanisms (Vasta, Quesenberry, Ahmed, and O'Leary 1999). Furthermore, the instructive roles of innate immunity on adaptive immunity are widely recognized now, and the functional interplay between lectins and their receptors implicated in various cellular/humoral functions, has been characterized in considerable detail (Akira, Takeda, and Kaisho 2001; Liu and Rabinovich 2005; Ludwig, Geijtenbeek, and van

Table 1. Classification of animal lectins

Lectin groups	Ca^{2+}	Specificity	Defining features
C-type	Yes	Variable	C-type sequence motif
Galectins (S-type)	No	β-galactosides	S-type sequence motif
P-type (M-6-P R)	Variable	Mannose-6-P	P-type sequence motif
I-type (Siglecs and sialoadhesins)	No	Variable	Ig-like domains
Pentraxins	Most	PC/Galactosides	Multimeric binding motif
Heparin-binding type	No	Heparin/Heparan-SO_4^{2-}	Basic amino acid clusters
F-type	Variable	L-fucose	F-type sequence motif
Calnexin	Yes	$Glc_1 Man_9$	Calnexin sequence motif
M-type	Yes	Man_8	M-type sequence motif
L-type	Yes	Variable	L-type sequence motif
R-type	No	Variable	R-type sequence motif
F-box	No	$GlcNAc_2$	F-box sequence motif
Ficolin	Yes	GlcNAc, GalNAc	Ficolin sequence motif
Chitinase-like (Chilectins)	No	Chito-ologosaccharides	triosephosphate isomerase (TIM) barrel-like structure
Intelectins (X-type)	Yes	Gal, galactofuranose, pentose	Intelectin sequence motif

Kooyk 2006). For example, C-type lectins, ficolins, siglecs, and galectins are not only key players in innate immune processes that lead to pathogen recognition, endocytosis, complement activation, and antigen processing, but are also involved in the regulation of adaptive immune functions, including B and T cell clonal selection, maturation, activation, and apoptosis.

In addition to acting as recognition molecules for potential pathogens, some humoral lectins, such as collectins, ficolins and pentraxins, function as effector factors by promoting their opsonization, phagocytosis, and activating the complement system (Fujita 2002). This activation pathway can be initiated by the binding of a lectin, such as the mannose-binding lectin (MBL) to microbial surfaces, and subsequent association with a serine protease [MBL-associated serine protease (MASP)] which, by activating C3, can lead to either phagocytosis of the opsonized target *via* the complement receptor, or its lysis *via* assembly of the membrane attack complex. Although initially described as mediators of developmental processes, galectins have been recently demonstrated to participate directly or indireectly in immune functions (Rabinovich, Rubinstein, and Toscano 2002). Galectins may directly bind microbial pathogens (John, Jarvis, Swanson, Leffler, Cooper, Huflejt, and Griffiss 2002), induce (galectin 1) or prevent (galectin 3) T cell apoptosis (Baum, Blackall, Arias-Magallano, Nanigian, Uh, Browne, Hoffmann, Emmanouilides, Territo, and Baldwin 2003), and modulate T cell responses (Rabinovich et al. 2002), and possibly the activity of the macrophage membrane receptor for complement components C3b and iC3b (Vasta et al. 1999). Lectins and other pattern-recognition molecules also play critical roles in quali- and quantitatively modulating other innate immune responses such as NK cell function, and adaptive immune responses through the requirement of co-stimulatory signals from cell surface proteins on antigen presenting cells to naïve T lymphocytes, via CD28 receptors (Becker and Lowe 2003). In summary, the functional interplay of various humoral and cell-associated lectin types in the acute phase response contributes not only to quickly recognize and neutralize the microbial challenge, but lead to an effective long term adaptive immunity.

Unlike vertebrates, however, invertebrates and protochordates lack adaptive immunity and rely solely on innate immunity for defense against microbial infection. Except for a few members of the immunoglobulin superfamily, such as hemolin (Lindstrom-Dinnetz, Sun, and Faye 1995; Mendoza and Faye 1999; Sun, Lindstrom, Boman, Faye, and Schmidt 1990) and mosaic FREP (Fibrinogen-related proteins) containing immunoglobulin domains (Adema, Hertel, Miller, and Loker 1997; Leonard, Adema, Zhang, and Loker 2001; Zhang and Loker 2003), no *bona fide* components of adaptive immunity are present in invertebrates or protochordates. Therefore, the structural and functional aspects of the innate immunite recognition and effector factors, particularly on lectins, toll-like receptors, and complement in invertebrates and protochordates have drawn particular attention (Fujita, Matsushita, and Endo 2004; Iliev, Roach, Mackenzie, Planas, and Goetz 2005; Khalturin, Panzer, Cooper, and Bosch 2004). Unlike immunoglobulins, T-cell receptors, and the recently identified variable lymphocyte receptors (VLRs) of agnathans (Pancer and Cooper 2006), lectins do not generate diversity in recognition by genetic recombination and therefore, great interest has arisen on the sources for diversity within the lectin repertoires, including the presence of multigene families, allelic variation or polymorphisms, alternative splicing, tandem gene duplications, and formation of chimeric structures by exon shuffling, as well as the structural basis for the potential "plasticity" of their carbohydrate binding sites (Garred, Larsen, Seyfarth, Fujita, and Madsen 2006; Vasta, Ahmed, Du, and Henrikson 2004a; Vilches and Parham 2002).

4 The use of non-mammalian models to assess the structural and functional diversity of lectin repertoires

In spite of the contributions of mammalian models to addressing basic question in adaptive immunity, the use of invertebrates [*Drosophila melanogaster* (Adams et al, 2000), *Caenorhabditis elegans* (Dodd and Drickamer 2001)] and ectothermic vertebrates such as zebrafish (*Danio rerio*) (Ahmed, Du, O'Leary, and Vasta 2004; Vasta et al. 2004a) as alternative model organisms amenable to genetic manipulation has proven equally valuable in that the use forward and reverse genetic approaches has provided further insight into both the structural and the functional aspects of innate immunity. Further, the dramatic increase in recent years in the availability of genome and EST public databases for many invertebrates and ectothermic vertebrates has greatly facilitated the search for sequences based on known lectin domains or sequence motifs of interest, significantly contributing to the full characterization of their lectin repertoires, and providing insight into their structural aspects, biological roles, and evolution (Dodd and Drickamer 2001; Zelensky and Gready 2004). Although representatives of the lectin families typical of mammals, such as C-type lectins, galectins and pentraxins have been described in these taxa, the detailed study of selected model species have yielded either novel lectin variants, or novel lectin families with unique sequence motifs, multidomain arrangements, and a new structural fold. Recent studies of lectin repertoires in invertebrates and lower vertebrates have resulted in the identification of novel variants of known lectin types, such as the galectins from *Drosophila*, *C. elegans*, the oyster *Crassostrea virginica*, the ascidian *Clavelina picta*, and zebrafish, or novel lectin families, most of them still undescribed at the structural level, including the rhamnose-binding lectins from trout (Tateno, Ogawa, Muramoto, Kamiya, and Saneyoshi 2002) and the "pufflectins", mannose-binding lectins from the pufferfish (*Takifugu rubripes*) with intriguing homology to those of monocotyledonous plants (Tsutsui, Tasumi, Suetake, Kikuchi, and Suzuki 2006). In invertebrates, whole-genome studies in *Drosophila* and *C. elegans* demonstrated that the C-type lectin domain (CTLD) superfamily is abundant and diverse (Drickamer and Dodd 1999; Zelensky and Gready 2005). A search of the genome of the protochordate *Ciona intestinalis* revealed not only a diversified lectin repertoire (at least 2 pentraxins and 120 C-type lectins) but other immune-related genes (Azumi, De Santis, De Tomaso, Rigoutsos, Yoshizaki, Pinto, Marino, Shida, Ikeda, Ikeda, Arai, Inoue, Shimizu, Satoh, Rokhsar, Du Pasquier, Kasahara, Satake, and Nonaka 2003). The zebrafish and pufferfish genomes have also revealed the presence of complex lectin repertoires, with multiple members of the C-, F-, and X-type lectins, galectin, pentraxin, and other lectin families (Ahmed et al. 2004; Bianchet, Odom, Vasta, and Amzel 2002; Kakinuma, Endo, Takahashi, Nakata, Matsushita, Takenoshita, and Fujita 2003; Lee, Baum, Moremen, and Pierce 2004; Odom and Vasta 2006; Vasta et al. 2004a; Yoder, Litman, Mueller, Desai, Dobrinski, Montgomery, Buzzeo, Ota, Amemiya, Trede, Wei, Djeu, Humphray, Jekosch, Hernandez Prada, Ostrov, and Litman 2004; Zelensky and Gready 2004) (zebrafish: zfin.org; fugu: Ensembl accession SINFRUT00000162546). Furthermore, intriguing mosaic proteins incorporating carbohydrate-binding and immunoglobulin domains, such as immulectins, FREP, and novel immune-type receptors are present in invertebrates, protochordates and fish (Cannon, Haire, and Litman 2002; Zhang, Ademam, Kepler, and Locker 2004). C-type lectin receptors containing immunoreceptor tyrosine-based inhibition motifs have been identified in the rainbow trout (Yoder, Mueller, Nichols, Ristow, Thorgaard, Ota, and Litman 2002). In summary, the mining of genomic and transcriptomic databases has revealed that each organism displays a highly diversified lectin repertoire, within each lectin family represented by multiple members, each including variable numbers of lectin isoforms (Bianchet et al. 2002; Kakinuma et al. 2003; Lee et al. 2004; Odom and Vasta 2006;

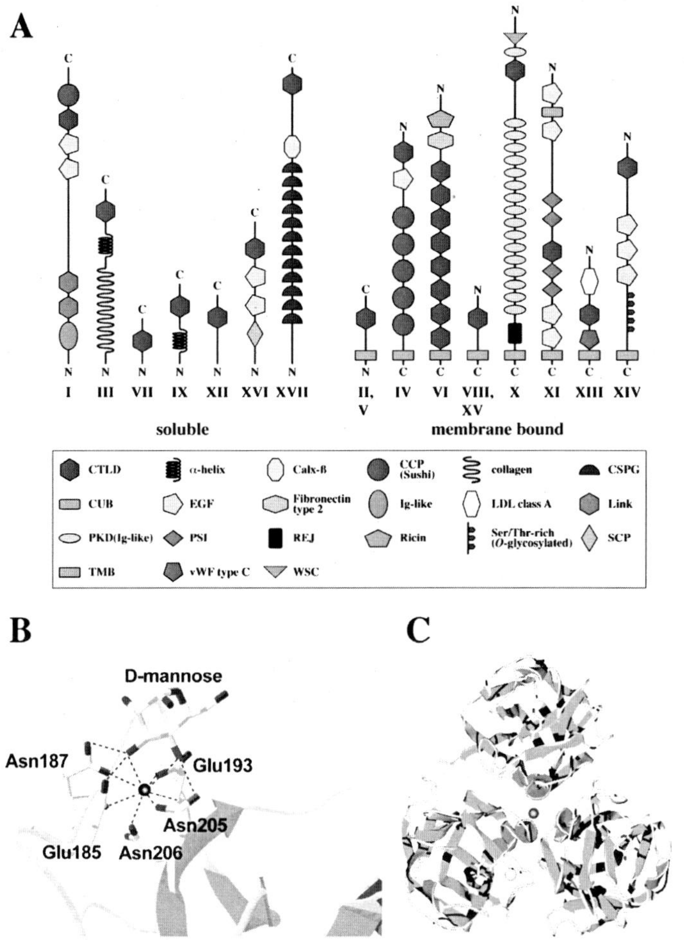

Fig. 1A. Domain architecture of the C-type lectin family. Group numbers indicate: I, lecticans; II, ASGR group; III, collectins; IV, selectins; V, NK receptors; VI, macrophage mannose receptor goup; VII, REG proteins; VIII, chondrolectin group; IX tetranectin group; X, polycystin 1; XI, attractin; XII, EMBP; XIII, DGCR2; XIV thrombomodulin group; XV, Bimlec; XVI, SEEC; XVII CBCP. N and C-terminus of protein are shown as "N" and "C" in this figure, respectively. This figure is adapted from Zelensky and Gready 2005.

B. Ca^{2+} -dependent carbohydrate binding of CTLD. The interactions of MBP-A amino acid residues and Ca^{2+} with D-mannose are shown by dashed line.

C. Quaternary structure of MBL.

Zelensky and Gready 2004). These studies have also revealed that other lectin families, such as P-type lectins, appear to be conspicuously absent in invertebrates and protochordates. This extends to the siglecs, representatives of which can be identified only as early as teleost fish (Lehmann, Gathje, Kelm, and Dietz 2004).

In the following sections of this review, the C-type lectins, and a novel family of fucose-binding lectins with a novel signature sequence motif and structural fold, the F-lectins, will be used to illustrate the diversification of lectin repertoires.

5 C-type lectins

Initially identified their Ca^{2+} requirement for ligand binding, C-type lectins are characterized by a conserved amino acid sequence motif, a structural fold, and the frequent association of CTLDs with other recognition or effector domains. The latter characteristic is evident in the collectins (MBLs, ficolin, conglutinin, pulmonary surfactant), proteoglycan core proteins, selectins, endocytic receptors, the mannose-macrophage receptor, and DC-SIGN (Weis, Taylor, and Drickamer 1998; Zhang, Robison, Thorgaard, and Ristow 2000) (Fig. 1 A). Collectins are lectins with a collagenous region linked to the CTLD that recognizes sugars on microbial surfaces, and upon binding to a serine protease (MBL-associated serine proteases; MASPs) may activate the complement cascade (Wallis 2002).

The CTLD fold forms a double-loop structure with its N- and C-terminal β strands (β1, β5) coming close together. The long second loop lies within the domain and it enters and exits the core domain at the same location. Four conserved cysteine residues (C1-C4) in the CTLD form disulfide bridges at the bases of the loops. The β5 strand and α1 helix (the whole domain loop) are linked by the residues C1 and C4, while β3 and β5 strands (the long loop region) are linked by the C2 and C3 residues. The long loop region is involved in Ca^{2+}-dependent carbohydrate binding, and in domain-swapping dimerization of some CTLDs (Liu and Eisenberg 2002; Weis et al. 1998). CTLDs are structurally two types: canonical and compact. The canonical CTLDs exhibit a long loop region, which is missing in the compact CTLD. The "Link" or "protein tandem repeat" (PTR) domains (Brissett and Perkins 1996) and bacterial CTLDs (Kelly, Prasannan, Daniell, Fleming, Frankel, Dougan, Connerton, and Matthews 1999) are examples of the compact CTLDs. CTLDs are also known as "long form" and "short form". The long form CTLD contains an N-terminal extension that forms a β-hairpin at the base of the domain. The hairpin is stabilized by an additional cystine bridge. The short form CTLD lacks this N-terminal extension. The CTLD structure contains four Ca^{2+}-binding sites –of which sites 1, 2 and 3 are located in the upper lobe of the structure. The Ca^{2+}-binding site 4 is involved in salt bridge formation between α2 and the β1/β5 sheet (Feinberg, Uitdehaag, Davies, Wallis, Drickamer, and Weis 2003; Loeb and Drickamer 1988). The Ca^{2+}-binding site 2 together with the amino acid residues form two characteristic motifs (galactose-type QPD motif and mannose-type EPN motif) in the CTLD sequence, and are directly involved in carbohydrate binding (Fig. 1 B). The replacement of the EPN sequence in MBP-A with the QPD sequence switches its specificity from mannose to galactose (Drickamer 1992). The CTLD/Ca^{2+}/monosaccharide complex is stabilized by a network of coordination and hydrogen bonds (Weis, Drickamer, and Hendrickson 1992): oxygen atoms from 4- and 3- hydroxyls of the mannose form two coordination

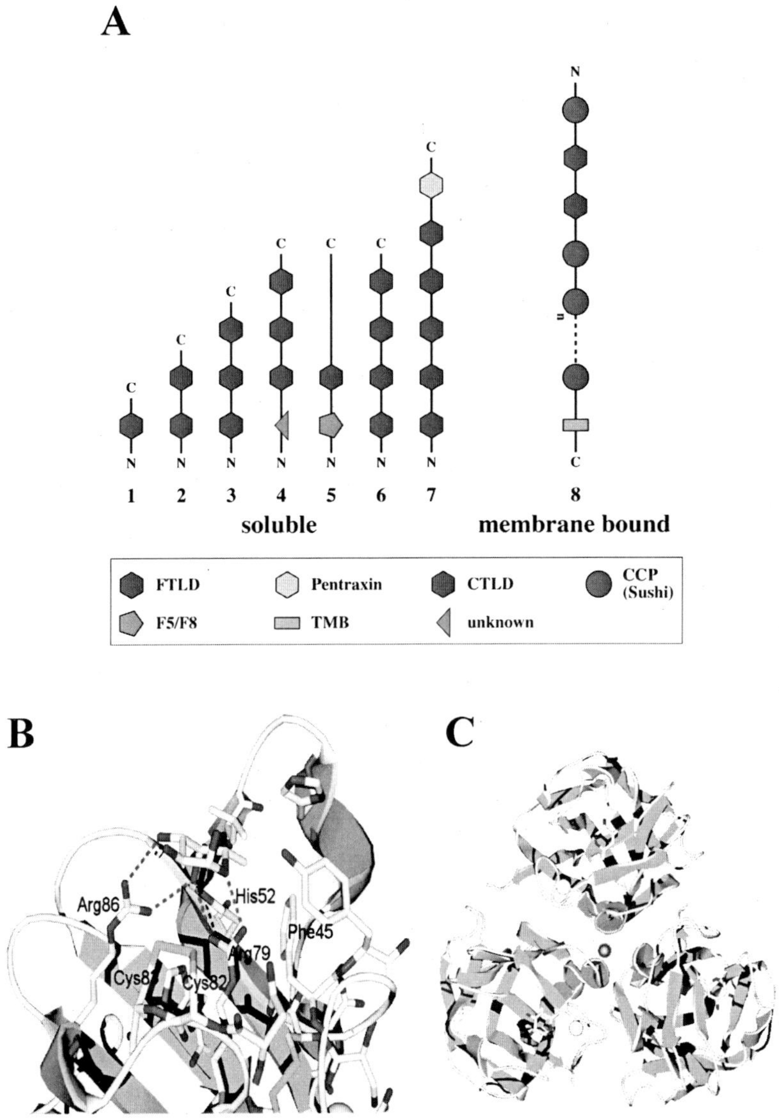

Fig. 2A. Domain architecture of the F-type lectin family. FBPL, FBP-like lectin; UNK, unknown; FAC5/8, coagulation factor V and VIII; PXN, pentraxin; CCP, complement control protein; TMB, transmembrane. The subscript "n" indicates the extended number of CCP domains present in *furrowed*. Numbers indicate: 1, *Anguilla anguilla* agglutinin (AAA); 2, MsaFBP32; 3, *X. tropicalis*; 4, *S. pneumoniae*; 5, *M. degradans*; 6, *O. mykiss*; 7, Xla-PXN-FBPL; 8, CG9095, *furrowed*.

B. Crystallographic model of AAA. View of α-Fuc bound to AAA. Broken green lines represent the hydrogen bonds involved in ligand binding. Oxygen is colored red. The exocyclic C6 is proximal to the face of Phe 45, which provides a hydrophobic environment to accommodate this group.

C. Quaternary structure of AAA. The single chlorine ion marking the three-fold axis of rotation is coordinated by a lysine (Lys16) from each subunit.

bonds with the Ca^{2+} ion and four hydrogen bonds with the carbonyl side chains that form the Ca^{2+}-binding site 2. The Ca^{2+}-binding site 2 is not only involved in carbohydrate-binding, but also in an interaction with a non-carbohydrate ligand as the antifreeze protein from Atlantic herring directly interacts with the ice crystal (Ewart, Li, Yang, Fletcher, and Hew 1998). Bound Ca^{2+} ions (at sites 1, 2, and 4) may have stabilizing effect on CTLD structure as removal of Ca^{2+} greatly increases CTLD susceptibility to proteolysis (Loeb and Drickamer 1988; Zelensky and Gready 2005).

Several lectins homologous of MBLs and ficolins, MASPs, and complement components have been identified in invertebrates and ectothermic vertebrates, suggesting that C-type lectins and the complement system played a pivotal role in innate immunity before the emergence of adaptive immunity in vertebrates. Immulectins have been identified in several insect and crustacean genomes, and may mediate defense responses in invertebrates in a manner similar to the mammalian MBLs. The immulectins isolated from the moth *Manduca sexta* hemolymph serve as a surveillance mechanism by binding to microbial surface molecules such as peptidoglycan, lipopolysaccharide, lipoteichoic acid, and β1,3-glucan (Ling and Yu 2006). The binding triggers diverse responses such as phagocytosis, nodule formation, encapsulation, melanization, and synthesis of anti-microbial peptides. At present, both collectins and group II/V receptors have been described from teleosts (Vitved, Holmskov, Koch, Teisner, Hansen, Salomonsen, and Skjodt 2000). Galactose-binding C-type lectins, were identified in the Japanese eel *Anguilla japonica* (Tasumi, Ohira, Kawazoe, Suetake, Suzuki, and Aida 2002) and rainbow trout (*Oncorhynchus mykiss*) (Zhang et al. 2000). More recent studies in rainbow trout revealed CTLDs associated with two extracellular immunoglobulin domains and cytoplasmic immunoreceptor tyrosine-based inhibition motifs (ITIMs) in the "novel immune-type receptors" (NITRs) (van den Berg, Yoder, and Litman 2004). In addition, CTLDs including selectin and macrophage mannose receptor gene sequences are detectable in the whole-shotgun genome assembly of zebrafish, fugu, *Drosophila* and *C. elegans* (Zelensky and Gready 2004; Zelensky and Gready 2005).

6 F-type lectins

The F-type is the most recent lectin family to be identified and structurally characterized, and its members are widely distributed from ectothermic invertebrates to protochordates and vertebrates (Bianchet et al. 2002; Odom and Vasta 2006; Vasta, Ahmed, and Odom 2004b). Like CTLDs, F-type lectin domains (FTLDs), are found in varying orders of tandemly arrayed repeats, yielding subunits of variable sizes, even within a single species (Odom and Vasta 2006; Vasta et al. 2004b) (Fig. 2A). The 3-D structure of the first F-type lectin from the European eel (*Anguilla anguilla*) agglutinin (AAA) in complex with the ligand α-L-fucose. α-Fuc) has provided not only a novel CRD sequence motif, but also a novel animal lectin fold, which consists of a β-jelly roll sandwich composed of three- and five-stranded β-sheets (Bianchet et al. 2002). The lectin binding to α-Fuc is mediated by hydrogen bonds from a trio of

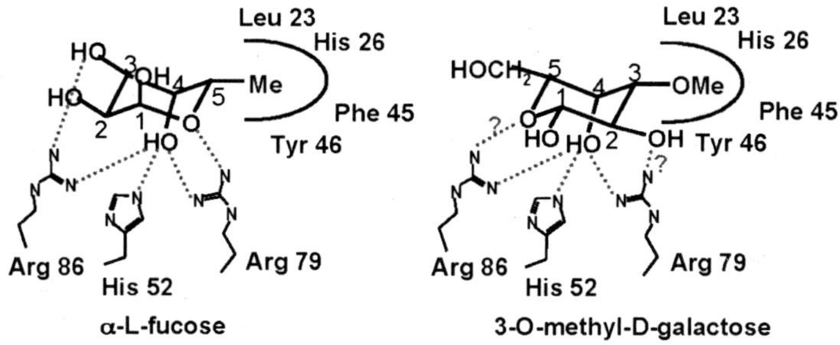

A

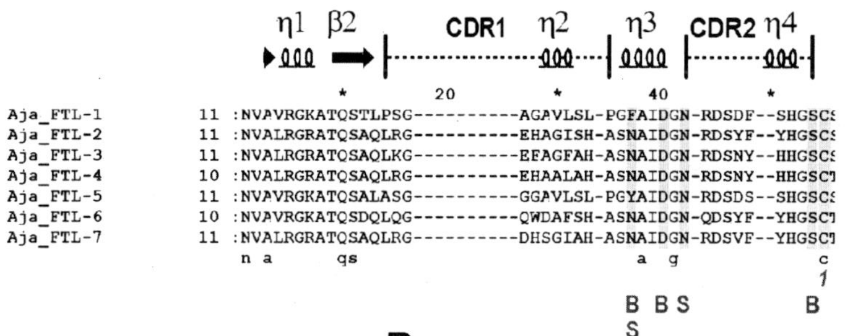

B

Fig. 3A. Docking of 3-O-methyl-D-galactose in AAA-α-Fuc complex. Additional interactions are possible with amino acid residues in the loops surrounding the binding pocket.
B. Amino acid sequence alignment of FTLD isoforms from the Japanese eel. The variability of critical residues in the binding pocket and surrounding loops in the multiple isoforms contribute alternative interactions with terminal and subterminal sugar units and thus expand the range of diverse oligosaccharides recognition by the lectin isoform repertoire. Substitutions are also present in critical positions of ligand binding loops of the AAA isoforms.

basic side chains that emerge from a shallow pocket, van der Waals contacts with a unique disulfide bridge formed by contiguous cysteines, and seclusion of the 6-deoxy methyl moiety in a hydrophobic cavity (Fig. 2B). Docking of H type 1 and Le[a] trisaccharides in AAA-α-Fuc complex revealed that additional interactions are evident with residues in the loops surrounding the binding pocket conferring fine discrimination of its cognate glycoconjugate ligands (Fig. 3 A). Interestingly, variability of critical residues in the binding pocket and surrounding loops in the multiple isoforms, as expressed in the Japanese eel (Fig. 3 B) (Honda, Kashiwagi, Miyamoto, Takei, and Hirose 2000), may contribute alternative interactions with terminal and subterminal sugar units and thus expand the range of diverse oligosaccharides recognition by the lectin isoform repertoire (Bianchet et al. 2002). This is an intriguing observation for a protein proposed to mediate the recognition of potential microbial pathogens. The homotrimeric structure and the three-fold cyclic symmetry of the native AAA (Fig. 2 C) is very similar to collectins (see Fig. 1 C) (Weis and Drickamer 1994), and would optimize the spacing and orientation of binding sites for recognition of glycoconjugates displayed on microbial surfaces. In AAA, calcium appears to play in structural stabilization, rather than participating in direct cation-saccharide interactions.

Although AAA possesses a single FTLD, a large number of F-type lectins of diverse domain topologies were identified in a variety of taxa from prokaryotes to amphibians (Honda et al. 2000; Odom and Vasta 2006; Saito, Hatada, Iwanaga, and Kawabata 1997) (see Fig. 2 A). For examples, most teleost F-type lectins contain either duplicate or quadruplicate tandem domains, whereas in *Xenopus* spp., these lectins are composed of either triplicate or quintuple tandem F-type domains. Clearly, the F-type fold with its joined N- and C-terminals favors the formation of concatenated CRD topologies in numbers that appear lineage-related. These tandem arrays may yield mosaic proteins by including pentraxin (*X. laevis*) or C-type domains (*D. melanogaster* CG9095, malarial mosquito, and honey bee).

Similarity searches in genomic databases revealed the presence of the FTLD sequence motif in both Gram positive and Gram negative bacteria, both lophotrochozoan (i.e. molluscs) and ecdysozoan protostomes (i.e. horseshoe crabs), invertebrate deuterostome (i.e. echinoderm), cartilaginous vertebrates (i.e skate), and the early branching clades of vertebrates, lobe-finned and ray-finned fish. In contrast, the FTLD sequence motif appears absent from protozoa, fungi, nematodes, ascidians, and higher vertebrates (i.e. reptiles, birds and mammals) suggesting that it may have been selectively lost even in relatively closely-related lineages. The paucity of bacteria possessing FTLDs suggests that it may have either been acquired through horizontal transfer from metazoans, or less likely, that most prokaryote lineages lost this CRD. The absence of the FTLDs in higher vertebrates is an evolutionary quandary that correlates with the appearance of cleidoic egg, and the colonization of land by vertebrates.

The structure of the AAA-α-Fuc complex enabled the thorough analysis of variations in the FTLD sequence motif focusing regions relevant for ligand or metal binding, or tertiary structure. Alignment of the primary structures of various F-type lectins (Odom and Vasta 2006) revealed that most FTLDs can bind α-Fuc as they share common α-Fuc-binding motif (HX24RXDX4(R/K). Some FTLDs, however, deviate from this α-Fuc-binding motif, suggesting that they may have either alternate carbohydrate-binding specificities or even lack carbohydrate-binding activity. For example, FTLD of *D. melanogaster* CG9095 is unable to produce the typical H-bonds due to the substitution of two amino acid residues in the α-Fuc-binding motif. Moreover, most duplicate tandem F-type lectins (i.e. striped bass, zebrafish) possess a unique combination of saccharide-binding motif and cystine bonds (N-terminal: CXHX24RGDCCXERXX16XX22C) and (C-terminal: CXHX24RDXXXERCX16CX22C). Specifically, one domain has lost the contiguous cystines that contact the saccharide ring while it gained the nested cystine.

Like other proteins, FTLDs contain sequence insertions or deletions (indels) in the loops where they have minimal effect on the core fold. Interestingly, the CDR1 loop, which interacts with saccharides subterminal to α-Fuc, shows great divergence suggesting that it might regulate binding to the wide diversity of glycoconjugates.

Although the FTLD fold is unique among animal lectins, a structure-based search identified several unrelated proteins, which shared the same jellyroll fold with AAA (Vasta et al. 2004b), despite the negligible sequence similarity to the F-type lectin. The C domains of human blood coagulation factor V (FA58C) and VIII (PDB 1CZT) (Macedo-Ribeiro, Bode, Huber, Quinn-Allen, Kim, Ortel, Bourenkov, Bartunik, Stubbs, Kane, and Fuentes-Prior 1999), the C-terminal domain of a bacterial sialidase (PDB 1EUT) (Gaskell, Crennell, and Taylor 1995), the NH_2-terminal domain of a fungal galactose oxidase (PDB 1GOF) (Firbank, Rogers, Wilmot, Dooley, Halcrow, Knowles, McPherson, and Phillips 2001; Ito, Phillips, Stevens, Ogel, McPherson, Keen, Yadav, and Knowles 1991), a subunit of the human APC10/DOC1 ubiquitin ligase (PDB 1XNA) (Wendt, Vodermaier, Jacob, Gieffers, Gmachl, Peters, Huber, and Sondermann 2001), the N-terminal domain of the XRCC1 single-strand DNA repair complex (PDB 1JHJ) (Marintchev, Mullen, Maciejewski, Pan, Gryk, and Mullen 1999), and a yeast allantoicase (PDB 1SG3) (Leulliot, Quevillon-Cheruel, Sorel, Graille, Meyer, Liger, Blondeau, Janin, and van Tilbeurgh 2004) are a few examples. The higher vertebrate F-lectin analogues may have originated from carbohydrate-binding domains, that later evolved new specificities. It also appears that all of these families share the placement of their binding-site among the loops at the opening of the β-barrel.

7 Conclusions

The use of non-mammalian models (invertebrates, protochordates and ectothermic vertebrates) for the assessment of the diversity and structural/functional characterization of the lectin repertoires in innate immunity has proved to be very useful. A variety of experimental *in vivo, in vitro,* and more recently, *in silico* approaches using available genomic databases have enabled greater insight into the diversity and complexity of their lectin repertoires. The identification of members from lectin families previously described in mammals has resulted in the discovery of novel structural features, most likely revealing functional adaptations along the lineages leading to the mammals. Further, the identification of novel lectin families such as the F-type lectins, underscores the fact that more research in non-mammalian model organisms will provide new information on both the structural, functional and evolutionary aspects of lectin repertoires that may not be as obvious in mouse or man. The elucidation of the detailed structural aspects of lectin-ligand binding plasticity, and the presence of multiple isoforms that confer substantial diversity in oligosaccharide binding, provides the structural basis for novel mechanisms for generating diversity for non-self recognition in innate immunity, that resembles those operative through adaptive immunity in higher vertebrates. Ongoing genome, transcriptome and proteome projects on model organisms representative of non-mammalian taxa will continue to reveal not only the extent of their full lectin repertoires, but coupled to the structural analysis of selected components has the potential to uncover novel structural features, on which a rigorous experimental assessment of their biological roles may be supported.

8 Acknowledgements

Work cited in this review was supported by grants R01 GM070589-01 from the National Institutes of Health, and MCB-00-77928 and IOB 0618409 from the National Science Foundation

9 Abbreviations

CRD, carbohydrate recognition domain; CTLD, C-type lectin-like domain; DC-SIGN, dendritic cell-specific ICAM-grabbing non-integrin; MBL, mannose-binding lectin; AAA, *Anguilla anguilla* agglutinin

References

Adams, M.D., Celniker, S.E., Holt, R.A., Evans, C.A., Gocayne, J.D., Amanatides, P.G., Scherer, S.E., Li, P.W., Hoskins, R.A., Galle, R.F., George, R.A., Lewis, S.E., Richards, S., Ashburner, M., Henderson, S.N., Sutton, G.G., Wortman, J.R., Yandell, M.D., Zhang, Q., Chen, L.X., Brandon, R.C., Rogers, Y.H., Blazej, R.G., Champe, M., Pfeiffer, B.D., Wan, K.H., Doyle, C., Baxter, E.G., Helt, G., Nelson, C.R., . Gabor, G.L., Abril, J.F., Agbayani, A., An, H.J., Andrews-Pfannkoch, C., Baldwin, D., Ballew, R.M., Basu, A., Baxendale, J., Bayraktaroglu, L., Beasley, E.M., Beeson, K.Y., Benos, P.V., Berman, B.P., Bhandari, D., Bolshakov, S., Borkova, D., Botchan, M.R., Bouck, J., Brokstein, P., .Brottier, P., Burtis, K.C., Busam, D.A., Butler, H., Cadieu, E., Center, A., Chandra, I., Cherry, J.M., Cawley, S., Dahlke, C., Davenport, L.B., .Davies, P., de Pablos, B., Delcher, A., Deng, Z., Mays, A.D., Dew, I., Dietz, S.M., Dodson, K., Doup, L.E., Downes, M., Dugan-Rocha, S., Dunkov, B.C., Dunn, P., Durbin, K.J., Evangelista, C.C., Ferraz, C., Ferriera, S., Fleischmann, W., Fosler, C., Gabrielian, A.E., Garg, N.S., Gelbart, W.M., Glasser, K., Glodek, A., Gong, F., Gorrell, J.H., Gu, Z., Guan, P., Harris, M., Harris, N.L., Harvey, D., Heiman, T.J., Hernandez, J.R., Houck, J., Hostin, D., Houston, K.A., Howland, T.J., Wei, M.H., Ibegwamm, C., Jalali, M., Kalush, F., Karpen, G.H., Ke, Z., Kennison, J.A., Ketchum, K.A., Kimmel, B.E., Kodira, C.D., Kraft, C., Kravitz, S., Kulp, D., Lai, Z., Lasko, P., Lei, Y., Levitsky, A.A., J., L., Z., L., Liang, Y., Lin, X., Liu, X., Mattei, B., McIntosh, T.C., McLeod, M.P., McPherson, D., G., M., V., M.N., C., M., J., M., Moshrefi, A., Mount, S.M., Moy, M., Murphy, B., Murphy, L., Muzny, D.M., Nelson, D.L., Nelson, D.R., Nelson, K.A., Nixon, K., Nusskern, D.R., Pacleb, J.M., Palazzolo, M., Pittman, G.S., Pan, S., Pollard, J., Puri, V., Reese, M.G., Reinert, K., Remington, K., Saunders, R.D., Scheeler, F., Shen, H., Shue, B.C., Siden-Kiamos, I., Simpson, M., Skupski, M.P., Smith, T., Spier, E., Spradling, A.C., Stapleton, M., Strong, R., Sun, E., Svirskas, R., Tector, C., Turner, R., Venter, E., Wang, A.H., Wang, X., Wang, Z.Y., Wassarman, D.A., Weinstock, G.M., Weissenbach, J., Williams, S.M., Woodage, T., Worley, K.C., Wu, D., Yang, S., Yao, Q.A., Ye, J., Yeh, R.F., Zaveri, J.S., Zhan, M., Zhang, G., Zhao, Q., Zheng, L., Zheng, X.H., Zhong, F.N., Zhong, W., Zhou, X., Zhu, S., Zhu, X., Smith, H.O., Gibbs, R.A., Myers, E.W., Rubin, G.M. and Venter, J.C. (2000) The genome sequence of *Drosophila melanogaster*. Science 287, 2185-2195.

Adema, C.M., Hertel, L.A., Miller, R.D. and Loker, E.S. (1997) A family of fibrinogen-related proteins that precipitates parasite-derived molecules is produced by an invertebrate after infection. Proc. Natl. Acad. Sci. U S A 94, 8691-8696.

Ahmed, H., Du, S.J., O'Leary, N. and Vasta, G.R. (2004) Biochemical and molecular characterization of galectins from zebrafish (*Danio rerio*): notochord-specific expression of a prototype galectin during early embryogenesis. Glycobiology 14, 219-232.

Akira, S., Takeda, K. and Kaisho, T. (2001) Toll-like receptors: critical proteins linking innate and acquired immunity. Nat. Immunol. 2, 675-680.

Azumi, K., De Santis, R., De Tomaso, A., Rigoutsos, I., Yoshizaki, F., Pinto, M.R., Marino, R., Shida, K., Ikeda, M., Ikeda, M., Arai, M., Inoue, Y., Shimizu, T., Satoh, N., Rokhsar, D.S., Du Pasquier, L., Kasahara, M., Satake, M. and Nonaka, M. (2003) Genomic analysis of immunity in a Urochordate and the emergence of the vertebrate immune system: "waiting for Godot". Immunogenetics 55, 570-581.

Barrionuevo, P., Beigier-Bompadre, M., Ilarregui, J.M., Toscano, M.A., Bianco, G.A., Isturiz, M.A. and Rabinovich, G.A. (2007) A novel function for galectin-1 at the crossroad of innate and adaptive immunity: galectin-1 regulates monocyte/macrophage physiology through a nonapoptotic ERK-dependent pathway. J. Immunol. 178, 436-445.

Baum, L.G., Blackall, D.P., Arias-Magallano, S., Nanigian, D., Uh, S.Y., Browne, J.M., Hoffmann, D., Emmanouilides, C.E., Territo, M.C. and Baldwin, G.C. (2003) Amelioration of graft versus host disease by galectin-1. Clin. Immunol. 109, 295-307.

Becker, D.J. and Lowe, J.B. (2003) Fucose: biosynthesis and biological function in mammals. Glycobiology 13, 41R-53R.

Bianchet, M.A., Odom, E.W., Vasta, G.R. and Amzel, L.M. (2002) A novel fucose recognition fold involved in innate immunity. Nat. Struct. Biol. 9, 628-634.

Bottazzi, B., Garlanda, C., Salvatori, G., Jeannin, P., Manfredi, A. and Mantovani, A. (2006) Pentraxins as a key component of innate immunity. Curr. Opin. Immunol. 18, 10-15.

Brissett, N.C. and Perkins, S.J. (1996) The protein fold of the hyaluronate-binding proteoglycan tandem repeat domain of link protein, aggrecan and CD44 is similar to that of the C-type lectin superfamily. FEBS Lett. 388, 211-216.

Cammarata, M., Benenati, G., Odom, E.W., Salerno, G., Vizzini, A., Vasta, G.R. and Parrinello, N. (2007) Isolation and characterization of a fish F-type lectin from gilt head bream (*Sparus aurata*) serum. Biochim. Biophys. Acta 1770, 150-155.

Cannon, J.P., Haire, R.N. and Litman, G.W. (2002) Identification of diversified genes that contain immunoglobulin-like variable regions in a protochordate. Nat. Immunol. 3, 1124-1125.

Dodd, R.B. and Drickamer, K. (2001) Lectin-like proteins in model organisms: implications for evolution of carbohydrate-binding activity. Glycobiology 11, 71-79.

Drickamer, K. (1992) Engineering galactose-binding activity into a C-type mannose-binding protein. Nature 360, 183-186.

Drickamer, K. and Dodd, R.B. (1999) C-Type lectin-like domains in *Caenorhabditis elegans*: predictions from the complete genome sequence. Glycobiology 9, 1357-1369.

Ewart, K.V., Li, Z., Yang, D.S., Fletcher, G.L. and Hew, C.L. (1998) The ice-binding site of Atlantic herring antifreeze protein corresponds to the carbohydrate-binding site of C-type lectins. Biochemistry 37, 4080-4085.

Feinberg, H., Uitdehaag, J.C., Davies, J.M., Wallis, R., Drickamer, K. and Weis, W.I. (2003) Crystal structure of the CUB1-EGF-CUB2 region of mannose-binding protein associated serine protease-2. EMBO J. 22, 2348-2359.

Firbank, S.J., Rogers, M.S., Wilmot, C.M., Dooley, D.M., Halcrow, M.A., Knowles, P.F., McPherson, M.J. and Phillips, S.E. (2001) Crystal structure of the precursor of galactose oxidase: an unusual self-processing enzyme. Proc. Natl. Acad. Sci. U S A 98, 12932-12937.

Fujita, T. (2002) Evolution of the lectin-complement activation pathway and its role in innate immunity. Nat. Rev. Immunol. 2, 346-353.

Fujita, T., Matsushita, M. and Endo, Y. (2004) The lectin-complement pathway--its role in innate immunity and evolution. Immunol. Rev. 198, 185-202.

Garred, P., Larsen, F., Seyfarth, J., Fujita, R. and Madsen, H.O. (2006) Mannose-binding lectin and its genetic variants. Genes Immun. 7, 85-94.

Gaskell, A., Crennell, S. and Taylor, G. (1995) The three domains of a bacterial sialidase: a beta-propeller, an immunoglobulin module and a galactose-binding jelly-roll. Structure 3, 1197-1205.
Honda, S., Kashiwagi, M., Miyamoto, K., Takei, Y. and Hirose, S. (2000) Multiplicity, structures, and endocrine and exocrine natures of eel fucose-binding lectins. J. Biol. Chem. 275, 33151-33157.
Iliev, D.B., Roach, J., C., Mackenzie, S., Planas, J.V. and Goetz, F.W. (2005) Endotoxin recognition: in fish or not in fish? FEBS Lett. 579, 6519-6528.
Ito, N., Phillips, S.E., Stevens, C., Ogel, Z.B., McPherson, M.J., Keen, J.N., Yadav, K:D. and Knowles, P.F. (1991) Novel thioether bond revealed by a 1.7 A crystal structure of galactose oxidase. Nature 350, 87-90.
John, C.M., Jarvis, G.A., Swanson, K.V., Leffler, H., Cooper, M.D., Huflejt, M.E. and Griffiss, J.M. (2002) Galectin-3 binds lactosaminylated lipooligosaccharides from *Neisseria gonorrhoeae* and is selectively expressed by mucosal epithelial cells that are infected. Cell Micorbiol. 4, 649-662.
Kakinuma, Y., Endo, Y., Takahashi, M., Nakata, M., Matsushita, M., Takenoshita, S. and Fujita, T. (2003) Molecular cloning and characterization of novel ficolins from *Xenopus laevis*. Immunogenetics 55, 29-37.
Kang, D., Liu, G., Lundstrom, A., Gelius, E. and Steiner, H. (1998) A peptidoglycan recognition protein in innate immunity conserved from insects to humans. Proc. Natl. Acad. Sci. U S A 95, 10078-10082.
Kelly, G., Prasannan, S., Daniell, S., Fleming, K., Frankel, G., Dougan, G., Connerton, I. and Matthews, S. (1999) Structure of the cell-adhesion fragment of intimin from enteropathogenic *Escherichia coli*. Nat. Struct. Biol. 6, 313-318.
Khalturin, K., Panzer, Z., Cooper, M.D. and Bosch, T.C. (2004) Recognition strategies in the innate immune system of ancestral chordates. Mol. Immunol. 41, 1077-1087.
Khan, A.I. and Kubes, P. (2003) L-selectin: an emerging player in chemokine function. Microcirculation 10, 351-358.
Kim, M.S., Byun, M. and Oh, B.H. (2003) Crystal structure of peptidoglycan recognition protein LB from *Drosophila melanogaster*. Nat. Immunol. 4, 787-793.
Lee, J.K., Baum, L.G., Moremen, K. and Pierce, M. (2004) The X-lectins: a new family with homology to the *Xenopus laevis* oocyte lectin XL-35. Glycoconj.J. 21, 443-450.
Lehmann, F., Gathje, H., Kelm, S. and Dietz, F. (2004) Evolution of sialic acid-binding proteins: molecular cloning and expression of fish siglec-4. Glycobiology 14, 959-968.
Leonard, P.M., Adema, C.M., Zhang, S.M. and Loker, E.S. (2001) Structure of two FREP genes that combine IgSF and fibrinogen domains, with comments on diversity of the FREP gene family in the snail *Biomphalaria glabrata*. Gene 269, 155-165.
Leulliot, N., Quevillon-Cheruel, S., Sorel, I., Graille, M., Meyer, P., Liger, D., Blondeau, K., Janin, J. and van Tilbeurgh, H. (2004) Crystal structure of yeast allantoicase reveals a repeated jelly roll motif. J. Biol. Chem. 279, 23447-23452.
Ley, K. and Kansas, G.S. (2004) Selectins in T-cell recruitment to non-lymphoid tissues and sites of inflammation. Nat. Rev. Immunol. 4, 325-335.
Lindstrom-Dinnetz, I., Sun, S.C. and Faye, I. (1995) Structure and expression of Hemolin, an insect member of the immunoglobulin gene superfamily. Eur. J. Biochem. 230, 920-925.
Ling, E. and Yu, X.Q. (2006) Cellular encapsulation and melanization are enhanced by immulectins, pattern recognition receptors from the tobacco hornworm *Manduca sexta*. Dev. Comp. Immunol. 30, 289-299.
Liu, F.T. and Rabinovich, G.A. (2005) Galectins as modulators of tumour progression. Nat. Rex. Cancer 5, 29-41.
Liu, Y. and Eisenberg, D. (2002) 3D domain swapping: as domains continue to swap. Protein Sci. 11, 1285-1299.
Loeb, J.A. and Drickamer, K. (1988) Conformational changes in the chicken receptor for endocytosis of glycoproteins. Modulation of ligand-binding activity by Ca^{2+} and pH. J. Biol. Chem. 263, 9752-9760.
Ludwig, I.S., Geijtenbeek, T.B. and van Kooyk, Y. (2006) Two way communication between neutrophils and dendritic cells. Curr. Opin. Pharmacol., 4.
Macedo-Ribeiro, S., Bode, W., Huber, R., Quinn-Allen, M.A., Kim, S.W., Ortel, T.L., Bourenkov, G.P., Bartunik, H.D., Stubbs, M.T., Kane, W.H. and Fuentes-Prior, P. (1999) Crystal structures of the membrane-binding C2 domain of human coagulation factor V. Nature 402, 434-439.

Marintchev, A., Mullen, M.A., Maciejewski, M.W., Pan, B., Gryk, M.R. and Mullen, G.P. (1999) Solution structure of the single-strand break repair protein XRCC1 N-terminal domain. Nat. Struct. Biol. 6, 884-893.

Mendoza, H.L. and Faye, I. (1999) Physiological aspects of the immunoglobulin superfamily in invertebrates. Dev. Comp. Immunol. 23, 359-374.

Odom, E.W. and Vasta, G.R. (2006) Characterization of a binary tandem domain F-type lectin from striped bass (*Morone saxatilis*). J. Biol. Chem. 281, 1698-1713.

Pancer, Z. and Cooper, M.D. (2006) The evolution of adaptive immunity. Annu. Rev. Immunol. 24, 497-518.

Rabinovich, G.A., Rubinstein, N. and Toscano, M.A. (2002) Role of galectins in inflammatory and immunomodulatory processes. Biochim. Biophys. Acta 1572, 274-284.

Rabinovich, G.A., Toscano, M.A., Ilarregui, J.M. and Rubinstein, N. (2004) Shedding light on the immunomodulatory properties of galectins: novel regulators of innate and adaptive immune responses. Glycoconj. J. 16, 565-573.

Saito, T., Hatada, M., Iwanaga, S. and Kawabata, S. (1997) A newly identified horseshoe crab lectin with binding specificity to O-antigen of bacterial lipopolysaccharides. J. Biol. Chem. 272, 30703-30708.

Sun, S.C., Lindstrom, I., Boman, H.G., Faye, I. and Schmidt, O. (1990) Hemolin: an insect-immune protein belonging to the immunoglobulin superfamily. Science 250, 1729-1732.

Tasumi, S., Ohira, T., Kawazoe, I., Suetake, H., Suzuki, Y. and Aida, K. (2002) Primary structure and characteristics of a lectin from skin mucus of the Japanese eel *Anguilla japonica*. J. Biol. Chem. 277, 27305-27311.

Tateno, H., Ogawa, T., Muramoto, K., Kamiya, H. and Saneyoshi, M. (2002) Rhamnose-binding lectins from steelhead trout (*Oncorhynchus mykiss*) eggs recognize bacterial lipopolysaccharides and lipoteichoic acid. Biosci. Biotechnol. Biochem. 66, 604-612.

Taylor, M.E. and Drickamer, K. (2003) Binding of oligosaccharide ligands to the selectins requires additional interactions with the carbohydrate-recognition domains. In: M.E. Taylor and K. Drickamer (Eds), Introduction of Glycobiology. Oxford University Press, Oxford; New York, p. 207.

Tsutsui, S., Tasumi, S., Suetake, H., Kikuchi, K. and Suzuki, Y. (2006) Carbohydrate-binding site of a novel mannose-specific lectin from fugu (*Takifugu rubripes*) skin mucus. Comp. Biochem. Physio. B Biochem. Mol. Biol. 143, 514-519.

van den Berg, T.K., Yoder, J.A. and Litman, G.W. (2004) On the origins of adaptive immunity: innate immune receptors join the tale. Trends Immunol. 25, 11-16.

Varki, A., Cummings, R., Esko, J., Freeze, H., Hart, G. and Marth, J. (1999) Essentials of Glycobiology. Cold Spring Harbor Laboratory Press, New York.

Vasta, G.R., Ahmed, H., Du, S.-J. and Henrikson, D. (2004a) Galectins in teleost fish: Zebrafish (*Danio rerio*) as a model species to address their biological roles in development and innate immunity. Glycoconj. J. 21, 503-521.

Vasta, G.R., Ahmed, H. and Odom, E.W. (2004b) Structural and functional diversity of lectin repertoires in invertebrates, protochordates and ectothermic vertebrates. Curr. Opin. Struct. Biol. 14, 617-630.

Vasta, G.R., Quesenberry, M., Ahmed, H. and O'Leary, N. (1999) C-type lectins and galectins mediate innate and adaptive immune functions: their roles in the complement activation pathway. Dev. Comp. Immunol. 23, 401-420.

Vasta, G.R., Quesenberry, M.S., Ahmed, H. and O'Leary, N. (2001) Lectins from tunicates: structure-function relationships in innate immunity. Adv. Exp. Med. Biol. 484, 275-287.

Vilches, C. and Parham, P. (2002) KIR: diverse, rapidly evolving receptors of innate and adaptive immunity. Annu. Rev. Immunol. 20, 217-251.

Vitved, L., Holmskov, U., Koch, C., Teisner, B., Hansen, S., Salomonsen, J. and Skjodt, K. (2000) The homologue of mannose-binding lectin in the carp family cyprinidae is expressed at high level in spleen, and the deduced primary structure predicts affinity for galactose. Immunogenetics 51, 955-964.

Wallis, R. (2002) Structural and functional aspects of complement activation by mannose-binding protein. Immunibiology 205, 433-445.

Weis, W.I. and Drickamer, K. (1994) Trimeric structure of a C-type mannose-binding protein. Structure 2, 1227-1240.

Weis, W.I., Drickamer, K. and Hendrickson, W.A. (1992) Structure of a C-type mannose-binding protein complexed with an oligosaccharide. Nature 360, 127-134.

Weis, W.I., Taylor, M.E. and Drickamer, K. (1998) The C-type lectin superfamily in the immune system. Immunol. Rev. 163, 19-34.
Wendt, K.S., Vodermaier, H.C., Jacob, U., Gieffers, C., Gmachl, M., Peters, J.M., Huber, R. and Sondermann, P. (2001) Crystal structure of the APC10/DOC1 subunit of the human anaphase-promoting complex. Nat. Struct. Biol. 8, 784-788.
Yoder, J.A., Litman, R.T., Mueller, M.G., Desai, S., Dobrinski, K.P., Montgomery, J.S., Buzzeo, M.P., Ota, T., Amemiya, C.T., Trede, N.S., Wei, S., Djeu, J.Y., Humphray, S., Jekosch, K., Hernandez Prada, J.A., Ostrov, D.A. and Litman, G.W. (2004) Resolution of the novel immune-type receptor gene cluster in zebrafish. Proc. Natl. Acad. Sci. U S A 101, 15706-15711.
Yoder, J.A., Mueller, M.G., Nichols, K.M., Ristow, S.S., Thorgaard, G.H., Ota, T. and Litman, G.W. (2002) Cloning novel immune-type inhibitory receptors from the rainbow trout, *Oncorhynchus mykiss*. Immunogenetics 54.
Zelensky, A.N. and Gready, J.E. (2004) C-type lectin-like domains in *Fugu rubripes*. BMC Genomics 5, 51.
Zelensky, A.N. and Gready, J.E. (2005) The C-type lectin-like domain superfamily. FEBS Lett. 272, 6179-6217.
Zhang, H., Robison, B., Thorgaard, G.H. and Ristow, S.S. (2000) Cloning, mapping and genomic organization of a fish C-type lectin gene from homozygous clones of rainbow trout (*Oncorhynchus mykiss*). Biochim. Biophys. Acta 1494, 14-22.
Zhang, S.-M., Ademam, C.M., Kepler, T.B. and Locker, E.S. (2004) Diversification of Ig superfamily genes in an invertebrate. Science 305, 251-254.
Zhang, S.M. and Loker, E.S. (2003) The FREP gene family in the snail *Biomphalaria glabrata*: additional members, and evidence consistent with alternative splicing and FREP retrosequences. Fibrinogen-related proteins. Dev. Comp. Immunol. 27, 175-187.

28

Hydrogen/Deuterium Exchange Mass Spectrometry: Potential for Investigating Innate Immunity Proteins

Michael C. Schuster,[1] Hui Chen[2] and John D. Lambris[3]

[1] University of Pennsylvania, Department of Medicine, Division of Rheumatology, schustmi@mail.med.upenn.edu
[2] University of Pennsylvania School of Medicine, Department of Pathology and Laboratory Medicine, jacobhui1208@gmail.com
[3] University of Pennsylvania School of Medicine, Department of Pathology and Laboratory Medicine, lambris@mail.med.upenn.edu

1 Introduction

Innate immunity has evolved to provide a rapid, non-specific defensive response to invading pathogens (Kenzel and Henneke 2006). As central participants in the innate immune response, proteins mediate both the initiation and propagation of the defense reaction through their interactions with pathogen-derived macromolecules and other host proteins. Although the complement system constitutes an important line of defense, unregulated or misdirected activation of the components of innate immunity is detrimental and is associated with disease states, including autoimmunity and immune-mediated tissue injury. (Carroll and Holers 2005). Therefore, great effort has been expended to structurally characterize the molecular determinants of proteins of the innate immunity system that dictate their functions. Understanding protein structure and intermolecular interactions are key steps not only in delineating the mechanisms of innate immunity but also in the development of therapeutics that can modulate these interactions. Despite active investigation, a precise structural characterization of many such proteins remains elusive.

The gold standard for describing the static structure of macromolecules is X-ray crystallography. Crystallographic models are capable of providing precise descriptions of protein structure and, in the case of macromolecular complexes, intermolecular interfaces in the solid state. Unfortunately, this technique is often hindered by the inability to generate well-diffracting crystals, particularly in the case of heavily glycosylated proteins, flexible modular proteins or protein complexes. Once obtained, these solid-state models are snapshots of the single protein conformation and they do not necessarily provide insight into the range of structural fluctuations that proteins may adopt in solution. Filling this void, hydrogen/deuterium exchange has emerged as a complementary and powerful technique. This approach is capable of providing not only insight into the structure of proteins in solution but also information regarding intermolecular interactions in the liquid phase (Busenlehner and Armstrong 2005; Hoofnagle, Resing, and Ahn 2003; Lanman and Prevelige 2004; Wales and Engen 2006).

The amide backbone of proteins provides a unique tool for investigating protein structure. The backbone hydrogen atoms, bound to the amide bond nitrogen atoms, are capable of exchanging with water-derived hydrogens in solution at neutral pH. Hydrogen/deuterium exchange is a method for measuring this exchange process and is accomplished by substituting deuterium oxide (D_2O) for water in the buffer of the protein. Amide hydrogen atoms exchange with solvent deuterium atoms in a time-dependent fashion, the rate of which is influenced by their local environment. Solvent-exposed amide hydrogens undergo exchange with an average rate of $k_i \approx 10\ s^{-1}$ at neutral pH, while amide hydrogens that are buried in the protein interior and/or participating in hydrogen bonds have a significantly slower exchange rate ($k_i \approx 10\ s^{-9}$ at neutral pH), or do not exchange at all (Busenlehner et al. 2005; Zhang and Smith 1993). Thus, the rate of exchange for backbone amide hydrogens provides information about the solvent accessibility of these amides, and thus insight into the structure, together with the dynamics of the protein.

Measuring the kinetics of hydrogen/deuterium exchange requires an analytical technique capable of providing exchange information about individual, or small stretches of, amide hydrogens on the protein backbone. Traditionally, nuclear magnetic resonance (NMR) spectroscopy was the technique of choice for following the exchange process (Englander and Kallenbach 1983; Englander, Mayne, Bai, and Sosnick 1997). However, NMR spectroscopy requires a relatively high concentration of protein, and assignments are difficult when protein masses exceed 50 kDa. With the recent advances in mass spectrometry, electrospray ionization (ESI) mass spectrometry and matrix-assisted laser desorptive-ionization (MALDI) mass spectrometry have emerged as alternate methods for monitoring hydrogen/deuterium exchange rates (Eyles and Kaltashov 2004). These techniques are capable of providing similar resolution of the exchange reaction and do not suffer from the protein size limits that are applicable to NMR spectroscopy.

In hydrogen/deuterium exchange mass spectrometry (HDXMS), proteins are exposed to D_2O for varying amounts of time; aliquots are removed and quenched at pre-defined time points and then analyzed by mass spectrometry (Fig. 1). In the quench step, the pH of the aliquot is adjusted to pH $\approx$ 2.4, and the temperature is reduced, both serving to slow down non-specific exchange and "lock on" the deuterium atoms already incorporated. To achieve spatial resolution during the mass spectrometry step, proteins are subjected to rapid proteolysis, often with an acid-stable enzyme such as pepsin, and then analyzed (Cravello, Lascoux, and Forest 2003). In MALDI mass spectroscopy, aliquots are typically analyzed directly after proteolysis, without further chromatography or purification. When ESI mass spectroscopy is utilized, the incorporation of a brief liquid chromatography step prior to mass spectrometric analysis can further improve the resolution of the analysis. The data obtained through this analysis allows one to measure the change in m/z

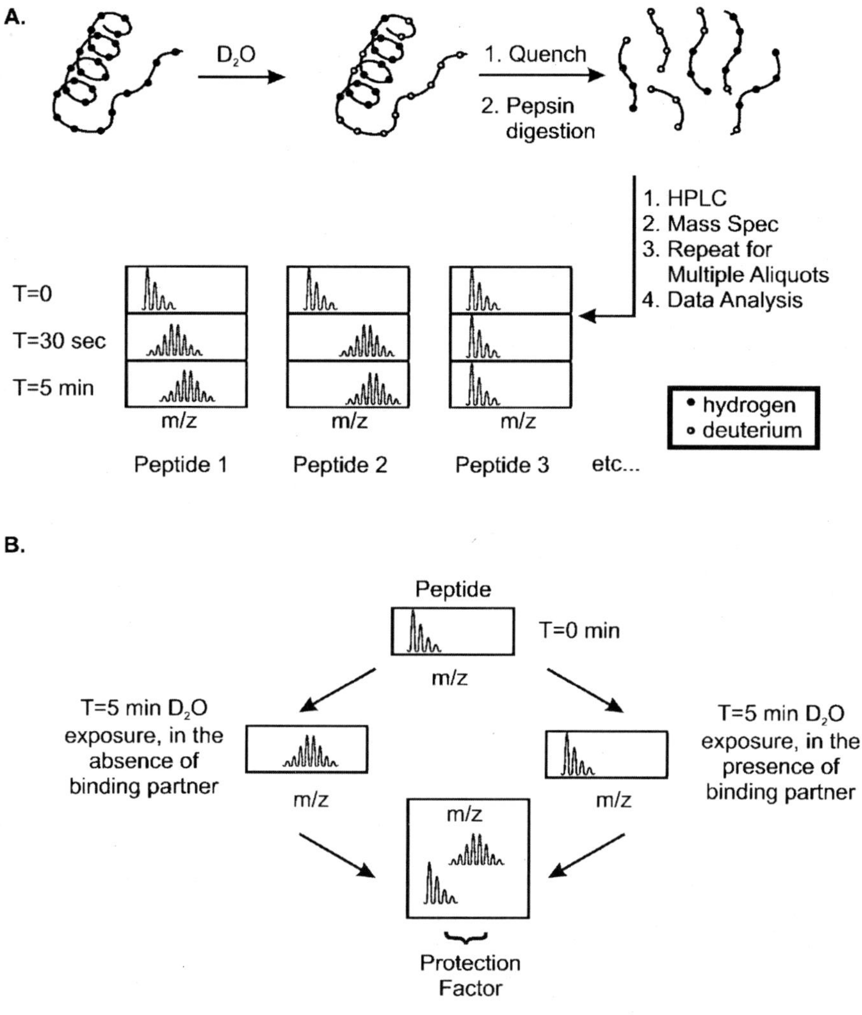

Fig. 1A. General overview of the hydrogen/deuterium exchange mass spectrometry method revealing expected mass envelope shifts for moderate (peptide 1), fast (peptide 2) and slow (peptide 3) exchanging peptides from a protein. **B.** Mass envelope shift for a single peptide derived from a protein involved in ligand binding. When the protein is incubated with D2O in the absence of a known ligand, there is an envelope shift for the peptide, suggesting deuterium exchange. When the protein is incubated with D2O in the presence of a known ligand, the same peptide is protected from deuterium exchange and does not readily undergo mass shift, suggesting that the peptide is shielded from solvent exposure as a result of ligand binding. This difference in mass shift between the bound and unbound states provides a quantitative measure of solvent protection, sometimes referred to as the protection factor.

(mass-to-charge ratio) for individual peptides as a function of time. Peptides that have a greater increase in m/z as a function of time, representing peptides with greater deuterium incorporation, are presumed to originate from less ordered and more solvent exposed environments. Although individual peaks yield the *average* deuterium uptake for a given peptide, the simultaneous analysis of multiple overlapping peptides provides resolution at nearly the amino acid level. {Peaks in the mass spectra are correlated with specific peptide stretches in the protein by performing MS/MS analysis of enzymatically cleaved proteins in their fully protonated state, using proteomic methods that are now routine (Domon and Aebersold 2006).}

HDXMS is a powerful technique that has been used to study a wide variety of proteins and macromolecules of varied sizes. Surprisingly, it has not found significant appreciation in the investigation of mechanisms of innate immunity, although it is capable of providing the precise structural information that is lacking for many of the proteins and protein complexes in this field. In this short review, we will highlight recent applications of HDXMS for the study of protein structure, protein-ligand interactions and protein-protein interactions in non-innate immunity systems. We will conclude by discussing our experience in using this technique to investigate the structure of complement protein C3 and its hydrolytic product C3(H_2O).

2 Applications of HDXMS

2.1 Protein Structure: HIV and SIV Nef Proteins

The Nef protein of the human (HIV-1 and HIV-2) and simian (SIV) immunodeficiency viruses is an important factor in the pathogenesis of infections caused by these viruses. By enhancing viral replication, downregulating receptors on immune cells and activating specific host signaling pathways, Nef promotes high viral loads and is necessary for progression to AIDS (Arold and Bauer 2001; Renkema and Saksela 2000). Although HIV Nef and SIV Nef both bind to the Hck tyrosine kinase, the mechanism of these interactions varies; HIV Nef interacts with the SH3 domain, whereas SIV Nef interacts with the SH2 domain of this tyrosine kinase. How the Nef proteins from these two viruses differ in structure is unknown, as no structural information is available for SIV Nef. Although HIV Nef has been investigated with a combination of techniques, including X-ray crystallography (Arold, Franken, Strub, Hoh, Benichou, Benarous, and Dumas 1997; Lee, Saksela, Mirza, Chait, and Kuriyan 1996) and NMR spectroscopy (Grzesiek, Bax, Clore, Gronenborn, Hu, Kaufman, Palmer, Stahl, and Wingfield 1996; Grzesiek, Bax, Hu, Kaufman, Palmer, Stahl, Tjandra, and Wingfield 1997), these studies have been hampered by the idiosyncrasies of the protein. Full-length HIV Nef could not be crystallized, and crystallographic data are available only for deletion mutants containing residues 54-205 and 58-206. Even then, residues 54-69 and 149-178 are disordered in both crystal structure models. Similarly, because Nef forms multimers at the concentrations required for NMR studies, NMR spectroscopy has been accomplished only for a deletion mutant of HIV Nef that is missing residues 2-39 and 159-173. Despite these limitations, a structural model of HIV Nef was developed from the available information (Geyer, Fackler, and Peterlin 2001). To better characterize the structure of HIV Nef and to gain insight into the structure of SIV Nef, Hochrein et al. took advantage of HDXMS to study both full-length proteins (Hochrein, Wales, Lerner, Schiavone, Smithgall, and Engen 2006).

Using an ESI-based HDXMS method with inline pepsin digestion, Hochrein and coworkers obtained deuterium exchange data for more than 85% of the primary sequence of both the HIV and SIV Nef proteins. Given the sensitivity of the instrumentation, they could utilize low concentrations of proteins, avoiding the aggregation that had prohibited NMR analysis of the full-length protein. With regard to HIV Nef, their results were in complete agreement with the pre-existing structure model; exchange was slow in areas shielded from the solvent and rapid in areas of predicted solvent exposure. It is particularly interesting that the exchange data from a central loop of the protein, obscured in the crystallographic map and deleted in the NMR experiments, suggested that this region was protected from solvent and likely possessed structure. When compared to HIV Nef, SIV Nef had a similar deuterium incorporation profile but appeared to have greater dynamics in solution.

This example of an HDXMS application confirms several key features of the method. Most importantly, it allowed for the structural characterization of full-sequence HIV Nef, which was not amenable to investigation by X-ray crystallography or NMR spectroscopy. Conversely, the HDXMS technique was itself validated by its agreement with the available structural model. Finally, using this technique, the authors were able to gain insight into a related but previously uncharacterized protein, SIV Nef.

2.2 Protein-Protein Interactions: Anthrax Lethal Factor

Bacillus anthracis, the organism that causes anthrax, secretes three monomeric proteins that mediate its toxicity toward host cells. Interestingly, these monomeric proteins participate in a series of intermolecular protein-protein interactions with one another, which facilitates entry of the toxins into host cells (Bradley, Mogridge, Mourez, Collier, and Young 2001; Molloy, Bresnahan, Leppla, Klimpel, and Thomas 1992; Scobie, Rainey, Bradley, and Young 2003). Assembly of the toxin begins when one of these secreted proteins, called protective antigen (PA), binds to host-cell surface receptors and is proteolytically cleaved. The cleaved, receptor-bound PA fragment can then bind additional PA proteins and spontaneously form the circular, homo-heptameric pre-pore. The pre-pore is capable of binding up to three copies of either of the two remaining secreted proteins, edema factor (EF) or lethal factor (LF). It has been suggested that EF and LF both bind to the pre-pore via their N-terminal 250 residue domains, termed EF_N and LF_N, in a competitive fashion (Elliott, Mogridge, and Collier 2000). The resulting pre-pore-EF or pre-pore-LF complexes are then internalized into the endosomal compartment, where the factors dissociate and the pre-pore forms a membrane-spanning pore. The factors EF and LF can then translocate into the cytosol, where they exert their toxic effects (Zhang, Finkelstein, and Collier 2004).

Understanding the protein-protein interactions that mediate this process has the potential to contribute to the development of anthrax-directed therapeutics. Previous work in this area, using mutagenesis, has indicated that a small patch of seven residues on EF_N/LF_N is responsible for binding to the pre-pore (Lacy, Mourez, Fouassier, and Collier 2002). However, complementary studies with the pre-pore revealed a much larger footprint on the pre-pore that was binding to EF_N/LF_N (Cunningham, Lacy, Mogridge, and Collier 2002). Melnyk et al. utilized HDXMS to better define the interacting surface between PA and LF_N, and with this information, they performed directed mutagenesis to identify critical residues in the pre-pore-LF_N interaction (Melnyk, Hewitt, Lacy, Lin, Gessner, Li, Woods, and Collier 2006).

Utilizing an ESI-based HDXMS technique with inline pepsin digestion, Melnyk and colleagues studied the termolecular complex of LF_N with two molar equivalents of trypsin-activated PA. Under optimized conditions, they obtained deuterium exchange data for over 75% of the primary sequence of LF_N. In the absence of PA, HDXMS analysis revealed rapid deuterium uptake in an area of LF_N that is known from the X-ray crystal structure to be surface-exposed. In the presence of two equivalents of activated PA, four discrete sites on LF_N (residues 95-120, 137-147, 177-189, and 225-235) were protected from the solvent. These residues formed one continuous surface on LF_N, which was larger than the protein-protein interface surface previously identified through mutagenesis studies. To better define the residues that are critical for binding, they prepared 27 mutants of LF_N by incorporating single amino acid substitutions into the observed interface region. When assayed for their ability to bind cells in the presence of PA, binding-defective mutants revealed the importance of multiple amino acids, including Asp-182 and Glu-135, which were separated by a larger-than-expected ~40 Å.

As in the case of the previously described study, the work by Melnyk et al. validates the HDXMS technique in terms of the crystal structure model of LF_N and the pattern of deuterium uptake. In addition, by comparing the deuterium uptake profile of the protein in the absence and then the presence of its binding partner, the authors were better able to describe the interaction surface between LF_N and its binding partner. Additional use of site-directed mutagenesis allowed them to define which residues that are important in binding and to obtain valuable structural information regarding this critical interaction.

2.3 Protein-Ligand Interactions: PPARγ

Peroxisome proliferator-activated receptor γ (PPARγ) is a ligand-dependant transcription factor whose actions influence glucose homeostasis and adipocyte differentiation (Gampe, Montana, Lambert, Miller, Bledsoe, Milburn, Kliewer, Willson, and Xu 2000; Yang, Rachez, and Freedman 2000). Because nuclear receptors such as PPARγ modulate fundamental biological processes such as metabolism and cellular differentiation, they are natural targets for drug development

(Olefsky and Saltiel 2000). PPARγ itself is a key regulator of glucose and lipid homeostasis and is a target of the anti-diabetes drugs rosiglitazone and pioglitazone. In addition, several agonists, partial agonists and antagonists of PPARγ have been synthesized (Berger et al. 2003; Brown et al. 1999; Leesnitzer et al. 2002).

It is known that nuclear receptors such as PPARγ undergo conformational changes upon ligand binding (Kallenberger, Love, Chatterjee, and Schwabe 2003). While the PPARγ full agonists rosiglitazone and pioglitazone are effective drugs in the treatment of diabetes mellitus, they cause undesirable side effects, such as edema and weight gain. Interestingly, a partial-agonist of PPARγ, nTZDpa, induces fewer side effects. While an X-ray crystal structure model of the ligand binding domain (LBD) of PPARγ with a partial agonist exists (Oberfield, Collins, Holmes, Goreham, Cooper, Cobb, Lenhard, Hull-Ryde, Mohr, Blanchard, Parks, Moore, Lehmann, Plunket, Miller, Milburn, Kliewer, and Willson 1999), such structural information for the LBD of PPARγ with agonists and antagonists is not available. Knowledge of the extent to which full agonists, partial agonists and antagonists influence PPARγ structure, and thus function, could be useful in the development and screening of new medications targeted toward nuclear receptors.

Hamuro and coworkers (Hamuro, Coales, Morrow, Molnar, Tuske, Southern, and Griffin 2006) have characterized the effects of agonists, partial agonists and antagonists on PPARγ structure using HDXMS. Utilizing an ESI-based approach with inline pepsin digestion, they characterized the deuterium exchange profile of the LBD of PPARγ alone and also in the presence of two full agonists (including rosiglitazone), the partial agonist nTZDpa, and a covalently binding antagonist. They identified 49 peptides for which they could monitor the deuterium uptake, spanning 97% of the PPARγ LBD sequence.

In the absence of ligand, the deuterium uptake profile of the PPARγ LBD was consistent with its known structure. When the PPARγ LBD was analyzed in the presence of either of the two full agonists tested, the deuterium uptake was remarkably different. Specifically, the exchange profile was dramatically slowed in nine segments, suggesting that multiple areas of the protein became less flexible/solvent-exposed, including the entire binding cavity, after binding the agonist. In contrast, data obtained with the PPARγ LBD in the presence of an antagonist revealed a slowing of exchange in only five segments. The partial agonist had the smallest structural perturbation, decreasing the exchange in only three segments. All four ligands slowed the exchange at residues 281-287; however, only the two agonists slowed the exchange at residues 445-451 and 472-477. From this information, the authors concluded that the full agonists stabilize the PPARγ LBD more fully than do the partial agonist or antagonist. In addition, given the unique ability of the full agonists to stabilize the two stretches (445-451 and 472-477) in the LBD, these data provide the foundation for future structure-function research with this important class of nuclear receptors.

3 HDXMS Utilized to Explore the Structures of C3 and C3(H_2O)

The complement system is an important component of innate and acquired immunity and a primary contributor to the inflammatory response. It consists of approximately 30 proteins that generally circulate in the blood in inactive forms (zymogens). Upon activation by one of three known pathways, the classical, lectin or alternative pathways, the active complement proteins assume varied roles, functioning as fluid-phase or cell-bound enzymes, chemoattractants, opsonins and cytolytic factors (the membrane attack complex) (Walport 2001a, 2001b). Complement protein C3 plays a central role in complement activation and is capable of binding to over 20 different proteins (Lambris 1988; Sahu and Lambris 2001). As a result of enzymatic activity occurring within the complement cascade, C3 undergoes significant conformational changes as it is activated and then degraded (Becherer, Alsenz, and Lambris 1990).

In addition to the changes in C3 structure that are mediated by its interactions with other proteins, C3 undergoes conformational changes during its non-enzymatic conversion to C3(H_2O). C3(H_2O) is the product of spontaneous hydrolysis of the C3 thioester bond. C3(H_2O), through the binding to factor B, forms the alternative complement pathway convertase C3(H_2O)Bb (Fishelson, Pangburn, and Mullereberhard 1984; Xu, Narayana, and Volanakis 2001). Because the alternative pathway convertase is implicated in several autoimmune diseases, there is interest in characterizing the molecular determinants that lead to its formation (Thurman and Holers 2006). X-ray crystal structures have been reported for C3 and its breakdown products C3c and C3d, but no crystal structure has been published for C3(H_2O) (Janssen, Huizinga, Raaijmakers, Roos, Daha, Nilsson-Ekdahl, Nilsson, and Gros 2005; Nagar, Jones, Diefenbach, Isenman, and Rini 1998).

To better characterize the structural changes that occur with the formation of C3(H_2O), our laboratory utilized MALDI-based HDXMS to investigate the structural differences between C3 and C3(H_2O) (Winters, Spellman, and Lambris 2005). To begin these studies, we utilized ESI mass spectrometry to identify 354 peptides produced from the pepsin-catalyzed digestion of C3. The identified peptides accounted for 80% of the total C3 primary sequence. Once this task was completed, deuteration studies and enzymatic digestions were performed *in silico*, and the resulting fragments were analyzed by MALDI-time of flight (MALDI-TOF) mass spectrometry. With this technique, we were able to monitor the deuterium incorporation for 31 peptides derived from the C3 and C3(H_2O) sequences, with a significant difference in uptake observed between the two proteins in the case of 17 peptides.

Analysis of these data in the context of the known interactions in which C3 and C3(H_2O) participate was provocative. As compared to C3, C3(H_2O) exhibited increased deuterium uptake at one of its factor I cleavage sites, via the spanning peptide 944-967. Factor I is responsible for degrading C3(H_2O) and C3b after their activation, thereby limiting excessive complement activation. Therefore, the data suggest that conformational changes in this region might predispose the molecule to digestion by factor I. With these data, 30% of the C3d region of C3 and C3(H_2O) could also be analyzed. There was a difference in deuterium incorporation for 7 out of

10 peptides identified from this region after conversion from C3 to C3(H_2O). This diffusely distributed and extensive change in deuterium incorporation suggested a significant alteration in the structure of this region of the protein. Interestingly, the recently published X-ray crystal structure of C3b, the protein with which C3(H_2O) shares significant functional similarities, does indeed exhibit a significant change in the C3d region of the protein.

4 Conclusions

Hydrogen/deuterium exchange, coupled with mass spectrometry, has emerged as a powerful technique for the structural analysis of proteins. It is capable of providing structural information, including data concerning solvent-exposed surface areas of previously uncharacterized proteins. For proteins whose crystal structures have been determined, the information obtained from HDXMS is additive and can provide clues to the structural dynamics of the protein in solution. Finally, and perhaps most significantly, HDXMS has been shown to be a powerful tool in the mapping of protein-protein and protein-ligand binding sites. Our laboratory has initiated studies utilizing HDXMS to investigate the interactions of C3 with other complement components. These studies are directed toward precise characterization of the mechanisms through which C3 interacts with its molecular partners within the complement cascade.

References

Alsenz, J., Becherer, J.D., Nilsson, B. and Lambris, J.D. (1990). Structural and Functional-Analysis of C3 Using Monoclonal-Antibodies. Current Topics in Microbiology and Immunology, 153, 235-248.

Arold, S., Franken, P., Strub, M.P., Hoh, F., Benichou, S., Benarous, R. and Dumas, C. (1997). The crystal structure of HIV-1 Nef protein bound to the Fyn kinase SH3 domain suggests a role for this complex in altered T cell receptor signaling. Structure, 5(10), 1361-1372.

Arold, S.T. and Bauer, A.S. (2001). Dynamic Nef and Nef dynamics: how structure could explain the complex activities of this small HIV protein. Trends in Biochemical Sciences, 26(6), 356-363.

Becherer, J.D., Alsenz, J. and Lambris, J.D. (1990). Molecular Aspects of C3 Interactions and Structural Functional-Analysis of C3 from Different Species. Current Topics in Microbiology and Immunology, 153, 45-72.

Berger, J.P., Petro, A.E., Macnaul, K.L., Kelly, L.J., Zhang, B.B., Richards, K., Elbrecht, A., Johnson, B.A., Zhou, G.C., Doebber, T.W., Biswas, C., Parikh, M., Sharma, N., Tanen, M.R., Thompson, G.M., Ventre, J., Adams, A.D., Mosley, R., Surwit, R.S. and Moller, D.E. (2003). Distinct properties and advantages of a novel peroxisome proliferator-activated protein gamma selective modulator. Molecular Endocrinology, 17(4), 662-676.

Bradley, K.A., Mogridge, J., Mourez, M., Collier, R.J. and Young, J.A.T. (2001). Identification of the cellular receptor for anthrax toxin. Nature, 414(6860), 225-229.

Brown, K.K., Henke, B.R., Blanchard, S.G., Cobb, J.E., Mook, R., Kaldor, I., Kliewer, S.A., Lehmann, J.M., Lenhard, J.M., Harrington, W.W., Novak, P.J., Faison, W., Binz, J.G., Hashim, M.A., Oliver, W.O., Brown, H.R., Parks, D.J., Plunket, K.D., Tong, W.Q., Menius, J.A., Adkison, K., Noble, S.A. and Willson, T.M. (1999). A novel N-aryl tyrosine activator of peroxisome proliferator-activated receptor-gamma reverses the diabetic phenotype of the Zucker diabetic fatty rat. Diabetes, 48(7), 1415-1424.

Busenlehner, L.S. and Armstrong, R.N. (2005). Insights into enzyme structure and dynamics elucidated by amide H/D exchange mass spectrometry. Archives of Biochemistry and Biophysics, 433(1), 34-46.
Carroll, M.C. and Holers, V.M. (2005). Innate autoimmunity. Advances in Immunology, 86, 137-157.
Cravello, L., Lascoux, D. and Forest, E. (2003). Use of different proteases working in acidic conditions to improve sequence coverage and resolution in hydrogen/deuterium exchange of large proteins. Rapid Communications in Mass Spectrometry, 17(21), 2387-2393.
Cunningham, K., Lacy, D.B., Mogridge, J. and Collier, R.J. (2002). Mapping the lethal factor and edema factor binding sites on oligomeric anthrax protective antigen. Proceedings of the National Academy of Sciences of the United States of America, 99(10), 7049-7053.
Domon, B. and Aebersold, R. (2006). Review - Mass spectrometry and protein analysis. Science, 312(5771), 212-217.
Elliott, J.L., Mogridge, J. and Collier, R.J. (2000). A quantitative study of the interactions of Bacillus anthracis edema factor and lethal factor with activated protective antigen. Biochemistry, 39(22), 6706-6713.
Englander, S.W. and Kallenbach, N.R. (1983). Hydrogen-Exchange and Structural Dynamics of Proteins and Nucleic-Acids. Quarterly Reviews of Biophysics, 16(4), 521-655.
Englander, S.W., Mayne, L., Bai, Y. and Sosnick, T.R. (1997). Hydrogen exchange: The modern legacy of Linderstrom-Lang. Protein Science, 6(5), 1101-1109.
Eyles, S.J. and Kaltashov, I.A. (2004). Methods to study protein dynamics and folding by mass spectrometry. Methods, 34(1), 88-99.
Fishelson, Z., Pangburn, M.K. and Mullereberhard, H.J. (1984). Characterization of the Initial C-3 Convertase of the Alternative Pathway of Human-Complement. Journal of Immunology, 132(3), 1430-1434.
Gampe, R.T., Montana, V.G., Lambert, M.H., Miller, A.B., Bledsoe, R.K., Milburn, M.V., Kliewer, S.A., Willson, T.M. and Xu, H.E. (2000). Asymmetry in the PPAR gamma/RXR alpha crystal structure reveals the molecular basis of heterodimerization among nuclear receptors. Molecular Cell, 5(3), 545-555.
Geyer, M., Fackler, O.T. and Peterlin, B.M. (2001). Structure-function relationships in HIV-1 Nef. Embo Reports, 2(7), 580-585.
Grzesiek, S., Bax, A., Clore, G.M., Gronenborn, A.M., Hu, J.S., Kaufman, J., Palmer, I., Stahl, S.J. and Wingfield, P.T. (1996). The solution structure of HIV-1 Nef reveals an unexpected fold and permits delineation of the binding surface for the SH3 domain of Hck tyrosine protein kinase. Nature Structural Biology, 3(4), 340-345.
Grzesiek, S., Bax, A., Hu, J.S., Kaufman, J., Palmer, I., Stahl, S.J., Tjandra, N. and Wingfield, P.T. (1997). Refined solution structure and backbone dynamics of HIV-1 Nef. Protein Science, 6(6), 1248-1263.
Hamuro, Y., Coales, S.J., Morrow, J.A., Molnar, K.S., Tuske, S.J., Southern, M.R. and Griffin, P.R. (2006). Hydrogen/deuterium-exchange (H/D-Ex) of PPAR gamma LBD in the presence of various modulators. Protein Science, 15(8), 1883-1892.
Hochrein, J.M., Wales, T.E., Lerner, E.C., Schiavone, A.P., Smithgall, T.E. and Engen, J.R. (2006). Conformational features of the full-length HIV and SIV Nef proteins determined by mass spectrometry. Biochemistry, 45(25), 7733-7739.
Hoofnagle, A.N., Resing, K.A. and Ahn, N.G. (2003). Protein analysis by hydrogen exchange mass spectrometry. Annual Review of Biophysics and Biomolecular Structure, 32, 1-25.
Janssen, B.J.C., Huizinga, E.G., Raaijmakers, H.C.A., Roos, A., Daha, M.R., Nilsson-Ekdahl, K., Nilsson, B. and Gros, P. (2005). Structures of complement component C3 provide insights into the function and evolution of immunity. Nature, 437(7058), 505-511.
Kallenberger, B.C., Love, J.D., Chatterjee, V.K.K. and Schwabe, J.W.R. (2003). A dynamic mechanism of nuclear receptor activation and its perturbation in a human disease. Nature Structural Biology, 10(2), 136-140.
Kenzel, S. and Henneke, P. (2006). The innate immune system and its relevance to neonatal sepsis. Current Opinion in Infectious Diseases, 19(3), 264-270.
Lacy, D.B., Mourez, M., Fouassier, A. and Collier, R.J. (2002). Mapping the anthrax protective antigen binding site on the lethal and edema factors. Journal of Biological Chemistry, 277(4), 3006-3010.
Lambris, J.D. (1988). The Multifunctional Role of C-3, the 3rd Component of Complement. Immunology Today, 9(12), 387-393.

Lanman, J. and Prevelige, P.E. (2004). High-sensitivity mass spectrometry for imaging subunit interactions: hydrogen/deuterium exchange. Current Opinion in Structural Biology, 14(2), 181-188.

Lee, C.H., Saksela, K., Mirza, U.A., Chait, B.T. and Kuriyan, J. (1996). Crystal structure of the conserved core of HIV-1 Nef complexed with a Src family SH3 domain. Cell, 85(6), 931-942.

Leesnitzer, L.M., Parks, D.J., Bledsoe, R.K., Cobb, J.E., Collins, J.L., Consler, T.G., Davis, R.G., Hull-Ryde, E.A., Lenhard, J.M., Patel, L., Plunket, K.D., Shenk, J.L., Stimmel, J.B., Therapontos, C., Willson, T.M. and Blanchard, S.G. (2002). Functional consequences of cysteine modification in the ligand binding sites of peroxisome proliferator activated receptors by GW9662. Biochemistry, 41(21), 6640-6650.

Melnyk, R.A., Hewitt, K.M., Lacy, D.B., Lin, H.C., Gessner, C.R., Li, S., Woods, V.L. and Collier, R.J. (2006). Structural determinants for the binding of anthrax lethal factor to oligomeric protective antigen. Journal of Biological Chemistry, 281(3), 1630-1635.

Molloy, S.S., Bresnahan, P.A., Leppla, S.H., Klimpel, K.R. and Thomas, G. (1992). Human Furin Is a Calcium-Dependent Serine Endoprotease That Recognizes the Sequence Arg-X-X-Arg and Efficiently Cleaves Anthrax Toxin Protective Antigen. Journal of Biological Chemistry, 267(23), 16396-16402.

Nagar, B., Jones, R.G., Diefenbach, R.J., Isenman, D.E. and Rini, J.M. (1998). X-ray crystal structure of C3d: A C3 fragment and ligand for complement receptor 2. Science, 280(5367), 1277-1281.

Oberfield, J.L., Collins, J.L., Holmes, C.P., Goreham, D.M., Cooper, J.P., Cobb, J.E., Lenhard, J.M., Hull-Ryde, E.A., Mohr, C.P., Blanchard, S.G., Parks, D.J., Moore, L.B., Lehmann, J.M., Plunket, K., Miller, A.B., Milburn, M.V., Kliewer, S.A. and Willson, T.M. (1999). A peroxisome proliferator-activated receptor gamma ligand inhibits adipocyte differentiation. Proceedings of the National Academy of Sciences of the United States of America, 96(11), 6102-6106.

Olefsky, J.M. and Saltiel, A.R. (2000). PPAR gamma and the treatment of insulin resistance. Trends in Endocrinology and Metabolism, 11(9), 362-368.

Renkema, G.H. and Saksela, K. (2000). Interactions of HIV-1 Nef with cellular signal transducing proteins. Frontiers in Bioscience, 5, D268-D283.

Sahu, A. and Lambris, J.D. (2001). Structure and biology of complement protein C3, a connecting link between innate and acquired immunity. Immunological Reviews, 180, 35-48.

Scobie, H.M., Rainey, G.J.A., Bradley, K.A. and Young, J.A.T. (2003). Human capillary morphogenesis protein 2 functions as an anthrax toxin receptor. Proceedings of the National Academy of Sciences of the United States of America, 100(9), 5170-5174.

Thurman, J.M. and Holers, V.M. (2006). The central role of the alternative complement pathway in human disease. Journal of Immunology, 176(3), 1305-1310.

Wales, T.E. and Engen, J.R. (2006). Hydrogen exchange mass spectrometry for the analysis of protein dynamics. Mass Spectrometry Reviews, 25(1), 158-170.

Walport, M.J. (2001a). Advances in immunology: Complement (First of two parts). New England Journal of Medicine, 344(14), 1058-1066.

Walport, M.J. (2001b). Advances in immunology: Complement (Second of two parts). New England Journal of Medicine, 344(15), 1140-1144.

Winters, M.S., Spellman, D.S. and Lambris, J.D. (2005). Solvent accessibility of native and hydrolyzed human complement protein 3 analyzed by hydrogen/deuterium exchange and mass spectrometry. Journal of Immunology, 174(6), 3469-3474.

Xu, Y., Narayana, S.V. and Volanakis, J.E. (2001). Structural biology of the alternative pathway convertase. Immunological Reviews, 180, 123-135.

Yang, W., Rachez, C. and Freedman, L.P. (2000). Discrete roles for peroxisome proliferator-activated receptor gamma and retinoid X receptor in recruiting nuclear receptor coactivators. Molecular and Cellular Biology, 20(21), 8008-8017.

Zhang, S., Finkelstein, A. and Collier, R.J. (2004). Evidence that translocation of anthrax toxin's lethal factor is initiated by entry of its N terminus into the protective antigen channel. Proceedings of the National Academy of Sciences of the United States of America, 101(48), 16756-16761.

Zhang, Z.Q. and Smith, D.L. (1993). Determination of Amide Hydrogen-Exchange by Mass-Spectrometry-a New Tool for Protein-Structure Elucidation. Protein Science, 2(4), 522-531.

Index

Printed in the USA